T0319461

**Antenna-in-Package Technology and Applications**

**IEEE Press**
445 Hoes Lane
Piscataway, NJ 08854

# Antenna-in-Package Technology and Applications

*Edited by*

*Duixian Liu*
IBM Thomas J. Watson Research Center
New York, USA

*Yueping Zhang*
School of Electrical and Electronic Engineering
Nanyang Technological University, Singapore

Published by John Wiley & Sons, Inc., Hoboken, New Jersey.
Published simultaneously in Canada.



*Library of Congress Cataloging-in-Publication data applied for*

ISBN: 9781119556633

Cover Design: Wiley
Cover Image: © windwheel/Shutterstock

Set in 9.5/12.5pt STIXTwoText by SPi Global, Chennai, India

10 9 8 7 6 5 4 3 2 1

# Contents

# List of Contributors

*Yueping Zhang*
School of Electrical and Electronic Engineering
Nanyang Technological University
Singapore

*Ning Ye*
Package Technology Development & Integration
Milpitas
USA

*Xiaoxiong Gu*
Thomas. J. Watson Research Center, IBM
New York
USA

*Pritish Parida*
Thomas. J. Watson Research Center, IBM
New York
USA

*A.C.F. Reniers*
Department of Electrical Engineering
Eindhoven University of Technology (TU/e)
Eindhoven
The Netherlands

*U. Johannsen*
Department of Electrical Engineering
Eindhoven University of Technology (TU/e)
Eindhoven
The Netherlands

*A.B. Smolders*
Department of Electrical Engineering
Eindhoven University of Technology (TU/e)
Eindhoven
The Netherlands

*Atif Shamim*
Computer, Electrical and Mathematical Sciences and Engineering Division
King Abdullah University of Science & Technology (KAUST)
KSA

*Haoran Zhang*
Computer, Electrical and Mathematical Sciences and Engineering Division
King Abdullah University of Science & Technology (KAUST)
KSA

*Frédéric Gianesello*
ST Microelectronics
Technology R&D
Silicon Technology Development
Crolles
France

*Diane Titz*
Université Nice Sophia Antipolis
Polytech'Lab
Biot
France

*Cyril Luxey*
Université Nice Sophia Antipolis
Polytech'Lab
Biot
France

*Maciej Wojnowski*
Infineon Technologies AG
Neubiberg
Germany

*Klaus Pressel*
Infineon Technologies AG
Regensburg
Germany

*Tong-Hong Lin*
The School of Electrical and Computer Engineering
Georgia Institute of Technology
Atlanta
USA

*Ryan A. Bahr*
The School of Electrical and Computer Engineering
Georgia Institute of Technology
Atlanta
USA

*Manos M. Tentzeris*
The School of Electrical and Computer Engineering
Georgia Institute of Technology
Atlanta
USA

*Duixian Liu*
Thomas. J. Watson Research Center, IBM
New York
USA

*Amin Enayati*
Emerson & Cuming Anechoic Chambers
Antwerp Area
Belgium

*Karin Mohammadpour-Aghdam*
School of Electrical and Computer Engineering
University of Tehran
Iran

*Farbod Molaee-Ghaleh*
School of Electrical and Computer Engineering
University of Tehran
Iran

# Preface

Rapid advances in semiconductor and packaging technologies have promoted the development of two system design concepts known as system on chip (SoC) and system-in-package (SiP). SoC integrates analog, digital, mixed-signal, and radio frequency (RF) circuits on a chip by a semiconductor process, while SiP implements separately manufactured functional blocks in a package by a packaging process. SoC yields improved system reliability and functionality at a much lower system cost. However, it degrades system performance and increase system power consumption due to unavoidable compromises in every circuit type in order to use the same material and process. On the contrary, SiP enhances system performance and reduces system power consumption but results in lower system reliability and higher system cost because of functional blocks and the fabrication of the package with different materials and processes.

Antennas are essential components for wireless systems. It is known that antennas are difficult to miniaturize, let alone integrate. Nevertheless, there have been attempts to integrate an antenna (or antennas) with other circuits in a die on a wafer using the back end of the line. An antenna realized in such a way is called an antenna on a chip (AoC) and is more suitable for terahertz applications for cost and performance reasons. In addition, there have been studies to integrate an antenna (or antennas) with a radio or radar die (or dies) into a standard surface-mounted device using a packaging process, which has created a new trend in antenna and packaging termed antenna-in-package (AiP). AoC and AiP are obviously subsets of the above SoC and SiP concepts, so why do we specifically differentiate them from SoC and SiP? The reason is to highlight their unique property of radiation.

AiP technology balances performance, size, and cost, hence it has been widely adopted by chipmakers for 60-GHz radios, augmented/virtual reality gadgets, and gesture radars. It has also found applications in 79-GHz automotive radars, 94-GHz phased arrays for imaging and data communications, 122-GHz, 145-GHz, and 160-GHz sensors, as well as 300-GHz wireless links. Recently, AiP technology has been under further development for millimeter-wave (mmWave) fifth-generation (5G) technology. Scalable large AiPs and multiple small AiPs have been successfully demonstrated in base stations, mobile phones, and networked cars at 28 GHz. We therefore believe that AiP technology will cause fundamental changes in the design of antennas for mobile communications for 5G and beyond operating in mmWave bands.

The development of mmWave AiP technology is particularly challenging because of the associated complexity in design, fabrication, integration, and testing. This book aims to face these challenges through disseminating relevant knowledge, addressing practical engineering issues, meeting immediate demands for existing systems, and providing the antenna and packaging solutions for the latest and emerging applications.

This book contains 11 chapters. The first five chapters lay some foundation and introduce fundamental knowledge. After the introductory chapter about how AiP technology has been developed as we know it today, several types of antennas are discussed in Chapter 2. An attempt is made to summarize the basic antennas and those antennas specifically developed for AiP technology. Emphasis is given to microstrip patch antennas and arrays, grid array antennas, Yagi–Uda antennas, and magneto-electric dipole antennas because of their dominance in AiP technology. Performance improvement techniques of antennas for AiP technology are also described. Chapter 3 describes today's mainstream packaging solutions with either wire-bond or flip-chip interconnects, wafer-level package, and fan-out wafer-level package. Chapter 4 focuses on the electrical, mechanical, and thermal co-design for AiP modules. More importantly, the thermal management considerations for next-generation heterogeneous integrated systems are reviewed in order to address the growing need for cooling the high-power devices of future radio systems. Chapter 5 presents the design and optimization of an anechoic test facility for testing mmWave integrated antennas. This facility can be used for both probe-based and connector-based measurements.

The next five chapters are related to the design, fabrication, and characterization of AiPs in different materials and processes for mmWave applications. Chapter 6 discusses low-temperature co-fired ceramic (LTCC)-based AiP. LTCC has unique properties for packaging mmWave circuits since it can provide a durable hermetic package with antennas, cavities, and integrated passive components. Chapter 7 illustrates how industrial organic packaging substrate technology used for classical integrated circuit (IC) packaging can support the development of innovative, efficient, and cost-effective mmWave AiPs from 60 GHz up to 300 GHz. Chapter 8 focuses on embedded wafer-level ball grid array (eWLB)-based AiP. Unlike LTCC or high-density integr (HDI), eWLB eliminates the need for a laminate substrate and replaces it with copper redistribution layers. Polymers are used for the electrical isolation between the metal layers. The metal routings are deposited by a combination of sputtering and electroplating with a thin film process. eWLB has historically been developed for mmWave automotive radar systems and therefore has naturally been used for mass production of mmWave AiPs. Chapter 9 presents surface laminar circuit (SLC)-based AiP. Compared to LTCC, HDI, and eWLB, SLC is more suitable for fabrication of very large or dense AiPs. The chapter describes SLC materials and design guidelines, and then addresses the design challenges and solutions for 8 × 8 dual-polarized phased arrays at 94 GHz for imaging and 28 GHz for 5G base station applications, respectively. Chapter 10 introduces different additive manufacturing technologies, methods to characterize three-dimensional (3D)-printed materials, a hybrid printing process by integrating 3D and inkjet printing, and a broadband 5G AiP realized with the hybrid process.

The last chapter turns to 3D AiP for power transfer, sensor nodes, and Internet of Things applications. This package has a cubic geometry with radiating antennas on its surrounding faces. The chapter highlights small antenna design and miniaturizing techniques as well as multi-mode capability as a way to achieve wideband antennas.

This book is the result of the joint efforts of the 21 authors in eight different institutions in Asia, Europe, and the United States. A book on an emerging topic like AiP technology would not have been possible without such collaborations. We thank all authors for their creative contributions and careful preparation of manuscripts. We are also pleased to acknowledge the professional cooperation of the publishers.

*Duixian Liu*
IBM Thomas J. Watson Research Center
Yorktown Heights, NY, USA

*Yueping Zhang*
School of Electrical and Electronic Engineering
Nanyang Technological University
Singapore

# Abbreviations

| | |
|---|---|
| 2D | two-dimensional |
| 3D | three-dimensional |
| 3GPP | 3rd Generation Partnership Project |
| 5G | fifth-generation |
| ABS | acrylonitrile butadiene styrene |
| ACE | Advanced Semiconductor Engineering, Inc. |
| ACP | aperture-coupled patch |
| ADS | Advanced Design System |
| AIA | active integrated antenna |
| AiM | antenna in a module |
| AiP | antenna-in-package |
| AM | additive manufacturing |
| AMC | artificial magnetic conductor |
| AoB | antenna on a board |
| AoC | antenna on a chip |
| AR | axial ratio |
| ARM | advanced reduced-instruction set-computer machine |
| ASIC | application-specific integrated circuit |
| ASUT | antenna system under test |
| AUT | antenna under test |
| Az | azimuth |
| BCB | benzocyclobutene |
| BC-SRR | broadside-coupled SRR |
| BER | bit error rate |
| BERT | BER tester |
| BiCMOS | bipolar complementary metal oxide semiconductor |
| BGA | ball grid array |
| BLE | Bluetooth low energy |
| BT | Bluetooth |

| | |
|---|---|
| BT | bismaleimide triazine |
| BW | bandwidth |
| C4 | controlled collapse chip connection |
| CATR | compact antenna test range |
| CCL | copper-clad laminate |
| CLIP | continuous liquid interface printing |
| CMA | characteristic mode analysis |
| CMF | conjugate match factor |
| CMG | conjugate match gain |
| CMOS | complementary metal-oxide semiconductor |
| CNC | computer numerical controlled |
| CP | circular polarization |
| CPS | coplanar strip |
| CPU | central processing unit |
| CPW | coplanar waveguide |
| CT | computer tomography |
| CTE | coefficient of thermal expansion |
| CUF | capillary underfill |
| DC | direct current |
| DLP | digital light projection |
| DMA | dynamic mechanical analysis |
| DRAM | dynamic random-access memory |
| DRIE | deep reactive ion etching |
| EBG | electromagnetic bandgap |
| EIRP | equivalent isotropic radiated power |
| El | elevation |
| EM | electromagnetic |
| EMI | electromagnetic interference |
| ESD | electrostatic discharge |
| ETS | embedded traces |
| eWLB | embedded wafer-level ball grid array |
| EZL | embedded Z line |
| FCC | Federal Communications Commission |
| FDM | fused-deposition modeling |
| FE | front end |
| FF | far-field |
| FMCW | frequency modulated continuous wave |
| FoM | figure of merit |
| FO PoP | fan-out package-on-package |
| FO-WLP | fan-out wafer-level packaging |
| FPGA | field-programmable gate array |

| | |
|---|---|
| FR4 | flame resistant 4 |
| FSS | frequency selective surface |
| GaAs | gallium arsenide |
| GaN | gallium nitride |
| Gb/s | gigabit per second |
| GPU | graphics processing unit |
| GSG | ground-signal-ground |
| GSGSG | ground-signal-ground-signal-ground |
| GSM | global system for mobile communications |
| HAST | highly accelerated stress test |
| HBM | high bandwidth memory |
| HDI | high-density integration |
| HDI | high-density interconnect |
| HFSS | high-frequency structure simulator |
| HPBW | half-power beam width |
| HTCC | high-temperature co-fired ceramics |
| IC | integrated circuit |
| IEEE | Institute of Electrical and Electronics Engineers |
| IF | intermediate frequency |
| InFO_PoP | integrated fan out package on package |
| I/O | input/output |
| IoT | Internet of Things |
| ISM | industrial, scientific and medical |
| ISSCC | International Solid-State Circuits Conference |
| JPL | jet propulsion laboratory |
| LCD | liquid crystal display |
| LCP | liquid crystal polymer |
| LGA | land grid array |
| LHCP | left-hand circular polarization |
| LNA | low-noise amplifier |
| LP | linearly polarized |
| LTCC | low-temperature co-fired ceramic |
| MACM | multiple amplitude component method |
| MAPCM | multiple amplitude phase component method |
| MCM | multi-chip module |
| MEMS | micro-electromechanical systems |
| MIM | metal–insulator–metal |
| MIMO | multiple input multiple output |
| MMIC | monolithic microwave integrated circuit |
| MMWAC | mmWave anechoic chamber |
| mmWave | millimeter-wave |

| | |
|---|---|
| mSAP | modified semi-additive process |
| MUF | molded underfill |
| NF | near-field |
| NF | noise figure |
| NIST | National Institute of Standards and Technology |
| NRE | non-recurring engineering |
| NRW | Nicholson–Ross–Weir |
| OTA | over-the-air |
| PAE | power-added efficiency |
| PAM | phase amplitude method |
| PBO | polybenzoxazoles |
| PCB | printed circuit board |
| PEC | perfect electric conductor |
| PER | packet error rate |
| PET | polyethylene terephthalate |
| pHEMT | pseudomorphic high electron mobility transistor |
| PI | polyimide |
| PLA | polylactic acid |
| PLL | phase-locked loop |
| PMC | perfect magnetic conductor |
| PP | polypropylene |
| PPM | polarization pattern method |
| PTH | plated-through-hole |
| p.u.l. | per unit length |
| QFN | quad flat non-leaded |
| QFP | quad flat package |
| R&D | research and development |
| RAM | random-access memory |
| RCC | resin-coated copper |
| RCS | radar cross-section |
| RDL | redistribution layer |
| RF | radio frequency |
| RFIC | radio frequency integrated circuit |
| RFID | radio-frequency identification |
| RHCP | right-hand circular polarization |
| RLCG | resistance, inductance, capacitance, and conductance |
| RMS | root mean square |
| RoHS | Restriction of Hazardous Substances Directive |
| RSM | rotating source method |
| RX | receiver |
| SAM | scanning acoustic microscopy |

| | |
|---|---|
| SAP | semi-additive process |
| SAR | synthetic aperture radar |
| SEM | scanning electron microscopy |
| SG | signal-ground |
| SiGe | silicon germanium |
| SiP | system-in-package |
| SISO | single input, single output |
| SIW | substrate integrated waveguide |
| SLA | stereolithography |
| SLC | surface laminar circuit |
| SLM | selective laser melting |
| SLS | selective laser sintering |
| SMA | sub-miniature version A |
| SNR | signal-to-noise ratio |
| SoC | system on chip |
| SoP | system-on-package |
| SPDR | split post dielectric resonator |
| SPP | surface plasmon polariton |
| SRR | split-ring resonator |
| SSD | solid state drive |
| SSMA | small SMA |
| SUB | subtractive process |
| TCB | thermocompression bonding |
| TE | transverse electric |
| TEM | transverse electro-magnetic |
| TEV | through encapsulant via |
| TFMSL | thin-film microstrip line |
| TIM | thermal interface material |
| TIV | through InFO via |
| TL | transmission line |
| TMA | thermomechanical analyzer |
| TM | transverse magnetic |
| TMV | through-mold vias |
| TPP | two-photo polymerization |
| TSMC | Taiwan Semiconductor Manufacturing Company |
| TSOP | thin small outline package |
| TSV | through silicon via |
| TX | transmitter |
| μvia | microvia |
| UBM | under bump metallurgy |
| UHF | ultra-high frequency |

| | |
|---|---|
| UV | ultraviolet |
| UWB | ultra-wideband |
| VCO | voltage-controlled oscillator |
| VGA | variable gain amplifier |
| VLSI | very-large-scale integration |
| VNA | vector network analyzer |
| VQFN | very thin quad flat no-lead |
| VSWR | voltage standing wave ratio |
| WGP | wave-guide port |
| WiGig | wireless gigabit alliance |
| WLCSP | wafer level chip scale package |
| WLP | wafer level package |
| WPAN | wireless personal area network |
| WPT | wireless power transfer |
| WSN | wireless sensor network |

## Symbols

| | |
|---|---|
| $Ae$ | effective antenna aperture |
| $c$ | speed of light |
| $D$ | directivity |
| $D = \max[D(\vartheta, \varphi)]$ | directivity function |
| $\eta$ | effciency |
| $\vec{e}$ | unit vector along the $r$ axis |
| $\vec{E}(\vec{r})$ | electric field in space frequency |
| $f$ | frequency |
| $G$ | gain |
| $G = \max[G(\vartheta, \varphi)]$ | gain function |
| $\Gamma$ | reflection coeffcient |
| $I^p$ | input current antenna |
| $k$ | coverage factor |
| $L$ | length of the radiating aperture |
| $\lambda_0$ | wavelength in free space |
| $M_v$ | measurement value |
| $P_{in}$ | input power |
| $P_t$ | total radiated power |
| $P(\vartheta, \varphi)$ | normalized radiation power |
| $R$ | radial distance from the antenna |
| $R_a$ | resistive |
| $R_L$ | radiated losses |
| $R_r$ | radiated resistance |

| | |
|---|---|
| $R_v$ | reference value |
| $\mu_{sys}$ | system uncertainty |
| $\vec{S}$ | Poynting vector |
| $V^p$ | input voltage antenna |
| $X_a$ | reactance |
| $Z_0$ | characteristic impedance |
| $Z_a$ | input impedance |
| $Z_g$ | generator impedance = $R_g$ |

# 1

# Introduction

*Yueping Zhang*

*School of Electrical and Electronic Engineering, Nanyang Technological University, Singapore*

## 1.1 Background

As the technology of choice for integration of digital circuitry, a complementary metal oxide semiconductor (CMOS) was proposed for the integration of analog circuitry for radio frequency (RF) applications in the mid-1980s, aiming for the ultimate goal of full integration of an entire wireless system on a chip [1]. In the mid-1990s, the first fully integrated CMOS transceiver for data communications in the 900-MHz industrial, scientific and medical (ISM) band was successfully demonstrated [2]. Since then, CMOS has been the enabler for wireless systems on chip (SoCs) operating from a few to tens of gigahertz. Figure 1.1 shows the die micrograph of the first wireless SoC, a 2.4-GHz CMOS mixed RF analog–digital Bluetooth radio announced at the International Solid-State Circuits Conference (ISSCC) 2001 [3]. The die size is 40.1 $mm^2$. It integrates on the same substrate a low intermediate frequency (IF) receiver, a Cartesian transmitter, a baseband processer, an advanced reduced-instruction set-computer machine (ARM) processor, flash memory, and random-access memory (RAM).

Full SoC integration is clearly not suitable in all cases. In fact, the radio chip is separate in many cases. Traditionally, silicon germanium (SiGe) seems to have been preferred to CMOS for analog RF. Figure 1.2 shows the die micrographs of the first SiGe 60-GHz transmitter and receiver disclosed at ISSCC 2006 [4]. The die sizes are $4.0 \times 1.6\,mm^2$ and $3.4 \times 1.7\,mm^2$, respectively. The level of integration achieved in these chips was high then for 60-GHz radios. The transmitter chip integrates a power amplifier, image-reject driver, IF-to-RF up-mixer, IF amplifiers, quadrature baseband-to-IF mixers, phase-locked loop (PLL), and frequency tripler. The receiver chip includes an image-reject low-noise amplifier, RF-to-IF

*Antenna-in-Package Technology and Applications*, First Edition.
Edited by Duixian Liu and Yueping Zhang.

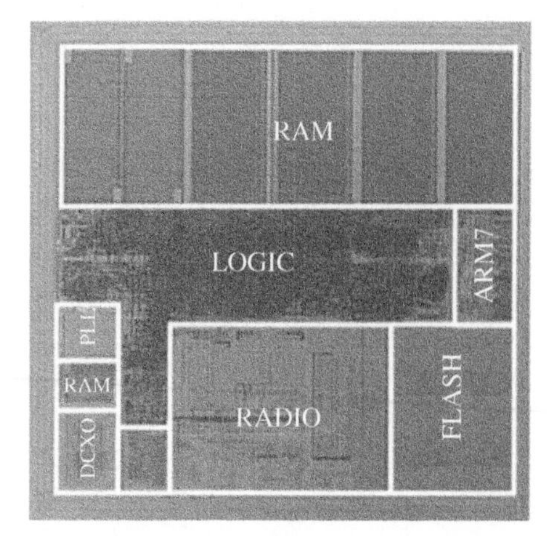


**Figure 1.1** Micrograph of the first 2.4-GHz CMOS wireless SoC, a Bluetooth radio (from [3], © 2001 IEEE, reprinted with permission).

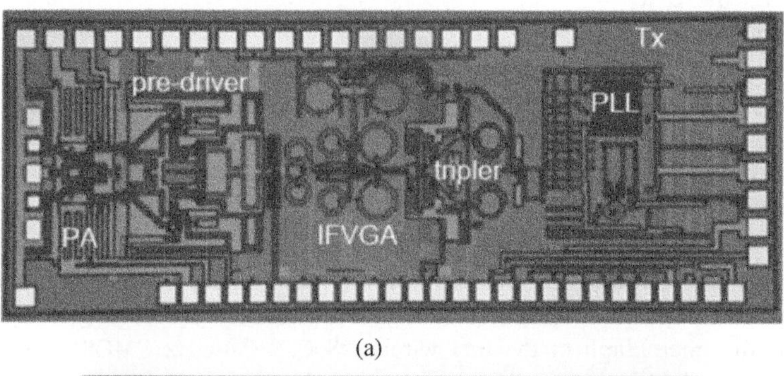


(a)

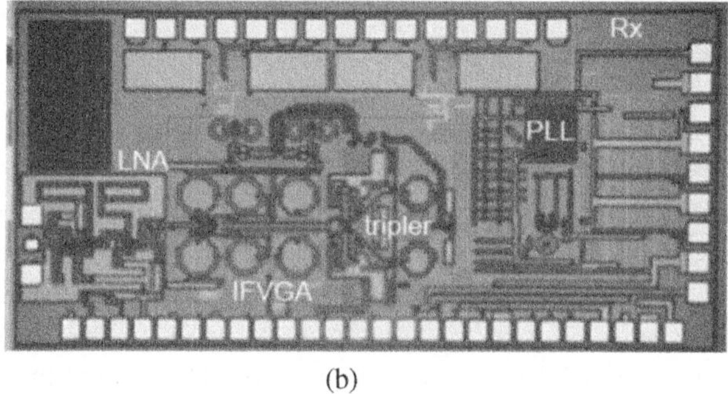


(b)

**Figure 1.2** Photographs of the first 60-GHz SiGe radio chipset: (a) transmitter and (b) receiver (from [4], © 2006 IEEE, reprinted with permission).

mixer, IF amplifiers, quadrature IF-to-baseband mixers, PLL, and frequency tripler. The input/output (I/O) pads are peripheral, with 60 on the transmitter and 53 on the receiver chips.

The emergence of wireless SoCs or single-chip radios called for compatible antenna solutions, which provided an excellent opportunity for researchers of prepared minds to seriously explore the feasibility of integrating an antenna in a chip package using packaging materials and processes in the late 1990s, leading to the development of antenna-in-package (AiP) technology [5]. This chapter recounts how AiP technology has been developed to its current state. Section 1.2 describes the idea of AiP with respect to the ideas of antenna on chip (AoC), antenna in module (AiM), antenna on board (AoB), and active integrated antenna (AIA). Section 1.3 reviews the early attempts to explore the idea of AiP. Section 1.4 reflects on the milestones in the development of the idea of AiP into a mainstream antenna and packaging technology. Finally, Section 1.5 gives concluding remarks.

## 1.2 The Idea

The idea of AiP was triggered by the demand for innovative antenna solutions to wireless SoCs [6]. It features using packaging technology to implement an antenna (or antennas) with a radio or radar die (or dies) in a chip package. It emphasizes only the addition of the unique function of radiation to the package. In this sense, it is different from the concept of system-in-package (SiP).

The idea of AoC sounds attractive [7]. It attempts to integrate an antenna (or antennas) with other circuits on a die directly using semiconductor technology. It is obviously a subset of the concept of SoC. Then why do we specifically differentiate it from SoC? The reason is to highlight the unique property of radiation, which is not necessarily being improved like digital circuits as the technology scales down. It is clear that AoC is more suitable for terahertz applications for cost and performance reasons.

The idea of AiM was proposed for multichip 60-GHz radios [8]. It uses micro-assembly technology to mount a few monolithic microwave integrated circuits (MMICs) and a small flat antenna in a hermetically sealed package. A window for the propagation of electromagnetic waves is formed above the antenna at the lid of the package. The window is also hermetically sealed.

The idea of AoB is similar to the idea of AiP. However, it relies on printed circuit board (PCB) technology to make an antenna (or antennas) on one surface of a board and to solder a packaged chip (or chips) on the other surface of the board. A few techniques, such as probe feeding or aperture coupling, are available to interconnect the packaged chip with the antenna. Of course, the antenna, the packaged chip, and the necessary feed networks can be contained on the same surface of the board. Recently, the idea of AoB has received considerable attention for millimeter-wave (mmWave) fifth-generation (5G) base stations [9].

A typical AIA consists of active devices such as Gunn diodes or transistors that form an active circuit and a planar antenna. The idea of AIA was proposed to eliminate the lossy and bulky interconnect between the active device and radiating element [10]. Later, the idea of AIA was employed for quasi-optical power combining. The output power from an array of many solid-state devices was combined in free space to overcome the power limitations of individual solid-state devices at mmWave frequencies.

Although the origin of the above ideas can be traced back to the invention of microstrip antennas in the early 1970s [11], it should be noted that they extended the concept of microstrip antennas to different levels of integration.

## 1.3 Exploring the Idea

In this section, the early attempts to explore the idea of AiP are reviewed. It should be mentioned that researchers in university labs devoted their efforts regarding Bluetooth radios to 2.4 GHz or other RF applications, while researchers in company labs focused on 60-GHz radios and other mmWave applications. At 2.4 GHz, a key challenge was how to miniaturize the antenna size, while at 60 GHz, it was how to minimize the interconnect loss between the die and antenna.

### 1.3.1 Bluetooth Radio and Other RF Applications

In 1998, Zhang started to work in the Division of Circuits and Systems at the School of Electrical and Electronic Engineering, Nanyang Technological University, Singapore. The division soon initiated a strategic research project entitled "Software radio on a chip." Zhang was tasked to develop an antenna technology for the project. Inspired by the structural similarity shared by a microstrip antenna and a microchip, shown in Figure 1.3, and foreseeing the outcome of an interesting antenna solution, Zhang immediately started to investigate the antenna performance of the microchip. First, Zhang did antenna experiments with used microchips, as shown in Figure 1.4a. Then, Zhang tried PCB mock-ups, as shown in Figure 1.4b. Encouraged by good antenna results from the microchips and PCB mock-ups, the research team led by Zhang realized more sophisticated designs, as shown in Figure 1.4c,d, with low-temperature co-fired ceramic (LTCC) technologies in 2004 [12]. It is interesting to note that differential microstrip patch and meander antennas were designed to suit high-level integration of wireless SoCs.

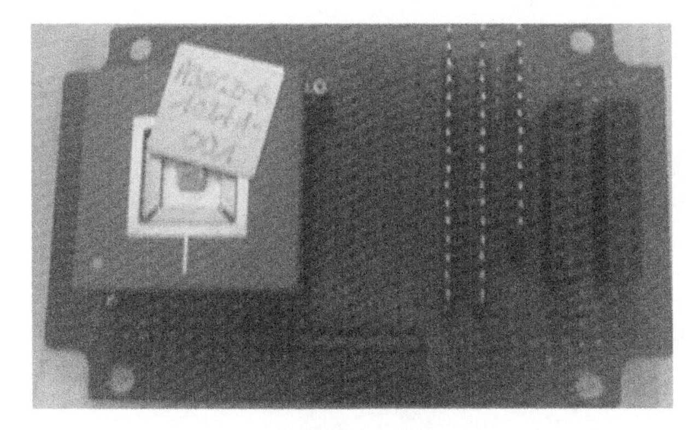

**Figure 1.3** Photograph of a microchip.

They were integrated on the top surfaces of two ball grid array (BGA) packages. Both packages had cavities to house wireless SoC dies. The interconnects from the die to the antenna were cascaded bond wires, traces, and vias. The interconnected die was encapsulated with epoxy.

In 2000, Song et al. at University of Birmingham presented an integrated antenna package [13]. An electrically small feed antenna was designed on a semiconductor substrate, which also supported the RF front-end circuits. The parasitic radiator placed above the feed antenna also acted as a top cover, sealing the entire package. Later, Song et al. presented another integrated antenna package [14]. A small antenna was embedded within the chip encapsulating material. A parasitic radiator was placed in close proximity to the embedded antenna, where it enhanced the poor gain and bandwidth of the packaged antenna.

In 2001, package engineers started to tackle the same problem. Lim et al. at the Georgia Institute of Technology managed to integrate RF passives, a patch antenna, and chips at the package level to enhance the overall performance of and to add more functionalities to an SoP paradigm [15]. Mathews et al. disclosed a package with an integral shield and antenna for a complete Bluetooth radio design [16].

In 2003, Ryckaert et al. at the Interuniversity Microelectronics Center, Belgium reported the co-design of circular-polarized slotted patch antenna with a wireless local area network (WLAN) transceiver in a multilayer package [17]. Popov et al. at the Institute of Microelectronics, Singapore reported the design of part of an RF chip package as a dielectric resonator antenna with a high dielectric constant when the antenna feed was integrated with the rest of the circuitry [18]. Leung at City University of Hong Kong independently proposed adding the chip package function to a dielectric resonator antenna in 2004 [19].

In 2005, Castany et al. at Fractus, Spain filed a patent about the integration of fractal antennas in a chip package [20]. They claimed that fractal antennas

Figure 1.4 Photographs showing the evolution of the integration of antenna in package: (a) an antenna on a used microchip package, (b) an antenna on a chip package mock-up, (c) a microstrip patch antenna as an LTCC package, and (d) a microstrip meander antenna as an LTCC package.

could provide very good antenna performance while allowing a high degree of miniaturization and an enhancement of isolation between the antenna and the die in the package.

In 2006, Brzezina et al. at Carleton University reported planar antennas with transceiver integration capability for ultra-wideband (UWB) radios [21]. Sun et al.

devised a novel technique that reduces the size of a conventional planar antenna by 40% and used it as a package to house a single-chip UWB radio [22].

In 2007, Wi et al. at Yonsei University, Korea presented an antenna-integrated package [23]. A modified U-shaped slot antenna was designed and measured, showing bandwidth of 180 MHz at 5.8 GHz. A parametric study was conducted with a full-wave electromagnetic solver to determine the critical factors in design and fabrication, as well as to estimate the performance accuracy of the antenna-integrated package.

### 1.3.2 60-GHz Radio and Other Millimeter-wave Applications

In early March 2005, Zhang met Brian P. Gaucher and Duixian Liu from the IBM Thomas J. Watson Research Center for the first time at the First International Workshop on Small Antenna Technology in Singapore and invited them to visit Nanyang Technological University. Gaucher gave a talk about the IBM 60-GHz radio SiGe chipsets, antennas, packages, and measurement setup. Figure 1.5 shows the chip-scale package for proof of concept [24]. The chip-scale package had a size of $13 \times 13 \times 1.25\,mm^3$. Note that the radio dies were flip-chip bonded with the antennas and assembled together as an land grid array (LGA) package. The package required a dedicated opening for an antenna and an encapsulation prior to the mounting of the antenna. Thus, it was difficult to fabricate and assemble this chip-scale package for mass production. With that in mind, Zhang briefed the visitors about antennas designed for 2.4 and 5 GHz as chip packages in LTCC, and a collaboration plan for a feasibility study for developing LTCC packages embedded with antennas for the IBM 60-GHz radio SiGe chipset was agreed. Zhang paired up with Mei Sun to lead the design effort, a team of researchers from Singapore Institute of Manufacturing Technology (Chee Wai Lu, Kai Meng Chua, and Lai Lai Wai) was responsible for fabrication, and Duixian Liu did the evaluation and feedback.

**Figure 1.5** Photograph of IBM chip-scale package with integrated chips and antennas.

In June 2005, Naoyuki Shino, Hiroshi Uchimura, and Kentaro Miyazato from the Kyocera research and development center reported the unification of antenna with a package based on laminated waveguides, coupling slots, and open cavity radiating elements in LTCC for short-range car radar operating in the 77-GHz band [25].

Even after all of the early attempts at integration of antennas onto or into packages, there is not yet an all-encompassing term to describe this process. Inspired by SiP, Zhang coined the term AiP to promote the idea [26]. In early March 2006, Zhang attended the Second International Workshop on Small Antenna Technology in New York and visited the IBM Thomas J. Watson Research Center. The visit strengthened the collaboration for integrating antennas in packages and accelerated AiP development. Figure 1.6 shows images captured in the design stage and photos of fabricated AiP samples for IBM 60-GHz SiGe transmitter and receiver dies. The AiPs for the transmitter and receiver dies measured $12.5 \times 8.6 \times 1.265\ mm^3$ and $12.5 \times 8 \times 1.265\ mm^3$, respectively. The transmitter AiP integrated a Yagi-Uda antenna and the receiver AiP a novel coplanar waveguide (CPW)-fed quasi-cavity-backed, guard-ring-directed, substrate-material-modulated slot antenna. The measured antenna results were promising and were recognized with the Best Paper Award in the third International Workshop on Small Antenna Technology held on 21–23 March 2007 in Cambridge, UK [27].

In August 2007, Sibeam put the first CMOS 60-GHz phased array radio in an LTCC AiP [28]. It made board and system design much easier by containing all high-frequency routing within the CMOS die and chip package, and enabled products in the market for wireless transmission of high-definition video content. It is worthwhile mentioning that after Zhang's keynote address at the Antenna Systems and Short-Range Wireless Conference 2005 held on 22–23 September in Santa Clara, California, USA, in the break Zhang enthusiastically discussed some issues about integration of antennas in LTCC with Doan, one of founders of SiBeam and pioneers in mmWave CMOS [29].

## 1.4 Developing the Idea into a Mainstream Technology

The success of the idea of AiP is largely due to the renewed interest in 60-GHz radios. In 2007 a new phase of 60-GHz radios began as the Institute of Electrical and Electronics Engineers (IEEE) initiated development of a new standard for the unlicensed 60-GHz band and many companies started to get involved in the development of 60-GHz radio chipsets with multiple antennas in chip packages. Since then, there have been numerous AiP designs reported by industry for

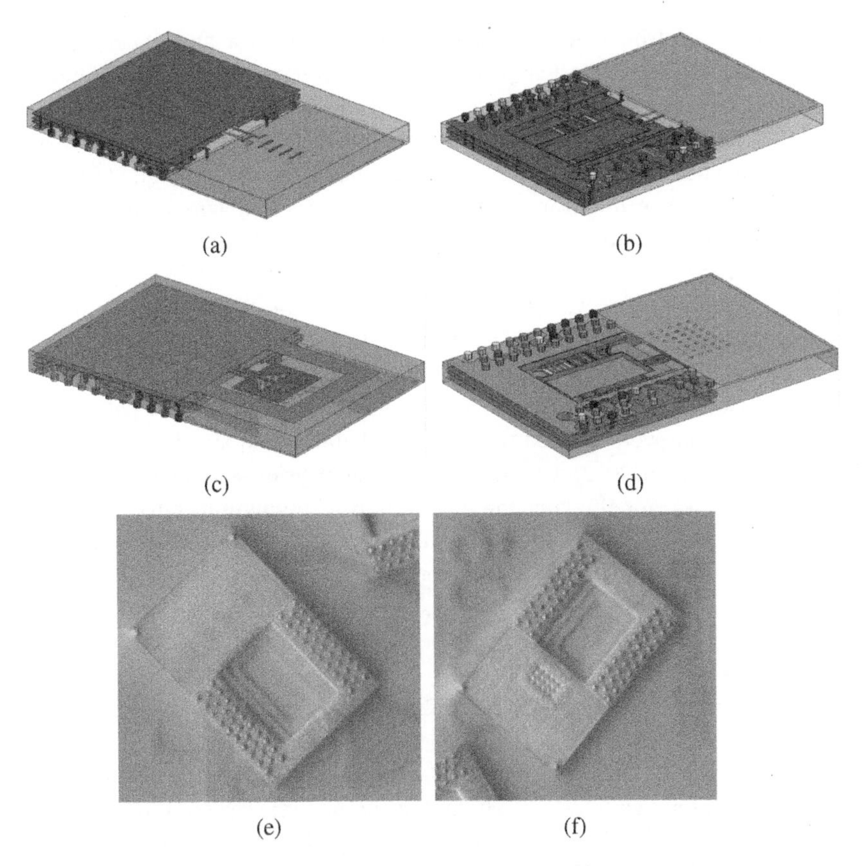


**Figure 1.6** Captured images and photographs of AiP for IBM 60-GHz SiGe chip sets: (a) top view of the captured image of the AiP for the transmitter, (b) bottom view of the captured image of the AiP for the transmitter, (c) top view of the captured image of the AiP for the receiver, (d) bottom view of the captured image of the AiP for the receiver, (e) photograph of the AiP for the transmitter, and (f) photograph of the AiP for the receiver.

applications at 60 GHz and other mmWave frequencies. Clearly, the idea of AiP has been developed into a mainstream antenna and packaging technology. The milestones in the development are described here.

The first and perhaps most important milestone was to break the boundaries between antenna and circuit fields, and to implement the design considerations for both antenna and circuit in a single design platform. It was stressed by the design strategy that the AiP should be co-simulated with the radio frequency integrated circuit (RFIC) in a circuit simulator with compact models extracted from full-wave electromagnetic simulations of the antenna and package for optimum results. The co-design platform with software tools from different vendors had been built before Cadence released the RF-SiP methodology kit in 2006 [30]. It runs the Advanced Design System from Agilent for two-dimensional (2D) electromagnetic simulation, the High Frequency Structure Simulator (HFSS) and ePhysics from Ansoft for three-dimensional (3D) electromagnetic simulation as well as 3D steady-state thermal, transient thermal and linear stress analyses coupling to HFSS.

The mass production of AiP began with LTCC technology. In 2005, the non-standard process to embed an air cavity in LTCC was developed to improve antenna performance [31] and an open cavity radiating element was created to relax LTCC fabrication tolerance. In 2008, a grid array antenna and patterned mesh ground plane were devised to enhance reliability and avoid warpage for AiP in LTCC [32]. In 2012, a paper entitled "Dual grid array antennas in a thin-profile package for flip-chip interconnection to highly-integrated 60-GHz radios" won the IEEE AP-S Sergei A. Schelkunoff Transactions Prize Paper Award [33]. In 2013, a step-profiled corrugated horn antenna was realized in LTCC for AiP to operate at 300 GHz [34]. High-density interconnect (HDI) technologies were developed specifically for the low-cost production of AiP. In 2012, Samsung developed FR4-based HDI technology for 60-GHz radios. Despite the relatively high loss tangent of the FR4, Samsung confirmed that unit loss was comparable with the LTCC-based AiP at 60 GHz [35]. In 2015, Intel developed liquid crystal polymer (LCP)-based HDI technology for 60-GHz radios. By limiting the number of metal layers to four, with the 60-GHz routing on the same layer as the die pads, Intel demonstrated that AiP achieved an ultra-thin profile at 60 GHz [36]. Unlike LTCC and HDI, the embedded wafer-level ball grid array (eWLB) technology eliminates the need for a laminate substrate and replaces it with a copper redistribution layer. It was developed by Infineon in 2006 and proved to be an alternative approach to fabricating AiP in high volume with low cost [37]. However, eWLB technology only produces a single redistribution layer (RDL), which limits the realization of antennas. To overcome this limitation, the Taiwan Semiconductor Manufacturing Company (TSMC) developed the InFO-AIP technology in 2018, which places the feeding line in the RDL at the bottom of the package, coupled to the patch antenna on the top side of the package. As a result, InFO-AIP yields a smaller form factor and a higher gain for 5G mmWave system applications [38]. In addition, a major concern with AiP is the risk of electromagnetic interference (EMI). In 2014, Advanced Semiconductor Engineering, Inc. (ASE) developed package-level conformal and compartment shielding techniques with metal coating to suppress EMI [39].

Accurate characterization of AiP was first made possible with a probe-based antenna measurement setup built by IBM in 2004 [40]. It used a ground-signal-ground (GSG) probe connected to one port of a vector network analyzer with a coaxial cable to feed the AiP. A special waveguide arm with a standard gain born antenna could be rotated around the AiP at a distance to ensure a far-field condition for radiation pattern measurement. In 2009, Toshiba built a setup to measure the radiation pattern of differential AiP with a ground-signal-ground-signal-ground (GSGSG) probe [41]. In 2011, Karim et al. introduced a setup to minimize the effect of the probe radiation and to enhance the dynamic range by implementing a backside probing technique [42]. In 2012, Diane et al. demonstrated a setup to measure 3D radiation patterns [43]. In 2015, Reniers et al. developed a bended probe to reduce both the blockage and the interference due to reflections from a conventional probe [44]. In 2018, fast testing of AiP for a production line was proved feasible with over-the-air (OTA) contactors [45]. An OTA patch antenna was embedded into the lead backer of a production pick and place handler, which offers a unique and reliable production solution testing an AiP device with 60-GHz RF signals both radiating out of an antenna array in the lid and connected through the ball grid array. OTA contactors have also been used for testing of 76–81-GHz AiP automotive radar devices and are being designed for 5G applications at 28 and 39 GHz.

A large number of needs are met by the use of AiP technology. including Internet of Things (IoT) devices at 2.4 GHz, 5G new radio and networked cars at 28 GHz, VR, axial ratio (AR), and gesture radars at 60 GHz, automotive radars at 79 GHz, imagers at 94 GHz, sensors at 122, 145, and 160 GHz, as well as 300-GHz wireless links. The advantages of AiP technology will continue to generate new applications, for example the adoption of AiP technology in the development of highly integrated micro-synthetic aperture radar (SAR) for deep-space exploration [46].

## 1.5 Concluding Remarks

AiP technology has broken the boundaries between antenna and circuit fields. The methodology and platform to co-design antenna and circuits is now available. AiP technology has justified developing new materials and processes, which is rare, to the best of my knowledge, and only microstrip patch antennas have received such attention. Testing has to be considered along the whole manufacturing cycle, including test strategies, verification and characterization, production testing, integration and system level testing. Probe-based measurement setups are suitable for AiP design verification and characterization. OTA antenna measurements are required for production testing, integration and system level testing. In the future, AiP technology will continue to provide direct antenna solutions to highly integrated wireless systems that will operate at even higher mmWave frequencies. It will also provide parasitic or distributive radiator functions to enhance the terahertz antenna performance of AoC technology.

## Acknowledgements

The author is grateful to his students Dr. Wang Junjun, Dr. Sun Mei, Dr. Zhang Bing, Dr. Chen Zihao, Mr. Lin Wei, and Mr. Xue Yang, his research staff Dr. Zhang Wenmei, Dr. Sang-Hyuk Wi, Dr. Tu Zhihong, and Ms. Zhang Lin, and his collaborators Dr. Duixian Liu and Mr. Brian P. Gaucher of the IBM Thomas J. Watson Research Center and Dr Albert Lu, Mr. Chua Kai Meng, and Ms. Wai Lai Lai of the Singapore Institute of Manufacturing Technology for their contribution in the development of AiP technology.

## References

1 B. Song, CMOS RF circuits for data communications applications. *IEEE Journal of Solid-State Circuits*, vol. SC-21, pp. 310–317, February 1986.
2 A. Abidi, A. Rofougaran, G. Chang et al., The future of CMOS wireless transceivers. In *ISSCC Digest of Technical Papers, IEEE*, February 1997.
3 F. Eynde, J.-J. Schmit, V. Charlier et al., A fully integrated single-chip SoC for Bluetooth. In *ISSCC Digest of Technical Papers, IEEE*, February 2001.
4 S.K. Reynolds, B.A. Floyd, U.R. Pfiffer et al., A silicon 60-GHz receiver and transmitter chipset for broadband communications. In *ISSCC Digest of Technical Papers*, IEEE, February 2001.
5 D. Liu, U. Pfeiffer, J. Grzyb, and B. Gaucher (eds.), *Advanced Millimeter-wave Technologies: Antennas, Packaging and Circuits*. Wiley, 2009.
6 Y.P. Zhang, Antenna-in-package technology: Its early development. *IEEE Antennas Propagation Magazine*, vol. 61, no. 3, pp. June 2019.
7 K.K. O, K. Kim, B.A. Floyd et al., On-chip antennas in silicon ICs and their applications. *IEEE Transactions on Electron Devices*, vol. 52, no. 7, pp. 1312–1323, July 2005.
8 Y. Hirachi, H. Nakano, and A. Kato, A cost-effective RF-module with built-in patch antenna for millimeter-wave wireless systems. In *Proceedings of the European Microwave Conference*, Munich, Germany, 5–7 October 1999.
9 https://www.anokiwave.com/.
10 Y.X. Qian and T. Itoh, Progress in active integrated antennas and their applications. *IEEE Transactions on Microwave Theory and Techniques*, vol. 46, no. 11, pp. 1891–1900, November 1998.
11 R.E. Munson, Conformal microstrip arrays and microstrip phased arrays. *IEEE Transactions on Antennas and Propagation*, vol. 22, no. 1, pp. 74–78, February 1974.
12 W. Lin, *Integrated Circuit Package Antenna*, M.Sc. thesis, School of Electrical and Electronic Engineering, Nanyang Technical University, Singapore, 2003.

**13** C.T.P. Song, P.S. Hall, H. Ghafouri-Shiraz et al., Packaging technique for gain enhancement of electrically small antenna designed on gallium arsenide. *Electronics Letters*, vol. 36, no. 18, pp. 1524–1525, 2000.

**14** C.T.P. Song, P.S. Hall, and H. Ghafouri-Shiraz, Novel RF front end antenna package. *IEE Proceedings – Microwaves, Antennas and Propagation*, vol. 150, no. 4, pp. 290–294, August 2003.

**15** K.T. Lim, A. Obatoyinbo, A. Sutono et al., A highly integrated transceiver module for 5.8 GHz OFDM communication system using multi-layer packaging technology. In *IEEE International Microwave Symposium Digest*, Denver, CO, USA, 8–13 June 2001.

**16** D.J. Mathews, R.J. Hill, M.P. Gaynor et al., Multi-chip semiconductor package with integral shield and antenna. US Patent 6686649B1.

**17** J. Ryckaert, S. Brebels, B. Come et al., Single-package 5 GHz WLAN RF module with embedded patch antenna and 20 dBm power amplifier. In *IEEE Microwave Symposium Digest*, Philadelphia, PA, USA, 8–13 June 2003.

**18** A.P. Popov and M.D. Rotaru, A novel integrated dielectric resonator antenna for circular polarization. In *Proceedings of the IEEE Electronic Components and Technology Conference* New Orleans, LA, USA, 27–30 May 2003.

**19** K.W. Leung, *The hollow DRA and its novel application as a packaging cover*. Hong Kong Research Grants Council, 2004.

**20** J.S. Castany, C. Puente, and J. Mumbruforn, Antenna-in-package with reduced electromagnetic interaction with on chip elements. US Patent 8330259B2.

**21** G. Brzezina, L. Roy, and L. MacEachen, Planar antennas in LTCC technology with transceiver integration capability for ultra-wideband application. *IEEE Transactions on Microwave Theory and Techniques*, vol. 54, no.6, pp. 2830–2839, June 2006.

**22** M. Sun, Y.P. Zhang, and Y.L. Lu, Ultra-wideband integrated circuit package antenna in LTCC. In *Proceedings of the Asia Pacific Microwave Conference*, Yokohama, Japan, 12–15 December 2006.

**23** S.H. Wi, J.-S. Kim, N.-K. Kang et al., Package-level integrated LTCC antenna for RF package application. *IEEE Transactions on Advanced Packaging*, vol. 30, no. 1, pp. 132–141, February 2007.

**24** U.R. Pfeiffer, J. Grzyb, D. Liu et al., A chip-scale packaging technology for 60-GHz wireless chipsets. *IEEE Transactions on Microwave Theory and Techniques*, vol. 54, no. 8, pp. 3387–3397, August 2006.

**25** N. Shino, H. Uchimura, and K. Miyazato, 77GHz band antenna array substrate for short range car radar. In *Proceedings of the IEEE International Microwave Symposium Digest*, Long Beach, CA, USA, 12–17 June 2005.

**26** Y.P. Zhang, Antenna-in-package (AiP) technology for modern radio systems. In *Proceedings of the IEEE International Workshop on Antenna Technology*, New York, USA, 6–8 March 2006.

**27** Y.P. Zhang, M. Sun, K.M. Chua et al., Antenna-in-package in LTCC for 60-GHz radio. In *Proceedings of the IEEE International Workshop on Antenna Technology*, Cambridge, UK, 21–23 March, 2007.

**28** http://www.sibeam.com/

**29** Y.P. Zhang, Integrated circuit package antenna: An elegant antenna solution for single-chip RF transceiver. In *Proceedings of the Antenna Systems and Short-Range Wireless Conference*, Santa Clara, CA, USA, 22–23 September 2005.

**30** https://www.cadence.com/

**31** A. Panther, A. Petos, M.G. Stubbs et al., A wideband array of stacked patch antennas using embedded air cavities in LTCC. *IEEE Microwave and Wireless Components Letters*, vol. 15, no. 12, pp. 916–918, December 2005.

**32** Y.P. Zhang and M. Sun, Grid array antennas and an integration structure, US Patent 8842054B2.

**33** https://www.ieeeaps.org/about-the-transactions/best-paper-awards.

**34** T. Tajima, H.J. Song, M. Yaita et al., 300-GHz LTCC horn antennas based on antenna-in-package technology. In *Proceedings of the European Microwave Conference*, Durenberger, Germany, 6–11 October 2013.

**35** W.B. Hong, K.H. Baek, and A. Goudelev, Multilayer antenna package for IEEE 802.11ad employing ultralow-cost FR4. *IEEE Transactions on Antennas and Propagations*, vol. 60, no. 12, pp. 5932–5938, December 2012.

**36** T. Kamgaing, A.A. Elsherbini, S.N. Oster et al., Low-profile fully integrated 60 GHz 18 element phased array on multilayer liquid crystal polymer flip chip package. In *Proceedings of the IEEE Electronic Components Technology Conference*, San Diego, CA, USA, 26–29 May 2015.

**37** M. Wojnowski, R. Lachner, J. Böck et al., Embedded wafer level ball grid array (eWLB) technology for millimeter-wave applications. In *Proceedings of the IEEE Electronics Packaging Technology Conference*, Singapore, 7–9 December 2011.

**38** https://www.tsmc.com/english/default.htm.

**39** http://ase.aseglobal.com/.

**40** T. Zwick, C. Baks, U.R. Pfeiffer et al., Probe based MMW antenna measurement setup. In *Proceedings of the IEEE Antennas and Propagation Symposium*, Monterey, CA, USA, 20–25 June 2004.

**41** T. Ito, Y. Tsutsumi, S. Obayashi et al., Radiation pattern measurement system for millimeter-wave antenna fed by contact probe. In *Proceedings of the European Microwave Conference*, Rome, Italy, 29 September–1 October 2009.

**42** K. Mohammadpour-Aghdam, S. Brebels, A. Enayati et al., RF probe influence study in millimeter-wave antenna pattern measurements. *International Journal of Radio Frequency and Microwave Computer-Aided Engineering*, vol. 21, no. 4, July 2011.

**43** D. Titz, F. Ferrero, and C. Luxey, Development of a millimeter-wave measurement setup and dedicated techniques to characterize the matching and radiation performance of probe-fed antennas. *IEEE Antennas and Propagation Magazine*, vol.54, no. 4, pp.188–203, August 2012.

**44** A.C.F. Reniers, A.R.V. Dommele, A.B. Smolders et al., The influence of the probe connection on mm-wave antenna measurements. *IEEE Transactions on Antennas and Propagation*, vol. 63, no. 9, pp. 3819–3825, September 2015.

**45** https://xcerra.com/multitest.

**46** J.G. Lu, *Synthetic Aperture Radar Design and Technology*. National Defence Industry Press, Beijing, 2016.

# 2

# Antennas

*Yueping Zhang*

*School of Electrical and Electronic Engineering, Nanyang Technological University, Singapore*

## 2.1 Introduction

An antenna-in-package (AiP) acts as an antenna. A good AiP design must incorporate not only fundamental antenna characteristics but also specific package constructions, for example turn one lead to a monopole or two leads as a dipole of a dual-in-line package. Unfortunately, the dual-in-line package, because of inherent large parasitic effects, is not suitable for many AiP designs. As a result, an antenna has to be integrated into a surface-mounted package. The antenna is usually chosen from popular antenna types such as dipole, monopole, loop, slot, Yagi-Uda, and patch, which can be quickly designed or modified for the application [1–22]. The choice of a grid array antenna for AiP design is an exception. The grid array antenna had not found any commercial application until it was adopted in AiP design for 60-GHz radios [23]. In addition, several unusual antennas have been specifically developed for AiP technology: laminated resonator, dish-like reflector, slab waveguide, horn-like, differentially fed aperture, and step-profiled corrugated horn antennas [24–30]. In this chapter an attempt is made to summarize these antennas. Special emphasis is given to microstrip patch antennas and arrays, grid array antennas, Yagi-Uda antennas, and magneto-electric dipole antennas. Performance improvement techniques of antennas for AiP technology are also described.

## 2.2 Basic Antennas

### 2.2.1 Dipole

A dipole is a basic antenna with such properties as simplicity, small size, and linear polarization. The dipole is called a printed dipole if printed on a substrate and

*Antenna-in-Package Technology and Applications,* First Edition.
Edited by Duixian Liu and Yueping Zhang.

a microstrip dipole if printed on a grounded substrate. The length of the dipole needs to be slightly less than half a wavelength for resonance. The width of the dipole can be chosen as one third of the length. At resonance, the resistance of the microstrip dipole is very small for thin substrate. Thus, it is difficult to match to a 100-Ω source. To overcome this matching problem, a folded dipole can be used instead since the resistance is about four times of that of a normal dipole. For example, Zwick et al. chose a folded dipole [1]. A single folded dipole was designed with metal bars for surface wave suppression. A dual folded dipole array was also designed. The half-wavelength separation of the two dipoles canceled the surface wave. The dipole was printed on the bottom of a fused silica superstrate with a ground plane below. The folded dipole and array both achieved more than 10% bandwidth and 8-dBi gain at 64 GHz.

### 2.2.2 Monopole

A monopole is another basic antenna. Its planar form has broad bandwidth and is suitable for ultra-wideband (UWB) radios [17–22]. A symmetrical beveled shape printed on the top surface of an ungrounded substrate yields a typical planar monopole. Openings can be cut at proper locations on the monopole so that they do not degrade the antenna electrical performance but improve the mechanical stability and avoid warpage. The monopole is fed with a coplaner waveguide (CPW). The monopole and the CPW share the same ground patches on the same surface. It is found that the gain of the antenna improves for the longer monopole length at lower frequencies over the wider bandwidth. A wider monopole width is beneficial to improve the gain and directivity of the antenna, in particular at higher frequencies. There is current distribution on the ground patches. The properties of the antenna are inevitably affected by the ground patch size. Furthermore, the current is mainly distributed on the upper edges of the ground patches. That means that this portion of the ground patches acts as part of the radiating structure. A narrower ground patch deteriorates the voltage standing wave ratio (VSWR) near the lower and upper edge frequencies. A wider ground patch improves the VSWR. The VSWR stays nearly constant for the ground width that exceeds the half-guided wavelength. A reasonable width can be set as the half-guided wavelength. The lower edge frequency is more sensitive to the ground length, showing a fluctuation with the length. The length should be longer than a quarter and shorter than a half-guided wavelength. When the length is longer than a half-guided wavelength, the fluctuation indicates the influence of higher-order modes on the dominant one.

### 2.2.3 Loop

A loop is a third basic antenna, which is like a dipole and has similar properties. The total length of a typical loop antenna is approximately one wavelength. In that case, the loop input impedance is low because the current intensity is strong at the feed point. Consequently, impedance matching to the 100-Ω source is difficult. To overcome this matching problem, a loop of one-and-half wavelengths can be used instead since the current intensity at the feed point is low and the loop input impedance is higher. Tsutsumi et al. selected a loop of one and half wavelengths for easy formation with bonding wires [2]. The loop was made with two bonding wires and two pads connected with a U-shaped metal strip in the package. It exhibited an impedance bandwidth from 56 to 64 GHz and gain from −2.4 to 4.9 dBi over the frequency range of 57–65 GHz.

### 2.2.4 Slot

A slot is also a basic antenna. It is cut in a ground plane. The length of the slot needs to be slightly less than half a wavelength for resonance. The width of the slot is typically much narrower than the length, although the bandwidth of the slot can be enhanced by making it wider. Unlike a dipole or a loop, differential feeding is not required. A microstrip line or a CPW can feed it. For example, Yoshida et al. used a coplanar waveguide to feed a slot in a ground plane on a substrate [3]. The slot obtained an impedance bandwidth of 26.3% and a peak realized gain of 7.1 dBi at 60 GHz.

## 2.3 Unusual Antennas

### 2.3.1 Laminated Resonator Antenna

A laminated resonator antenna consists of an opened cavity and a feeding slot [24]. The opened cavity is realized with a metallic fence of vias and floor embedded in a dielectric body. The pitch between two adjacent vias should be kept small to make the fence of vias act as an electrical wall. The opened cavity is designed to resonate at the fundamental frequency, which can be calculated using the formula based on the dimensions and dielectric constant of the corresponding closed cavity resonator. The dimension height is critical. It is doubled in the calculation formula to account for the open-ended effect. It should also be a quarter wavelength for better radiation. The slot is cut in the floor to couple the energy to the opened cavity to radiate into the air. The size, shape, location, and orientation of the slot should be designed together with the opened cavity for optimum antenna performance [25]. It has been shown that the laminated resonator antenna is a promising candidate for 77-GHz automotive applications and ideally suited for low-temperature co-fired ceramic (LTCC)-based AiP technology [26].

### 2.3.2 Dish-like Reflector Antenna

A dish-like reflector antenna consists of a driving patch, a ground plane, a primary reflector, and a dish-like reflector [27]. The driving patch and the ground plane form a typical microstrip antenna. The primary reflector is a metallic annulus on the top surface of the whole structure. The dish-like reflector is composed of stacked metallic rings within multiple laminated layers. It is sandwiched between the primary reflector and the ground plane. The wave launched from the microstrip antenna is first reflected by the primary reflector and then reflected again by the dish-like reflector into the air. The gain and bandwidth of the dish-like reflector antenna can be adjusted through varying the size and placement of the various reflective conductors. A dish-like reflector antenna in an LTCC achieved a gain of more than 10 dBi in the 60-GHz band.

### 2.3.3 Slab Waveguide Antenna

A slab waveguide antenna is composed of a post-wall waveguide, a dielectric slab waveguide, and a feeding network [28]. The feeding network is a cascaded interconnect from the ground-signal-ground (GSG) pads through a microstrip line to a microstrip line/post-wall waveguide transition. To achieve broadband characteristics, the dielectric slab waveguide is arranged in front of the open-ended post-wall waveguide. Two metal side walls made of vias are built on both sides of the dielectric slab waveguide. The transition between the post-wall waveguide and the dielectric slab waveguide should be designed to realize lower reflection and not to excite evanescent modes. The conversion of the transverse electric ($TE_{10}$) mode in the post-wall waveguide to the transverse magnetic ($TM_0$)-like mode in the dielectric slab waveguide benefits integration and improves the isolation between slab waveguide antennas. For example, two slab waveguide antennas were integrated side by side into an antenna package. Measured results showed that excellent isolation between the two antennas of more than 38.1 dB (average 46 dB) was achieved from 59 to 66 GHz.

### 2.3.4 Differentially Fed Aperture Antenna

A differentially fed aperture antenna is devised with an H-shaped opening cavity in which there is a long cross-shaped patch fed differentially by two grounded coplanar waveguides [29]. The opened cavity is designed to have both the width and length larger than one wavelength and the height a quarter wavelength. The field distribution within the cavity is not in the resonant modes. The energy goes into the cavity, splits into two parts, and propagates on the patch as a traveling wave. The patch excites a uniform field distribution on the aperture. The field around the edges of the cavity also contributes to the radiation and helps increase the electrical size of the aperture and thus the gain. Experiment results showed that the differentially fed aperture antenna achieved a −15-dB impedance bandwidth from 56.2 to 69.7 GHz, peak gain of 15.3 dBi at 60 GHz, and 3-dB gain bandwidth from 54.0 to 67.5 GHz.

### 2.3.5 Step-profiled Corrugated Horn Antenna

A step-profiled corrugated horn antenna can be created by flaring the dimensions of a rectangular waveguide along both the E-plane and the H-plane [30]. The flare of the step-profiled corrugated horn antenna is realized by the stepped cavity surrounded by the fence of vias and metal layers. The step-profiled corrugated horn antenna is ideally suited for LTCC-based AiP designs to operate at 300 GHz because the hollow structure can reduce dielectric loss of LTCC and enhance antenna efficiency. A step-profiled corrugated horn antenna with size $5 \times 5 \times 2.8\,\text{mm}^3$ in an LTCC achieved 18-dBi peak gain at 300- and 100-GHz bandwidth with more than 10-dB return loss.

## 2.4 Microstrip Patch Antennas

Microstrip patch antennas find widespread application in AiP technology [6–16] due to the availability of abundant design techniques and versatility to meet design requirements. This section presents some techniques considered important for AiP design and new research results not covered in any other book.

### 2.4.1 Basic Patch Antennas

Figure 2.1 shows the structure of a basic patch antenna. It is printed on the top surface of a grounded substrate of relative permittivity $\varepsilon_r$, loss tangent $\delta$, length $a$, width $b$, and thickness $h$. The metal patch of width $w$ and length $l$ can be excited at feed 1 or 2 or both.

Consider a general case where two feeds at $(x_1, y_1)$ and $(x_2, y_2)$ are excited simultaneously. The antenna becomes a two-port network with the self-impedances $Z_{11}$ and $Z_{22}$, and mutual impedances $Z_{12}$ and $Z_{21}$. For an electrically thin substrate, the self- and mutual impedances $Z_{11}$ and $Z_{12}$ are given by the cavity model [31] as:

$$Z_{11} = j\omega\mu_0 h \sum_{m,n=0}^{\infty} \frac{\varphi_{mn}^2(x_1, y_1) j_0^2\left(\frac{m\pi d}{2w}\right)}{k_{mn}^2 - k_e^2} = \sum_{m,n=0}^{\infty} Z_{11,mn} \tag{2.1}$$

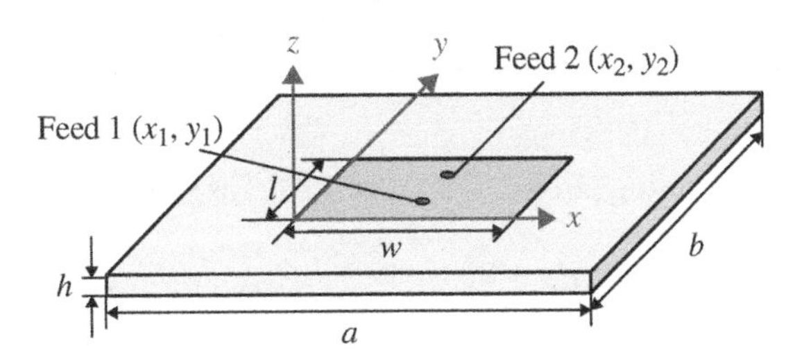


**Figure 2.1** Structure of a basic patch antenna and coordinate system.

and

$$Z_{12} = j\omega\mu_0 h \sum_{m,n=0}^{\infty} \frac{\varphi_{mn}(x_1, y_1)\varphi_{mn}(x_2, y_2) j_0^2\left(\frac{m\pi d}{2w}\right)}{k_{mn}^2 - k_e^2} = \sum_{m,n=0}^{\infty} Z_{12,mn} \tag{2.2}$$

where

$$\varphi_{mn}(x, y) = \frac{\chi_{mn}}{\sqrt{wl}} \cos\left(\frac{m\pi}{w}\right) \cos\left(\frac{n\pi}{l}\right), \tag{2.3}$$

$$j_0(x) = \frac{\sin(x)}{x}, \tag{2.4}$$

$$k_{mn}^2 = \left(\frac{m\pi}{w}\right)^2 + \left(\frac{n\pi}{l}\right)^2, \tag{2.5}$$

$$k_e^2 = \varepsilon_r(1 + j\delta_e)k_0^2, \tag{2.6}$$

$$\chi_{mn} = \begin{cases} 1, m = 0 \ and \ n = 0 \\ \sqrt{2}, m = 0 \ or \ n = 0 \\ 2, m \neq 0 \ and \ n \neq 0 \end{cases}, \tag{2.7}$$

$$k_0 = \frac{2\pi}{\lambda}. \tag{2.8}$$

The parameter $d$ is the feed probe diameter and $\delta_e$ is the effective loss tangent. The self- and mutual impedances $Z_{22}$ and $Z_{21}$ can be expressed in a similar way.

The input impedance of the differential patch antenna $Z_d$ can be obtained [32] as:

$$Z_d = Z_{11} - Z_{12} - Z_{21} + Z_{22}. \tag{2.9}$$

Due to symmetry, $Z_d$ is simplified as:

$$Z_d = 2(Z_{11} - Z_{12}) = 2(Z_{22} - Z_{21}). \tag{2.10}$$

The input impedance of the single-ended patch antenna $Z_s$ can be obtained as:

$$Z_s = Z_{11} - \frac{Z_{12}^2}{Z_{11}Z_L} \tag{2.11}$$

where $Z_L$ represents the load impedance connected to feed 2, which can be changed for tuning, matching, or modifying patterns. In the simplest case, we consider $Z_L$ to be infinitely large, that is, feed 2 is opened or there would not be feed 2, so $Z_s$ is simplified as:

$$Z_s = Z_{11}. \tag{2.12}$$

Substituting (2.1) and (2.2) into (2.10) as well as (2.1) into (2.12), we get

$$Z_d = 2(Z_{11} - Z_{12}) = 2 \sum_{m,n=0}^{\infty} (Z_{11,mn} - Z_{12,mn}), \tag{2.13}$$

$$Z_s = \sum_{m,n=0}^{\infty} Z_{11,mn}. \tag{2.14}$$

From (2.1) and (2.2), we find

$$\frac{Z_{11,mn}}{Z_{12,mn}} = \frac{\varphi_{mn}(x_1, y_1)}{\varphi_{mn}(x_2, y_2)} = \frac{\cos\left(\frac{m\pi x_1}{w}\right)\cos\left(\frac{n\pi y_1}{l}\right)}{\cos\left(\frac{m\pi x_2}{w}\right)\cos\left(\frac{n\pi y_2}{l}\right)}. \tag{2.15}$$

Since $x_1 = x_2$ and $y_2 = l - y_1$,

$$\frac{Z_{11,mn}}{Z_{12,mn}} = \frac{\varphi_{mn}(x_1, y_1)}{\varphi_{mn}(x_2, y_2)} = \frac{\cos\left(\frac{n\pi y_1}{l}\right)}{\cos\left(\frac{n\pi(l - y_1)}{l}\right)} = \begin{cases} 1, & n = 2, 4, 6 \ldots \\ -1, & n = 1, 3, 5 \ldots \end{cases}. \tag{2.16}$$

Consequently,

$$Z_d = 2\sum_{m,n=0}^{\infty} (Z_{11,mn} - Z_{12,mn}) = 4 \sum_{m=0,n=2i+1}^{\infty} Z_{11,mn} \tag{2.17}$$

where $i = 0, 1, 2, 3, \ldots$. It is important to note that $Z_d$ has no term where $n$ is an even number [33]. As for the single-ended impedance $Z_s$, it is given by:

$$Z_s = \sum_{m,n=0}^{\infty} Z_{11,mn}. \tag{2.18}$$

Note that $Z_s$ has terms with $n$ either odd or even. It is interesting to examine the ratio of $Z_d$ to $Z_s$:

$$\frac{Z_d}{Z_s} = 4\frac{\sum_{m=0,n=2i+1}^{\infty} Z_{11,mn}}{\sum_{m,n=0}^{\infty} Z_{11,mn}} \approx 4. \tag{2.19}$$

The ratio is approximately 4 because the basic patch antenna operates at the fundamental $TM_{01}$ mode under differential or single-ended excitation. The ratio shows that if the single-ended patch antenna matches to a 50-Ω source, the differential patch antenna will match to a 200-Ω source. It should also be noted that the patch size in the above derivation remains the same for both differential and single-ended excitations. Hence, the size ratio is 1. In fact the size ratio can be larger than 1, which will become clear shortly.

The quality factor is an important parameter for the basic patch antenna, which is given by:

$$Q = \left[\frac{120\lambda h G_r}{\varepsilon_r w l\left(1 - \frac{3.4h\sqrt{\varepsilon_r - 1}}{\lambda}\right)} + \frac{1}{\pi h}\sqrt{\frac{\lambda}{120\sigma_c}} + \tan\delta\right]^{-1} \tag{2.20}$$

for $h/\lambda < 0.06$, where the first term is related to power loss due to radiation and surface waves, and the last two terms are related to power loss due to conductor and dielectric materials [34]. The quality factor due to radiation only can be expressed as:

$$Q_r = \frac{\varepsilon_r wl}{120\lambda h G_r} \tag{2.21}$$

where $G_r$ is the radiation conductance of the basic patch antenna, which is given by:

$$G_r = \frac{1}{60}\frac{w}{\lambda} - \frac{1}{30\pi^2} \tag{2.22}$$

for differential and

$$G_r = \frac{1}{45}\left(\frac{w}{\lambda}\right)^2 \tag{2.23}$$

for single-ended excitations, respectively. Substituting (2.22) and (2.23) into (2.20) and (2.21), respectively, we can obtain the quality factors $Q_d$ and $Q_s$ as well as $Q_{rd}$ and $Q_{rs}$ with patch widths $w_d$ and $w_s$ for the differential and single-ended patch antennas, respectively. Therefore, we can find the impedance bandwidth radio of $B_d$ to $B_s$:

$$\frac{B_d}{B_s} = \frac{Q_s}{Q_d} \approx \frac{w_d}{w_s}. \tag{2.24}$$

Similarly, we can find the radiation efficiency ratio of $\eta_d$ to $\eta_s$:

$$\frac{\eta_d}{\eta_s} = \frac{Q_d}{Q_s}\frac{Q_{rs}}{Q_{rd}} \geq 1. \tag{2.25}$$

The reason why the radiation efficiency ratio can be more than 1 is that the surface wave can be suppressed by the differential excitation.

In the design of a differential patch antenna, the electrical separation $ES$, defined as the ratio of physical distance between the two feeds to the free-space wavelength, plays an important role [35, 36]. The resonance of the differential patch antenna requires:

$$ES > 0.1. \tag{2.26}$$

It was found that the design formula for a single-ended patch antenna could be used for the design of a differential patch antenna. However, the $ES$ must be observed and satisfied. Usually, the $ES$ can be satisfied with an increased patch width $w$. For the single-ended patch antenna, $w$ is usually $1.5l$, while for the differential patch antenna, $w$ can be equal to or larger than $2l$. In other words, $w_d$ can be larger than $w_s$, which shows from (2.24) that the impedance bandwidth of the differential patch antenna is wider than that of the single-ended patch antenna. Therefore, together with (2.25), we can conclude that the differential patch antenna outperforms the single-ended patch antenna.

Finally, we turn to the design of the basic patch antenna for dual linear polarizations only under single-ended excitation. This requires the basic patch to be a square shape $w = l$, feed 1 to be at $(x_1, y_1) = (0.5w, y_1)$, feed 2 to be at $(x_2, y_2) = (x_2, 0.5l)$, and the coordinate values for $y_1$ and $x_2$ to be the same. Thus, if feeds 1 and 2 are excited with the same signal, the antenna will radiate for dual linear polarizations. If feeds 1 and 2 are excited with the signals of the same amplitude but 90° phase difference, the antenna will radiate for single circular polarization. Four feeds will be needed for the antenna to radiate for dual circular polarizations.

### 2.4.2 Stacked Patch Antennas

A fundamental drawback of basic patch antennas is narrow impedance bandwidth, typically in the order of a few percent for a VSWR of less than 2. To enhance impedance bandwidth, a large body of research has been conducted, resulting in several bandwidth enhancement techniques. The technique utilizing the double-resonance phenomenon is probably the most common one. A double resonance can be easily created by mounting a parasitic patch on a driven patch, leading to a stacked patch antenna. It was found that considerable bandwidth improvement could be achieved when the size of the parasitic patch was equal to or slightly larger than that of the driven patch [37]. As well as enhanced impedance bandwidth, stacked patch antennas have additional advantages such as filter characteristics, easy implementation of electrostatic discharge (ESD), and facilitation to meet the metal density requirement and to remove heat [38–43]. Hence, AiP designers favor stacked patch antennas.

Figure 2.2 shows the exploded view of a stacked patch antenna. As can be seen from the figure, the driven patch is embedded between a grounded substrate and a covered superstrate of relative permittivity $\varepsilon_r$, loss tangent $\delta$, length $a$, and width $b$, while the parasitic patch is printed on the top surface of the superstrate. Both patches are centered relative to the ground plane. The driven patch of width $w_1$ and length $l_1$ is fed by a coaxial feed through the ground plane. The parasitic patch

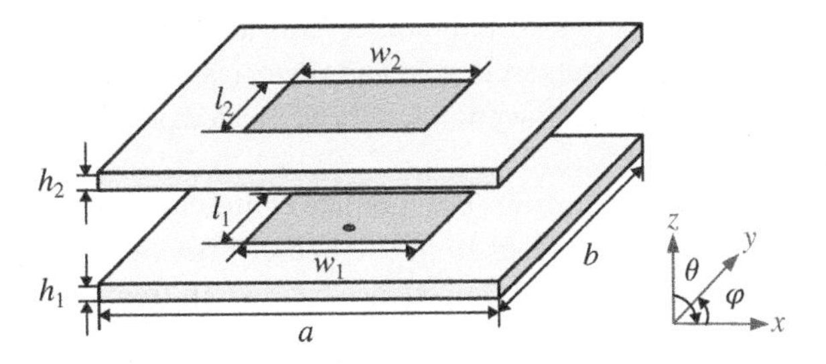


**Figure 2.2** Structure of a stacked patch antenna and coordinate system.

**Figure 2.3** Photograph of a stacked patch antenna.

has width $w_2$ and length $l_2$. The thicknesses for the substrate and superstrate are denoted by $h_1$ and $h_2$, respectively.

The stacked patch antenna can be considered as two coupled resonators with one formed by the driven patch and the ground plane and the other by the driven and parasitic patches. The design involves determining the resonant frequencies of the two resonators and the coupling level between them. Generally, the resonant frequencies and coupling level are obtained by parameter scan with a full-wave simulator. The designer often finds that the matching is good around the two resonant frequencies but poor in a band between the two resonant frequencies. The poor matching is due to the inappropriate coupling level. In fact, formulas exist for calculation of the resonant frequencies [41]. The coupling between the two resonators can be monitored with a method proposed by Wu et al. based on the theory of a coupling matrix for the design of band pass filters [42]. The method treats the design of the stacked patch antenna as the synthesis, diagnosis, and tuning of a two-order band pass filter. The input port of the filter is the input port of the antenna, while the output port of the filer is the far-field radiation of the antenna.

A stacked patch antenna was designed and fabricated on an LTCC substrate of relative permittivity $\varepsilon_r = 5.9$ and loss tangent $\delta = 0.002$ in the 28-GHz band for 45° polarization [43]. The antenna used six ceramic layers and the thickness of each layer was 96 μm. There were two ceramic layers between the driven patch and the ground plane and four ceramic layers between the driven and parasitic patches. In addition, there were four ceramic layers beneath the ground plane and two metal layers to create a stripline to feed the antenna from a CPW port. The outer and inner metal layers were gold and silver, respectively, with thicknesses of 8–10 μm. Figure 2.3 shows the photos of the fabricated prototype. The antenna total size was $6.0 \times 6.0\,\text{mm}^2$.

Figure 2.4 compares the simulated with the measured $|S_{11}|$ of the stacked patch antenna. It is evident from the figure these values are in good agreement. The antenna achieved an impedance bandwidth from 26.43 to 29.32 GHz or 10.3% at 28 GHz.

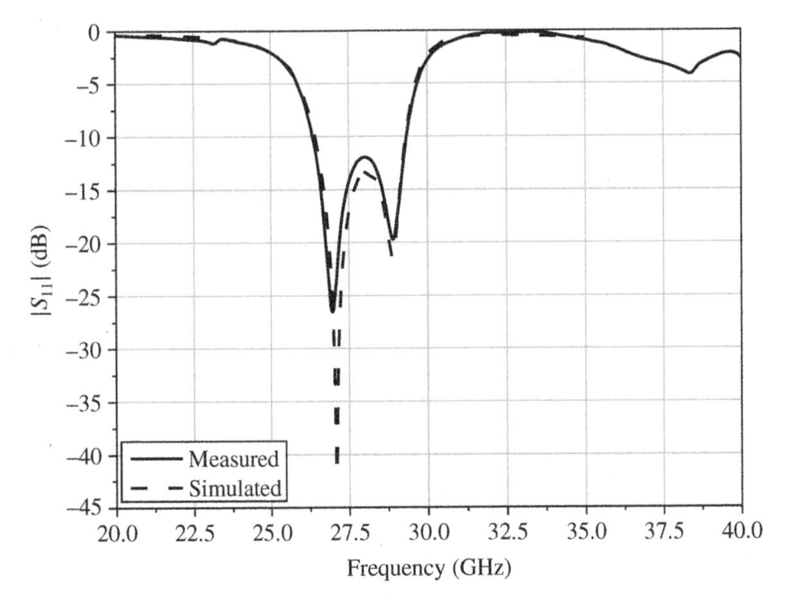


**Figure 2.4** Simulated and measured $|S_{11}|$ of a stacked patch antenna.

Figure 2.5 compares the simulated with the measured realized gain of the stacked patch antenna at the boresight direction. As can be seen, they are in good agreement. The antenna achieved a gain of 5 dBi at 28 GHz and a 3-dB gain bandwidth from 25.5 to 30 GHz or 16.1% at 28 GHz.

### 2.4.3 Patch Antenna Arrays

The maximum gain of a basic or stacked patch antenna is less than 10 dBi. To achieve higher gains, patch antenna arrays are required. Patch antenna arrays are designed as fixed-beam, switched-beam, and phased arrays.

Several fixed-beam patch antenna arrays have been developed as 60-GHz AiPs. Figure 2.6 shows a linearly polarized $2 \times 4$ patch antenna array constructed on a fused-silica substrate of size $10.49 \times 12.6 \times 0.3\,\text{mm}^3$ [44]. It is fed by a network involving both a coplaner strip (CPS) and CPW, and is enhanced by a metal cavity to achieve a measured impedance bandwidth of more than 9 GHz and maximum gain of about 15 dBi at 60 GHz. A circularly polarized patch antenna array reduces the sensitivity to antenna orientation, and it makes the overall system more resistant to multipath effects. Figure 2.7 shows a circularly polarized $4 \times 4$ patch antenna array on a multilayer LTCC substrate of size $13 \times 13 \times 0.9\,\text{mm}^3$ [45]. By applying a stripline sequential rotation feed network, the antenna array exhibits a wide impedance bandwidth of 12 GHz for $\text{VSWR} \leq 2$ and 3-dB axial

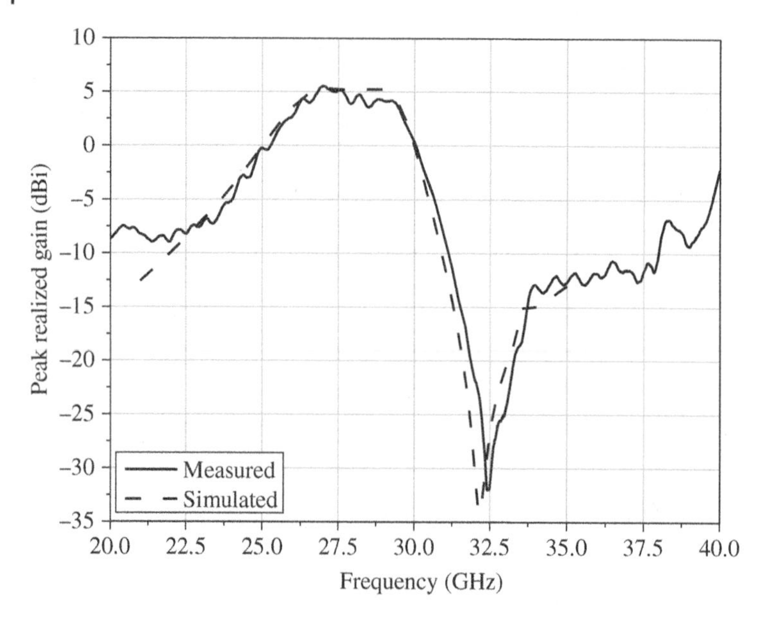


**Figure 2.5** Simulated and measured gain of a stacked patch antenna.

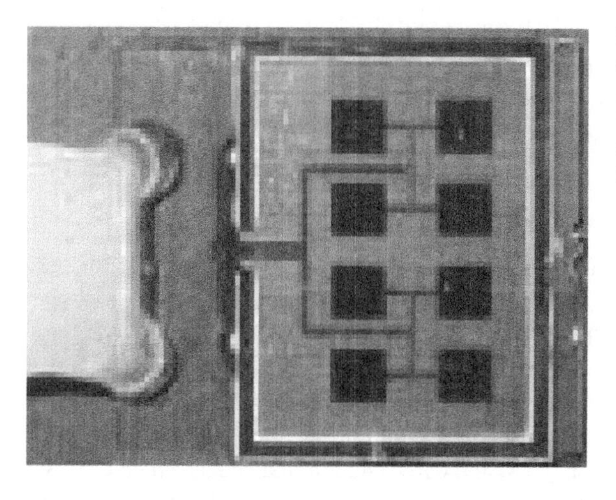

**Figure 2.6** Photograph of a linearly polarized 2 × 4 patch antenna array (from [44], © 2008 IEEE, reprinted with permission).

ratio bandwidth of 8.9 GHz, and a peak gain of 17.1 dBi at 63 GHz. The gains of patch antenna arrays can be further enhanced using soft surface structures. It was found that the gain enhancement can be up to 2 dB at 60 GHz [46].

A switched-beam antenna array uses a switch circuit in front of a fixed beam-forming network such as a Rotman lens or a Butler matrix to select the best signal path. It may be a cost-effective approach to implementing steerable antennas.

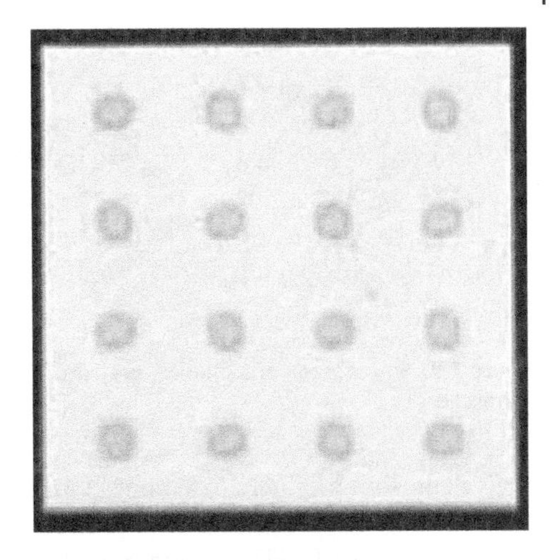

**Figure 2.7** Photograph of a circularly polarized 4×4 patch antenna array (from [45], © 2013 IEEE, reprinted with permission).

**Figure 2.8** Photograph of a switched-beam patch antenna array (from [47], © 2008 IEEE, reprinted with permission).

Figure 2.8 shows a switched-beam patch antenna array using advanced thin-film and flip-chip technology on a silicon mother board. A 4×8 Rotman lens was designed on a 30-μm-thick benzocyclobutane substrate to form four beams in the ±6° and ±18° directions at 60 GHz. The flip-chip structure was optimized for heat dissipation and compensated to minimize the parasitic effects in the 60-GHz band. The fabricated switched-beam array antenna shown in Figure 2.8 produced

**Figure 2.9** Photograph of a phased array (from [11], © 2011 IEEE, reprinted with permission).

equivalent isotropically radiated power (EIRP) higher than 17.3 dBm in the ±6° and ±20° directions [47].

When compared with a fixed-beam, and even a switched-beam, antenna, phased arrays provide higher EIRP for the transmitter and higher signal-to-noise ratio (SNR) for the receiver. Figure 2.9 shows a phased array in a multilayer LTCC substrate of size $13 \times 20 \times 1.4\,\text{mm}^3$ [11]. A radio frequency integrated circuit (RFIC) is attached to the backside of the substrate with 16 antennas. The $4 \times 4$ patch antennas are organized in a triangular lattice. Beside antenna gain, impedance matching, polarization, and mutual coupling, antenna element gain variation is also an important issue for small phased array designs. When the coupling between array elements is weaker than −15 dB, the return loss of an antenna will be almost unaffected by the nearby antennas. However, the antenna element radiation pattern or gain is still strongly affected by the nearby elements so the element gain is dependent on its location. This is unlike an infinite array where all antenna elements have the same return loss, radiation patterns, and gains. It was found though extensive simulations that there is about a 2-dB gain variation depending on an antenna location for the array size and layout. Dummy antenna elements can be added to the packages outside the array to reduce the variation but that will increase the substrate size and manufacturing cost.

## 2.5 Microstrip Grid Array Antennas

The grid array antenna was invented by Kraus in 1962 as a travelling-wave (non-resonant) antenna with the main lobe of radiation in a backward angle-fire direction [48]. The grid array antenna was implemented in microstrip technology by Conti et al. in 1981 as a standing-wave (resonant) antenna with the main beam of radiation in the broadside direction. They showed that the grid array antenna in microstrip technology has all the usual benefits of conventional patch-type radiators plus broad bandwidth, high gain, and adequate cross-polarization control [49]. However, the grid array antenna had not found any commercial application until Sun and Zhang first applied it to 60-GHz radios in 2008 [23], followed by Menzel et al. to 79 GHz [50], Chen et al. to 94 GHz [51], Zwick et al. to 122-GHz radars [52], and Zhang to 145-GHz systems [53, 54].

### 2.5.1 Basic Configuration

Figure 2.10 shows the structure of a microstrip grid array antenna. The grid array radiator is printed on the top surface of a substrate of relative permittivity $\varepsilon_r$, loss tangent $\delta$, length $a$, width $b$, and thickness $h$. The grid array radiator consists of rectangular loops whose long sides are $l$ long and $w_l$ wide and whose short sides are $s$ long and $w_s$ wide. The grid array radiator is fed by two coaxial probes that have their outer conductors connected to the ground plane and their inner conductors connected to the two ends of one short side or two short sides.

### 2.5.2 Principle of Operation

The radiation properties of the microstrip grid array antenna are uniquely determined by grid geometry. The individual loop dimensions set the center frequency of operation. Loop impedance, interconnections, and feed point determine the amplitude distribution on the array, while the resonant behavior of the multiple loop array determines the overall frequency response [49]. For the design frequency of interest, the loops are sized such that the current on each short side is essentially in phase, while each long side supports a full wavelength current element. Thus, the radiation of the microstrip grid array antenna is essentially from the short sides with the long sides acting mainly as guiding or transmission lines. However, unlike the short sides (or radiating elements) in [48] and [50], they are arranged to have the same number in the E- and H-planes so as to realize the pencil beam [55]. Also, unlike the dual feeds in [50] that use two coplanar strip lines to make it only possible for differential operation, the dual feeds here that use two coaxial probes to make it possible not only for differential but also for

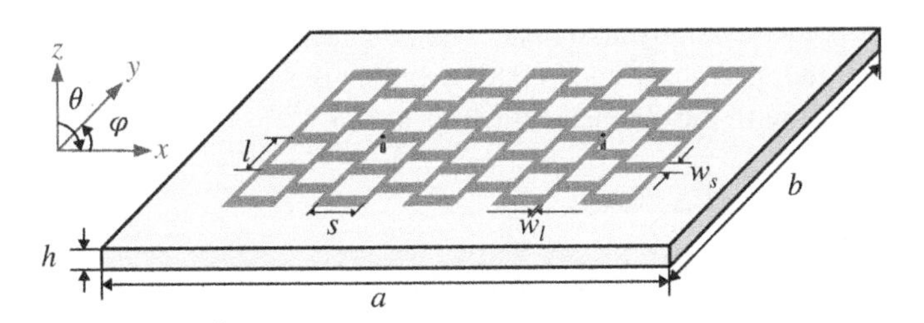


**Figure 2.10** Structure of a microstrip grid array antenna and coordinate system.

single-ended operations. In addition, the capability of the dual-feed microstrip grid array antenna to function as two single-ended antennas is particularly useful from the system design perspective because it only takes up one antenna real estate, which definitely helps the system to reduce cost and size.

### 2.5.3 Design Formulas with an Example

The design of a microstrip grid array antenna involves the choice of substrate type and size, the determination of the number of loops and feed location, and the optimization of the dimensions of a loop. In the following, we show how to design the microstrip grid array antenna for 24-GHz automotive radars to monitor blind spots. Based on the specifications in [56] of directivity $D \geq 20$ dB, angular resolution of the single-shaped beam $\Phi \leq 15°$, and size $a \times b \leq 100 \times 100\ \text{mm}^2$, we start the design with the RT/Duroid 5880 substrate for its low dielectric constant ($\varepsilon_r = 2.2$) and low loss tangent ($\delta = 0.0009$). We choose the substrate thickness to be $h = 0.787$ mm for mechanical strength.

We then implement the square-shaped substrate for pencil-beam patterns and choose the dimensions to satisfy the angular resolution as:

$$a = b = 60^0 \frac{\lambda}{\Phi} \geq 4\lambda \tag{2.27}$$

where $\lambda$ is the free space wavelength at 24 GHz. We determine the substrate size to be smaller than $100 \times 100\ \text{mm}^2$ as $a \times b = 4.8\lambda \times 4.8\lambda = 60 \times 60\ \text{mm}^2$.

Next, we determine the length $l$ and width $w_l$ for the long sides and the length $s$ and width $w_s$ for the short sides of the rectangular meshes. The resonance of the grid array radiator requires the length

$$l = \frac{\lambda}{\sqrt{\varepsilon_{eff}}} \tag{2.28}$$

where $\varepsilon_{eff}$ is the effective dielectric constant, which can be estimated as

$$\varepsilon_{eff} = \frac{\varepsilon_{r+1}}{2} + \frac{\varepsilon_{r-1}}{2} \left\{ \left[1 + 12\frac{h}{w_l}\right]^{-\frac{1}{2}} + 0.02(\varepsilon_r - 1)\left[1 - \frac{w_l}{h}\right]^2 \right\} \tag{2.29}$$

for $w_l/h < 1$ and the third term in (2.29) is zero for $w_l/h > 1$ [57]. It is known that the width $w_l$ controls the loss of transmission lines and the radiation of cross-polar components. To lower them, as suggested in [58], we adopt the same width $w_l$ for all long sides for $w_l/h < 1$ with the characteristic impedance

$$Z_c = \frac{60}{\sqrt{\varepsilon_{eff}}} \ln \left[\frac{8h}{w_l} + \frac{w_l}{4h}\right] \leq 200\Omega \tag{2.30}$$

Considering realizability from fabrication and reliability for use in automobiles, we prefer $Z_c = 125\ \Omega$. Substituting (2.29) into (2.30), we get $w_l \approx 0.4$ mm, which leads to $\varepsilon_{eff} \approx 1.7$ from (2.29) and $l \approx 9.6$ mm from (2.28).

Similarly, the resonance of the grid array radiator also requires the length $s$ to be half of the length $l$, that is, $s \approx 4.8$ mm. It is more involved to determine the widths $w_s$ because the width $w_s$ sets the array radiation coefficients with the first-order approximation of the coefficient being proportional to width [58]. For simplicity, we assign the same radiation coefficient for all radiating elements, leading to the same width $w_s$ for all short sides. To enhance the radiation of co-polar components, the choice of $w_s$ should be

$$w_s = \frac{\lambda}{4\sqrt{\varepsilon_{eff}}} \tag{2.31}$$

or $w_s \approx 2.4$ mm.

After that, we lay out as many rectangular loops as possible on the substrate to form the grid array radiator to achieve the directivity. Assuming that the grid array radiator has $N$ radiating elements or short sides and the current is uniformly distributed on them, the directivity can be estimated by

$$D = 5.16 + 10 \log N \text{ dB} \tag{2.32}$$

where 5.16 dB accounts for the directivity due to the ground plane and one radiating element. There are 32 loops in this example, so the number of short sides (or radiating elements) is 41. This indicates that directivity >20 dB is achievable.

Finally, we locate the dual feeds as illustrated in Figure 2.10 and simulate the performance of the dual-feed microstrip grid array antenna with the high-frequency structure simulator (HFSS$^{TM}$) simulator. In this design, the dual feeds are two sub-miniature version A (SMA) coaxial probes with Teflon ($\varepsilon_r = 2.1$) filling, inner diameter 0.5 mm, and outer diameter 2 mm. They should be located symmetrically with respect to the broadside direction, taking the differential operation into account. The simulations are conducted mainly to yield the optimized length $l$ and width $w_l$ for the long sides and length $s$ and width $w_s$ for the short sides of rectangular meshes. They are found to be $l = 9.8$ mm, $w_l = 0.4$ mm, $s = 4.9$ mm, and $w_s = 2.5$ mm, which are quite close to our calculated values.

The designed dual-feed microstrip grid array antenna was fabricated. Figure 2.11 shows the photo of the fabricated dual-feed microstrip grid array antenna. It has a size of $60 \times 60 \times 0.787$ mm$^3$. There are four holes each with radius of 2 mm at the four corners of the substrate reserved for stabilizing the radiator in order to flatten the whole antenna structure. The $S$-parameter measurements of the fabricated dual-feed microstrip grid array antenna were made with a Rohde-Schwarz ZVA 67 four-port vector network analyzer, which can be used for differential and single-ended excitations up to 67 GHz, respectively.

Figure 2.12a shows the simulated and measured $|S_{d11}|$ results for the dual-feed microstrip grid array antenna under differential excitation at both feeds. The measured 10-dB impedance bandwidth is 1.17 GHz from 23.75 to 24.92 GHz (or 4.85% at 24.15 GHz) and the simulated 10-dB impedance bandwidth is 0.7639 GHz

**Figure 2.11** Photograph of a dual-feed microstrip grid array antenna.

from 23.8894 to 24.6533 GHz (or 3.16% at 24.15 GHz). The difference between the simulated and measured $|S_{d11}|$ results is mainly caused by soldering. The irregular solder joints cannot be accurately modelled and were not included in simulations. Nevertheless, both simulated and measured $|S_{d11}|$ results demonstrate that the dual-feed microstrip grid array antenna under differential excitation has achieved very good matching and enough impedance bandwidth. Figure 2.12b shows the simulated and measured $|S_{11}|$ results for the dual-feed microstrip grid array antenna excitation at one feed and the other feed opened. The measured 10-dB impedance bandwidth is 1.46 GHz from 23.23 to 24.69 GHz (or 6.05% at 24.15 GHz) for the antenna excited at feed 1. The simulated 10-dB impedance bandwidth is 1.3136 GHz from 23.2864 to 24.60 GHz (or 5.44% at 24.15 GHz) for the antenna excited at the feed. The slight difference between the measured and simulated $|S_{11}|$ results is due to the soldering. Nevertheless, both simulated

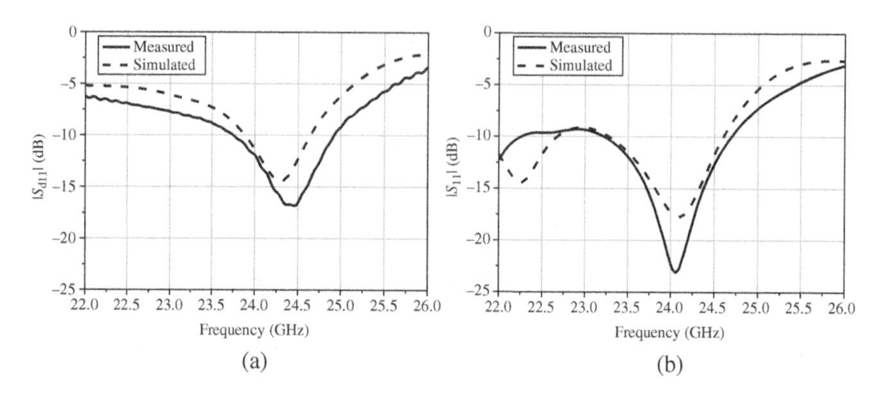


**Figure 2.12** Simulated and measured results (a) $|S_{d11}|$ under differential excitation and (b) $|S_{11}|$ under single-ended excitation.

and measured $|S_{11}|$ results demonstrate that the dual-feed microstrip grid array antenna under single-ended excitation has achieved very good matching and enough impedance bandwidth.

The radiation patterns of the dual-feed microstrip grid array antenna were characterized in an anechoic chamber by utilizing a two-port Agilent N5230A PNA-L network analyzer, a 180° hybrid coupler, and a Satimo SGH-820 ridged horn antenna. Figure 2.13 shows the simulated and measured radiation patterns of the dual-feed microstrip grid array antenna at 24.15 GHz under differential excitation. As expected, both E- and H-plane patterns have the main beam of radiation in the broadside direction. The simulated and measured co-polarized radiation patterns agree well. The simulated half-power beam widths (HPBWs) of the radiation beam are both 14° in the E- and H-planes. The measured HPBWs of the radiation beam are also both 14° in the E- and H-planes. However, the simulated and measured cross-polarized radiation patterns deviate largely because it is difficult to generate an ideal differential signal. Practically, the differential signal has an amplitude imbalance and a phase difference. The amplitude imbalance and phase difference can be up to 1 dB and 20°. Our simulations show that such imperfect differential excitation slightly tilts the main beam of radiation from the broadside direction in the E plane and enhances cross-polarized radiation in the H-plane. Figure 2.14 shows the simulated and measured radiation patterns of the dual-feed microstrip grid array antenna at 24.15 GHz under single-ended excitation at one feed and the other feed opened. It is interesting to note that the main beam of radiation in both simulated and measured E planes is deviated 6° from the broadside direction. This is because the phase center of the dual-feed microstrip grid array antenna under single-ended excitation is near the feed

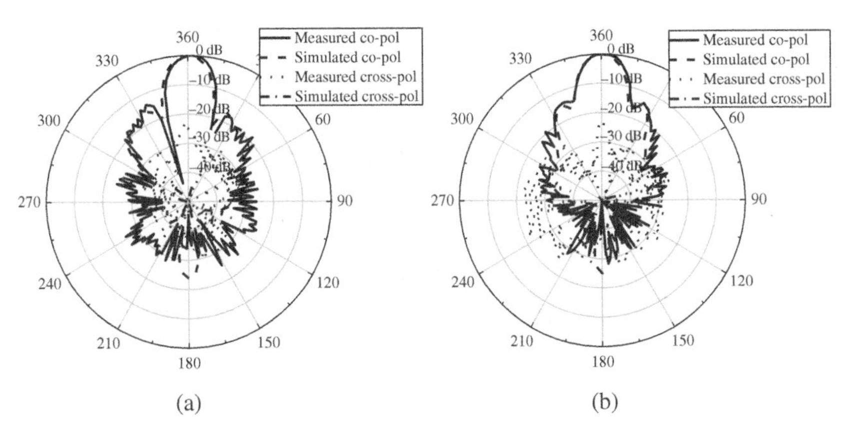


**Figure 2.13** Simulated and measured radiation patterns of a dual-feed grid array at 24.15 GHz under differential excitation (a) in the E-plane and (b) in the H-plane.

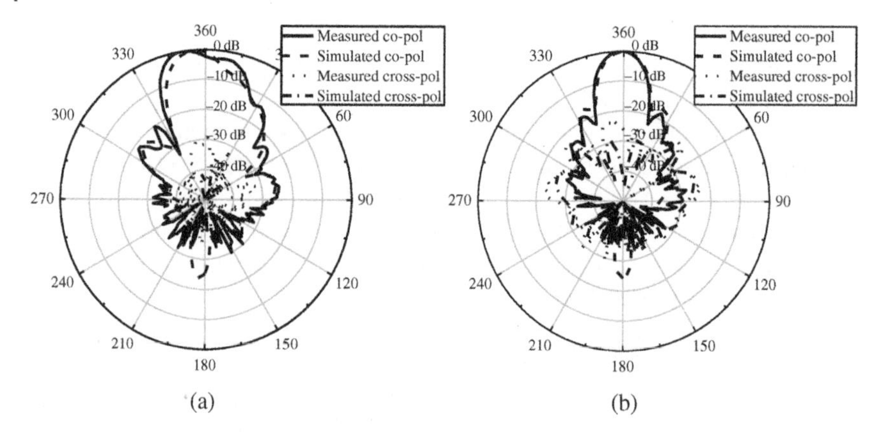


**Figure 2.14** Simulated and measured radiation patterns of a dual-feed grid array at 24.15 GHz under single-ended excitation (a) in the E-plane and (b) in the H-plane.

point rather than the geometrical center of the antenna. The simulated HPBWs of the radiation beam are 17° and 15°, respectively, in the E- and H-planes. The measured HPBWs of the radiation beam are 18° and 15°, respectively, in the E- and H-planes. Figures 2.13 and 2.14 clearly demonstrate that the dual-feed microstrip antenna radiates pencil-beam patterns under either differential or single-ended excitation.

Figure 2.15 shows the simulated and measured gain values of the dual-feed microstrip grid array antenna under differential and single-ended excitations. Under differential operation, the measured highest gain is 21.0 dBi at 24.28 GHz and the measured 3-dB gain bandwidth is from 23.09 to 25.27 GHz (or 9.03% at

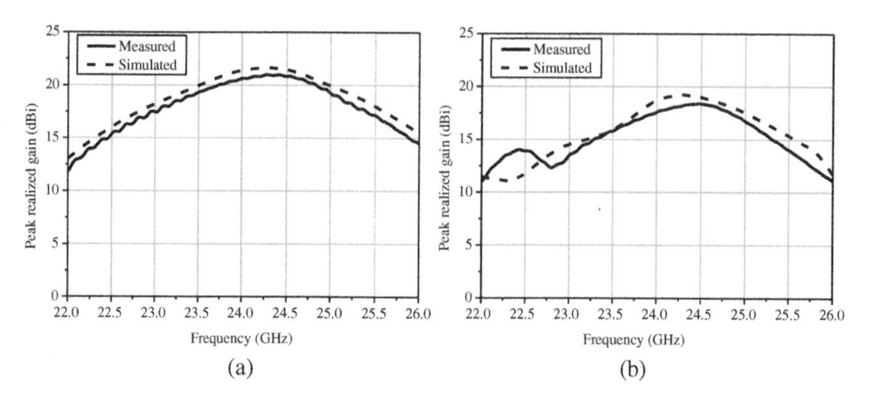


**Figure 2.15** Simulated and measured gain values of a dual-feed microstrip grid array antenna under (a) differential and (b) single-ended excitations.

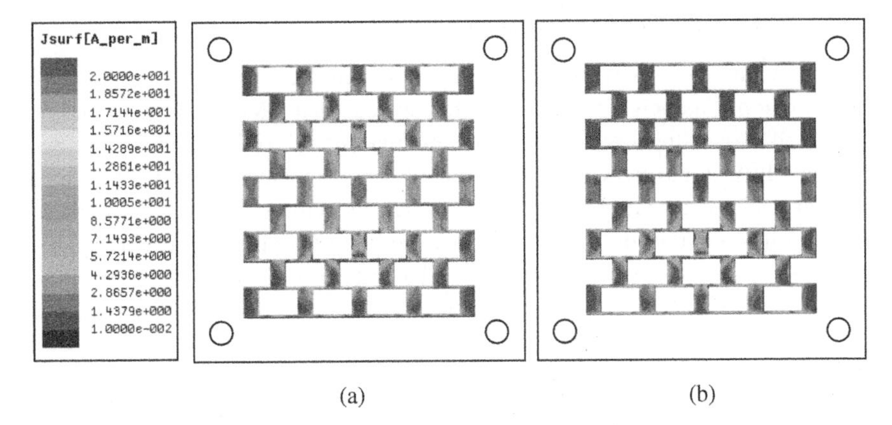


**Figure 2.16** Simulated current distributions of a dual-feed microstrip grid array antenna for (a) differential and (b) single-ended excitations.

24.15 GHz), the simulated highest gain is 21.6 dBi at 24.25 GHz, and the simulated 3-dB gain bandwidth is from 23.12 to 25.35 GHz (or 9.23% at 24.15 GHz). The simulations also show that the imperfect differential excitation slightly affects the gain. Under single-ended operation, the measured highest gain is 18.4 dBi at 24.48 GHz and the measured gain bandwidth is from 23.39 to 25.26 GHz (or 7.74% at 24.15 GHz), the simulated highest gain is 19.3 dBi at 24.25 GHz and the simulated 3-dB gain bandwidth is from 23.60 to 25.15 GHz (or 6.42% at 24.15 GHz).

Figure 2.16 shows the simulated current distributions on the dual-feed microstrip grid array radiator at 24.15 GHz under either differential or single-ended excitation. The higher gain and wider gain bandwidth under differential excitation are mainly attributed to the improved current distribution and phase synchronization on radiating elements. The simulated and measured aperture efficiencies of the dual-feed microstrip grid array antenna are 51% and 44% under differential operation and 32% and 24% under single-ended excitation.

## 2.6 Yagi-Uda Antennas

Yagi-Uda antennas are used in AiP technology for their unique end-fire radiation characteristics [4, 5, 39, 40, 59, 60]. The Yagi-Uda antenna by Kim et al. is particularly interesting [5] because they embedded it vertically in rather than horizontally on a substrate. In such a manner, they made good use of the ground plane according to the image theory. In addition, the design of horizontal and vertical Yagi-Uda antennas in the vicinity of each other has been done with the aim of providing dual polarizations [61].

### 2.6.1 Horizontal Yagi-Uda Antenna

Figure 2.17 shows the structure of a horizontal Yagi-Uda antenna that consists of three elements, namely a driver, a reflector, and a director. More commonly, there is more than one director to improve the antenna gain. The driver and director are horizontally printed on the top surface of a substrate of relative permittivity $\varepsilon_r$, loss tangent $\delta$, length $a$, width $b$, and thickness $h$. The reflector is usually printed on the bottom surface of the substrate, which has another function as the ground plane.

The driver is a half-wave dipole. It is fed by a uniplanar microstrip-to-coplanar strip transition as a broadband balun. The transition is realized as an impedance-matched T-junction with one side of microstrip line longer than the other side by a half-wavelength at the desired frequency. The driver should be designed to launch the $TE_0$ mode surface wave in the substrate [62]. The $TE_0$ mode surface wave radiates at the end edge of the substrate into free space. The $TM_0$ mode surface wave should not be excited because it degrades the antenna performance with an increase in cross-polarization radiation and a reduction in gain and front-to-back ratio. The reflector or the grounded substrate region can be designed to completely cut off the $TE_0$ mode surface wave by correctly choosing the substrate thickness $h$. The director shares the same field polarization as the driver and the $TE_0$ mode surface wave is strongly coupled to the director. For this reason, the director guides the electromagnetic field towards the end-fire direction while simultaneously acting as an impedance matching element.

The first horizontal Yagi-Uda antenna was demonstrated using a substrate of relative permittivity $\varepsilon_r = 10.2$ and thickness $h = 0.635$ mm at X-band frequencies [62]. The antenna total size, including the microstrip-to-coplanar strip transition, was less than $0.5\lambda_0$ in width and $0.6\lambda_0$ in length. It achieved a bandwidth of 17%, a gain of 6.5 dBi, a front-to-back ratio of 18 dB, and a cross-polarization level lower than 15 dB at 10 GHz. It was also found that the horizontal Yagi-Uda antenna could be scaled linearly to any frequency band of interest while retaining wideband characteristics [63].

### 2.6.2 Vertical Yagi-Uda Antenna

Figure 2.18 shows the structure of a vertical Yagi-Uda antenna that consists of three elements, namely a driver, a reflector, and a director. The driver, reflector,

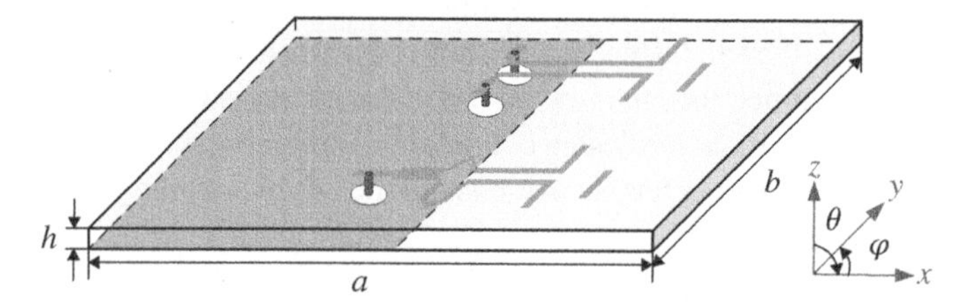


**Figure 2.17** Structure of a horizontal Yagi-Uda antenna and coordinate system.

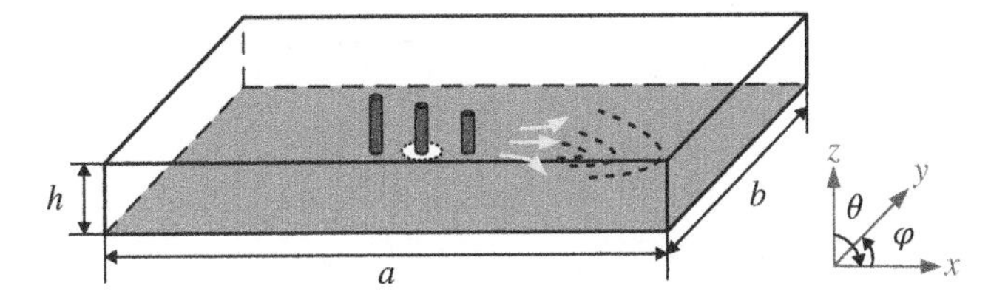


**Figure 2.18** Structure of a vertical Yagi-Uda antenna and coordinate system.

and director are vertically embedded in a substrate of relative permittivity $\varepsilon_r$, loss tangent $\delta$, length $a$, width $b$, and thickness $h$.

Because of the ground plane, the driver is a quarter-wave monopole. The driver launches the $TM_0$ mode surface wave in the substrate. The $TM_0$ mode surface wave radiates at the end edge of the substrate into free space. The director of the vertical Yagi-Uda antenna has the same wave-guiding effect as the director in the horizontal Yagi-Uda antenna does. The reflector of the vertical Yagi-Uda antenna, however, cannot be designed as that in the horizontal Yagi-Uda antenna to completely reflect the surface wave.

The vertical Yagi-Uda antenna was demonstrated using an LTCC substrate of relative permittivity $\varepsilon_r = 5.9$ and loss tangent $\delta = 0.0035$ in the 60-GHz band [5]. The substrate had two layers separated by a ground plane. The upper layer was 0.25 mm thick and the lower layer 0.6 mm thick. The driver was a metal via with a height of 0.55 mm. The reflector was metal vias all of height 0.6 mm. These vias were grounded and arranged as two fences. There were seven and 11 vias in the first and second fences, respectively. The pitch between two adjacent vias of the same fence was 0.3 mm. The distance from the driver to the first fence was 0.8 mm and from the first fence to the second fence was 1.2 mm. There were three directors using three metal vias to enhance the antenna gain. The first and second directors had the same height of 0.45 mm, and the height of the third director was 0.35 mm. The directors were equally separated at 0.4 mm. The antenna total size was $5.7 \times 3.7\,\text{mm}^2$ or $1.178\lambda_0 \times 0.765\lambda_0$.

Figure 2.19 shows the simulated and measured $S_{11}$ of the vertical Yagi-Uda antenna as a function of frequency from 50 to 67 GHz. It is evident from the figure that although there are differences between the simulated and measured values, the antenna achieved acceptable matching from 57 to 67 GHz, satisfying all 60-GHz standards. Figure 2.20 shows the simulated and measured radiation patterns at 60 GHz. As expected, the antenna radiates an end-fire beam. The measured gain at 60 GHz is 7.3 dBi. The simulated radiation efficiency is 90% at 60 GHz.

### 2.6.3 Yagi-Uda Antenna Array

A Yagi-Uda antenna is suitable to form a linear array because of its small size, broad pattern, and wide bandwidth [39, 40]. In antenna array design, the effect

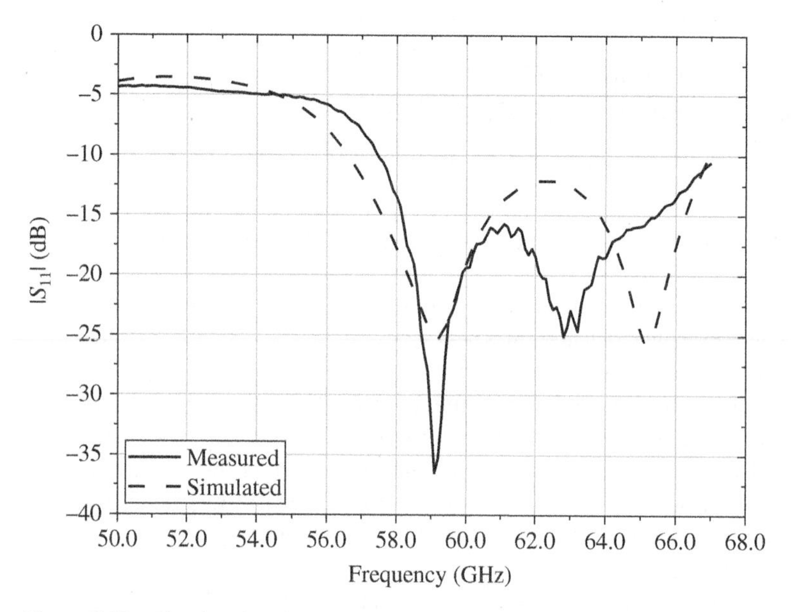


**Figure 2.19** Simulated and measured $|S_{11}|$ of a vertical Yagi-Uda antenna.

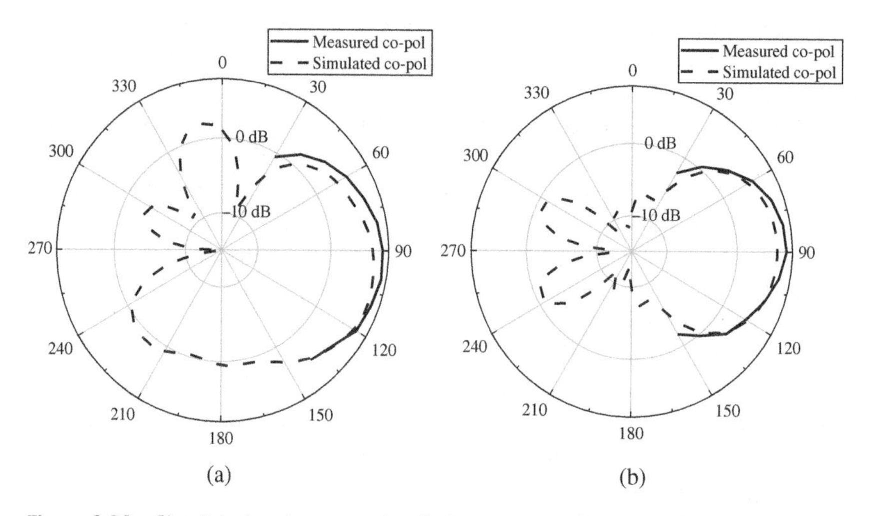


**Figure 2.20** Simulated and measured radiation patterns of a vertical Yagi-Uda antenna: (a) E-plane and (b) H-plane.

of mutual coupling between elements within the array must be considered. The mutual coupling is undesirable because strong mutual coupling may cause scan blindness to limit the actual beam scanning range of the antenna array. The mutual coupling between two horizontal Yagi-Uda antennas was measured at X-band frequencies. The two antennas were fabricated on the same substrate of relative permittivity $\varepsilon_r = 10.2$ and thickness $h = 0.635$ mm. At a spacing of 15.0 mm, which corresponds to $0.5\lambda_0$ at 10.0 GHz, the mutual coupling was measured to be −32.0 dB [57]. The mutual coupling between two vertical Yagi-Uda antennas was measured for frequencies from 30 to 42 GHz. The two antennas were embedded in the same substrate of relative permittivity $\varepsilon_r = 3.38$ and thickness $h = 1.93$ mm. At a spacing of 7.6 mm, which corresponds to $0.864\lambda_0$ at 36.0 GHz, the mutual coupling was measured to be about −17.0 dB, which is too strong and must be significantly suppressed. It was shown that the mutual coupling could be reduced by nearly 20 dB with horn-shaped edge walls composed of metal vias [61].

A linear 1 × 4 Yagi-Uda antenna array has been demonstrated. The length of microstrip line connecting the array and the chip was about 5 mm and its loss was 0.5 dB at 28 GHz [39]. The single antenna and the array gains were 5.3 and 11.3 dBi, respectively. Beam patterns were measured in the azimuth direction for phase shifts of 0–180° with ±45°. The 3-dB beam coverages were wide as much as 105° for the array. It was expected that the linear array could be applied to smart phones.

## 2.7 Magneto-Electric Dipole Antennas

The magneto-electric dipole antenna was invented by Luk and Wong in 2006 [64]. It consists of a planar electric dipole and a vertically oriented patch antenna as a magnetic dipole. The planar electric dipole should be designed as a half-wavelength dipole and the vertically oriented patch antenna as a quarter-wavelength patch antenna. The electric dipole and magnetic dipole are placed orthogonally. They are excited simultaneously with equal amplitude and proper phase. The resulting radiation pattern will be identical in the E- and H-planes, and furthermore the back radiation will be suppressed significantly. The magneto-electric dipole antenna is a new type of complementary antenna. It exhibits a wide impedance bandwidth, a stable gain, and a stable radiation pattern with low cross-polarization and back radiation levels over the operating frequencies [65]. The magneto-electric dipole antenna has been explored recently for AiP technology, showing promising results [66].

### 2.7.1 Single-polarized Microstrip Magneto-electric Dipole Antenna

Figure 2.21 shows the structure of a single-polarized microstrip magneto-electric dipole antenna [66]. Note that the planar electric dipole is realized on the top surface of a grounded substrate of relative permittivity $\varepsilon_r = 3.85 \pm 0.05$ and loss tangent $\delta = 0.007 \pm 0.002$ at 28.5 GHz. The magnetic dipole is realized with the gap formed by the two halves of the planar electric dipole, the shorting vias, and the ground plane between the shorting vias. Due to the diameter of vias being limited to 50 μm, three feeding vias are optimized for impedance matching to a 50-Ω source. The antenna has a small form factor of $10 \times 10\,\text{mm}^2$ in slightly less than 1.0-mm thickness. The antenna geometry and its feature dimensions are given in Table 2.1.

Figure 2.22 shows a photograph of a fabricated single-polarized microstrip magneto-electric dipole antenna.

### 2.7.2 Dual-polarized Microstrip Magneto-electric Dipole Antenna

Figure 2.23 shows the structure of a dual-polarized magneto-electric dipole antenna [66]. Note that the two planar electric dipoles are realized with four square patches on the top surface of a grounded substrate of relative permittivity

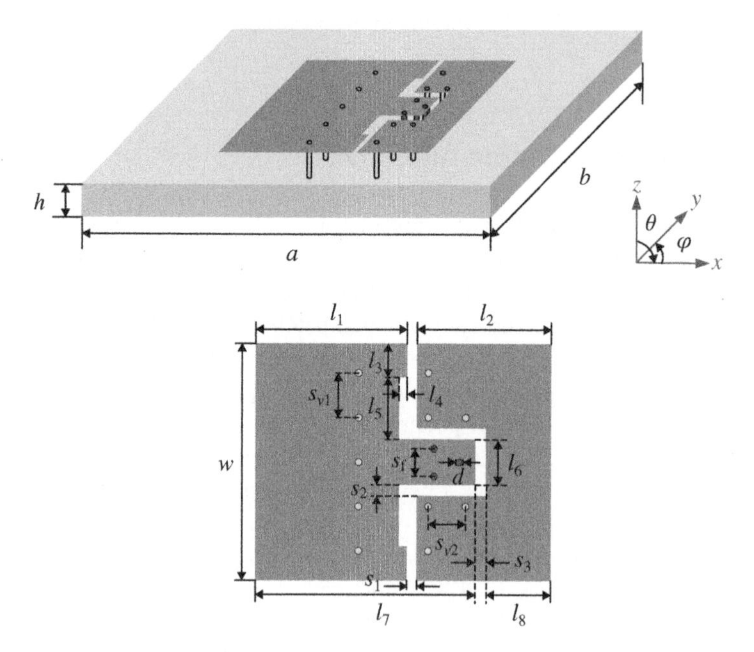


**Figure 2.21** Structure of a single-polarized magneto-electric dipole and coordinate system: upper diagram, 3D view; lower diagram, top view.

**Table 2.1** Dimensions of magneto-electric dipole antennas.

| Symbol | Value (mm) | Symbol | Value (mm) |
|---|---|---|---|
| W | 2.00 | Lx | 3.40 |
| L1 | 1.40 | Ly | 3.40 |
| L2 | 1.05 | Lx1 | 1.40 |
| L3 | 0.22 | Ly1 | 1.40 |
| L4 | 0.05 | C1 | 0.50 |
| L5 | 0.58 | C2 | 0.10 |
| L6 | 0.40 | Wf1 | 0.45 |
| L7 | 1.94 | Wf2 | 0.45 |
| L8 | 0.54 | Sv | 0.30 |
| S1 | 0.10 | Sf | 0.23 |
| S2 | 0.08 | D | 0.05 |
| S3 | 0.08 | Sf | 0.23 |
| Sv1 | 0.40 | Lf1 | 1.45 |
| Sv2 | 0.25 | Lf2 | 1.45 |

**Figure 2.22** Photograph of a single-polarized magneto-electric dipole antenna.

$\varepsilon_r = 3.85 \pm 0.05$ and loss tangent $\delta = 0.007 \pm 0.002$ at 28.5 GHz. The two magnetic dipoles are realized with the gaps formed by the four square patches, the 12 shorting vias, and the ground plane between the shorting vias. The two coupling strips, in the center of the antenna on different metal layers, are each connected at one end with feeding vias. Due to the diameter of vias being limited to 50 μm, three feeding vias are optimized for impedance matching to a 50-Ω source for one polarization. The antenna has a small form factor of $10 \times 10\ \text{mm}^2$ in slightly more

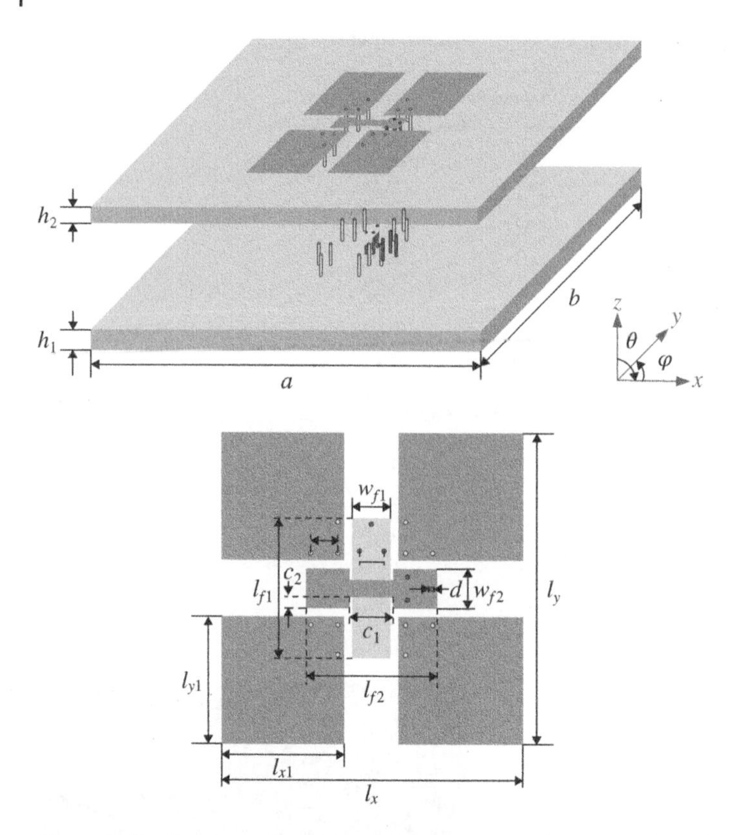


**Figure 2.23** Structure of a dual-polarized magneto-electric dipole and coordinate system.

**Figure 2.24** Photograph of a dual-polarized magneto-electric dipole antenna.

**Table 2.2** Measured gain of magneto-electric dipole antennas.

| Item | 24 GHz | 28 GHz | 39 GHz |
|---|---|---|---|
| Single polarization | 4.0 dBi | 5.1 dBi | 5.5 dBi |
| Dual polarization | 3.8 dBi | 5.0 dBi | 5.4 dBi |

than 1.0-mm thickness. The antenna geometry and its feature dimensions are given in Table 2.1. Figure 2.24 shows a photograph of a fabricated dual-polarized microstrip magneto-electric dipole antenna.

### 2.7.3 Simulated and Measured Results

Figure 2.25 shows the simulated and measured $|S_{11}|$ of the single- and dual-polarized magneto-electric dipole antennas. The single-polarized magneto-electric dipole antenna achieves a measured impedance bandwidth from 24.5 to 40 GHz (or 48% at 28 GHz) for $|S_{11}| \leq -10$ dB and the dual-polarized magneto-electric dipole antenna attains a measured impedance bandwidth from 25 to 43 GHz (or 52% at 28 GHz).

Figure 2.26 shows the measured and simulated radiation patterns of the single-polarized magneto-electric dipole antenna in the E- and H-planes at 24 and 39 GHz. It can be seen that there are acceptable agreements between the measurements and simulations. The small differences may be caused by fabrication tolerance, measurement error, and other uncertainties.

Figure 2.27 shows the measured and simulated radiation patterns of the dual-polarized magneto-electric dipole antenna in the E- and H-planes at 24 and 39 GHz for horizontal polarization. It can be seen that there are acceptable agreements between the measurements and simulations. The small differences may be caused by fabrication tolerance, measurement error, and other uncertainties. The radiation patterns for vertical polarization are omitted due to page limitations. They are quite similar to those for horizontal polarization.

The measured peak gains of the single- and dual-polarized magneto-electric dipole antennas are presented in Table 2.2. It is interesting to note that the measured gains vary within 1.7 dB from 3.8 to 5.5 dBi over such a wideband bandwidth.

## 2.8 Performance Improvement Techniques

Performance improvement of antennas for AiP technology is strongly related to the enhancement or suppression of surface waves. As stated in Section 2.6, the horizontal and vertical Yagi-Uda antennas rely on $TE_0$ and $TM_0$ mode surface waves, respectively, to operate. Hence, the enhancement of proper surface waves

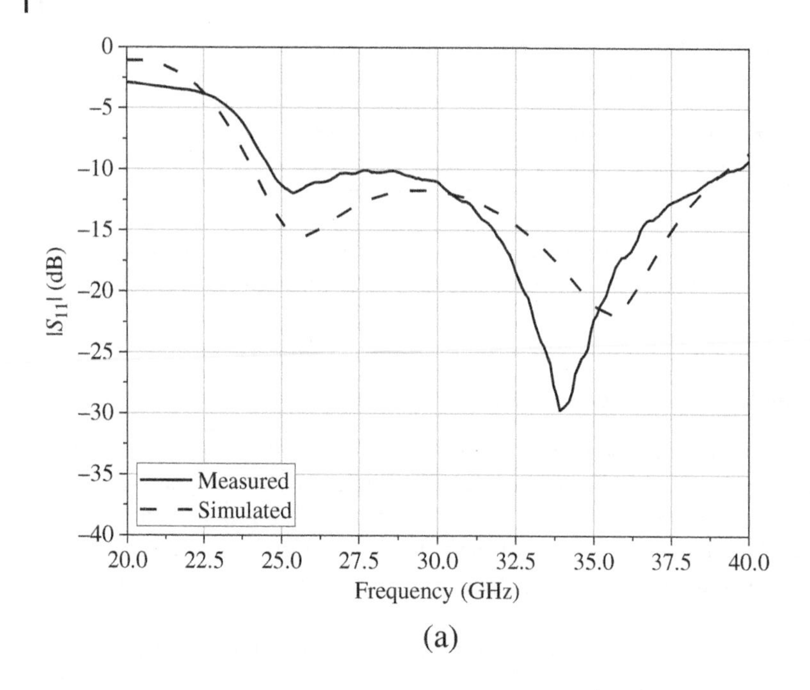


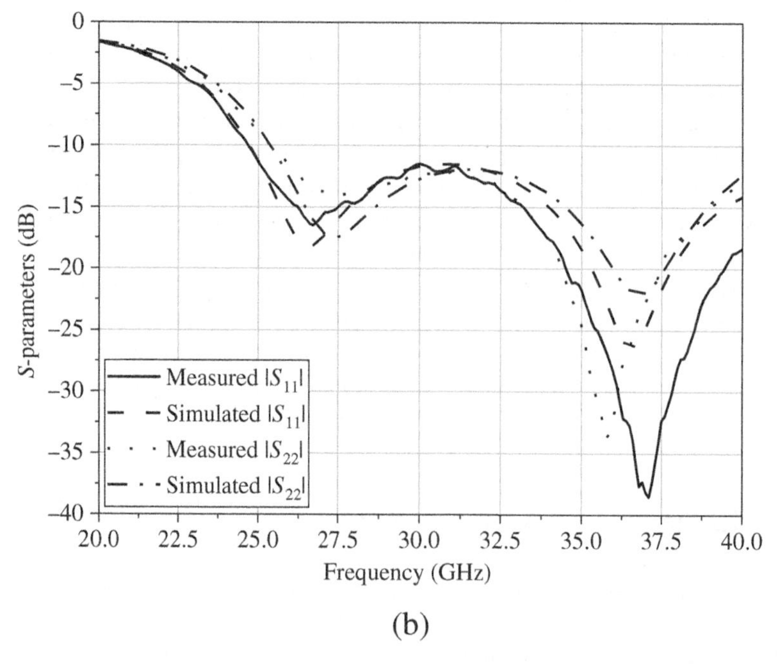


**Figure 2.25** Simulated and measured $|S_{11}|$ results for (a) single-polarized and (b) dual-polarized magneto-electric dipole antennas.

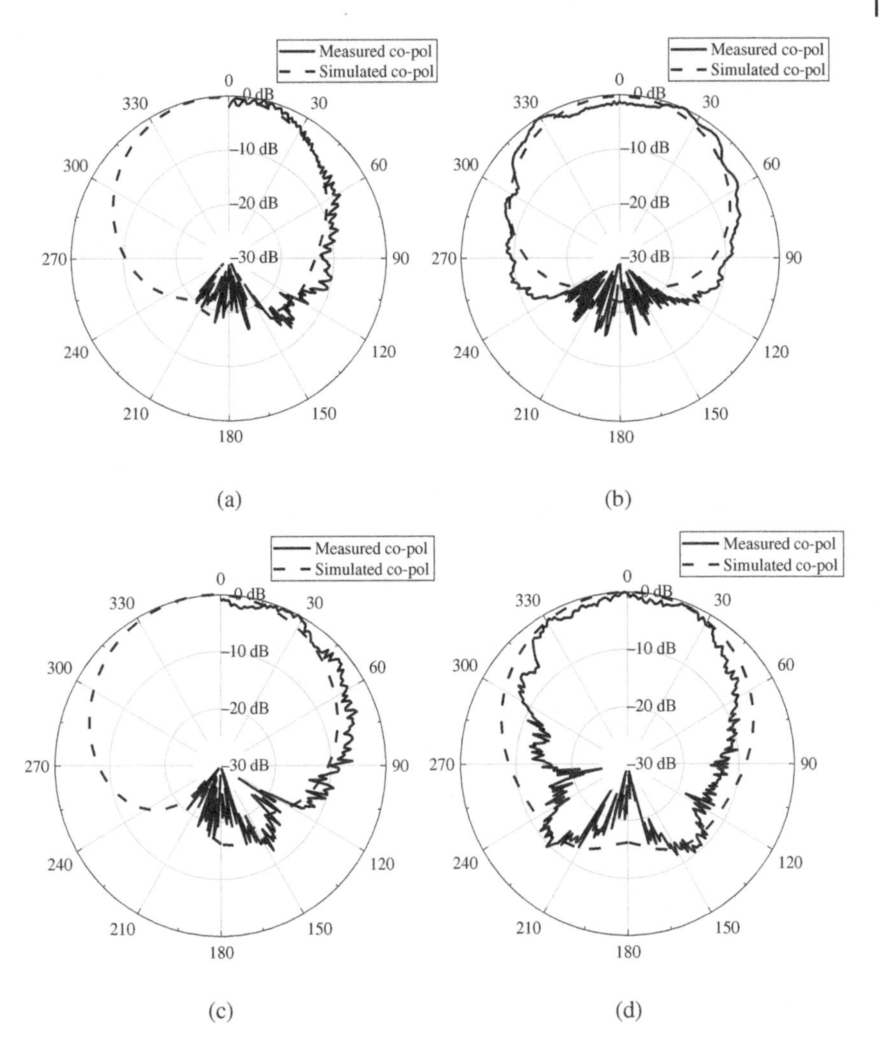


**Figure 2.26** Simulated and measured radiation patterns of a single-polarized magneto-electric dipole antenna: (a) E- and (b) H-planes at 24 GHz and (c) E- and (d) H-planes at 39 GHz.

using a thicker substrate of higher permittivity helps to improve the performance of Yagi-Uda antennas. For other types of antennas, such as microstrip patch and magneto-electric dipole antennas, surface waves should be suppressed otherwise they will reduce the radiation efficiency and distort the radiation pattern of the antenna. Moreover, surface waves increase unwanted mutual coupling between antenna elements in an array, which causes impedance and pattern anomalies associated with the blind angle.

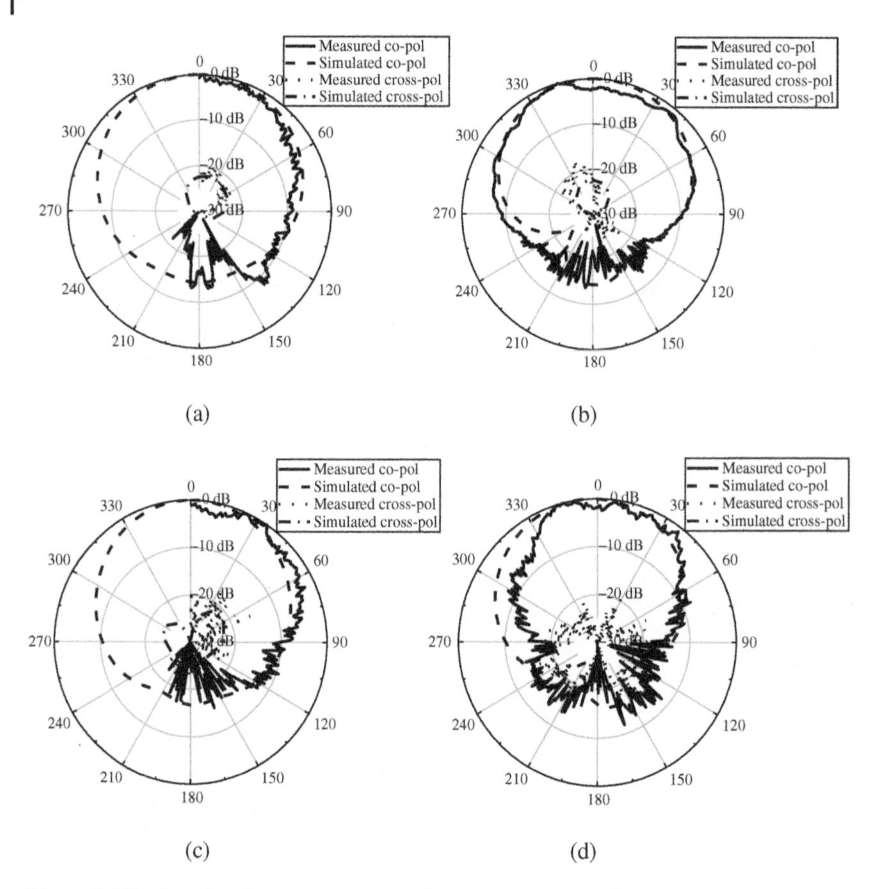


**Figure 2.27** Simulated and measured radiation patterns of a dual-polarized magneto-electric dipole antenna for horizontal polarization: (a) E- and (b) H-planes at 24 GHz and (c) E- and (d) H-planes at 39 GHz.

Soft and hard surfaces, electromagnetic bandgap (EBG), and artificial magnetic conductors (AMCs) can be used to suppress surface waves [67–69]. The use of uniplanar compact EBG and soft surface structures to suppress surface waves to improve the performance of microstrip patch antennas and arrays for 60-GHz AiP technology will be addressed in Chapter 7. The use of AMCs to suppress surface waves to improve the performance of microstrip dipole antennas and arrays for 60-GHz AiP technology is described in this section. The AMC is based on spiral unit cells devised by Kyriazidou, Contopanagos, and Alexopoulos [70–72]. It has a single metallization layer with no vias, which allows simultaneous projection of AMC behavior directly on the antenna metallization layer, while eliminating surface waves and obeying package design rules.

### 2.8.1 Single-layer Spiral AMC

A generic simplified package capturing the essential features of realistic package technologies is considered. It uses three copper metallization layers, M1, M2, and M3, of 15 μm thickness each, separated by dielectrics of thicknesses $h_1 = 60$ μm and $h_2 = 190$ μm, and complex permittivity $\varepsilon = 4(1 - j0.01)$. M1 is for the antenna, M2 for the AMC, and M3 for the ground plane, respectively. The die lies underneath M3 where further metallization and dielectric layers may exist for other functions such as signal routings and bump pads etc. Since the AMC design terminates at the ground plane, any package layer below M3 is disregarded.

The unit cell of the AMC is a square spiral. The spiral layout is represented by the notation ($A$, $L$, $W$, $S$; $N$) where $A$ is the side length of the unit cell, $L$ is the longest spiral arm length, $W$ is the spiral trace width, $S$ is the nearest-edge spacing, and $N$ is the number turns in multiples of 0.25. The unit cell of the AMC is designed with the HFSS for normal plane-wave incidence and $S_{11}$ is de-embedded up to the surface of M1. The port and boundary setup in the HFSS simulation can refer to [69]. The AMC bandwidth $B_{AMC}$ is defined as the frequency range for which $\text{Re}(S_{11}) \geq 0$. $B_{AMC}$ should include the operation band $B_o$. There will occur band-limited surface plasmon polaritons (SPP) resonances below and above $B_{AMC}$, which should be kept far away from $B_o$. The optimized unit cell has the layout (430 μm, 400 μm, 75 μm, 30 μm; 1.25). The AMC response has a 20% bandwidth $B_{AMC} = (54, 67)$ GHz covering the whole 60-GHz band $B_o = (57, 66)$ GHz. SPP resonances occur at 39 and 90 GHz, respectively, which are far away from $B_o$.

### 2.8.2 Design Guidelines

To achieve integration of an antenna with an AMC, three additional bandwidths need to be defined and determined. They are (1) the radiation band $B_A$ of the antenna designed to be integrated in the absence of the AMC, (2) the band $B_{A+AMC}$ of the coupled frequency response of the integrated antenna with the AMC, and (3) the matching and radiating band $B_{RAD}$ of the antenna. The lower and higher frequency edges that set $B_{RAD}$ are given by

$$\text{Re}[S_{11}(f_l)] \approx 1 \tag{2.33}$$

and

$$\text{Re}[S_{11}(f_h)] = \text{Im}[S_{11}(f_h)] < 0. \tag{2.34}$$

The general design principles are as follows:

(1) Design the AMC on M2 as that given in section 2.8.1 so that $B_o$ is within $B_{RAD}$ where the band $B_{RAD}$ of the AMC is computed from (2.33) and (2.34) on the antenna metallization layer M1.

(2) Design the dipole on M1 in the absence of AMC on M2, so that $B_A \approx B_{RAD}$. Make sure that good impedance matching exists throughout $B_A$. If $B_{RAD}$ is not much broader than $B_o$ but significantly narrower than $B_A$, one can take $f_h \approx f_a$, the resonant frequency of the antenna at which $|S_{11}|$ goes to minimum.
(3) On printing the dipole (2) on the AMC (1), the resulting antenna will have a radiation band $B_{A+AMC} \subseteq B_{RAD} \subset B_o$ where good impedance matching will exist.

### 2.8.3 A Design Example

Given $B_o = (57, 66)$ GHz, $B_{RAD}$ is slightly wider than $B_o$ at (56, 68) GHz. The unit cell of the AMC is optimized to have the layout (480 μm, 450 μm, 100 μm, 30 μm; 1.25). Design the center-fed dipole in the absence of AMC to have a radiation band $B_A \subseteq B_{RAD}$ and to resonate at $f_a \approx f_h = 68$ GHz. The dipole has dimensions of length 1.25 mm and width 0.1 mm on a package of size $1.44 \times 2.4 \times 0.25$ mm$^3$. Integrate $3 \times 5$ cells of the AMC with the dipole in the package. The simulation shows that the antenna has $B_{A+AMC} = (56, 76)$ GHz. The antenna radiates with broadside gain values of 5.5–6 dBi over the whole band $B_{A+AMC}$, making it ideal for array design in the 60-GHz band for AiP technology.

## 2.9 Summary

This chapter has presented antennas for AiP technology. First, basic antennas such as dipole, monopole, loop, and slot were described. It was noted that these basic antennas are mainly used for proof of concept. Then, unusual antennas, for example laminated resonator, dish-like reflector, and slab waveguide, were described. It was stated that these unusual antennas were specifically proposed to suit some packaging materials and processes. Next, special emphasis was given to microstrip patch antennas and arrays, grid array antennas, and Yagi-Uda antennas and arrays for their wide applications. It was also highlighted that magneto-electric dipole antennas have achieved promising results, showing potential for use in AiP technology. Finally, antenna performance improvement techniques based on AMC were discussed. Although obtaining enhancements in antenna performance, the AMC design has not been well adopted, probably due to the increased cost of fabrication.

## Acknowledgements

The author is grateful to Mr. Z. J. Shao and Mr. Y. L. Fang of Shanghai Jiao Tong University for drawing the figures. The author also is appreciative of Professor C. S.

Park of the Korea Advanced Institute of Science and Technology, Daejeon, South Korea and Dr. B. Yu of Speed Wireless Technology Co. Ltd, Guangdong, China for providing some simulated and measured data.

## References


1 T. Zwick, D. Liu, and B.P. Gaucher, Broadband planar superstrate antenna for integrated millimeter wave transceivers. *IEEE Transactions on Antennas and Propagation*, vol. 54, no. 10, pp. 2790–2796, October 2006.

2 Y. Tsutsumi, T. Ito, K. Hashimoto et al., Bonding wire loop antenna in standard ball grid array package for 60-GHz short-range wireless communication. *IEEE Transactions on Antennas and Propagation*, vol. 60, no. 12, pp. 1557–1563, April 2013.

3 S. Yoshida, H. Oguma, S. Kameda et al., 60-GHz-band planar slot antenna using organic substrates for ultra-small WPAN modules. In *Proceedings of the International Conference on Communications and Electronics*, Nha Trang, Vietnam, 11–13 August 2010.

4 M. Sun, Y.P. Zhang, K.M. Chua et al., Integration of Yagi antenna in LTCC package for differential 60-GHz radio. *IEEE Transactions on Antennas and Propagtion*, vol. 56, no. 8, pp. 2780–2783, August 2008.

5 H.J. Kim, J.H. Lee, I.S. Song et al., Compact LTCC Yagi-Uda type end-fire antenna-in-package for 60 GHz wireless communications. In *IEEE MTT-S International Microwave Symposium Digest*, Orlando, FL, USA, 1–6 June 2014.

6 Y.P. Zhang, Integration of microstrip antenna on ceramic ball grid array package. *Electronics Letters*, vol. 38, no. 5, pp. 207–208, 2002.

7 Y.P. Zhang, Integration of microstrip antenna on cavity-down ceramic ball grid array package. *Electronics Letters*, vol. 38, no. 22, pp. 1307–1308, 2002.

8 R.L. Li, G. DeJean, M. Maeng et al., Design of compact stacked-patch antennas in LTCC multilayer packaging modules for wireless applications. *IEEE Transactions on Advanced Packaging*, vol. 27, no. 4, pp. 581–589, November 2004.

9 S.H. Wi, J.-S. Kim, N.K. Kang et al., Package-level integrated LTCC antenna for RF package application. *IEEE Transactions on Advanced Packaging*, vol. 30, no. 1, pp. 132–141, February 2007.

10 Y.P. Zhang, Enrichment of package antenna approach with dual feeds, guard ring, and fences of vias. *IEEE Transactions on Advanced Packaging*, vol. 32, no. 3, pp. 612–618, August 2009.

11 D.G. Kam, D. Liu, A. Natarajan et al., LTCC packages with embedded phased-array antennas for 60-GHz communications. *IEEE Microwave Wireless Computation Letters*, vol. 21, no. 3, pp. 142–144, March 2011.

**12** D.G. Kam, D. Liu, A. Natarajan et al., Organic packages with embedded phased-array antennas for 60-GHz wireless chipsets. *IEEE Transactions on Components, Packaging and Manufacturing Technology*, vol. 1, no. 11, pp. 1806–1814, November 2011.

**13** W.B. Hong, A. Goudelev, K.H. Baek et al., 24-element antenna-in-package for stationary 60-GHz communication scenarios. *IEEE Antennas and Wireless Propagation Letters*, vol. 10, pp. 738–741, 2011.

**14** W.B. Hong, K.H. Baek, and A. Goudelev, Multilayer antenna package for IEEE 802.11ad employing ultralow-cost FR4. *IEEE Transactions on Antennas and Propagation*, vol. 60, no. 12, pp. 5932–5938, December 2012.

**15** E. Cohen, M. Ruberto, M. Cohen et al., A CMOS bidirectional 32-element phased-array transceiver at 60 GHz with LTCC antenna. *IEEE Transactions on Microwave Theory and Techniques*, vol. 61, no. 3, pp. 1359–1375, March 2013.

**16** W.B. Hong, K.H. Baek, and A. Goudelev, Grid assembly-free 60-GHz antenna module embedded in FR-4 transceiver carrier board. *IEEE Transactions on Antennas and Propagation*, vol. 60, no. 12, pp. 1573–1580, April 2013.

**17** Y. Chen and Y.P. Zhang, Integration of ultra-wideband slot antenna on LTCC substrate. *Electronics Letters*, vol. 40, no. 11, pp. 645–646, May 2004.

**18** Y. Chen and Y.P. Zhang, A planar antenna in LTCC for single-package UWB radio. *IEEE Transactions on Antennas and Propagation*, vol. 53, no. 9, pp. 3089–3093, September 2005.

**19** G. Brzezina, L. Roy, and L. MacEachen, Planar antennas in LTCC technology with transceiver integration capability for ultra-wideband application. *IEEE Transactions on Microwave Theory and Techniques*, vol. 54, no.6, pp. 2830–2839, June 2006.

**20** M. Sun, Y.P. Zhang, and Y.L. Lu, Miniaturization of planar monopole antenna for ultrawide-band radios. *IEEE Transactions on Antennas and Propagation*, vol. 58, no. 7, pp. 2420–2425, July 2010.

**21** W.X. Wu and Y.P. Zhang, Analysis of ultra-wideband printed planar quasi monopole antennas using the theory of characteristic modes. *IEEE Antennas and Propagation Magazine*, vol. 52, no. 6, pp. 67–77, December 2010.

**22** S.H. Wi, Y.P. Zhang, H. Kim et al., Integration of antenna and feeding network for UWB single-chip radio. *IEEE Transactions on Components, Packaging and Manufacturing Technology*, vol. 1, no. 1, pp. 111–118, January 2011.

**23** M. Sun and Y.P. Zhang, Design and integration of 60-GHz grid array antenna in chip package. In *Proceedings of the Asia Pacific Microwave Conference*, Hong Kong, China, 16–20 December 2008.

**24** H. Uchimura and T. Takenoshita, A ceramic planar 77 GHz antenna array. In *IEEE International Microwave Symposium Digest*, Anaheim, CA, USA, 13–19 June 1999.

25 F. Bauer and W. Menzel, A 79-GHz planar antenna array using ceramic-filled cavity resonators in LTCC. *IEEE Antennas Wireless Propagation Letters*, vol 12, pp. 910913, 2013.

26 N. Shino, H. Uchimura, and K. Miyazato, 77GHz band antenna array substrate for short range car radar. In *IEEE International Microwave Symposium Digest*, Long Beach, CA, USA, 12–17 June 2005.

27 http://www.ntt.co.jp/news2012/1201e/120117a.html.

28 R. Suga, H. Nakano, Y. Hirachi et al., A small package with 46-dB isolation between Tx and Rx antennas suitable for 60-GHz WPAN module. *IEEE Transactions on Microwave Theory and Techniques*, vol. 60, no.3, pp. 640–646, March 2012.

29 S. Liao, P. Wu, K.M. Shum et al., Differentially fed planar aperture antenna with high gain and wide bandwidth for millimeter-wave application. *IEEE Transactions on Antennas and Propagation*, vol. 63, no. 3, pp. 966–977, March 2015.

30 T. Tajima, H.J. Song, M. Yaita et al., 300-GHz step-profiled corrugated horn antennas integrated in LTCC. *IEEE Transactions on Antennas and Propagation*, vol. 62, no. 11, pp. 5437–5444, November 2014.

31 W.F. Richards, Y.T. Lo, and D.D. Harrison, An improved theory for microstrip antennas and applications. *IEEE Transactions on Antennas and Propagation*, vol. 29, no. 1, pp. 38–46, January 1981.

32 Y.P. Zhang and J.J. Wang, Theory and analysis of differentially-driven microstrip antenna. *IEEE Transactions on Antennas and Propagation*, vol. 54, no. 4, pp. 1092–1099, April 2006.

33 Z.Q. Tong, A. Stelzer, and W. Menzel, Improved expressions for calculating the impedance of differential feed rectangular microstrip patch antennas. *IEEE Microwave and Wireless Components Letters*, vol. 22, no. 9, pp. 441–443, August 2012.

34 S.S. Zhong, *Microstrip Antenna Theory and Applications*. Xidian University Press, Xian, 1991.

35 Y.P. Zhang, Design and experiment on differentially-driven microstrip antennas. *IEEE Transactions on Antennas and Propagation.*, vol. 55, no. 10, pp. 2701–2708, October 2007.

36 Y.P. Zhang, Electrical separation and fundamental resonance of differentially-driven integrated-circuit antennas. *IEEE Transactions on Antennas and Propagation*, vol. 59, no. 4, pp. 1078–1084, April 2011.

37 P.S. Hall and C. Wood, Wide bandwidth microstrip antennas for circuit integration. *Electronics Letters*, vol. 15, no. 15, pp. 458–460, July 1979.

38 D.X. Liu, X.X. Gu, C.W. Baks et al., Antenna-in-package design considerations for Ka-Band 5G communication applications. *IEEE Transactions on Antennas and Propagation*, vol. 65, no. 12, pp. 6372–6379, December 2017.

39 H.T. Kim, B.-S. Park, S.-S. Song et al., A 28-GHz CMOS Direct conversion transceiver with packaged 2 times 4 antenna array for 5G cellular system. *IEEE Journal of Solid-State Circuits*, vol. 53, no. 5, pp.1245–1259, May 2018.

40 J.D. Dunworth, A. Homayoun, B.-H. Ku et al., A 28GHz bulk-CMOS dual-polarization phased-array transceiver with 24 channels for 5G user and base station equipment. In *ISSCC Digest of Technical Papers*, February 2018.

41 Z.F. Liu, P.S. Kooi, L.W. Li et al., A method for designing broad-band microstrip antennas in multilayered planar structures. *IEEE Transactions on Antennas and Propagation*, vol. 47, no. 9, pp. 1416–1420, September 1999.

42 H.H. Zhou, L.S. Wu, L.F. Qiu et al., Diagnosis and tuning of filtering antenna based on extracted coupling matrix. In *Proceedings of the IEEE Electrical Design of Advanced Packaging Systems*, Honolulu, HI, USA, 14–16 December 2016.

43 G. Guo, L.S. Wu, Y.P. Zhang et al., Stacked patch array in LTCC for 28 GHz antenna-in-package applications. In *Proceedings of the IEEE Electrical Design of Advanced Packaging Systems*, Haining, Zhejiang, China, 14–16 December 2017.

44 D. Liu and R. Sirdeshmukh, A patch array antenna for 60 GHz package applications. In *Proceedings of the IEEE Antenna and Propagation Symposium*, Honolulu, HI, 10–15 June 2008.

45 W.M. Zhang, Y.P. Zhang, M. Sun et al., A 60-GHz circularly-polarized array antenna-in-package in LTCC technology. *IEEE Transactions on Antennas and Propagation*, vol. 61, no. 12, pp. 6228–6232, December 2013.

46 L. Wang, Y.X. Gao, and W.X. Sheng, Wideband high-gain 60-GHz LTCC L-probe patch antenna array with a soft surface. *IEEE Transactions on Antennas and Propagation*, vol. 61, no. 4, pp. 1802–1809, April 2013.

47 S. Lee, S. Song, Y. Kim et al., A V-band beam-steering antenna on a thin-film substrate with a flip-chip interconnection. *IEEE Microwave Wireless Components Letters*, vol. 18, no. 4, pp. 287–289, April 2008.

48 J.D. Kraus, A backward angle-fire array antenna. *IEEE Transactions on Antennas and Propagation*, vol. 12, pp. 48–50, January 1964.

49 R. Conti, J. Toth, T. Dowling et al., The wire-grid microstrip antenna. *IEEE Transactions on Antennas and Propagation*, vol 29, pp. 157–166, January 1981.

50 F. Bauer, X. Wang, W. Menzel et al., A 79-GHz radar sensor in LTCC technology using grid array antennas. *IEEE Transactions on Microwave Theory and Techniques*, vol. 61, no. 6, pp. 2514–2521, June 2013.

51 Z.H. Chen, Y.P. Zhang, A. Bisognin et al., An LTCC microstrip grid array antenna for 94-GHz applications. *IEEE Antennas Wireless Propagation Letters*, vol. 14, pp. 1279–1281, 2015.

52 S. Beer, C. Rusch, B. Göottel et al., D-band grid-array antenna integrated in the lid of a surface-mountable chip-package. In *Proceedings of the European*

*Conference on Antennas and Propagation*, Gothenburg, Sweden, 8–12 April 2013.

**53** B. Zhang, H. Gulan, T. Zwick et al., Integration of a 140 GHz packaged LTCC grid array antenna with an InP detector. *IEEE Transactions on Components, Packaging and Manufacturing Technology*, vol. 5, no. 8, pp. 1060–1068, August 2015.

**54** B. Zhang, C. Karnfelt, H. Gulan et al., A D-band packaged antenna on organic substrate with high fault tolerance for mass production. *IEEE Transactions on Components, Packaging and Manufacturing Technology*, vol. 6, no. 3, pp. 359–365, March 2015.

**55** H. Nakano, T. Kawano, H. Mimaki et al., A fast MoM calculation technique using sinusoidal basis and testing functions for a wire on a dielectric substrate and its application to meander loop and grid array antennas. *IEEE Transactions on Antennas and Propagation*, vol. 53, no. 10, pp. 3300–3307, October 2005.

**56** K. Solbach and R. Schneider, Review of antenna technology for millimeter wave automotive sensors. In *Proceedings of the IEEE Antennas and Propagation Symposium*, Spokane, Washington, USA, 3–8 July 2011.

**57** E.O. Hammerstad, Equations for microstrip circuit design. In *Proceedings of the Fifth European Microwave Conference*, Hamburg, Germany, 1–4 September 1975.

**58** K.D. Palmer and J.H. Cloete, Synthesis of the microstrip wire grid array. In *Proceedings of the 10th International Conference on Antennas and Propagation*, vol. 1, pp. 114–118, 1997.

**59** X.X. Gu, D.X. Liu, C. Baks et al., A multilayer organic package with four integrated 60 GHz antennas enabling broadside and end-fire radiation for portable communication devices. In *Proceedings of the IEEE Electronic Components Technology Conference*, San Diego, CA, USA., 26–29 May 2015.

**60** C.Y. Ho, M.F. Jhong, P.C. Pan et al., Integrated antenna-in-package on low-cost organic substrate for millimeter-wave wireless communication applications. In *Proceedings of the IEEE Electronic Components Technology Conference* Orlando, FL, USA., 30 May–2 June 2017.

**61** Y.W. Hsu, T.C. Huang, H.S. Lin et al., Dual-polarized quasi Yagi–Uda antennas with endfire radiation for millimeter-wave MIMO terminals. *IEEE Transactions on Antennas and Propagation*, vol. 65, no. 12, pp. 6282–6289, December 2017.

**62** Y. Qian, W.R. Deal, N. Kaneda et al., Microstrip-fed quasi-Yagi antenna with broadband characteristics. *Electronics Letters*, vol. 34, no. 23, pp. 2194–2196, November 1998.

63 Y. Qian, W.R. Deal, N. Kaneda et al., A uniplanar quasi-Yagi antenna with wide bandwidth and low mutual coupling characteristics. In *Proceedings of the IEEE Antennas and Propagation Symposium*, Orlando, FL, 11–16 July 1999.
64 K.M. Luk and H. Wong, A new wideband unidirectional antenna element. *International Journal of Microwave and Optical Technology*, vol. 1, no. 1, pp. 35–44, June 2006.
65 K.B. Ng, H. Wang, K.K. So et al., 60 GHz plated through hole printed magneto-electric dipole antenna. *IEEE Transactions on Antennas and Propagation*, vol. 60, no. 7, pp. 3129–3136, July 2012.
66 B. Yu, Z.Y. Qian, C.C. Lin et al., Wideband antennas in fan-out wafer level packaging for 5G wireless communications. *IEEE Transactions on Antennas and Propagation* (submitted).
67 L. Wang, Y.X. Gao, and W.X. Sheng, Wideband high-gain 60-GHz LTCCL-probe patch antenna array with a soft surface. *IEEE Transactions on Antennas and Propagation*, vol. 61, no. 4, pp. 1802–1809, April 2013.
68 H.Y. Jin, W.Q. Che, K.S. Chin et al., 60-GHz LTCC differential-fed patch antenna array with high gain by using soft-surface structures. *IEEE Transactions on Antennas and Propagation*, vol. 65, no. 1, pp. 206–215, January 2017.
69 A.E. Lamminen, A.R. Vimpari, and J. Saily, UC-EBG on LTCC for 60-GHz frequency band antenna applications. *IEEE Transactions on Antennas and Propagation*, vol. 57, no. 10, pp. 2904–2912, 2009.
70 C. Kyriazidou, H. Contopanagos, and N.G. Alexopoulos, Space-frequency projection of planar AMCs on integrated antennas for 60 GHz radios. *IEEE Transactions on Antennas and Propagation*, vol. 60, no. 4, pp. 1899–1909, 2012.
71 H.F. Contopanagos, C.A. Kyriazidou, A.P. Toda et al., On the projection of curved AMC reflectors from physically planar surfaces. *IEEE Transactions on Antennas and Propagation*, vol. 63, no. 2, pp. 646–658, 2015.
72 H.F. Contopanagos, C.A. Kyriazidou, F. De Flaviis et al., Planar spiral AMCs integrated on 60 GHz antennas. In *Proceedings of the IEEE Antennas and Propagation Symposium*, Chicago, IL, USA, 8–14 July 2012.

# 3

# Packaging Technologies

*Ning Ye*

*Package Technology Development & Integration, Western Digital, 951 SanDisk Drive, Milpitas, CA 95035, USA*

## 3.1 Introduction

This chapter provides an overview of semiconductor packages. At the beginning of the chapter a historical viewpoint of semiconductor package is briefly explored. Then focus is placed on packaging solutions applied in today's mainstream products: lead-frame-based quad flat no-lead packages, laminate-based plastic ball grid array packages with either a wire bond or flip chip as the interconnect, wafer level packages, and fan-out wafer-level packages. The package assembly process is described together with major packaging materials required in the flow. Process steps covered include wafer thinning, wafer dicing, die attaching, wire bonding, thermo-compression bonding, mass reflow, molding, and package singulation. Packaging materials covered include lead frames, laminate substrates, redistribution layers, mold compounds, die attach materials, bonding wires, and solder balls. Some emerging trends in packaging technology are briefly touched upon in the summary.

## 3.2 Major Packaging Milestones

Looking back at the history of semiconductor packaging, there are several pivotal moments worth remembering. The invention of the transistor in 1947 marked the beginning of the semiconductor revolution [1]. 1957 was the year when the first paper was published on wire bonding as a technique for establishing an electrical connection between a semiconductor and external leads [2]. Flip chip was introduced as a new microelectronics packaging technique in 1964 [3]. In 1967, the lead frame package was developed as the number of package leads grew beyond 10 [4].

*Antenna-in-Package Technology and Applications*, First Edition.
Edited by Duixian Liu and Yueping Zhang.

Plastic ball grid array (BGA) package based on an organic laminate substrate was developed in 1989 [5]. In 1998, plastic quad flat non-leaded (QFN) package was introduced to the market [6], and wafer-level chip-scale package (WLCSP), also known as wafer-level package (WLP), was patented [7]. Fan-out wafer-level package (FO-WLP) was patented in 2001 [8].

## 3.3 Packaging Taxonomy

Packaging is also referred to as back-end processing in a semiconductor manufacturing flow. Front-end processing means wafer fabrication, whereas back-end processing, or packaging, converts wafers into packaged (enclosed) dies that are ready to be used in an electronic end application.

There are several functions enabled by a semiconductor package. It connects the die mechanically to the system-level printed circuit board (PCB). It protects the die from environmental impacts, such as mechanical stress, dust, contamination, and moisture, in the board assembly process and usage environment. It also provides heat dissipation paths for the die. Most importantly, the package offers electrical connections between the die and the PCB.

Tracing the electrical pathway from the die to the PCB, it can be observed that there is a disparity between the input/out (I/O) pad pitch on the die, on the order of tens of micrometers, and the package mounting pad pitch on the PCB, on the order of hundreds of micrometers. This pitch gap is bridged by a routing layer in the package.

There are many ways to classify the various types of packages. In this chapter, routing layer and die-to-routing layer interconnects are the focal points of the package categories.

### 3.3.1 Routing Layer in Packages

There are three choices for the routing layer in packages in high-volume production: lead frame, laminate, and redistribution.

#### 3.3.1.1 Lead Frame

A lead frame, as shown in Figure 3.1, is a strip of thin metal sheet that serves as carrier for many rows and columns of individual dies. It has a die paddle, to which an integrated circuit (IC) die is attached. In the periphery of the package there are a number of leads connecting the package to the system level PCB. The lead frame is typically made of copper (Cu) alloy manufactured by either stamping or etching. Stamping is better suited for high-volume products from a unit cost perspective, but requires a higher upfront non-recurring engineering (NRE) cost for die and

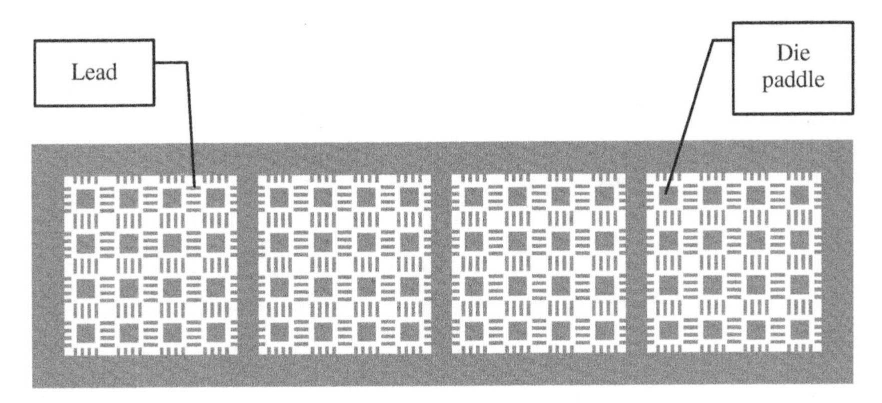


**Figure 3.1** Lead frame.

punch tooling. Etching has a lower NRE cost and a shorter lead time for tooling, but its volume production unit cost is higher. Some of the major lead frame manufacturers are ASM Pacific Technology, Chang Wah, Haesung DS, Mitsui, Shinko, and SDI.

The lead frame is single layer only and its external leads are limited to the package periphery. This constrains its routing capability and I/O counts. It is still widely used as a mainstream choice of packaging in many applications due to its low cost, proven reliability, and improved performance. Typical applications of lead frame packages include analog devices, application-specific integrated circuits (ASIC), digital signal processors, flash, logic devices, memory, microcontrollers, and optoelectronics.

#### 3.3.1.2 Laminate

Laminate, also known as organic laminate substrate, is a type of die carrier that consists of multiple layers of conductor and glass fiber reinforced epoxy resin. Laminate enables a package to PCB connection scheme that is different from that used by a lead frame. In a lead frame package, the package is connected to the PCB through leads that are constrained on the package periphery. The multiple conductor layers in laminate allow the routing freedom to place solder balls, which are the connection between the package and the PCB, across the entire bottom surface of the package.

A cross-section of laminate substrate is illustrated in Figure 3.2a. The conductor layers are etched Cu layers, which are separated by dielectric layers. A two-layer substrate is generally 130 μm thick. The middle dielectric layer is the core. One of the most popular core materials is bismaleimide triazine (BT) resin reinforced with glass fibers because it has a high glass transition temperature and low dielectric constant and loss factor. BT was invented by Mitsubishi Gas Chemical

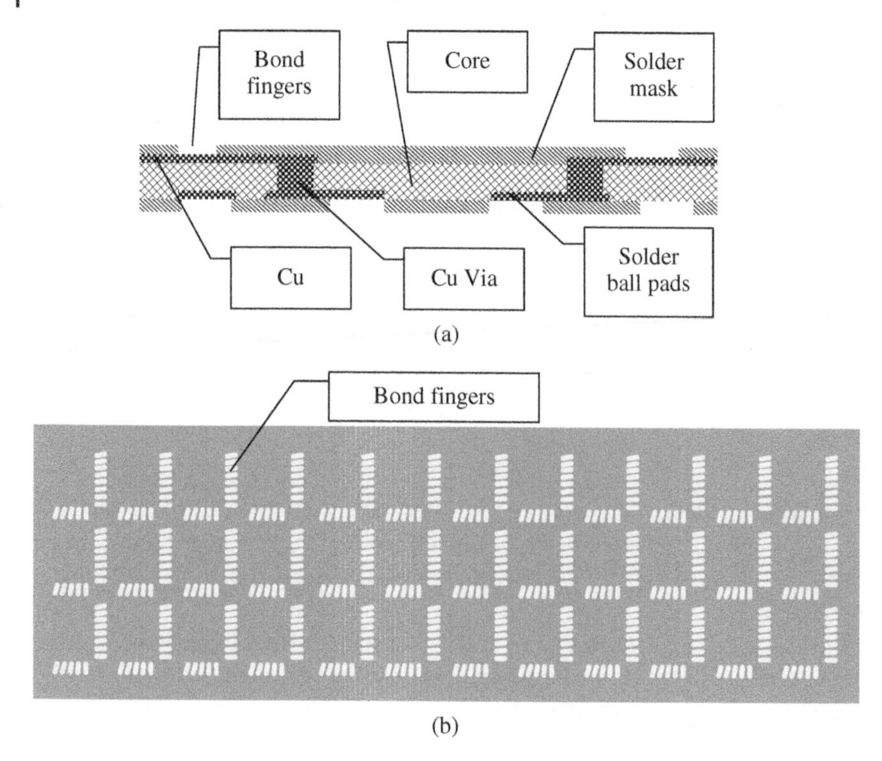


**Figure 3.2** Laminate substrate: (a) cross-section view of laminate substrate and (b) top view of laminate substrate strip with three rows and 12 columns of package substrates.

in 1974 [9]. Other core suppliers include Hitachi Chemical, Doosan Corporation Electro-Materials, Panasonic, LG Chem, and Sumitomo Bakelite.

The top metal layer has bond fingers connecting to the die, and the bottom metal layer has pads to mount solder balls connecting to the PCB. The bond fingers are plated with nickel/gold (Ni/Au). The surface finish of the solder ball pads can be either Ni/Au or organic solderability preservatives. The top and bottom metal layers are connected with Cu vias. For four or higher layer count laminate substrate design, designers use dedicated Cu layers as the power and ground planes in the inner layers for better electrical performance.

On the top side of the substrate, an opening in the solder mask, also known as a solder resist, is made where soldering pads for surface mount components and wire bond fingers are located. On the bottom side, solder mask openings define the location of the mounting pads for solder balls. The solder mask serves as the protection layer covering the Cu trace, and it can also prevent solder from bridging across adjacent Cu traces/pads. Among the several types of solder resist, liquid and dry film photo imageable solder resists are the most widely used in organic laminate package substrates. The main ingredients of solder resists are resin, inorganic fillers, and solvent. Major suppliers of solder resist include Taiyo Ink MFG and Hitachi Chemical.

Organic laminate substrate producers include ASE Materials, Daeduck Electronics, Daisho Denshi, Eastern, IBIDEN, Kinsus, Korean Circuit Company, Nan Ya PCB, SEMCO, SHINKO, Shennan Circuits, SIMMTECH, and Unimicron. Laminate substrates are shipped to package assembly factories in strip format, which contains multiple package substrates, as shown in Figure 3.2b.

Laminate packages are suitable for high-performance devices, e.g. ASICs, central processing units (CPUs), dynamic random-access memory (DRAM), flash memory, and field-programmable gate arrays (FPGAs).

#### 3.3.1.3 Redistribution Layer

In a WLP, the routing layer comes in the form of a redistribution layer (RDL), which consists of one or multiple conductor layers sandwiched by dielectric layers. These layers can be directly "grown" on top of the die in a wafer or panel format through a sequence of layering/patterning process steps similar to those used in wafer fabrication [10].

An example of RDL process flow is shown in Figure 3.3 and includes the following process steps: spin coating and patterning a polymer dielectric layer, sputtering a seed layer with Ti/Cu, electroplating with Cu, and depositing and patterning another polymer dielectric layer. It can be observed that the essential function of an RDL is to move/reroute the I/O pad on the die to a new location. The requirement on the RDL trace line and space (L/S) is driven by routing complexity, which depends on the number of I/Os and package size. For most applications L/S requirements are between 5 and 10 μm. This can be supported by the low curing temperature polymer approach, as illustrated in Figure 3.3. Common choices of polymer dielectric materials include polyimide (PI), polybenzoxazoles (PBO), and benzocyclobutene (BCB) [11]. To reduce warpage and improve reliability, Taiyo Ink developed a new photosensitive dielectric material, called PDM-1, with a curing temperature reduced to 180 °C and coefficient of thermal expansion (CTE) lowered to 30–35 ppm/°C. Cu traces with 2 μm L/S were demonstrated. The high resolution and low CTE were achieved by applying nano-sized fillers [12, 13]. For high interconnect density applications requiring finer L/S, a damascene process is required [14]. One of the concerns associated with finer L/S designs is a broken Cu trace due to electromigration caused by high current density. A study has shown that covering Cu traces with inorganic dielectrics can significantly increase electromigration resistance [15].

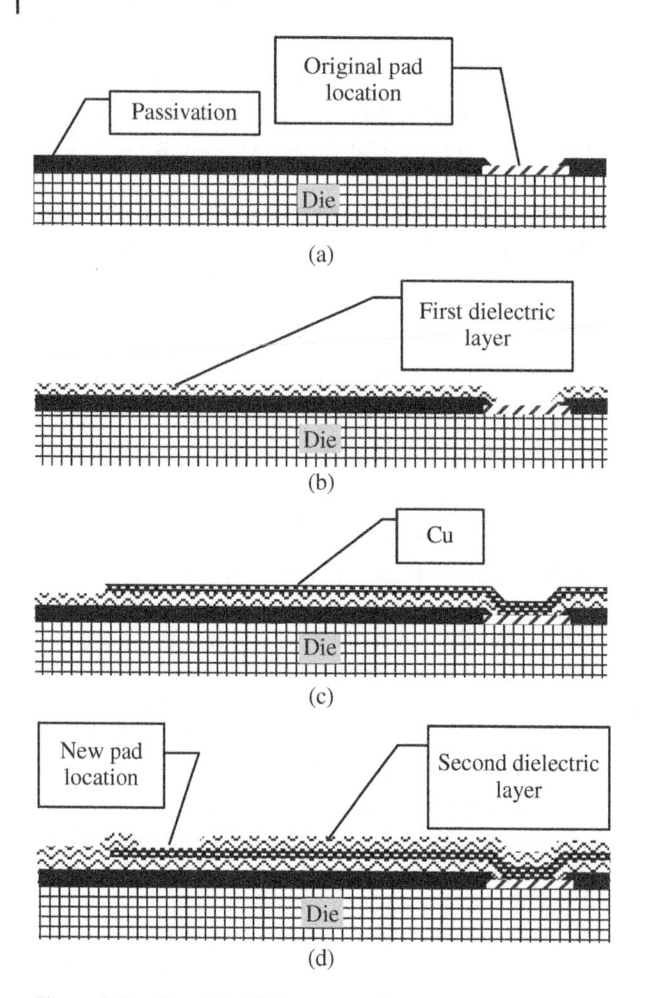


**Figure 3.3** Simplified RDL process flow: (a) incoming wafer, (b) spin coat and pattern first polymer dielectric layer, (c) sputter seed layer Ti/Cu and electroplate Cu, and (d) deposit and pattern second polymer dielectric layer.

### 3.3.2 Die to Routing Layer Interconnect

Wire bonds and flip chips are two major types of interconnect between the die and the routing layer.

#### 3.3.2.1 Wire Bonds

Wire bonds use fine metal wires as the interconnect between the I/O pads on a die and the routing layer, such as a lead frame or organic laminate substrate. The wire materials include Au, Cu, silver (Ag), and aluminum (Al). The wire diameter can be as thin as 15 μm, and its length can be as long as 5 mm. The geometry of the bond wire has a strong impact on the performance of the circuit, especially for high-speed designs. Commercial electromagnetic simulators, such as Advanced Design System (ADS), COMSOL, and ANSYS HFSS, are available to support circuit simulation. There is also ongoing effort in the research community to further enhance the methodology. One method is to model bond wires with arbitrary shapes using lumped element models [16]. The wire geometry is approximated by breaking the wire into small segments. The inductance and capacitance of each segment are modeled using traditional equations, and then they are entered into a circuit simulator, such as SPICE [17], as lumped elements. The advantage is that this algorithm can be realized with any scripting language. Another proposal is an analytical approach that models the self and mutual inductances of bond wires without segmenting them by adding a bonding-wire-geometric-profile-dependent term into the self-inductance formula [18]. The computation of this analytical approach can be completed in a few milliseconds, which is much faster than a numerical approach. An alternative approach to simplify the behavioral model of bonding wires is to introduce a frequency-dependent resistor, which cuts four design parameters from the original nine [19].

Gold used to be the preferred material for bonding wire. Due to the drive to reduce cost, Cu and Ag are gaining popularity. To overcome the bondability issue and corrosion failures sometimes observed for Ag and Cu wires during biased highly accelerated stress testing (HAST), noble metal coated Ag wire [20] and Au flash palladium coated Cu [21] wires have been developed and demonstrated with improved performance. Wire suppliers include Heraeus, MK Electron, Nippon Micrometal, and Tanaka. As shown in Figure 3.4a, the I/O pads designed for a wire bond die are located on the periphery of the die. Pad pitch can be pushed to be around 40 μm.

#### 3.3.2.2 Flip Chips

In a flip-chip design, the interconnect between the die and the routing layer is through metal bumps on the I/O pads located across the die surface, as illustrated in Figure 3.4b. The name "flip chip" comes from the fact that die is flipped, with the I/O pads of the die facing down toward the routing layer. This is opposite to the die attach orientation in a wire bond.

Instead of the bonding wires in a wire bond BGA package, a flip-chip package uses solder bumps or Cu pillar bumps (a recent development to reduce bump pitch) as the interconnect between the die and the laminate substrate. In wire bond package, the bond pads are limited on the die periphery only; in a flip-chip package, the bump pads can be located anywhere on the active side of the die surface. Therefore, flip-chip design lends itself to high I/O count packages. The shortened interconnect path in a flip-chip package offers better electrical performance than

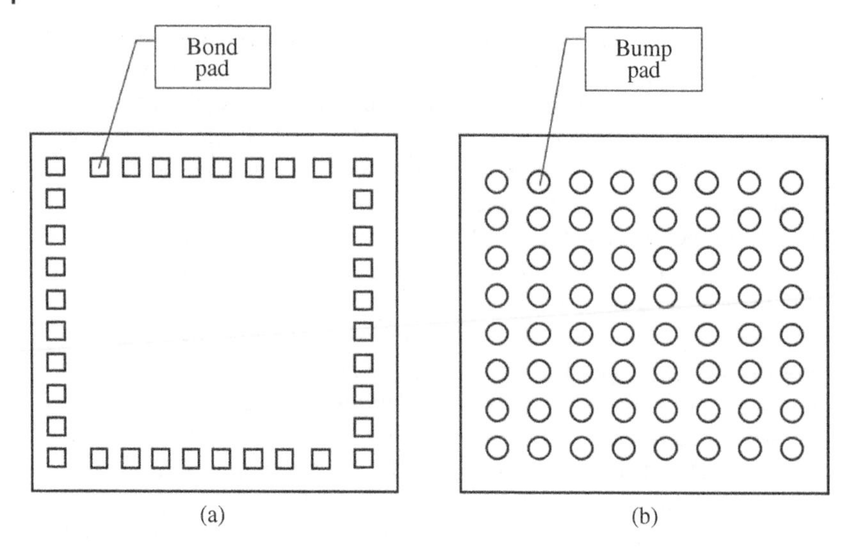


**Figure 3.4** Wire bond versus flip chip die: (a) wire bond die and (b) flip chip die.

in a wire bond package due to low inductance loss. This has made flip chip the preferred choice of interconnect for high-performance applications such as microprocessors, graphics processing units (GPUs), high-speed memory, basebands, and high-performance ASICs.

## 3.4 Packaging Process for Several Major Packages

The key packaging process steps are described here for several mainstream packages: wire bond plastic BGA, wire bond QFN, flip-chip plastic BGA, WLP, and FO-WLP.

### 3.4.1 Wire Bond Plastic Ball Grid Array

The wire bond plastic BGA package, illustrated in Figure 3.5, is plastic molded package with a wire bond connecting the die to the organic laminate substrate. It is called a ball grid array because solder balls, which connect the package to the next level of the PCB, are laid out in a grid pattern on the bottom side of the package. The advantage over lead frame packages is higher I/O counts, and better electrical and thermal performance. A high-level flow chart of assembly process for wire bond plastic BGA is shown in Figure 3.6.

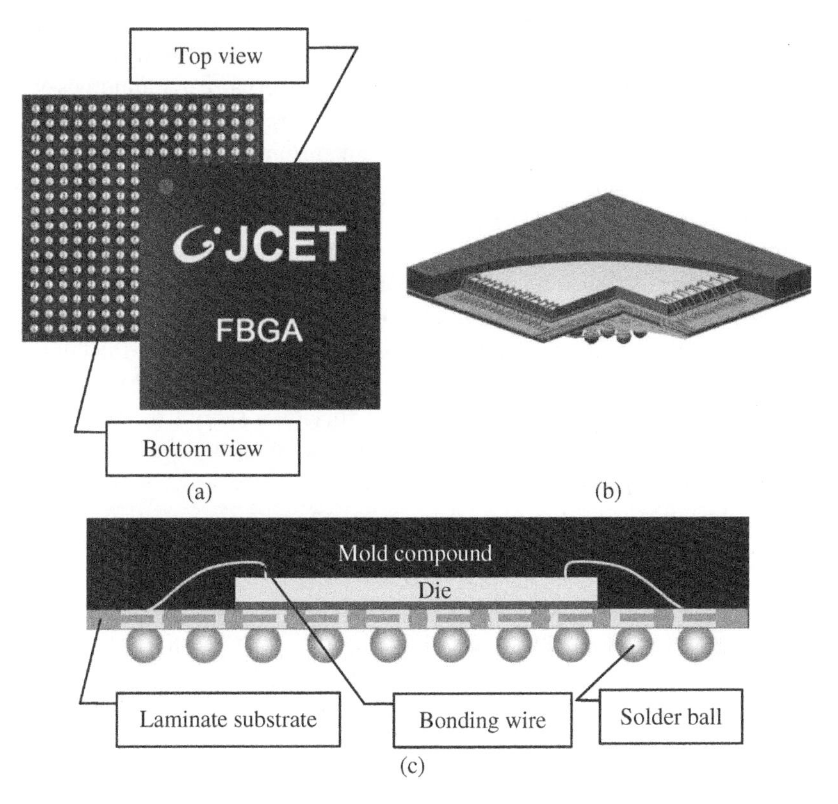


**Figure 3.5** Wire bond plastic BGA package: (a) top and bottom view, (b) isometric view with mold compound partially removed, and (c) cross-section view (from [22], © 2018 STATS ChipPAC Pte. Ltd, reprinted with permission).

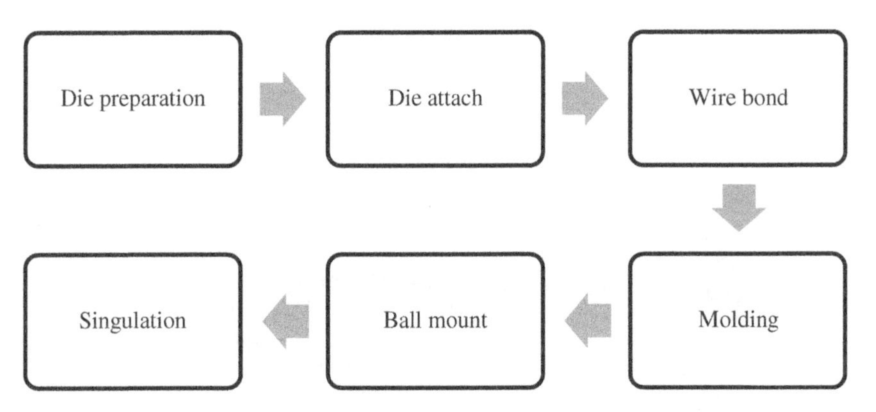


**Figure 3.6** Assembly process flow for wire bond plastic BGA.

#### 3.4.1.1 Die Preparation

ICs are fabricated on a wafer, which is a thin disk of semiconductor, with silicon being the dominant material used in the industry. One wafer can carry hundreds or thousands of individual ICs, also known as a dies or chips, that are manufactured to the same electrical specification and physical dimensions. In the process step of die preparation, wafers are converted into dies. Typical wafer sizes used by wafer fabs are 300, 200, and 150 mm in diameter.

Die preparation, as illustrated in Figure 3.7, comprises two major steps: wafer thinning and wafer dicing. The nominal thickness of a 300 mm wafer is 775 μm, which is too thick to fit in most packages. A wafer-thinning process is therefore required to reduce the wafer thickness to meet the required thickness specification. One common method of wafer thinning is mechanical grinding, which removes silicon material from the back (inactive) side of the wafer. A grinding wheel contains abrasive grains, usually diamond, bonded with resin material [23]. During grinding, the front (active) side of the wafer is protected with tape against physical damage and contamination.

Wafer dicing is commonly through a mechanical dicing blade, which spins at high speed and cuts the wafer along the saw street/scribe line, the spacing between adjacent dies, separating the wafer into individual dies. The mechanical dicing blade also relies on diamond grits bonded typically with nickel using an electroforming process. During dicing, the back side of the wafer is mounted on a wafer frame through a tape. Other dicing methods use lasers or plasma.

Wafer-thinning and dicing equipment manufacturers include Accretech, ASM Pacific, Disco, SPTS Technologies, and Plasma-Therm.

#### 3.4.1.2 Die Attach

Die attach is the process that establishes a mechanical connection between the die and the routing layer. In this process step, the die is picked up from the wafer frame and bonded to either a organic laminate substrate or a metal

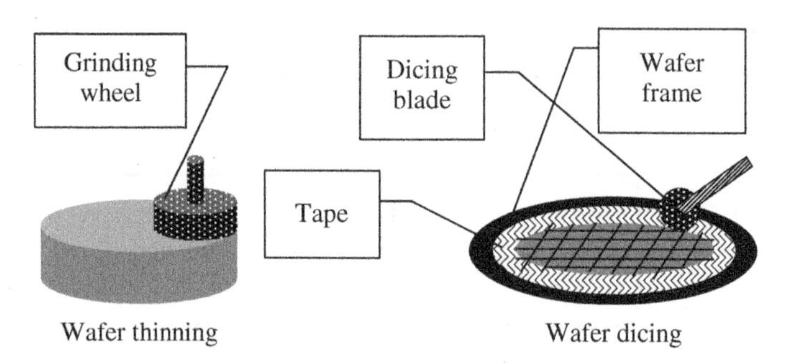


**Figure 3.7** Die preparation.

lead frame using adhesive materials. The thermoset resins used as die attach adhesives include epoxy, acrylate, bismaleimide, and polyimide. There are two common forms of die attach adhesives: paste and film. Die attach paste, usually seen in single die packages, is available in syringes and dispensed on a laminate substrate/lead frame. Die attach film [24], laminated on the back side of a wafer after thinning, is preferred in thin die or stacked die packages because film eliminates or reduces the process complexity of paste, such as bleeding or voiding. Air bubbles along the silicon/die attach film interface could still occur due to die and/or substrate warpage. Air voids trapped in the die attach material may cause die crack during wire bonding [25] or delamination during reflow of packaging mounting onto the PCB. Moisture absorption and elevated temperature exposure can degrade the interfacial adhesion strength of die attach film [26].

Die attach is performed at elevated temperatures, generally in the range 80–150 °C. One key step in die attach, called "die pick," picks the die away from the tape. During die pick, as shown in Figure 3.8, the die is pushed upward underneath the tape by an ejection tool, which can be an array of needles for thicker dies. However, for thinner dies, a more complex ejection tool, such as a multistep ejector, is required [27].

Die attach adhesive suppliers include Henkel, Hitachi Chemical, Lintec, LG Chem, and Nitto. Major die attach equipment makers include Besi, Fasford Technology, Kulicke & Soffa, and Panasonic.

#### 3.4.1.3 Wire Bonding

In wire bonding, electrical connections between the die and the routing layer are established through metal wires, as shown in Figure 3.9. Among the several kinds of wire bond processes, the most widely used one is thermosonic ball bonding with Au, Cu, or Ag wire.

During the wire bond process, the laminate substrate is held at a raised temperature, e.g. 150 °C for Au wire. The bonding process starts with forming a ball by melting the tip of the wire using a plasma discharge, a process known as electronic flame-off [28]. The ball is then pressed against the Al bond pad on a die while ultrasonic energy is applied. With the aid of force, elevated temperature, and ultrasonic energy, an intermetallic compound layer composed of the wire metal and the bond pad metal is formed. Next, the wire is lifted and dragged toward the bond finger on the organic laminate substrate following a series of specific motions designed to form a wire shape, called a "loop." Loop control is important as it can prevent wires from shorting to the die or to each other during wire bonding and the subsequent process step, molding. The second bond on the organic substrate does not involve the formation of a ball, but similar to the first bond, it also requires force, heat, and ultrasonic energy. Wire bonding is a sequential process, which means a wire bonder (wire bond machine) bonds one wire at a time. However, it is a very

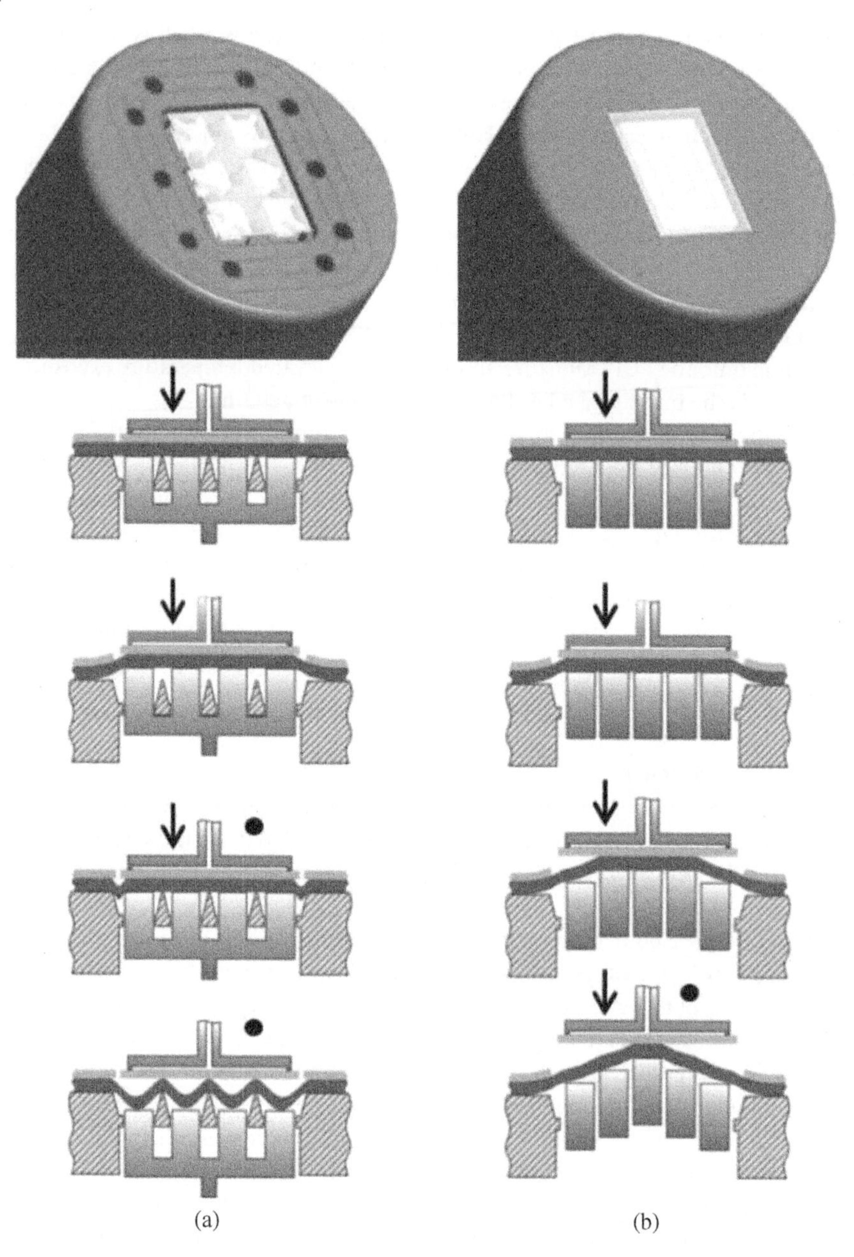


**Figure 3.8** Pick process flow: (a) piston (needle) ejector and (b) multi-step ejector (from [27], © 2018 IEEE, reprinted with permission).

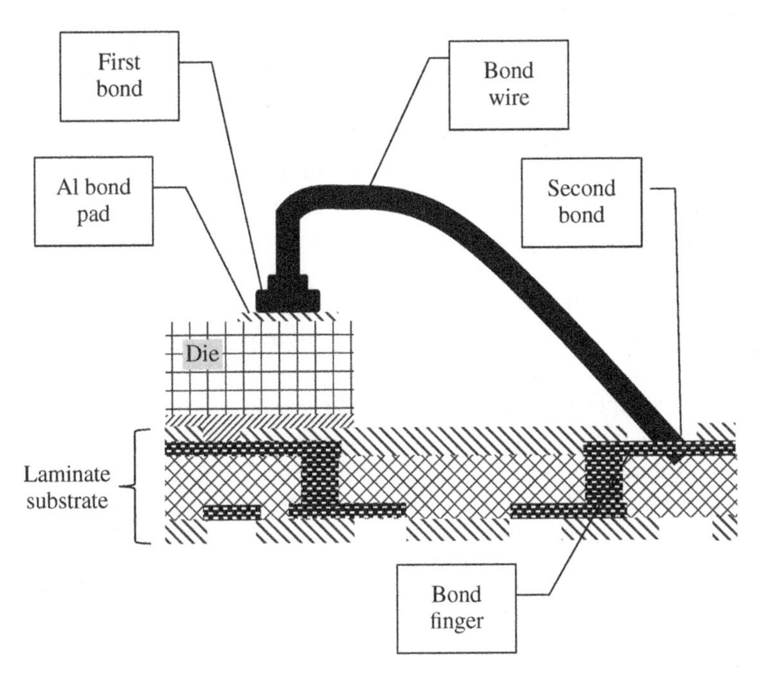


**Figure 3.9** Wire bond between die and laminate substrate.

fast process, as an advanced ball bonder can bond more than 20 wires per second. Wire bonder manufacturers include Kulicke & Soffa, Besi, and Shinkawa.

#### 3.4.1.4 Molding

In the molding process, the die is encapsulated in the mold compound, which provides mechanical and environmental protection to the die. This is especially important for wire bond packages. With the active side of the die facing up and fragile thin bonding wires, the wire bond package will not survive the stress of the mechanical handling in the subsequent assembly and test process steps without the mold compound.

Some leading mold compound suppliers are Panasonic, Sumitomo Bakelite, Kyocera Chemical, and Hitachi Chemical. Mold compound is mainly composed of epoxy resin, such as cresol novolac or biphenyl, and silica fillers, which are added to lower the CTE. The CTE of mold compound is important in managing package warpage [29], which is especially critical for low profile thin packages. Warpage prediction methodology leveraging both experimental study and numerical simulation has been developed to select the proper mold compound CTE in order to optimize package warpage at both room temperature and component mounting temperature [30]. Another key characteristic of the mold compound is its moisture sensitivity. If too much moisture is absorbed by the mold compound, delamination along material interfaces within package can be induced by vapor pressure during solder reflow for component mounting [31]. Moisture desorption and absorption of a mold compound have been experimentally studied to determine the diffusion constant and the saturated moisture concentration [32]. Multiple approaches have been proposed to numerically model moisture diffusion in electronic packages. One such proposal applied water activity-based theory to overcome the discontinuity issue at the material interface, covering both linear and non-linear diffusion behaviors [33].

As illustrated in Figure 3.10, two main methods are used in the package molding process: transfer molding and compression molding. In transfer molding, a substrate strip is first placed in the mold chase. The mold compound is then heated to a molten state at, for example, 175 °C and flows into the mold cavity under the pressure of the plunger. In compression molding, mold compound is dispensed into the mold cavity and heated up. The substrate is then immersed in the molten mold compound by compressing the top and bottom mold chases toward each other. Compression molding minimizes lateral mold flow, therefore the risk of excessive

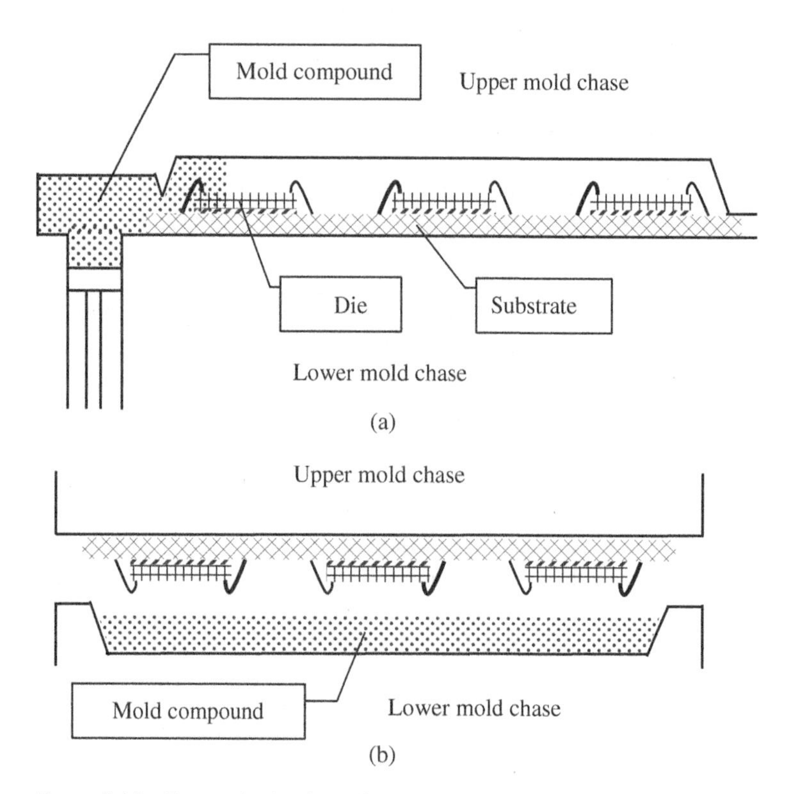


**Figure 3.10** Two methods of molding: (a) transfer molding and (b) compression molding.

wire sweep, i.e. lateral movement of bond wires during molding, is reduced. The mold compound hardens as curing or crosslinking occurs in the mold. Its full cure after molding, also known as post mold cure, takes between 4 and 16 h at 175 °C. Molding equipment manufacturers include APIC YAMADA, Besi, and Towa.

#### 3.4.1.5 Ball Mounting

In ball mounting, solder balls, which are the interconnect between the BGA package and the PCB, are mounted on the bottom side of the laminate substrate by (i) fluxing the ball pads of the substrate, (ii) placing the solder balls, e.g. by dropping the balls through a metal stencil, and (iii) reflowing the solder balls. The reflow process is performed in a reflow oven, which has multiple temperature zones. The conveyor belt in the oven carries the substrate strips through the multiple temperature zones to heat the solder balls to above the melting temperature of the solder and then cool them back down. After the reflow, the solder balls are bonded to the ball pad on the substrate through an intermetallic compound layer formed at the interface. As a result of the European Union's Restriction of Hazardous Substances Directive (RoHS), implemented in 2006, most solder balls are now lead-free and normally made of tin (Sn)-Ag-Cu alloys. Because of the CTE mismatch between the package and the PCB, solder joint fatigue is of critical consideration for board level reliability. Adding Sn-Ag-Cu with dopant, such as bismuth (Bi) for SACQ solder, has been shown to improve performance in drop and temperature cycling tests [34, 35]. To lower solder ball cost, a low-Ag solder (less than 0.7wt%) has been developed with Bi and Ni added, which shows better solderability, thermal resistance, electro-migration resistance, and mechanical properties [36].

Solder ball suppliers include Accurus Scientific, Indium, MK Electron, Nippon Micrometal, and Senju Metal Industry. Heller Industries and Senju Metal Industry Co. Ltd are among the manufactures of reflow oven.

#### 3.4.1.6 Package Singulation

In package singulation, molded strips are cut into individual packages using saw blades with diamond grits. The strips are mounted on tape or a vacuum jig to hold the packages during cutting. Tape is ideal for small volume production as there is no need to tool up a vacuum jig, which is package size dependent. Another advantage of tape is that it can handle larger strip warpage. However, if strip warpage is under control and production volume is high, a vacuum jig is more cost effective as it avoids the consumable cost associated with tape. Package singulation machine suppliers include Advanced Dicing Technologies, Besi, and Disco.

### 3.4.2 Wire Bond Quad Flat No-Lead Packages

Wire bond QFN packages, shown in Figures 3.11 and 3.12, are plastic molded packages with a wire bond connecting the die and the Cu/lead frame. The QFN name

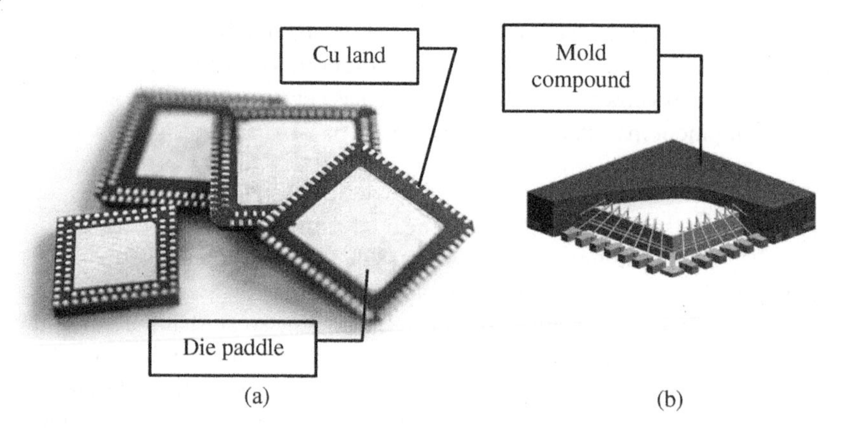


**Figure 3.11** QFN package: (a) bottom view and (b) isometric view with mold compound partially removed (from [37], © 2018 STATS ChipPAC Pte. Ltd, reprinted with permission).

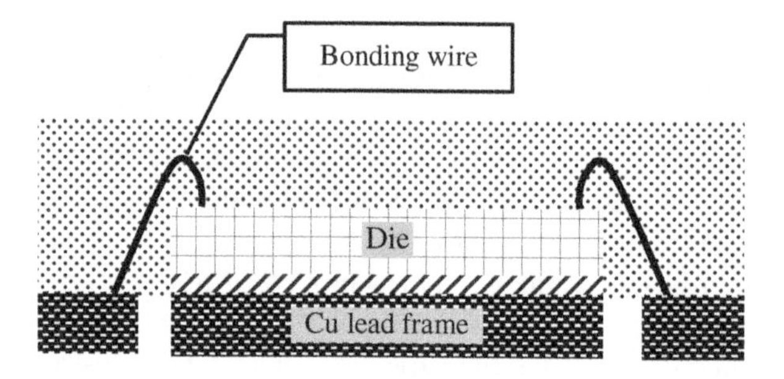


**Figure 3.12** QFN package cross-section.

originates from the fact that no leads extend beyond the package body, which is a major distinction from other traditional lead frame packages with external leads, such as thin small outline packages (TSOPs) and quad flat packages (QFPs).

The assembly flow of a QFN package, as illustrated in Figure 3.13, is very similar to that of a wire bond BGA package except it has an optional lead finish step. On the bottom side of a QFN package there are exposed metal lands on the periphery and die paddle in the middle that connect the package to the PCB. Without any protection for the lands and the die paddle, oxidation on the exposed Cu hinders the soldering process during surface mounting onto the PCB. Thus, a lead finish step is required to plate a protection layer such as matte Sn. If the lead frame is pre-plated with Ni/palladium (Pd)/Au, this step can be skipped.

A typical QFN pin count is below 200. QFN's small form factor and lower cost compared to BGA make it an ideal choice for low I/O dies in portable applications.

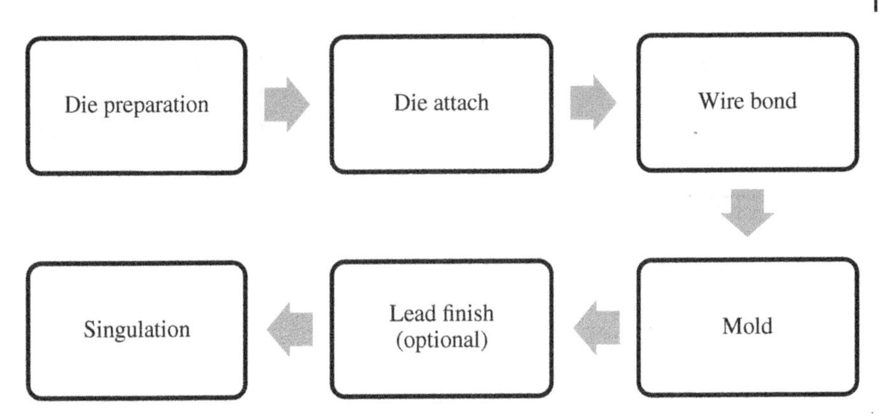


**Figure 3.13** QFN package assembly process flow.

### 3.4.3 Flip-chip Plastic Ball Grid Arrays

The flip chip plastic BGA package, shown in Figure 3.14, is similar to a wire bond plastic BGA, except that the die to laminate substrate interconnect is flip chip instead of wire bond. In flip chip, the inactive back side of the die faces up, so stress from mechanical handling in the subsequent assembly and test process steps is much less important. The molding process is therefore skipped in some flip-chip BGA designs. Figure 3.15 illustrates an assembly flow of a flip-chip plastic BGA. There are three steps that are different from the assembly flow of a wire bond plastic BGA: wafer bumping, flip-chip attach, and underfill.

#### 3.4.3.1 Flip-chip Bumping

There are two mainstream flip-chip bumping technologies: solder bumping and Cu pillar bumping. The bump layout on a die can be array or periphery. The

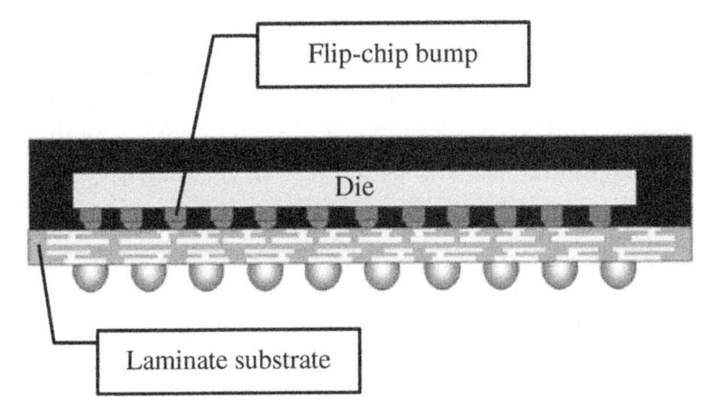


**Figure 3.14** Flip chip BGA package cross-section (from [38], © 2018 STATS ChipPAC Pte. Ltd, reprinted with permission).

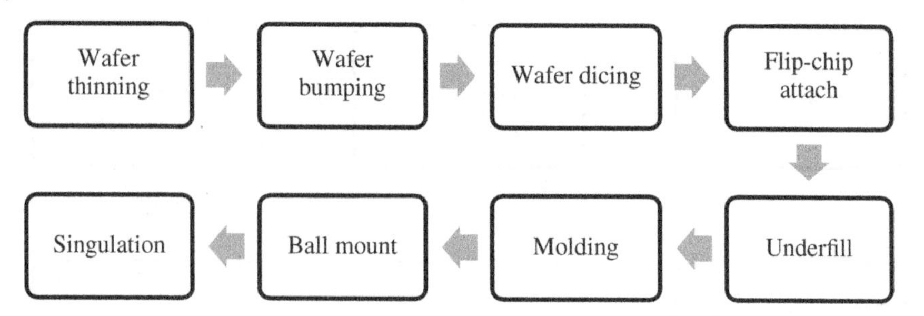


**Figure 3.15** Flip chip plastic BGA.

advantage of periphery is layer count reduction in the package substrate as the routing complexity is significantly decreased in contrast to an array. The bump pitch of a solder bump can be as small as 130 μm for array design and 80 μm for in-line periphery design. For finer pitch designs, a Cu pillar is preferred over a solder bump. This is because after flip-chip bonding to the package substrate, the solder bump collapse increases the shorting risk among adjacent bumps, while Cu pillars retain their shape. With Cu pillar bumps, the pitch can be pushed down to 100 μm for array design and 30 μm/60 μm staggered for in-line periphery design. Another benefit of Cu pillars is their stronger resistance against electromigration. Among various types of bumping process, electroplating is the most widely used. Flip-chip bumping is processed at wafer level. Equipment manufacturers for IC package electroplating include Applied Materials, Lam Research, and Tokyo Electron Ltd.

#### *3.4.3.1.1 Solder Bumping*

Solder bumps are made of eutectic Sn/Pb solder or Pb-free Sn/Ag solder. Typically, bump height is in in the range 60–100 μm and bump diameter is between 80 and 125 μm.

A solder bumping flow is shown in Figure 3.16. It starts with sputtering under bump metallurgy (UBM), e.g. less than 1 μm of titanium tungsten (TiW)/Cu, which serves as diffusion barrier between solder and Al on the bond pad. It then goes through photoresist coating and patterning, Cu plating, and solder plating steps. The next steps strip the photoresist and etch the UBM. Finally, solder bumps are formed after going through solder reflow.

#### *3.4.3.1.2 Cu Pillar Bumping*

The solder cap on top of a Cu pillar is mostly Pb-free Sn/Ag solder. Usually, total pillar height (including solder cap) is in in the range 30–60 μm and its diameter is between 20 and 50 μm. As illustrated in Figure 3.17, the Cu pillar bumping process is very similar to solder bumping flow. The major difference is the ratio of Cu and solder plating.

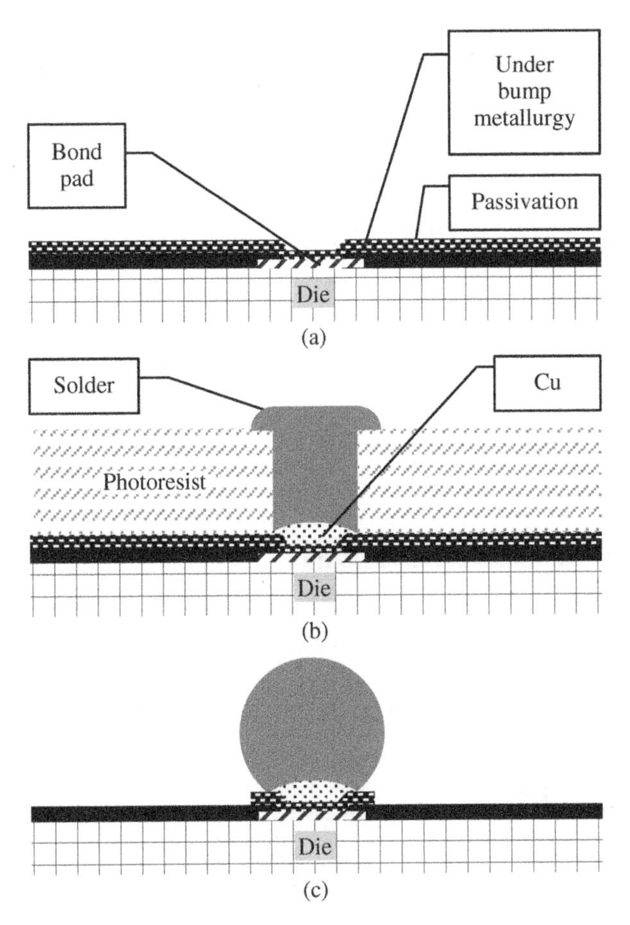


**Figure 3.16** Solder bumping process flow: (a) deposit UBM, (b) coat and pattern photoresist, deposit Cu, then repeat coating and patterning photoresist, deposit solder, and (c) remove photoresist, etch UBM, reflow.

#### 3.4.3.2 Flip-chip Attach

In the flip-chip attach process, a flip-chip die is attached to the package substrate through solder or Cu pillar bumps by mass reflow or thermocompression bonding (TCB). Mass reflow is the common method of flip-chip attach for Cu pillar bumps with pitch above 40 μm and solder bumps.

Mass reflow includes the following steps: (i) flux the bump pads on the substrate, (ii) pick and place the flip-chip die onto the substrate, and (iii) reflow. It is worth noting that reflow is performed after pick and place is completed for all the package substrate sites on a substrate strip, therefore all flip-chip dies on a substrate strip go through the reflow oven at the same time.

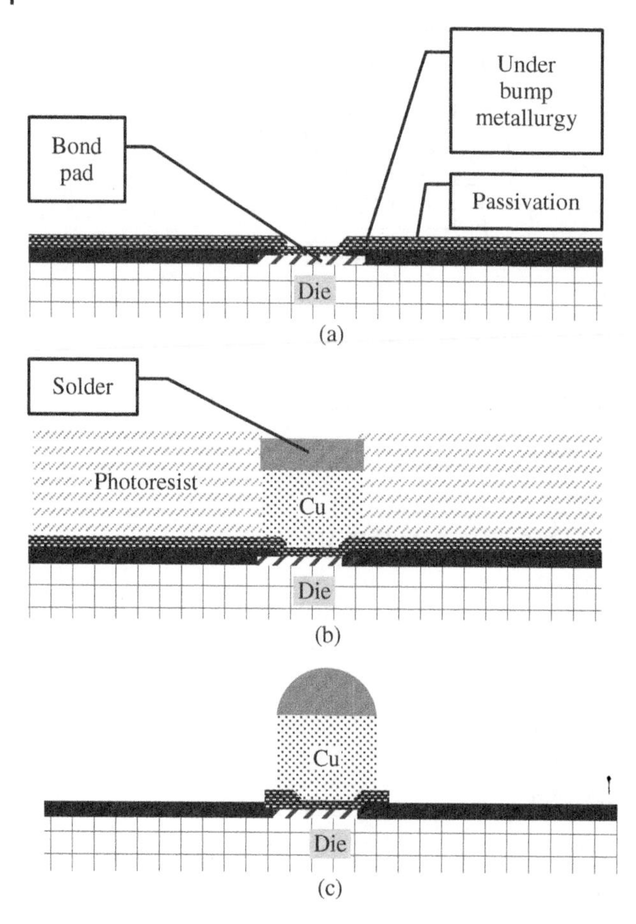


**Figure 3.17** Cu pillar bumping process flow: (a) deposit UBM, (b) coat and pattern photoresist, deposit Cu and solder, and (c) remove photoresist, etch UBM, reflow.

For Cu pillar bumps with pitch at or below 40 μm, TCB is required. The steps in a TCB process are (i) flux the bump pads on the substrate, (ii) pick and place the flip-chip die onto the substrate, and (iii) apply pressure on the die while heating the solder to above its reflow temperature. TCB is different from mass reflow in that solder heating is applied one die at a time. This is the reason why the throughput of TCB is lower than that of mass reflow [39]. Thermocompression bonder manufacturers include ASM Pacific Technology, Besi, Kulicke & Soffa, and Torray.

#### 3.4.3.3 Underfill

Underfill is applied between the die and the substrate, wrapping around the bumps, to reduce the risk of solder joint fatigue failure and cracking in extreme low $K$ dielectric layer or die caused by the CTE mismatch between silicon and substrate. For non-stacking flip chip, there are two types of underfill material: capillary underfill (CUF) and molded underfill (MUF).

CUF is based on epoxy [40] or cyanate ester [41] resin and is loaded with silica fillers to lower the CTE. CUF requires a dispensing process step along the die edge(s). It is then pulled by capillary force to spread underneath the entire die. Finally, CUF is thermally cured to form a solid bond between the die and the substrate. CUF suppliers include Henkel, Lord, Master Bond, NAMICS Technologies, Yincae Advanced Material, and Zymet.

MUF is a modified version of mold compound optimized to flow through narrow gaps. It eliminates the underfill dispensing step, but it is more expensive than normal mold compound. Depending on the die and package size, a trade-off analysis is required to determine whether CUF or MUF is the better choice.

### 3.4.4 Wafer Level Packaging

As shown in Figure 3.18, a WLP can be considered as a flip-chip die directly attached onto a PCB using package level solder balls (normally 250 μm in diameter or larger) without a laminate substrate. In contrast, flip-chip solder bumps are usually smaller than 130 μm in diameter. Due to the large CTE mismatch between silicon and PCB, WLP can only be used for small die sizes, typically smaller than $10 \times 10\,\text{mm}^2$. WLP is an ideal solution for low I/O devices, as large I/O pitch can fit within the ball grid array laid out across the die surface matching the PCB pad pitch.

The simplest WLP construction is shown in Figure 3.19. The process flow is illustrated in Figure 3.20. All major steps of WLP packaging are at the wafer level: (i) form the UBM, (ii) apply flux, (iii) mount the ball, and (iv) reflow the solder ball [42]. This simple process significantly lowers packaging cost. The WLP package

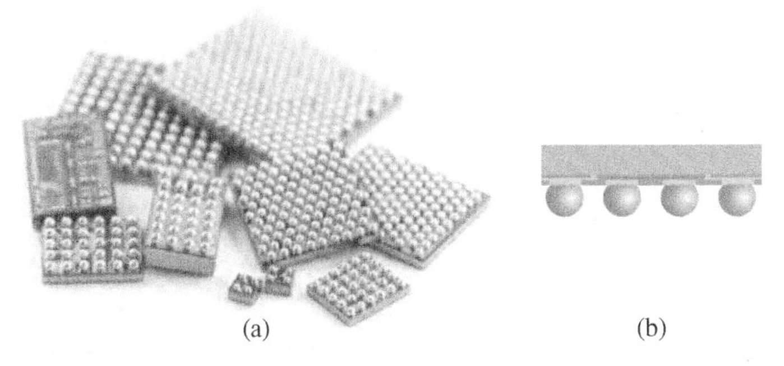


**Figure 3.18** Wafer level package: (a) bottom view and (b) cross-section view (from [43], © 2018 STATS ChipPAC Pte. Ltd, reprinted with permission).

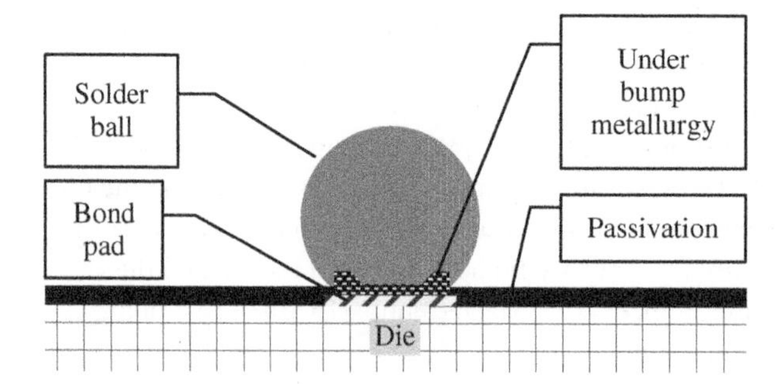


**Figure 3.19** Cross-section of solder ball on wafer level package.

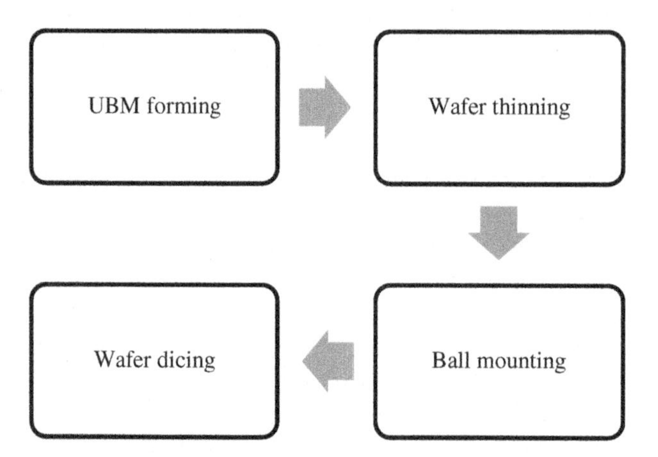


**Figure 3.20** Wafer level package assembly process flow.

size is exactly the same as the die size in length and width. Thanks to its low cost and small form factor, it is the perfect choice of packaging for smartphones and wearables.

If the bond pads of the die are originally designed for wire bonding and located on the periphery, a RDL process can be added to move the bond pads to redistribute across the entire die and form a grid array for ball mounting.

### 3.4.5 Fan Out Wafer Level Packaging

FO-WLP, shown in Figure 3.21, is a variation of regular WLP. FO-WLP has mold compound surrounding the die. Both the die surface and the mold compound surface at the bottom side of the package are utilized for package level ball pad placement, therefore FO-WLP can support larger number of I/Os than standard WLP for a given die size. FO-WLP uses RDL as the routing layer connecting the

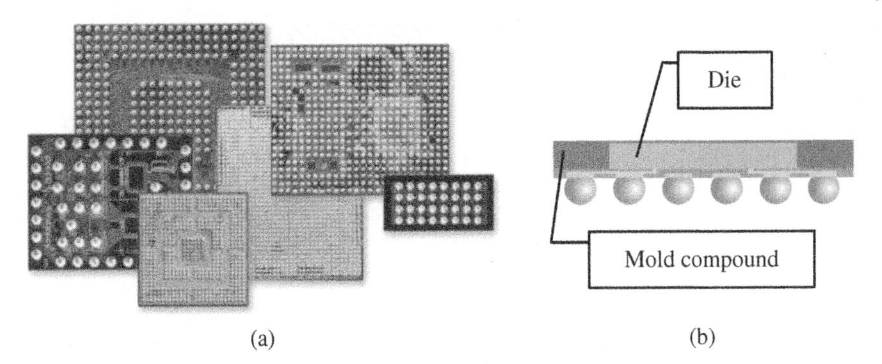


**Figure 3.21** Fan-out wafer level package: (a) bottom view and (b) cross-section view (from [44], © 2018 STATS ChipPAC Pte. Ltd, reprinted with permission).

bond pads on the die to package level ball pads either within the die surface or on the mold compound surface. An RDL with two Cu layers can be less than 20 μm thick, which is much thinner than a two-layer laminate substrate. Besides package thickness reduction, this also gives FO-WLP better thermal and electrical performance than laminate-based BGA.

One mainstream FO-WLP process is called chip-first/face-down, and is illustrated in Figure 3.22. Wafer is thinned and diced into individual dies. A thermal release tape is laminated to a temporary carrier, which is made of silicon, glass, or stainless steel. Dies are then attached to the temporary carrier with extra spacing added between adjacent dies. This allows the finished package to have enough space to accommodate all the package-level solder balls at the required pitch. Next, the dies are encapsulated in mold compound. After the temporary carrier is removed, the RDL is formed. The final steps are ball mounting and package singulation.

## 3.5 Summary and Emerging Trends

There are two main themes in the history of packaging evolution: increasing I/O count and shrinking the form factor.

Area array packaging enabled by a laminate substrate or RDL increases the I/O count by moving the package-to-PCB interconnects from the peripheral leads on a lead frame package to underneath the package. In a similar fashion, flip chip increases the I/O count by moving the die-to-package interconnects from the peripheral bond pads on a wire bond die to the entire active area of a die.

BGA and QFN reduce the package footprint by eliminating the leads extending beyond the package molded body, as is the case for conventional lead frame packages.

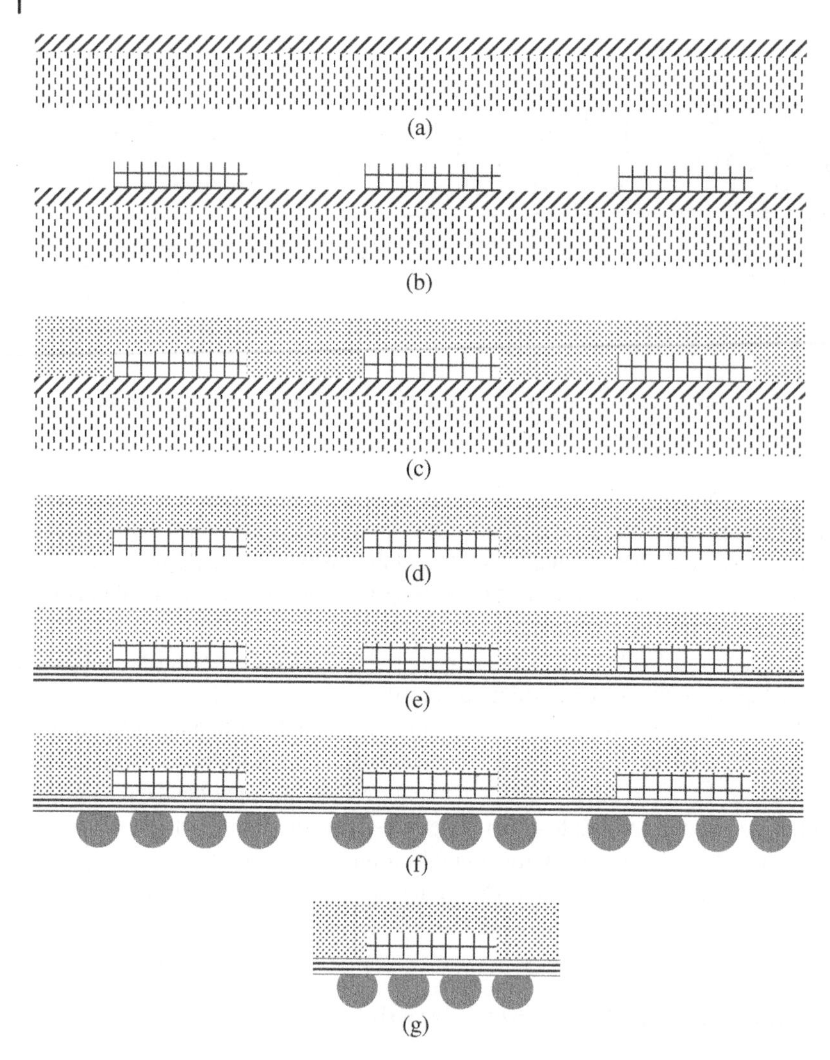


**Figure 3.22** Key steps in FO-WLP assembly process flow: (a) apply tape on temporary carrier, (b) attach die onto carrier with active side facing down, (c) encapsulate die with mold compound, (d) remove temporary carrier, (e) construct RDL, (f) mount solder balls, and (g) dice into individual package.

In the past decade, several new trends have emerged: 2.5D/3D integration with through Si via, 3D integration with fan-out package on package, and 3D FO-WLP.

2.5D is a packaging technology that enables homogeneous or heterogeneous integration using Si interposers with through-silicon vias (TSVs). This process is reserved for high-end products that require tight L/S in the routing layer that

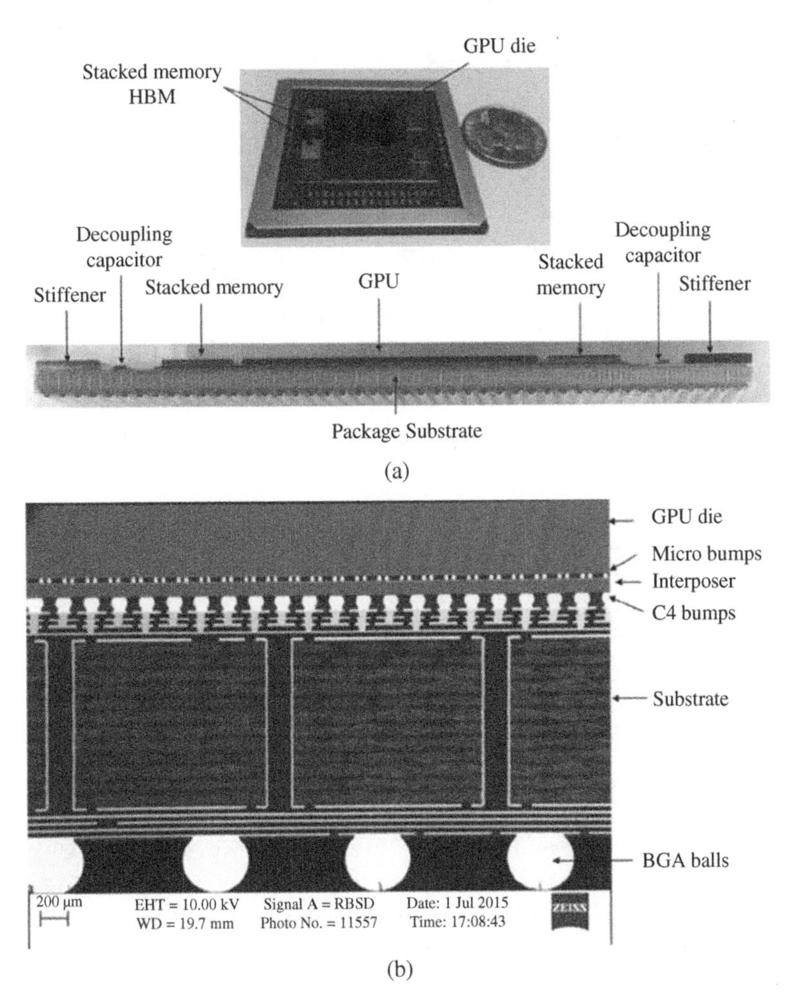


**Figure 3.23** AMD Radeon™ Fury GPU: (a) optical microscope photography from top view and cross-section and (b) scanning electron microscopy photo for micro-structure (from [45], © 2016 IEEE, reprinted with permission).

cannot be supported by an organic laminate substrate. One excellent example for both 2.5D and 3D integration with TSVs is Radeon Fury graphics cards shipped by AMD in 2015 [45]. As shown in Figures 3.23 and 3.24, the integration between GPUs (or CPUs) and high bandwidth memory (HBM) is 2.5D, as they are mounted on an Si interposer in a side-by-side fashion and assembled into one single package. In contrast to packaging GPUs and HBM separately as standalone components and connecting them through a system level PCB, 2.5D packaging offers a much shorter interconnect length, therefore higher bandwidth and lower power

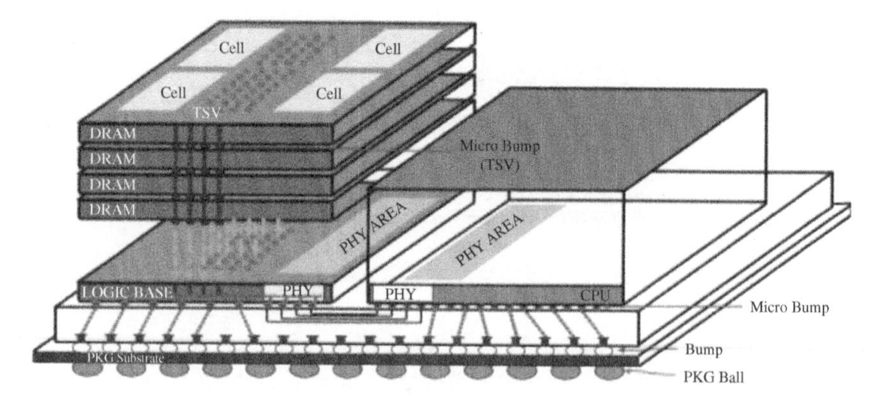


**Figure 3.24** System overview of general system in package using HBM DRAM (from [46], © 2016 IEEE, reprinted with permission).

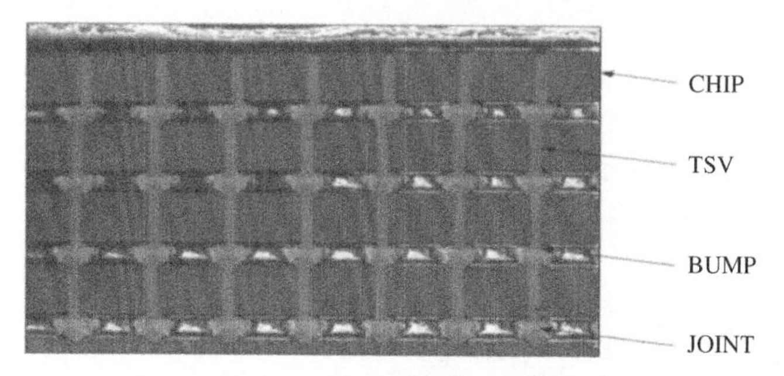


**Figure 3.25** TSV cross-section view (from [46], © 2016 IEEE, reprinted with permission).

at the system level are realized. HBM contains multiple DRAM core dies vertically, or in 3D, stacked on top of a base die connected with TSVs [46]. TSVs, as shown in Figure 3.25, offer a shorter interconnect path than bonding wires and connection through PCB traces. Thus, 3D stacking with TSVs increases speed and lowers the power of the system. 3D integration enabled by TSVs has also been proposed on solid state drives (SSDs) [47] to reduce SSD cost, as NAND memory die size is shrunk by moving the charge pumps to the boost converter shared by all the NAND dies on an interposer die, as illustrated in Figure 3.26.

Fan-out package-on-package (FO-PoP) offers an alternate means of 3D integration without using TSVs. A leading example of FO-PoP is TSMC's InFO_PoP (integrated fan-out package-on-package), which is the packaging solution for Apple's A10 processor, launched in 2016 [48]. As shown in Figure 3.27, InFO_PoP stacks a memory wire bond BGA on top of a system on chip (SoC) FO-WLP using vias through mold compound or Through InFO Via (TIV) in TSMC's terminology. InFO_PoP has clear advantage over flip-chip PoP (with the bottom package as a

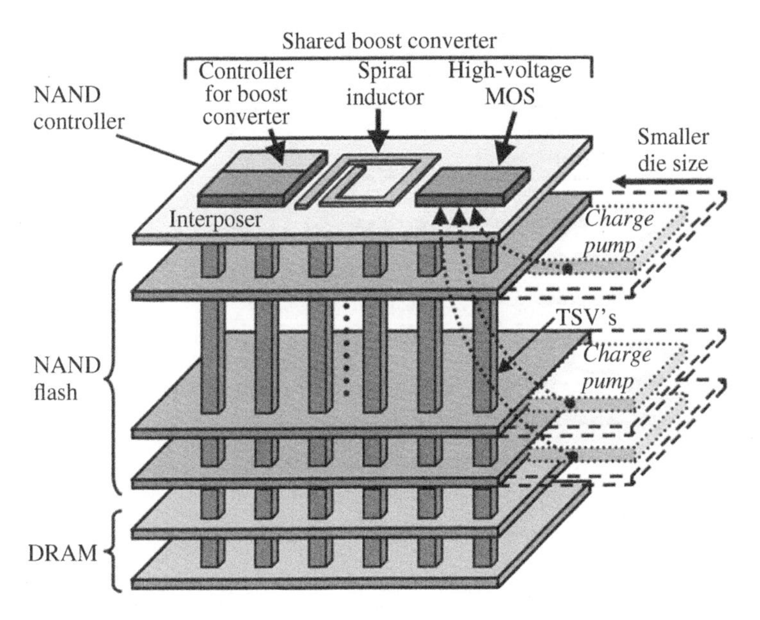


**Figure 3.26** Proposed 3D SSD with boost converter (from [47], © 2009 IEEE, reprinted with permission).

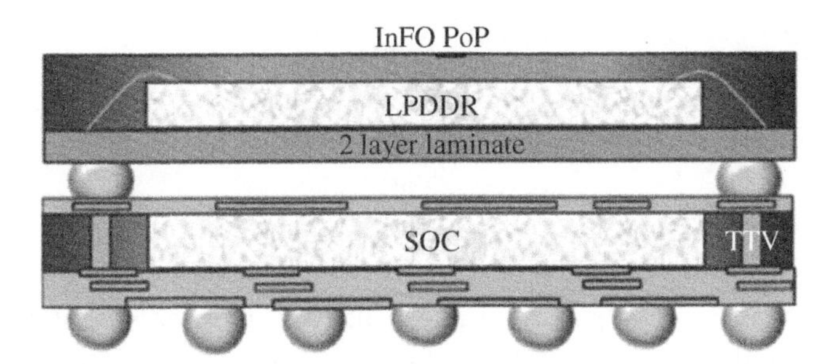


**Figure 3.27** Cross-section view of single chip InFO_PoP with TIV (Through InFO Via) (from [49], © 2016 IEEE, reprinted with permission).

flip-chip BGA) because InFO eliminates flip-chip bump and its RDL is thinner than the substrate in flip-chip BGA. 3D integration enabled by the InFO_PoP structure optimizes product performance and power while maintaining cost effectiveness by avoiding the relatively high process cost associated with TSVs [49].

Multilayer fan-out packaging, as illustrated in Figure 3.28, has been proposed as another flexible solution for 3D heterogeneous integration, targeted at wearable and portable devices. A low-profile six-layer stacked fan-out package, shown in Figure 3.29, has been demonstrated with acceptable warpage and good thermal performance [50].

**Figure 3.28** Sketch of 3D fan-out stacking. Note that the vertical interconnects are not shown (from [50], © 2018 IEEE, reprinted with permission).

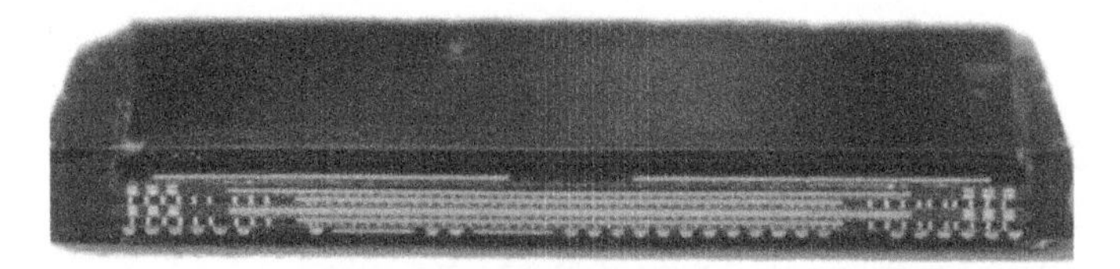

**Figure 3.29** Cross-section view of six-layer 3D fan-out stacking package (from [50], © 2018 IEEE, reprinted with permission).

As Moore's law slows down, packaging innovations allowing a higher level of integration are moving to the focal point of the microelectronics technology landscape as we continue to drive along the road to faster performance and more functionality in a miniaturized form.

## References


**1** J. Bardeen and W.H. Brattain, The transistor, a semi-conductor triode. *Physical Review*, vol. 74, no. 2, pp. 230–231, July 1948. Available: https://doi.org/10.1103/PhysRev.74.230.

**2** L. Anderson, H. Christensen, and P. Andreatch. (1957 May). Technique for connecting electrical leads to semiconductors. *Journal of Applied Physics*, vol. 28, no. 8, pp. 923, July 1948. Available: https://doi.org/10.1063/1.1722893.

**3** E.M. Davis, W.E. Harding, R.S. Schwartz et al., Solid logic technology: versatile, high performance microelectronics. *IBM Journal of Research and*

*Development*, vol .8, no. 2, pp. 102–114, April 1964. Available: https://doi.org/10.1147/rd.82.0102.

**4** J.E. Kauffman, Method of packaging integrated circuits. Patent 3436810, April 1969. Available: http://patft1.uspto.gov/netacgi/nph-Parser?patentnumber=3436810.

**5** B. Freyman and R. Pennisi, Overmolded plastic pad array carriers (OMPAC): a low cost, high interconnect density IC packaging solution for consumer and industrial electronics. In *Proceedings of the IEEE Electronic Components and Technology Conference*, pp. 176–182, 1991.

**6** Y. Yamaguchi, Resin encapsulated semiconductor device having a reduced thickness and improved reliability. Patent 6081029, June 2000. Available: http://patft1.uspto.gov/netacgi/nph-Parser?patentnumber=6081029.

**7** P. Elenius and H. Hollack, Method for forming chip scale package. Patent 6287893, September 2001. Available: http://patft1.uspto.gov/netacgi/nph-Parser?patentnumber=6287893.

**8** H. Hedler, T. Meyer, and B. Vasquez, Transfer wafer level packaging. Patent 6727576, April 2004. Available: http://patft1.uspto.gov/netacgi/nph-Parser?patentnumber=6727576.

**9** M. Gaku, K. Suzuki, and K. Nakamichi, Curable resin compositions of cyanate esters. Patent 4110364, August 1978. Available: http://patft1.uspto.gov/netacgi/nph-Parser?patentnumber=4110364.

**10** R.C. Jaeger, An overview of microelectronic fabrication. In *Introduction to Microelectronic Fabrication*, pp. 1–15. Prentice Hall, Upper Saddle River, NJ, 1998.

**11** M. Töpper, T. Fischer, T. Baumgartner et al., A comparison of thin film polymers for wafer level packaging. In *Proceedings of the IEEE Electronic Components and Technology Conference*, pp. 769–776, 2010.

**12** X. Wei and Y. Shibasaki, A novel photosensitive dry-film dielectric material for high density package substrate, interposer and wafer level package. In *Proceedings of the IEEE Electronic Components and Technology Conference*, pp. 159–164, 2016.

**13** D. Okamoto, Y. Shibasaki, D. Shibata et al., An advanced photosensitive dielectric material for high-density RDL with ultra-small photo-vias and ultra-fine line/space in 2.5D interposers and fan-out packages. In *Proceedings of the IEEE Electronic Components and Technology Conference*, pp. 1543–1548, 2018.

**14** F. Liu, C. Nair, A. Kubo et al., Organic damascene process for 1.5-μm panel-scale redistribution layer technology using 5-μm-thick dry film photosensitive dielectrics. *IEEE Transactions on Components, Packaging and Manufacturing Technology*, vol. 8, no. 5, pp. 792–801, April 2018. Available: https://doi.org/10.1109/TCPMT.2018.2821000.

**15** H. Kudo, R. Kasai, J. Suyama et al., Demonstration of high electromigration resistance of enhanced sub-2 micron-scale Cu redistribution layer for advanced fine-pitch packaging. In *Proceedings of the IEEE CPMT Symposium Japan*, pp. 5–8, 2017.

**16** H. Xue, C.R. Benedik, X. Zhang et al., Numerical solution for accurate bond-wire modeling. *IEEE Transactions on Semiconductor Manufacturing*, vol. 31, no. 2, pp. 258–265, March 2018. Available: https://doi.org/10.1109/TSM.2018.2818168.

**17** Electrical Engineering and Computer Sciences Department of the University of California at Berkeley,. “The Spice Page”. Available: http://bwrcs.eecs.berkeley.edu/Classes/IcBook/SPICE/.

**18** T.V. Dinh, J. Pagazani, D. Lesénéchal et al., Bonding wire geometric profile dependent model for mutual coupling between two bonding wires on a glass substrate. *IEEE Transactions on Components, Packaging and Manufacturing Technology*, vol. 5, no. 1, pp. 119–127, January 2015. Available: https://doi.org/10.1109/TCPMT.2014.2366730.

**19** T. Wang and Y. Lu., Fast and accurate frequency-dependent behavioral model of bonding wires. *IEEE Transactions on Industrial Informatics*, vol. 13, no. 5, pp. 2389–2396, August 2017. Available: https://doi.org/10.1109/TII.2017.2737525.

**20** B.S. Kumar, S. Murali, K.I. Tae et al., Novel coated silver (Ag) bonding wire: bondability and reliability. In *Proceedings of the IEEE 19th Electronics Packaging Technology Conference*, pp. 1–6, 2017.

**21** S. Na and S. Jeon, A study on Pd distribution effect on the reliability of Au coated PCC wire bonding. In *Proceedings of the 19th International Conference on Electronic Packaging Technology*, pp. 1346–1350, 2018.

**22** STATS ChipPAC Pte. Ltd. “FBGA Fine Pitch Ball Grid Array.” [Online]. Available: http://www.statschippac.com/~/media/Files/Package%20Datasheets/FBGA.ashx

**23** J.H. Liu, Z.J. Pei, and G.R. Fisher, Grinding wheels for manufacturing of silicon wafers: a literature review. *International Journal of Machine Tools and Manufacture*, vol. 47, no. 1, January 2007. Available: https://doi.org/10.1016/j.ijmachtools.2006.02.003.

**24** P. Krishnan, Y.K. Leong, F. Rafzanjani et al. Die attach film (DAF) for break-through in manufacturing (BIM) application. In *Proceedings of the 36th International Electronics Manufacturing Technology Conference*, pp. 2389–2396, 2014.

**25** N. Ye, Q. Li, H. Zhang et al. Challenges in assembly and reliability of thin NAND memory die. In *Proceedings of the IEEE 66th Electronic Components and Technology Conference*, pp. 1840–1846, 2016.

**26** C. Guan, M. Li, K. Chen et al., Effects of moisture absorption and temperature on the adhesion strength between die attach film (DAF) and silicon die. In

*Proceedings of the 14th International Conference on Electronic Materials and Packaging*, pp. 1–4, 2012.

**27** A. Marte, U. Ernst, H. Clauberg et al., Advances in memory die stacking. In *Proceedings of the IEEE 68th Electronic Components and Technology Conference*, pp. 407–418, 2018.

**28** W. Qin, I.M., Cohen, and P.S. Ayyaswamy, Fixed wand electronic flame-off for ball formation in the wire bonding process. *Journal of Electronic Packaging*, vol. 116, no. 3, pp. 212–219, September 1994. Available: https://doi.org/10.1115/1.2905688.

**29** S. Huang, Z. Ji, Y. Liu et al., Parametric optimization and yield probability prediction of package warpage. In *Proceedings of the IEEE 68th Electronic Components and Technology Conference*, pp. 243–248, 2018.

**30** P. Chen, Z. Ji, Y. Liu et al., Warpage prediction methodology of extremely thin package. In *Proceedings of the IEEE 67th Electronic Components and Technology Conference*, pp. 2080–2085, 2017.

**31** Y. Liu, Z. Ji, P. Chen et al., Moisture induced interface delamination for EMI shielding package. In *Proceedings of the IEEE CPMT Symposium Japan*, pp. 217–220, 2016.

**32** X. Fan and V. Nagaraj, In-situ moisture desorption characterization of epoxy mold compound. In *Proceedings of the 13th International Thermal, Mechanical and Multi-Physics Simulation and Experiments in Microelectronics and Microsystems*, pp. 1/6–6/6, 2012.

**33** L. Chen, Y. Liu, and X. Fan, Application of water activity-based theory for moisture diffusion in electronic packages using ANSYS. In *Proceedings of the 19th International Conference on Thermal, Mechanical and Multi-Physics Simulation and Experiments in Microelectronics and Microsystems*, pp. 1–6, 2018.

**34** T. Yeung, H. Sze, K. Tan et al., Material characterization of a novel lead-free solder material – SACQ. In *Proceedings of the IEEE 64th Electronic Components and Technology Conference*, pp. 518–522, 2014.

**35** P. Lall, V. Yadav, J. Suhling et al., High strain rate mechanical behavior of SAC-Q solder. In *Proceedings of the 16th IEEE Intersociety Conference on Thermal and Thermomechanical Phenomena in Electronic Systems*, pp. 1447–1455, 2017.

**36** S. Fenglian and L. Yang, A new low-Ag lead free solder alloy and the reliability. In *Proceedings of the 9th International Forum on Strategic Technology*, pp. 440–443, 2014.

**37** STATS ChipPAC Pte. Ltd. "QFN Quad Flat No-Lead Package." [Online]. Available: http://www.statschippac.com/~/media/Files/Package%20Datasheets/QFN.ashx

**38** STATS ChipPAC Pte. Ltd. "Flip Chip CSP." [Online]. Available: http://www.statschippac.com/~/media/Files/Package%20Datasheets/fcCSP.ashx

**39** N. Islam, M.C. Hsieh, K.K. Take et al., Fine pitch Cu pillar assembly challenges for advanced flip chip package. In *Proceedings of the International Wafer-Level Packaging Conference*, San Jose, CA, USA, 24–25 October 2017.

**40** Z. Zhang, P. Zhu, and C.P. Wong, Flip-chip underfill: materials, process, and reliability. In *Materials for Advanced Packaging*, 2nd edn, D. Lu and C. P. Wong (eds). Springer, Cham, pp. 331–371, 2017.

**41** Q.H. Tat and I.J. Rasiah, The study of cyanate ester underfill adhesives under typical flip chip assembly process conditions. In *Proceedings of the 3rd Electronics Packaging Technology Conference*, pp. 228–233, 2000.

**42** Y.H. Lin, F. Kuo, Y.F. Chen et al., Low-cost and fine-pitch micro-ball mounting technology for WLCSP. In *Proceedings of the IEEE 62nd Electronic Components and Technology Conference*, pp. 953–958, 2012.

**43** STATS ChipPAC Pte. Ltd. "Innovations in Wafer Level Technology." [Online]. Available: http://www.statschippac.com/~/media/Files/DocLibrary/brochures/STATSChipPAC_Wafer_Level_Technology.ashx

**44** STATS ChipPAC Pte. Ltd. "eWLB Embedded Wafer Level Ball Grid Array." [Online]. Available: http://www.statschippac.com/~/media/Files/Package%20Datasheets/eWLB.ashx

**45** C. Lee, C. Hung, C. Cheung et al., An overview of the development of a GPU with integrated HBM on silicon interposer. In *Proceedings of the IEEE 66th Electronic Components and Technology Conference*, pp. 1439–1444, 2016.

**46** J.C. Lee, J. Kim, K.W. Kim et al., High bandwidth memory (HBM) with TSV technique. In *Proceedigns of the International SoC Design Conference*, pp. 181–182, 2016.

**47** T. Yasufuku, K. Ishida, S. Miyamoto et al., Effect of resistance of TSV's on performance of boost converter for low power 3D SSD with NAND flash memories. In *Proceedings of the IEEE International Conference on 3D System Integration*, pp. 1–4, 2009.

**48** AspenCore, Inc. TSMC Likely to Lock up Apple A10, A11 Orders. Available: https://www.eetimes.com/document.asp?doc_id=1330677.

**49** C. Tseng, C. Liu, C. Wu et al., InFO (wafer level integrated fan-out) technology. In *Proceedings of the IEEE 66th Electronic Components and Technology Conference*, pp. 1–6, 2016.

**50** F. Hsu, J. Lin, S. Chen et al., 3D heterogeneous integration with multiple stacking fan-out package. In *Proceedings of the IEEE 68th Electronic Components and Technology Conference*, pp. 337–342, 2018.

# 4

# Electrical, Mechanical, and Thermal Co-Design

*Xiaoxiong Gu and Pritish Parida*

*Thomas. J. Watson Research Center, IBM, 1101 Kitchawan Rd. Route 134, Yorktown Heights, NY 10598, USA*

## 4.1 Introduction

As antenna element size and pitch are in the order of millimeters, i.e. comparable to wavelength at millimeter-wave (mmWave) frequencies, antenna-in-package (AiP) solutions are more feasible for design and manufacturing than at other frequencies. The implementation of AiP modules demands co-design of antenna elements and integration of the active mmWave radio frequency integrated circuit (RFIC). In addition, the increasing complexity of AiP module packaging poses significant challenges for thermal management and thermomechanical integrity. It is often desirable to integrate more RF functions in an AiP module (e.g. transmit and receive modes, support of different polarization, etc.), which raises the overall heat density. All of these issues need to be studied and addressed in a holistic manner during the design and development phases.

In this chapter, we first discuss the needs and challenges of performing electrical, mechanical, and thermal co-design for AiP modules with a particular focus on mmWave front-end module implementation in 28-GHz high-speed wireless data transmission applications. Furthermore, we review the thermal management considerations for next-generation heterogeneous integrated systems that are also applicable to phased array front-end modules. Different cooling methodologies are discussed for future communication devices under various power dissipation conditions.

Figure 4.1 illustrates key considerations for AiP module design and implementation, including substrate technology choice, physical and electrical design, thermal management, interconnection, electrical and RF testing, reliability, and rework (repair) capabilities.

*Antenna-in-Package Technology and Applications*, First Edition.
Edited by Duixian Liu and Yueping Zhang.

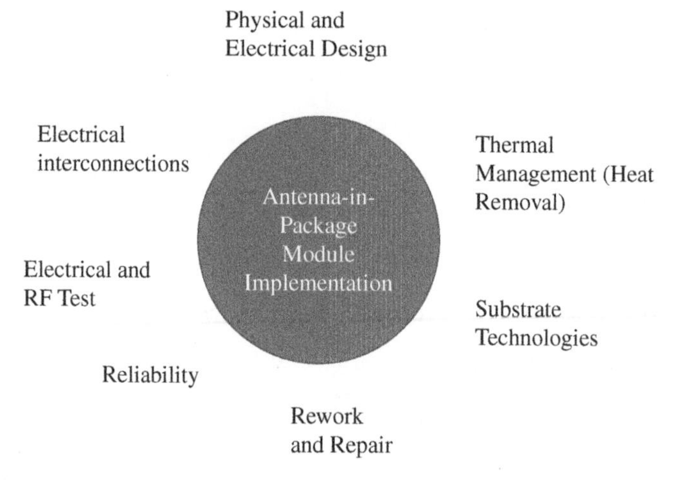


**Figure 4.1** Important parameters and considerations for AiP design and implementation.

A wide range of substrate technologies and the corresponding implementations of AiP approaches are covered in this book. In this chapter, we describe our approach, from an electrical, mechanical, and thermal co-design point of view, to tackle the challenges of integrating multiple phased array ICs and antenna elements using multilayer organic buildup substrate technology. As shown in Figure 4.2, the phased-array module integrates antennas on a first-level package and supports multiple RFICs by routing large numbers of RF, intermediate frequency (IF) and/or baseband signals on the package. Together with the flip-chip bonded ICs, the module can then be attached to a second-level printed circuit board (PCB) through ball grid arrays (BGAs). A digital control interface, e.g. via a nearby application-specific integrated circuit (ASIC) or field-programmable gate array (FPGA), as well as a thermal management solution such as a cooling heat sink can be implemented at the board level.

A conceptual breakdown of layers in an organic antenna array package is illustrated in Figure 4.2. The multilayer substrate typically includes radiation elements such as patch structures on the top layer and interconnect pads for RFICs [e.g. via controlled collapse chip connections (C4s)] and board (e.g. via BGAs) interfaces on the bottom layer, respectively. The inner layers are assigned for power supplies and ground planes, as well as the routing for RF [e.g. antenna feed, IF, local oscillator (LO)], baseband, and digital signals. The choice of layer stack-up depends on the detailed system design requirements. It is also important to characterize dielectric properties in advance for both antenna performance and signal/power integrity of the package interconnects. With RFICs and BGA solder balls attached,

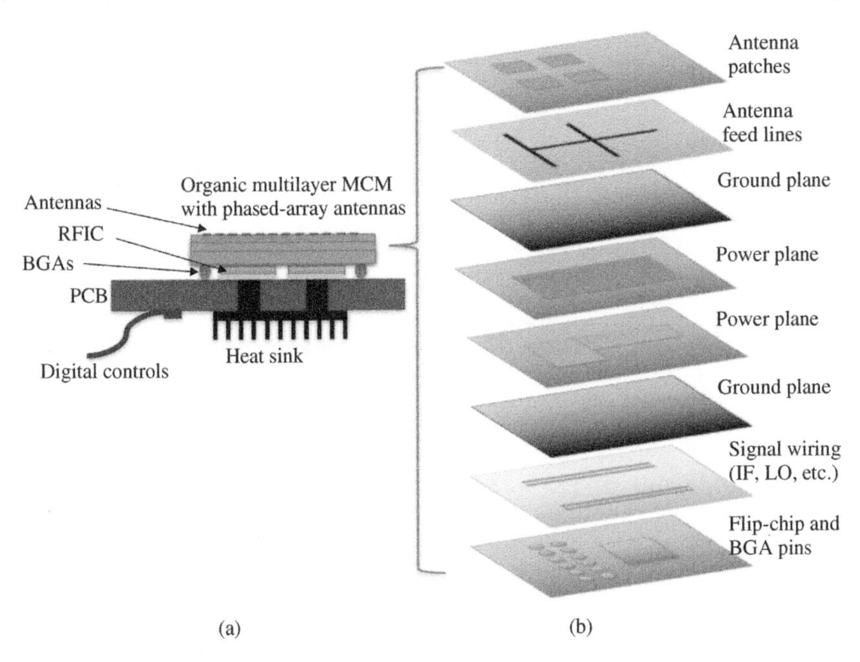


**Figure 4.2** Illustration of the multilayered antenna package concept: (a) cross-sectional view of an organic multichip phased array antenna module on a PCB and (b) layer breakdown view of the AiP module.

a mmWave AiP module is complete and can be readily tested and screened before being mounted in a system board-level assembly.

A typical co-design and test flow for implementing mmWave AiP modules is illustrated in Figure 4.3. The interactions between key design steps include substrate and material selection, antenna prototyping, warpage and thermo-mechanical integrity assessment, and package-circuit-antenna co-design for module integration and layout. During the design iterations, simulation and measurement of key aspects such as antenna performance (bandwidth, gain, radiation patterns), cooling solutions, and warpage conditions need to be evaluated and optimized to ensure that performance meets the overall system specifications.

To discuss the details of such co-design steps and the associated implementation considerations, one example of a 28-GHz phased array antenna module and its associated antenna prototype and thermomechanical test vehicle for high-speed radio access applications is described in detail in the subsequent sections. In this case, the module design needs to optimize antenna performance, i.e. gain, bandwidth, and radiation pattern, while taking into account system trade-offs in terms of substrate material properties, array size (i.e. the number of elements and pitch),

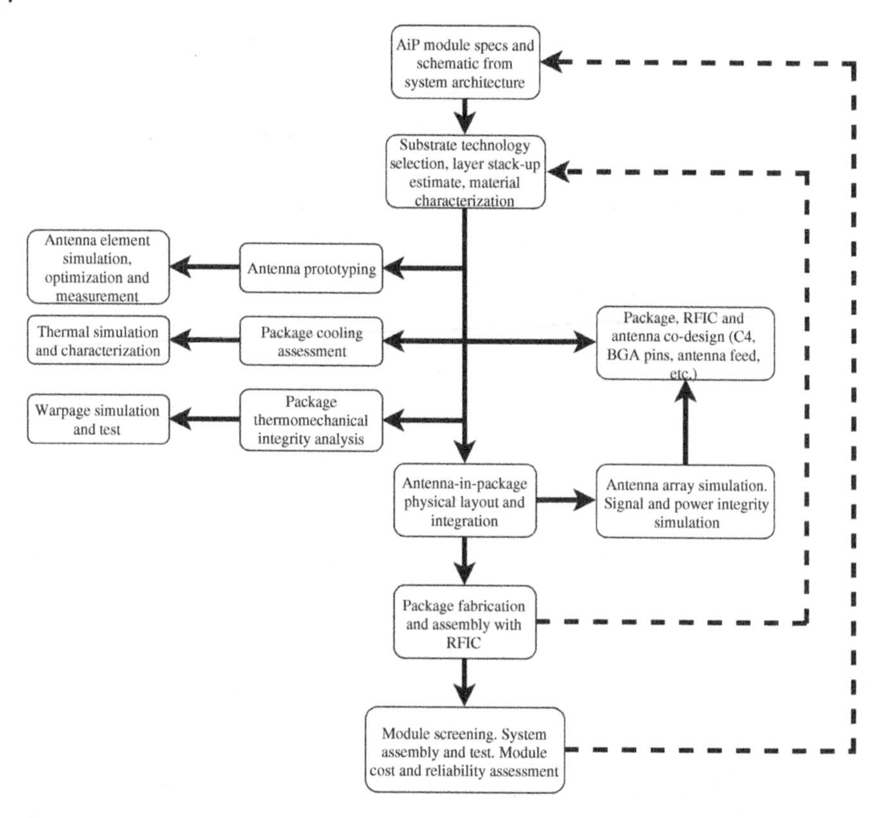


**Figure 4.3** Design and test flow example for implementing an AiP module.

interconnection flexibility (e.g. for wiring power supplies and control signals), thermo-mechanical compatibility, IC assembly and board integration, and so on.

## 4.2 Electrical, Warpage, and Thermomechanical Analysis for AiP Co-design

### 4.2.1 28-GHz Phased Array Antenna Module Overview

The fast-emerging next generation (5G) wireless communication system demands highly integrated radio access solutions that incorporate advanced phased-array transceiver and antenna front-end technologies to support high radiated power, large signal to noise ratio, and rapid and accurate beam forming and scanning in three dimensions within a wide range [1, 2]. Silicon integration of mmWave transceivers with AiP is one key enabling technology for the miniaturization and mass adoption of 5G wireless applications [3, 4]. Here, the silicon integration and packaging involve co-designs of electrical, mechanical, and thermal aspects of the phased array module to ensure that the AiP module can achieve not only the desired array beam forming and steering performance with sufficient power gain and spatial resolution, but also the manufacturability and yield required for high-volume production and the reliability to operate in various environment conditions.

Based on a multilayer organic buildup package concept originating from the design of a 94-GHz W-band phased-array unit tile module [5, 6], a 28-GHz functional AiP module has been designed and implemented. Figure 4.4 illustrates the signaling schemes of the 28-GHz four-chip antenna array module. It supports four 28-GHz flip-chip mounted transceiver ICs that feature dual-polarized operation in transmitter (TX) and receiver (RX) modes, incorporates 64 embedded antenna elements, and enables a reliable board-level assembly via a BGA interface with direct heat sink attachment for effective thermal management. Two of the four ICs are rotated 180° so that the digital and control signal pins are in closer proximity to each other and can be more easily connected with one bus. The external LO

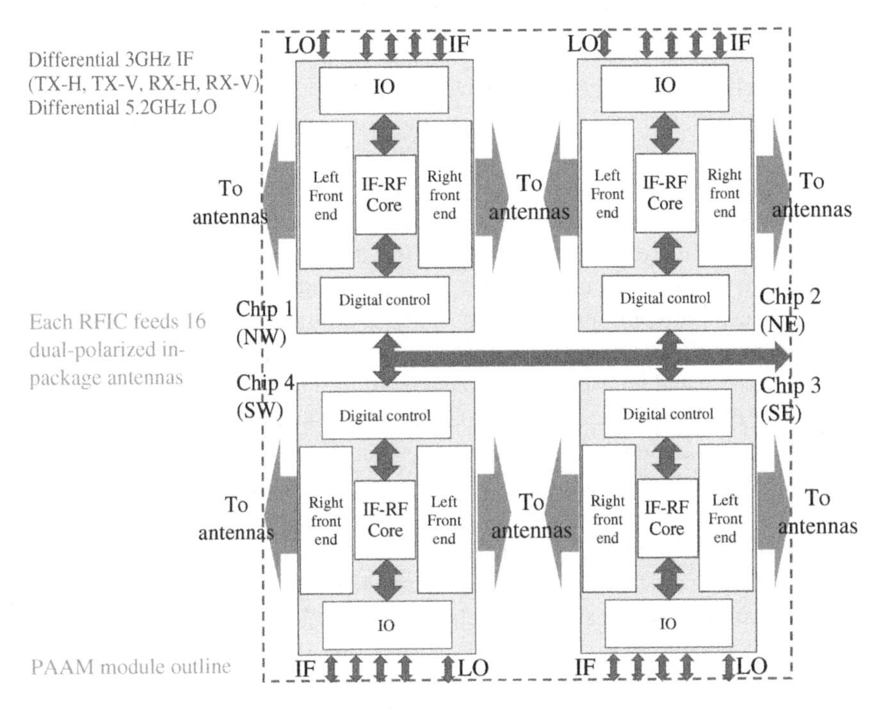


**Figure 4.4** Illustration of signaling schemes on the 28-GHz four-chip antenna array module.

and four IF signals (H- and V- for transmit and receive, respectively) are located on the outer perimeter and can be routed to the BGA with less complexity and lower loss. The 28-GHz RF signals are symmetrically fanned out from both sides of each IC and the symmetry is further maintained at the module level for the four ICs. Details of the package stack-up, antenna design, and implementation, as well as passive antenna characterization and active module-level measurement, are further discussed in Chapter 11 (e.g. the 28-GHz module photo is shown in Figure 11.14). Module design, characterization, and test results, including radiation pattern measurement and over-the-air communication link measurement, can be found in [8, 9].

This 28-GHz AiP module supports 128 front-end (FE) elements per single package. Enabled by the four phased array transceiver ICs, the AiP module also supports simultaneous dual-polarized beams in both TX and RX. Moreover, antenna feed lines with equal delay, uniform antenna radiation patterns across the array, and orthogonal amplitude and phase control per element in the IC enable high-precision beam steering without calibration. Details of the IC-level implementation are provided in [10].

### 4.2.2 Thermomechanical Test Vehicle Overview

The active 28-GHz phased-array packaging approach was first established based on a comprehensive study of thermal, mechanical, and electrical aspects of an antenna prototype test vehicle, along with the results of integration experiments with multiple ICs and a second-level PCB. Table 4.1 summarizes the main considerations of implementing a multichip phased array AiP module using organic buildup substrates, including the issues related to dielectric material, interconnect loss and ground rules, layer stack-up optimization, warpage control, antenna performance, thermal management and packaging joint reliability, etc.

An antenna prototyping and thermomechanical multi-chip module (MCM) test vehicle is described below that was intended to address the various integration challenges associated with module implementation. Figure 4.5 illustrates a bottom view of the test vehicle. The module dimensions are 68.5 × 68.5 mm. There are 100 (10 by 10) antenna prototype structures distributed evenly on the substrate at a 5.9 mm pitch. Probe launch sites are located on the bottom side of the module for direct antenna measurement. In addition, four 10 × 12 mm thermal test chips are flip-chip attached to the package along with 1-mm pitch solder balls for second-level board connection. These thermal chips have built-in resistor networks for multiple voltage supplies such that a controlled experiment with programmable temperature variation can be carried out, thus enabling cooling and thermomechanical stress testing.

To support a complex layout of the MCM, the stack-up of the package needs to have a sufficient number of layers to implement antenna patch structures, feed

**Table 4.1** Co-design considerations of implementing a multilayer organic AiP MCM for phased-array applications.

| Challenges | Implementation considerations and challenges |
|---|---|
| Substrate material | Dielectric electrical loss and mechanical stiffness properties |
| Interconnect | Design rules for fine-pitch wiring (trace and via); electrical loss for RF and antenna feed signals |
| Layer stack-up | Layer count and designation to meet overall module wiring needs |
| Substrate warpage | Coefficient of thermal expansion mismatch; mitigation of warpage for large-size modules |
| Antenna performance | Gain, frequency and bandwidth, radiation pattern, isolation between elements, and different polarizations |
| Thermal management | Control of RFIC junction temperature under different ambient and cooling conditions |
| Chip and BGA joints | Joint reliability with different underfill; mechanical stress on board level |

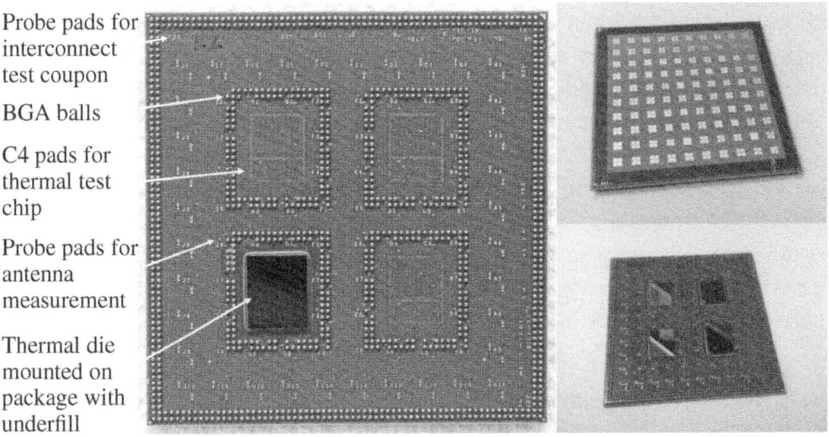

Probe pads for interconnect test coupon
BGA balls
C4 pads for thermal test chip
Probe pads for antenna measurement
Thermal die mounted on package with underfill

**Figure 4.5** Thermomechanical and antenna prototype package with a top view of antenna patches and a bottom view of four attached thermal test chips and BGA balls (from [9], © 2019 IEEE, reprinted with permission).

line routing, multiple power supply domains and ground planes, as well as various signal wiring including IF, LO, baseband, and digital control. For instance, it is common to have more than 10 layers for a complex phased-array package with a large number of antenna elements such as 100 elements [6–8]. The antenna design, including patch placement, isolation between elements, and feed line routing, needs to be optimized for the desired gain, bandwidth, and radiation pattern.

The prototypes shown in Figure 4.5 were characterized and validated by probe measurement and full-wave simulation to ensure performance in the frequency band of interest. Details of antenna implementation and characterization using such organic buildup substrate technology are discussed in Chapter 11.

### 4.2.3 Antenna Prototyping and Interconnect Characterization

During the design and layout, there are several important issues to be considered for the package from a manufacturability point of view. First, the dielectric materials, including buildup and core, need to have good electrical properties, i.e. low dielectric constant and loss tangent, and at the same time maintain a low profile (i.e. smooth surface) for the substrate to achieve tight ground rules due to the fast flash-etching process. Second, for organic buildup substrate, the wiring pitch is currently on the order of 10 μm for buildup layers in terms of achievable metal line width and spacing (https://global.kyocera.com/prdct/organic/prdct/package/fcbga/std/), which gives the desired flexibility for signal routing. In addition, vertical interconnects between layers keep improving in terms of their density and flexibility. For example, laser-drilled vias can be placed with a fine pitch less than 150 μm and can be also stacked directly for vertical escape of signals. The test vehicle contains additional interconnect coupon sites to measure broadband *S* parameters of trace and via structures, in addition to the main antenna test sites. Direct probing measurement using a network analyzer was performed on the package to analyze signal integrity and gather antenna radiation pattern data with a motor-controlled receiving horn antenna inside a chamber. The antenna prototypes in the package consist of both probe-fed and aperture-coupled dual-polarized stacked patch antennas with different variations. Measurement results from both the antenna prototypes after fabrication are reported in [7, 9] and in Chapter 11.

### 4.2.4 Warpage Analysis and Test

One important requirement that must be considered in package design is maintaining the overall flatness of the substrate. This is crucial, especially for a large MCM, as any warpage of the package can cause reliability issues and yield loss during the chip join process when the reflow temperature is elevated up to 260 °C [11]. For conventional C4 (controlled collapse chip connection) bumps, which are commonly used in flip-chip BGA and flip-chip chip-scale packages, the solder ball pitch and size are around 130 and 100 μm, with a 30 μm space between bumps. Any significant warpage can change the gap between chip and substrate causing bump disconnection or inducing solder bridging between adjacent bumps. For micro bump or chip connection (C2) bumps with a smaller pitch (e.g, 40–130 μm), a copper pillar with a thinner solder tip is generally produced to meet the finer pitch requirements. Here, the collapsing process during soldering faces steeper challenges in compensating for warpage effects due to the reduced solder volume. To make things worse, the warpage issue is also more pronounced for organic-based substrates due to the significantly different coefficients of thermal expansion (CTEs) between the effective overall package dielectric and silicon (e.g. 2.6 ppm/°C for silicon versus over 10 ppm/°C for the package). More thermomechanical stress will be added to the module during and after the chip attach process, as well as after the module is mounted to a PCB.

As a precaution to prevent warpage-related module failure, the substrate needs to maintain sufficient stiffness with an appropriate core layer thickness. For a large MCM, as shown in Figure 4.5, it is desirable to have over 1 mm core layer thickness in the stack-up. Furthermore, the metal content of all layers can be balanced for the upper portion versus the lower portion of the substrate to mitigate the thermomechanical stress on the package. However, this approach may impose difficult constraints on the antenna design as more metal fill may be required for the upper layers where antenna patches are located. Consequently, iterations of antenna simulation are required when metal patterns are adjusted near the antenna elements to avoid any detrimental effects on the radiation pattern.

The actual warpage of the package can be measured using thermal shadow moiré (TSM) techniques [12]. TSM is the most commonly employed metrology for conducting elevated temperature warpage measurements. It measures the topography of the surface of a solid object, i.e. its deviation from a planar surface. This measurement is made by placing a Ronchi ruled grating and sample of interest into a thermally insulated enclosure, as shown in Figure 4.6. A heat source is then used to ramp up the temperature of the sample under test. A shadow of the reference grating is cast onto the surface of the specimen below by projecting a beam of white light at a specified angle through the grating. Moiré fringe patterns are produced as a result of the geometric interference pattern created between the reference grating and the shadow grating. For the phased-array test vehicle, TSM measurements were successfully conducted and recorded as the temperature of the sample was increased to the peak reflow temperature and returned to near room temperature. Actual warpage numbers were subsequently calculated based on the TSM patterns. For example, Figure 4.6 (bottom) shows an example of the calculated overall warpage (excluding the chip area) on the bottom (BGA) side of the package at an elevated temperature. In this case, the warpage is less than 150 μm across the area surrounding four chips, indicating a very good possibility of supporting reliable 1-mm pitch BGA connections to the second-level PCB.

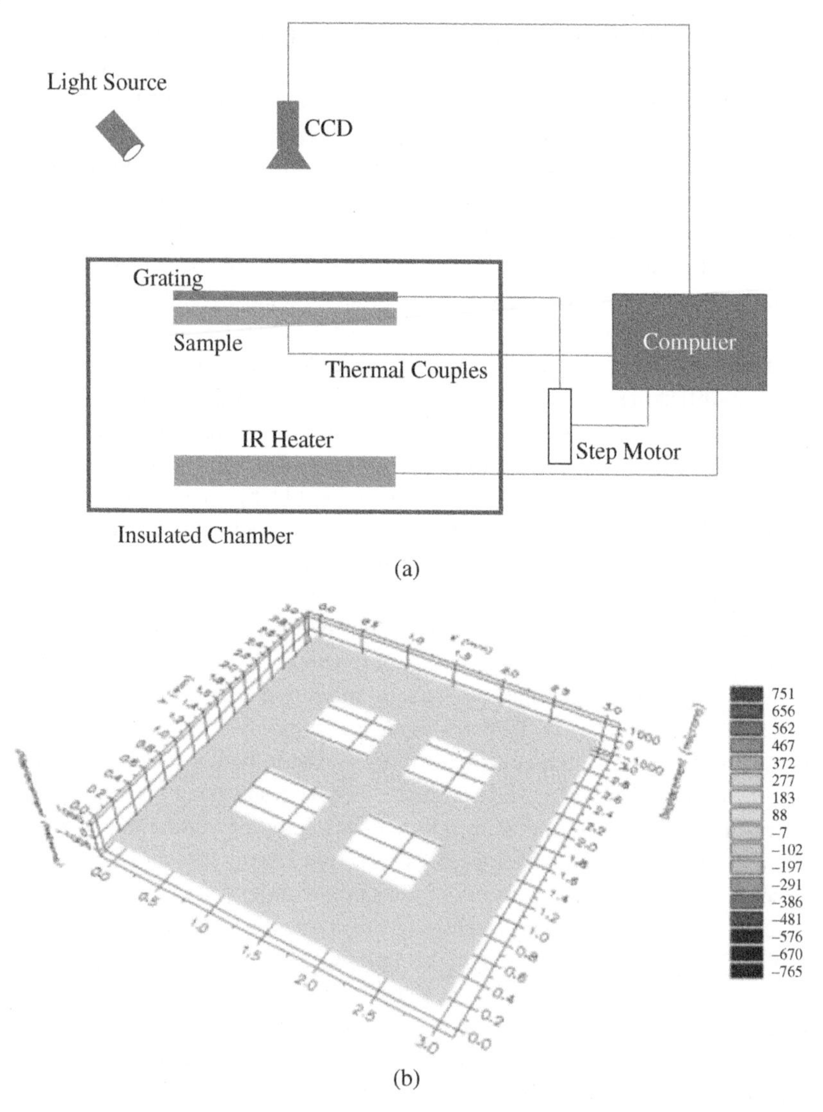


**Figure 4.6** (a) Thermal shadow moiré setup illustration for warpage measurement. (b) Measured warpage of a four-chip thermomechanical and antenna prototype package.

### 4.2.5 Thermal Simulation and Characterization

Thermal management is another key aspect of phased-array module integration. Table 4.2 summarizes RFIC power consumption and corresponding peak IC power density in recent silicon-based phased array AiP module demonstrations.

**Table 4.2** Peak chip power density of reported silicon-based mmWave phased array AiP modules.

| AiP module reference | RFIC technology | RFIC power consumption (W) | Die size ($mm^2$) | Peak chip power density ($W/cm^2$) |
|---|---|---|---|---|
| LG 28 GHz [13] | 28 nm CMOS | 0.4 (TX), 0.68 (RX) | 7.28 | 9.3 |
| IBM 28 GHz [10] | 0.13 μm SiGe | 10.2 (TX), 6.6 (RX) | 166 | 6.1 |
| UCSD 28 GHz [14] | 0.18 μm SiGe | 0.8 (TX), 0.42 (RX) | 11.7 | 6.8 |
| Qualcomm 28 GHz [18] | 28 nm CMOS | 0.36 (TX), 0.167 (RX) | 27.8 | 1.3 |
| SiBeam 60 GHz [15] | 65 nm CMOS | 2.16 (TX), 1.54 (RX) | 72.7 (TX), 77.1 (RX) | 3.0 |
| IBM 60 GHz [16, 17] | 0.13 μm SiGe | 6.4 (TX), 1.8 (RX) | 43.9 (TX), 37.7 (RX) | 14.6 |
| UCSD 60 GHz [19] | 0.13 μm SiGe | 32 (TX only) | 1740 | 1.8 |
| Intel 60 GHz [21] | 90 nm CMOS | 1.2 (TX), 0.85 (RX) | 29 | 4.1 |
| Broadcom 60 GHz [25] | 40 nm CMOS | 1.2 (TX), 0.96 (RX) | 26.3 | 4.6 |
| Broadcom 60 GHz [26] | 40 nm CMOS | 8.4 (TX), 6.6 (RX) | 292 | 2.9 |
| Intel 73 GHz [22] | 22 nm CMOS | 0.59 (TX), 0.672 (RX) | 5.04 | 13.3 |
| IBM 94 GHz [5] | 0.13 μm SiGe | 2.7 (TX), 3.4 (RX) | 44 | 7.7 |
| IBM 94 GHz [23] | 0.13 μm SiGe | 3 (TX), 4.6 (RX) | 37.5 | 12.2 |
| UC Berkeley 94 GHz [20] | 0.13 μm SiGe | 0.424 (TX), 0.364 (RX) | 7.4 | 5.7 |
| Nokia Bell Labs 94 GHz [24] | 0.18 μm SiGe | 5.5 (TX), 4.5 (RX) | 24.5 | 22.5 |

a) CMOS, complementary metal oxide semiconductor; SiGe.

The reported peak chip power density numbers vary between 1 W/cm$^2$ and 22 W/cm$^2$. This variation is the result of different levels of functionality in the IC, as well as different output power targets and operating efficiencies.

Multiphysics co-simulations with active IC power characteristics are required in the design phase to accurately evaluate on-die temperature based on different

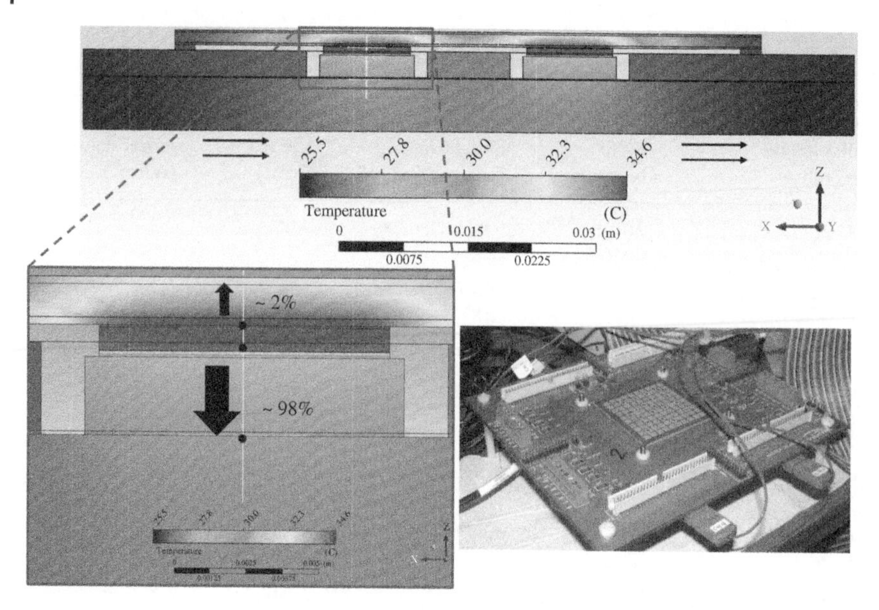


**Figure 4.7** Thermal simulation and measurement for the 5G prototype package (from [9], © 2019 IEEE, reprinted with permission).

cooling options and the projected ambient environment. For example, Figure 4.7 illustrates a temperature contour plot at multiple interfaces based on a multi-physics thermal simulation. The phased array thermomechanical test module is mounted to a PCB with cutouts directly underneath the four ICs. A heat sink is attached to the board from the back side with four pedestals going through the cutouts and touching the ICs with thermal interface material (TIM) in between ICs and heat sinks. The simulation analyzes the temperature drop starting from the IC junction assuming different boundary conditions, e.g. a certain ambient temperature with a determined air flow speed. Table 4.3 lists the geometry and thermal conductivity parameters of the thermal model. At 25 °C ambient environment, the simulation results indicate that most of the heat (approximately 98%) conducts down through the silicon and TIM toward the heat sink. The contour plots also show the non-uniform heating on the core and buildup layers. The simulated temperature difference between active devices and ambient environment is 9.5 °C. In this case, the cooling efficiency is sensitive to the gap of the TIM between ICs and the heat sink, highlighting the importance of tightening the gap tolerance for system board assembly.

Guided by simulation, thermal experiments were further carried out by taking advantage of the programmable thermal sensor chips on the test module. These thermal chips have built-in resistor networks for multiple voltage supplies to

**Table 4.3** Thermal simulation specifications for the AiP prototype.

| Components | Layer thickness (μm) | Thermal conductivity (W/mK) |
|---|---|---|
| Buildup FC | 240 | 2 |
| Core layer | 1134 | 0.2 |
| Buildup BC | 240 | 2 |
| C4 with underfill | 82 | 6.35 |
| RFIC | 800 | 140 |
| TIM1 | 200 | 2.5 |
| BGA connection | 600 | 0.2 |
| PCB | 3000 | 2 |
| TIM2 (air gap between PCB and heat sink) | 125 | 0.027 |
| Heat sink | 6000 | 385 |

a) FC, top layers; BC, bottom layers.

control temperature variation. A custom test board, a heat spreader with pedestals, as well as additional thermocouples are integrated together in a controlled setup to measure the temperature profiles of the module, as shown in Figure 4.7 (right). Figures 4.8 and 4.9 illustrate the detailed views of the thermocouple placement for direct temperature measurement. The on-chip thermal sensors were calibrated with respect to the embedded thermocouples based on an accurate resistance measurement over temperature. Both the on-chip temperature and the temperature at the bottom of the heat spreader were recorded when varying the power of the thermal chips. Table 4.4 lists the measured temperature numbers under different conditions. The average temperature difference for one thermal test chip

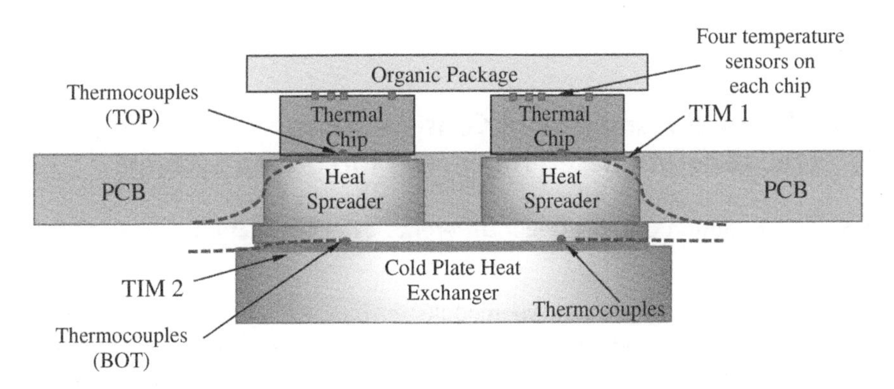


**Figure 4.8** Illustration of the cross-section view of the thermal experiment.

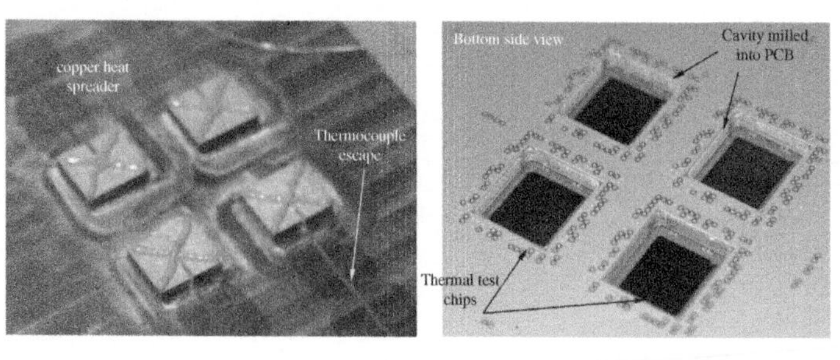


"TOP" thermo-
couple senses
intermediate
temperature
between x and y

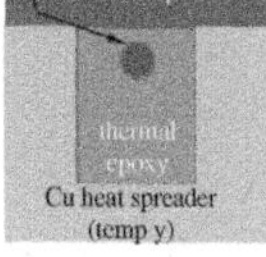


**Figure 4.9** Detailed views of heat spreader (left) and PCB cutout (middle) for thermocouple placement (right).

**Table 4.4** Temperature measurement of one active chip for thermal characterization.

| Power (W) | Chip max. temperature (°C) | Heat spreader bottom temperature (°C) | Temperature difference (°C) | Temperature difference scaled to 10 W (°C) |
|---|---|---|---|---|
| 3.327 | 25.3 | 23.1 | 2.2 | 6.6 |
| 5.099 | 32.3 | 28.9 | 3.4 | 6.7 |
| 7.257 | 29.7 | 24.9 | 4.8 | 6.6 |
| 9.750 | 32.8 | 26.0 | 6.8 | 6.9 |
| 12.532 | 36.2 | 27.3 | 8.9 | 7.1 |
| 18.913 | 44.0 | 30.7 | 13.3 | 7.1 |
| 26.245 | 52.5 | 33.2 | 19.0 | 7.2 |

with a scaled 10 W power setting is 7 °C, which is consistent with the thermal simulations since the actual TIM1 layer may be thinner than assumed in the simulations.

## 4.3 Thermal Management Considerations for Next-generation Heterogeneous Integrated Systems

### 4.3.1 AiP Cooling Options Under Different Power Dissipation Conditions

The electronic packaging of the AiP module provides both electrical connections to the multiple RFICs and a thermal structure to remove the heat generated by the RFICs. A typical AiP module would incorporate multiple RFICs, as illustrated in Figure 4.10, where the chips are mounted on a package (e.g. an organic substrate)

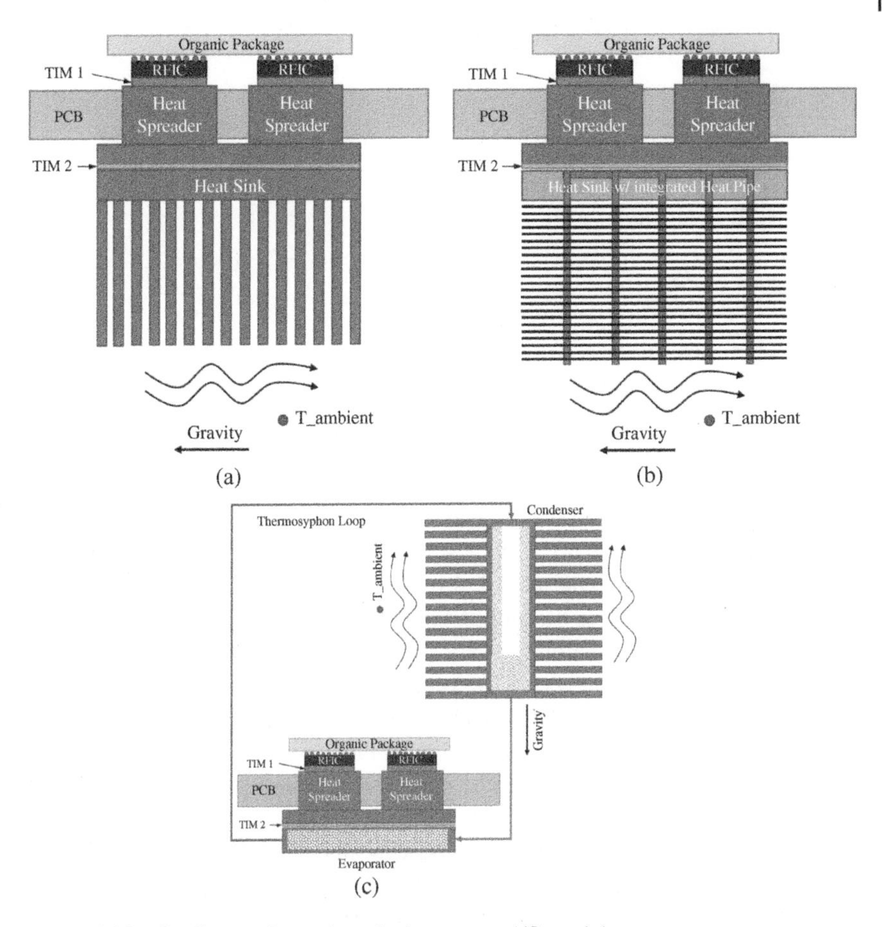


**Figure 4.10** Cooling configurations for low-power AiP modules.

to provide electrical connections for power and signals. The heat from the chips is removed by attachment of a heat spreader and a heat sink, which are thermally coupled to the back side of the chip with thermal interface materials. The temperature rise within the chip is a function of the thermal resistance between the active semiconductor devices on the RFIC and the coolant being utilized to remove the heat generated. The thermal resistances, as shown in Figure 4.11, include (a) a silicon RFIC chip, which also acts as a heat spreader, (b) a first thermal interface material (TIM1) between the silicon and the heat spreader, (c) the heat spreader, (d) a second thermal interface material (TIM2) between the heat spreader and the heat sink, and (e) the heat-sink to ambient air.

In general, thermal management approaches aim to address two key issues. The first of these is reducing the total thermal resistance from the device(s)/junction

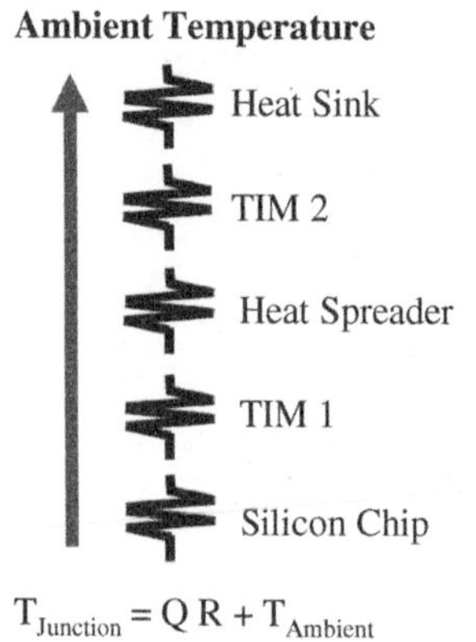


**Figure 4.11** A thermal resistance path.

to the coolant and coolant to the outdoor ambient temperature. This issue can be expressed as reducing $\Delta T_{ja}/Q = (\Delta T_{jc} + \Delta T_{ca})/Q$, where $\Delta T_{jc} = T_j - T_{coolant}$ and $\Delta T_{ca} = T_{coolant} - T_{ambient}$. The second of these issues is the transfer of heat from the device(s) being cooled to the outdoor ambient environment while minimizing the cooling power consumption.

Figure 4.10 illustrates a few cooling configurations that are commonly used in practice for low power (~1 W) devices having low heat densities (~1 W/cm$^2$). In such configurations the heat transfer to the ambient air is achieved through natural convection phenomena where the cooling fluid (air, in this case) motion is caused by the density differences in the fluid occurring naturally due to temperature gradients in proximity to the heat sink. This mode of cooling is called passive cooling where the fluid motion (in a direction opposite to the direction of gravity) occurs naturally and there is no need for external energy input for generating fluid motion. The associated heat transfer rates are low (~10 W/m$^2$ K), resulting in a large temperature change ($\Delta T_{ja}$) between the junction and ambient air per watt of heat dissipation ($Q$) from the device. Thus, this approach is suitable for only low power devices [27]. Figure 4.10b shows a second passive cooling configuration where the heat transfer rate across the heat sink is improved by integrating it with either a heat pipe or a vapor chamber [28]. The improved heat transfer rate results in a reduced $\Delta T_{ja}/Q$. Alternately, relatively higher (~25% more) heat dissipation from the ICs can be safely managed. Figure 4.10c illustrates a third passive cooling configuration, commonly referred to as the thermosyphon loop, where the fluid in the evaporator extracts the heat from the device and undergoes phase change from liquid to vapor [29]. The vapors that are generated rise to the condenser as they have a very low density compared to the liquid phase. In the condenser, which could be placed several feet away from the evaporator, the vapor condenses and returns to the evaporator, flowing along the direction of gravity.

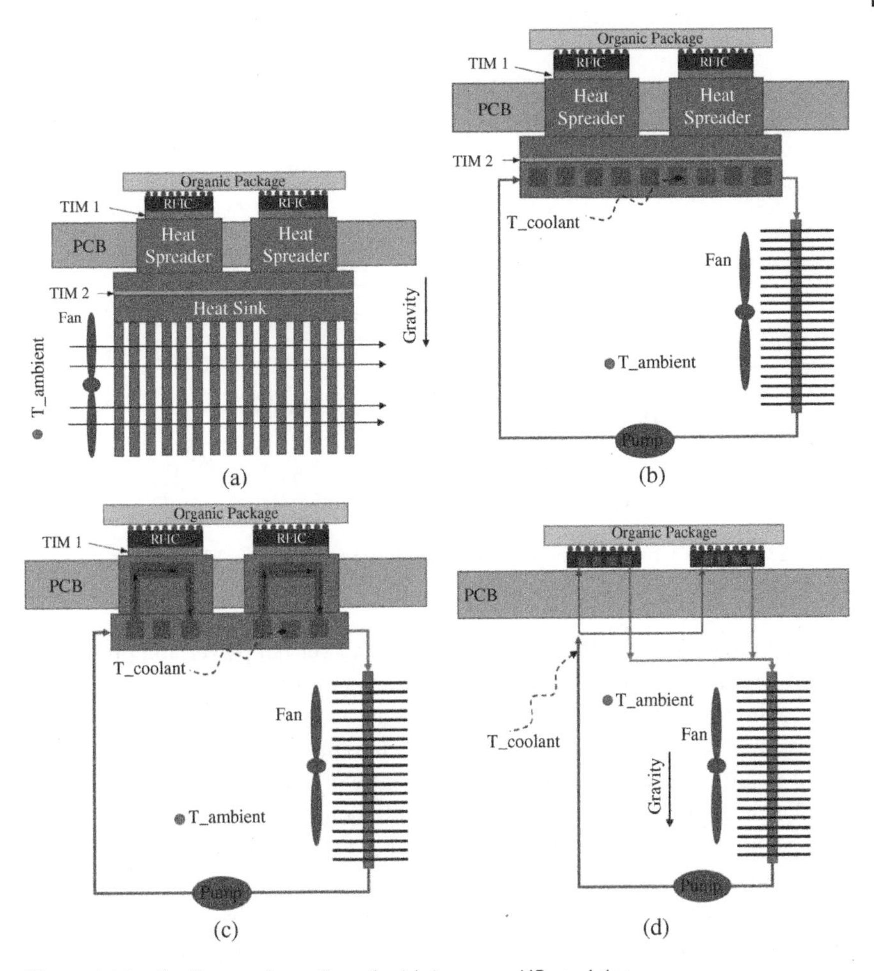


**Figure 4.12** Cooling configurations for high-power AiP modules.

In the passive mode of operation, the heat transfer in the condenser is achieved through natural convection.

For relatively higher heat dissipation (~10 W or more) from the ICs, forced cooling approaches are required. Figure 4.12a shows a traditional forced-air cooling configuration. The package is identical to that illustrated in Figure 4.10a except for the fans that are required to force the air at desired velocities across the heat sink. In general, the fan power required to force the air is a cubic function of air flow rate and can vary from 5% to 20% of the chip heat dissipation. That is, for 1 W heat dissipation from the device, fan power consumption can range from 0.05 to 0.2 W depending upon the ambient air temperature and heat dissipation from the

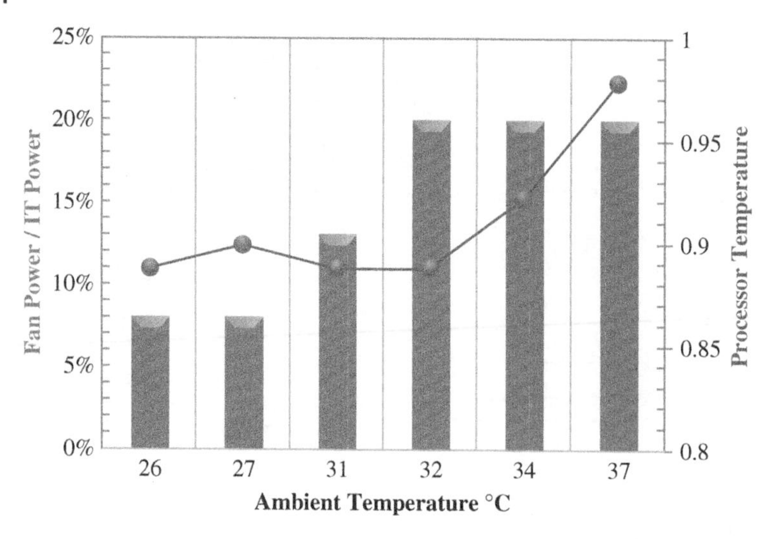


**Figure 4.13** Server fan power and processor junction temperature variation with ambient temperature.

device. Figure 4.13 shows the measurement of a server fan power and processor junction temperature variation with ambient temperature [30].

The cooling performance and energy efficiency can be significantly improved by utilizing liquid cooling to remove heat from high power components within the AiP module. In this case, as shown in Figure 4.12b, for an AiP module the air-cooled heat sink is replaced with a liquid cooled cold-plate while keeping the same heat spreader and both a TIM1 and TIM2. A pump is required to provide the desired liquid flow rate. The heat extracted by the flowing liquid must be transferred to the ambient air, which is achieved by blowing air over a liquid-to-air heat exchanger. The cooling power consumption in this configuration (Figure 4.12b) is a function of both liquid-coolant and air flow rates. However, because the capacity of liquid to extract heat is far superior to that of the air, a reduced amount of cooling power is required. Past studies on similar liquid cooling configurations have shown that for 1 W of heat dissipation from the device, the total cooling power (fan + pump) consumption can be ~0.04 W while maintaining the device temperature well below the limits [31].

In a more advanced liquid cooled AiP module, as illustrated in Figure 4.12c, the heat spreader and both TIMs can be replaced with a single high-performance liquid metal TIM (LMTIM) [32], which has a thermal conductivity an order of magnitude higher than commonly used TIMs. In this direct-attach approach, the cold-plate is attached directly on the back side of the chip, which eliminates the thermal resistance of the heat spreader and second TIM2, significantly lowering the thermal resistance path to the coolant [30, 33]. The LMTIM comprises an

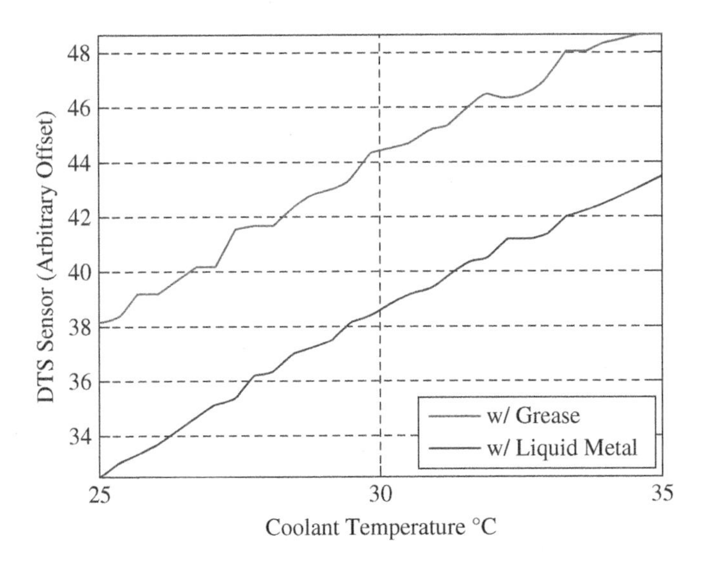


**Figure 4.14** Comparison between the conventional liquid-cooled cold-plate and direct attach cold-plate approaches.

indium-gallium-tin alloy that is liquid at all normal operating temperatures. Figure 4.14 shows the comparison between the conventional liquid-cooled cold-plate approach shown in Figure 4.12b and the direct attach cold-plate shown in Figure 4.12c with LMTIM for a 100 W processor versus coolant temperature. The results showed an estimated temperature improvement of nearly 5 °C over a range of coolant temperatures. Alternatively, this allows an increase in coolant temperature of 5 °C to achieve the same chip junction temperature at reduced cooling power consumption [33].

Figure 4.12d illustrates a possible next-generation chip-embedded cooling configuration in which a benign nonconductive fluid is made to flow through microscopic gaps, some no wider than a single strand of hair (~100 μm), etched on a high-power chip or between the stacked high-power active devices. Utilizing a dielectric fluid as coolant does away with the need for a barrier between the chip electrical signals and the fluid. The dielectric coolant is pumped into the chips, where it removes the heat from the chip by boiling from liquid-phase to vapor-phase. It then re-condenses, dumping the heat to the ambient environment where the process begins again, as shown in Figure 4.12d. As this cooling system does not need a compressor, it can operate at much lower power compared to typical refrigeration systems. Key elements of the approach and results are available in references [30, 34–39]. Past studies on a thermal test vehicle having 120 μm deep radially expanding micro channels with pin-fin arrays demonstrated that core power density of 350+ W/cm$^2$ and hot spot density of 2000+ W/cm$^2$

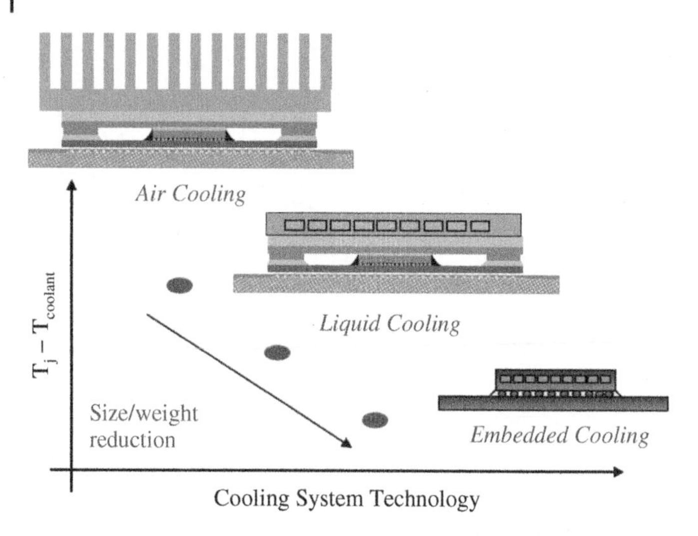


**Figure 4.15** Summary of cooling system technology improvement.

can be safely managed while achieving a chip junction to ambient temperature difference of roughly 35 °C [35]. The results demonstrate that this volumetrically efficient cooling solution (which is compatible with 3D heterogeneous chip packages) can manage three times the core power density of today's high-power processor while maintaining the device temperature well within limits. Although there are certain packaging and design challenges associated with this approach [38], this cooling approach (as summarized in Figure 4.15) not only delivers a lower device junction temperature ($T_j$), but also reduces system size, weight, and power consumption (SWaP).

### 4.3.2 Thermal Management for Heterogeneous Integrated High-power Systems

Enabled by high levels of integration of RF, analog, and digital functions, many silicon-based solutions have been developed and presented for emerging mmWave applications such as in 28-GHz 5G communication [10, 13, 14, 18]. At the same time, there is significant technology research and development in parallel on gallium arsenide (GaAs) and gallium nitride (GaN) devices due to their significantly higher output power level. For example, a fully integrated Ka-band (28-GHz) front end on a 150-nm GaAs pseudomorphic high electron mobility transistor (pHEMT) process is presented in [41]. The TX path achieves an output power at 1 dB compression (OP1dB) of 24–25 dBm from 27.5 to 28.35 GHz with a maximum power-added efficiency (PAE) of 27–32%. The RX achieves 18 dB gain with 2.9–3.3 dB noise figure (NF). A GaN-based 39-GHz FE monolithic microwave integrated circuit (MMIC) is described in [40]. The transmit path achieved an average output power of 26 dBm with 9% PAE and the receive path achieved 16 dB gain with 4 dB NF. On the system array level level, a 28-GHz 480-element digital active antenna system using a GaN high electron mobility transistor (HEMT) amplifier is shown in [42]. It has a ±60° steering range in azimuth and 68 dBm equivalent isotropically radiated power (EIRP) after back-off for long-range base station applications. In particular, GaN devices offer higher power density and bring system benefits including size reduction, lower current consumption, and higher system efficiency. We have already seen these benefits of using GaN technology in power amplifier designs of 4G base station systems [43]. For emerging 5G applications, GaN's ability to work in the high-frequency range allows it to evolve from base stations to small cell applications and, potentially, into mobile devices.

One major piece of the puzzle to enable the use of GaN for 5G involves advanced packaging techniques and thermal management. GaN devices for highly reliable military applications have traditionally been available in ceramic or metallic packages; this is also an active research area to develop heterogeneous integration of high-power GaN devices along with other components such as silicon-based RFIC modules with advanced digital functions. Inadequate thermal management in such packages will cause RF performance degradation in terms of reduced gain, output power, and efficiency, leading to premature failures and reliability issues [44]. Here, the challenge for infrastructure is to develop packaging solutions that maintain RF performance through proper thermal management. GaN's higher power densities present very challenging thermal and mechanical problems for the subsystem package. While single GaN HEMTs operating at over 30 W/mm output power have been demonstrated in controlled test environments [45], current fielded systems utilizing conventional cooling techniques can only handle devices operating at a small fraction (10–15%) of that dissipation before becoming thermally limited [44].

Recent studies have shown that advanced microfluidic cooling (like that illustrated in Figure 4.12c) has the potential to remove thermal limits from GaN devices and allow them to reach their full RF output potential [44]. For some of the high-power amplifiers (HPAs) tested, researchers have shown that a microfluidic cooled HPA can provide 4.5 dB more gain, 8 dB more RF output power, and a 3–5% increase in PAE compared to the same HPA when conventionally cooled (Figure 4.12b) [44]. For emerging radio access and handset application, the projected power density target is expected to be in the 2–4 W/mm range (https://www.qorvo.com/design-hub/blog/enabling-5g-with-gan-technology-setting-the-table-for-success). With the advanced thermal solutions described in this section, such power density targets can be adequately cooled.

## Acknowledgment

The 28-GHz phased array module was co-developed by IBM Research and Ericsson. The authors thank Ericsson's Ola Tageman, Joakim Hallin, Leonard Rexberg, Sandro Vecchiattini, Richard Lindman, Elena Pucci, Anatoli Deleniv, Eva Hellgren, Benny Gustafson, Agneta Ljungbro, Bo Eriksson, Tony Josefsson, Mikael Wahlen, Ali Ladjemi, and Adam Malmcrona for technical and management support, IBM's Duixian Liu, Christian Baks, Bodhisatwa Sadhu, Young Kwark, Alberto Valdes-Garcia, Nicolas Boyer, Eric Giguere, and Mark Ferriss for technical support, and IBM's Daniel Friedman for management support.

## References

1 J. Thompson et al., 5G wireless communication systems: prospects and challenges [Guest Editorial]. *IEEE Communications Magazine*, vol. 52, no. 2, pp. 62–64, February 2014.

2 T.S. Rappaport et al., Millimeter Wave Mobile Communications for 5G Cellular: It Will Work! *IEEE Access*, vol. 1, pp. 335–349, 2013.

3 F. Boccardi, R.W. Heath Jr., A. Lozano et al., "Five disruptive technology directions for 5G." *IEEE Communications Magazine*, pp. 74–80, February 2014.

4 W. Boh et al., "Millimeter-wave beamforming as an enabling technology for 5G cellular communications: theoretical feasibility and prototype results." *IEEE Communications Magazine*, pp. 106–113, February 2014.

5 A. Valdes-Garcia et al., A fully-integrated dual-polarization 16-element W-band phased-array transceiver in SiGe BiCMOS. *2013 IEEE Radio Frequency Integrated Circuits Symposium*, Seattle, WA, 2013, pp. 375–378.

6 X. Gu et al., A compact 4-chip package with 64 embedded dual-polarization antennas for W-band phased-array transceivers. *2014 IEEE 64th Electronic Components and Technology Conference*, Orlando, FL, 2014, pp. 1272–1277.

7 D. Liu, X. Gu, C.W. Baks et al., Antenna-in-package design considerations for Ka-band 5G communication applications. *IEEE Transactions on Antennae and Propagation*, vol. 65, no. 12, pp. 6372–6379, December 2017.

8 X. Gu, D. Liu et al., A multilayer organic package with 64 dual-polarized antennas for 28 GHz 5G communication. *IEEE MTT-S International Microwave Symposium*, Honololu, HI, 4–9 June 2017, pp. 1899–1901.

9 X. Gu et al., Development, implementation, and characterization of a 64-element, dual-polarized, phased array antenna module for 28-GHz high-speed data communications. *IEEE Transactions on Microwave Theory Techniques*, early access, 2019 (print ISSN: 0018-9480, online ISSN: 1557-9670, DOI: 10.1109/TMTT.2019.2912819).

10 B. Sadhu et al., A 28-GHz 32-element TRX phased-array IC with concurrent dual-polarized operation and orthogonal phase and gain control for 5G communications. *IEEE Journal of Solid-State Circuits*, vol. 52, issue 12, pp. 3373–3391, December 2017.

11 W.S. Tsai, C.Y. Huang, C.K. Chung et al., Generational changes of flip chip interconnection technology. *2017 12th International Microsystems, Packaging, Assembly and Circuits Technology Conference*, Taipei, 2017, pp. 306–310.

12 JEDEC Standard, JESD22-B112A, "Package warpage measurement of surface-mounted integrated circuits at elevated temperature. JEDEC Solid State Technology Association", October 2018.

13 H. Kim et al., A 28GHz CMOS direct conversion transceiver with packaged antenna arrays for 5G cellular system. *2017 IEEE Radio Frequency Integrated Circuits Symposium*, Honolulu, HI, 2017, pp. 69–72.

14 K. Kibaroglu, M. Sayginer, and G.M. Rebeiz, A scalable 64-element 28 GHz phased-array transceiver with 50 dBm EIRP and 8–12 Gbps 5G Link at 300 meters without any calibration. *2018 IEEE/MTT-S International Microwave Symposium*, Philadelphia, PA, 2018, pp. 496–498.

15 S. Emami et al., A 60 GHz CMOS phased-array transceiver pair for multi-Gb/s wireless communications. *2011 IEEE International Solid-State Circuits Conference*, San Francisco, CA, 2011, pp. 164–166.

16 A. Valdes-Garcia et al., A fully integrated 16-element phased-array transmitter in SiGe BiCMOS for 60-GHz communications. *IEEE Journal of Solid-State Circuits*, vol. 45, no. 12, pp. 2757–2773, December 2010.

17 A. Natarajan et al., A fully-integrated 16-element phased-array receiver in SiGe BiCMOS for 60-GHz communications. *IEEE Journal of Solid-State Circuits*, vol. 46, no. 5, pp. 1059–1075, May 2011.

18 J.D. Dunworth et al., A 28 GHz bulk-CMOS dual-polarization phased-array transceiver with 24 channels for 5G user and base station equipment. *2018 IEEE International Solid-state Circuits Conference*, San Francisco, CA, 2018, pp. 70–72.

19 S. Zihir, O.D. Gurbuz, A. Kar-Roy et al., 60-GHz 64- and 256-elements wafer-scale phased-array transmitters using full-reticle and subreticle stitching techniques. *IEEE Transactions on Microwave Theory and Techniques*, vol. 64, no. 12, pp. 4701–4719, December 2016.

20 A. Townley et al., A 94 GHz 4TX–4RX phased-array for FMCW radar with integrated LO and flip-chip antenna package. *2016 IEEE Radio Frequency Integrated Circuits Symposium*, San Francisco, CA, 2016, pp. 294–297.

21 E. Cohen, M. Ruberto, M. Cohen et al., A CMOS bidirectional 32-element phased-array transceiver at 60 GHz with LTCC antenna. *IEEE Transactions on Microwave Theory and Techniques*, vol. 61, no. 3, pp. 1359–1375, March 2013.

**22** S. Pellerano et al., 9.7 A scalable 71-to-76 GHz 64-element phased-array transceiver module with 2x2 direct-conversion IC in 22 nm FinFET CMOS technology. *2019 IEEE International Solid- State Circuits Conference*, San Francisco, CA, 2019, pp. 174–176.

**23** W. Lee et al., Fully integrated 94-GHz dual-polarized TX and RX phased array chipset in SiGe BiCMOS operating up to 105 °C. *IEEE Journal of Solid-State Circuits*, vol. 53, no. 9, pp. 2512–2531, September 2018.

**24** S. Shahramian, M.J. Holyoak, and Y. Baeyens, A 16-element W-band phased-array transceiver chipset with flip-chip PCB integrated antennas for multi-gigabit wireless data links. *IEEE Transactions on Microwave Theory and Techniques*, vol. 66, no. 7, pp. 3389–3402, July 2018.

**25** M. Boers et al., A 16TX/16RX 60 GHz 802.11ad chipset with single coaxial interface and polarization diversity. *IEEE Journal of Solid-State Circuits*, vol. 49, no. 12, pp. 3031–3045, December 2014.

**26** T. Sowlati et al., A 60-GHz 144-element phased-array transceiver for backhaul application. *IEEE Journal of Solid-State Circuits*, vol. 53, no. 12, pp. 3640–3659, December 2018.

**27** C. Lasance, Advances in high performance cooling for electronics. *Electronics Cooling*, November 2005. Available at: https://www.electronics-cooling.com/2005/11/advances-in-high-performance-cooling-for-electronics/.

**28** T. Nguyen, M. Mochizuki, K. Mashiko et al., "Use of heat pipe heat sink for thermal management of high performance CPUs." *Proceedings of 16th Annual IEEE Semiconductor Thermal Measurement and Management Symposium*, San Jose, CA, pp.76–79, March 2000.

**29** D. Khrustalev, Loop thermosyphons for cooling of electronics. *18th IEEE SEMI-THERM Symposium*, no. 717, pp. 145–150, 2002.

**30** T.J. Chainer, M.D. Schultz, P.R. Parida et al., "Improving data center energy efficiency with advanced thermal management." *IEEE Transactions on Components, Packaging and Manufacturing Technology*, vol. 7, issue 8, pp. 1228–1239, 2017.

**31** M. David, M. Iyengar, P.R. Parida et al., "Impact of operating conditions on a chiller-less data center test facility with liquid cooled servers." In *Proceedings of the 13th IEEE ITherm Conference 2012*, San Diego, CA, 2012.

**32** Y. Martin and T. Van-Kessel, "High performance liquid metal thermal interface for large volume production. *IMAPS Thermal and Power Management*, San Jose CA, 11–15 November 2007.

**33** M. Schultz, M. Gaynes, P. Parida et al., "Experimental investigation of direct attach microprocessors in a liquid-cooled chiller-less data center." *Proceedings of ITHERM*, Orlando, FL, 27–30 May 2014.

**34** P.R. Parida, A. Vega, A. Buyuktosunoglu et al., "Embedded two-phase liquid cooling for increasing computational efficiency. *Proceedings of the 15th IEEE ITherm Conference 2016*, Las Vegas, NV, 31 May–3 June 2016.

**35** M. Schultz, F. Yang, E. Colgan et al., "Embedded two-phase cooling of large 3D compatible chips with radial channels." *Journal of Electronic Packaging*, vol. 138, issue 2, IPACK2015-48348, 2016.

**36** P.R. Parida, A. Sridhar, A. Vega et al., "Thermal model for embedded two-phase liquid cooled microprocessor." *Proceedings of the 16th IEEE ITherm Conference 2017*, Orlando, FL, 30 May–2 June 2017.

**37** B. Dang, E. Colgan, F. Yang et al., "Integration and packaging of embedded radial micro-channels for 3D chip cooling." *Proceedings of the IEEE ECTC Conference 2016*, Las Vegas, NV, 31 May–3 June 2016.

**38** M. Schultz, P.R. Parida, M. Gaynes et al., "Microfluidic two-phase cooling of a high power microprocessor. Part A: Design and fabrication." *Proceedings of the 16th IEEE ITherm Conference 2017*, Orlando, FL, 30 May–2 June 2017.

**39** M. Schultz, P.R. Parida, M. Gaynes et al., "Microfluidic two-phase cooling of a high power microprocessor. Part B: Test and characterization." *Proceedings of the 16th IEEE ITherm Conference 2017*, Orlando, FL, 30 May–2 June 2017.

**40** B. Kim and V. Z. Li, 39 GHz GaN front end MMIC for 5G applications. *2017 IEEE Compound Semiconductor Integrated Circuit Symposium*, Miami, FL, 2017, pp. 1–4.

**41** J. Curtis, H. Zhou, and F. Aryanfar, A fully integrated Ka-band front end for 5G transceiver. *2016 IEEE MTT-S International Microwave Symposium*, San Francisco, CA, 2016, pp. 1–3.

**42** T. Kuwabara, N. Tawa, Y. Tone et al., A 28 GHz 480 elements digital AAS using GaN HEMT amplifiers with 68 dBm EIRP for 5G long-range base station applications. *2017 IEEE Compound Semiconductor Integrated Circuit Symposium*, Miami, FL, 2017, pp. 1–4.

**43** K. Kato et al., A 83-W, 51% GaN HEMT Doherty power amplifier for 3.5-GHz-band LTE base stations. *2016 46th European Microwave Conference*, London, 2016, pp. 572–575.

**44** J. Ditri, R. R. Pearson, R. Cadotte, J. W. Hahn, D. Fetterolf, M. McNulty et al., "GaN unleashed: the benefits of microfluidic cooling. *IEEE Transactions on Semiconductor Manufacturing*, vol. 29, no. 4, November 2016.

**45** Y.-F. Wu et al., "30-W/mm GaN HEMTs by field plate optimization." *IEEE Electron Device Letters*, vol. 25, no. 3, pp. 117–119, March 2004.

# 5

# Antenna-in-Package Measurements

*A.C.F. Reniers, U. Johannsen, and A.B. Smolders*

*Department of Electrical Engineering, Eindhoven University of Technology (TU/e), 5600 MB Eindhoven, The Netherlands*

## 5.1 General Introduction and Antenna Parameters

### 5.1.1 Antenna Measurement Concepts

Antenna measurements are usually categorized by the location of the measurement facility, either indoor or outdoor. The various measurement concepts which are used are as follows:

i. *Far-field range, outdoor and indoor*: In this case the antenna characteristics are directly measured in the far-field region of the *antenna-under-test* (AUT). Especially at lower frequencies (<1 GHz) or for very large antenna structures, it is often not practical to perform this type of measurement within an indoor facility. Usually, the AUT and receiving antenna are placed in a high location to minimize the impact of obstacles. A radiation pattern is obtained by rotating the test antenna and registering the received signal from the reception antennas as a function of the rotation angle. The advantage of this method is that the far-field radiation pattern is immediately available, without the need for any post-processing. The disadvantage is that this method is very sensitive for environmental disturbances and varying weather conditions.
ii. *Compact antenna test range, indoor*: The single-offset *compact antenna test range* (CATR) uses a parabolic reflector and is illuminated by a feed horn. In the near-field region of the parabola the reflected field behaves as a plane wave within the so-called "quiet zone". By taking care of edge diffraction from the large reflector, a small amplitude and phase ripple can be obtained within the quiet zone, which allows relative large antenna structures to be tested in a CATR. The advantage of a CATR is that instantaneous far-field measurement data is obtained. The disadvantage is that the CATR is expensive and bulky.

*Antenna-in-Package Technology and Applications*, First Edition.
Edited by Duixian Liu and Yueping Zhang.

iii. *Near-field range, indoor*: In this case a small probe antenna (often an open-ended waveguide) is used to sample the near-field of the AUT at a distance of $3\lambda_0$ to $5\lambda_0$ from the AUT, where $\lambda_0$ is the free-space wavelength. In this way, the test range can be very compact. The scanning surface of the probe can be planar, cylindrical or spherical. The acquired near-field data is post-processed to transform from near-field to far-field. The complexity of the transformation and the acquisition time increase from planar to cylindrical to spherical surfaces.

### 5.1.2 Field Regions

The radiation properties of antennas are generally expressed in terms of spherical coordinates. The spherical coordinate system is shown in Figure 5.1. The unit vectors in this spherical coordinate system along the $r$, $\theta$, and $\phi$ directions are denoted by $\vec{e}_r$, $\vec{e}_\theta$, and $\vec{e}_\phi$, respectively. The location of point $P(x, y, z)$ is indicated by the position vector $\vec{r} = x\vec{e}_x + y\vec{e}_y + z\vec{e}_z = \sqrt{x^2 + y^2 + z^2}\vec{e}_r = |\vec{r}|\vec{e}_r$.

The electric field $\vec{E}(\vec{r})$ in the space-frequency domain can now be written in terms of the spherical coordinates according to:

$$\vec{E}(\vec{r}) = E_r(\vec{r})\vec{e}_r + E_\theta(\vec{r})\vec{e}_\theta + E_\phi(\vec{r})\vec{e}_\phi. \tag{5.1}$$

The main function of an antenna is to transform a guided electromagnetic (EM) wave along a transmission line into radiated waves that propagate in free space. The transition from transmission line to the antenna is illustrated in Figure 5.2.

The EM field radiated by a transmitting antenna can be split up in three regions: (i) the *near-field* region, (2) the transition or *Fresnel* region, and (iii) the *far-field*

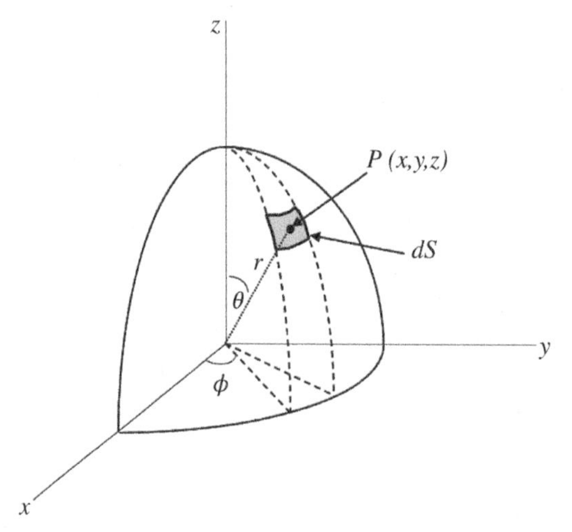


**Figure 5.1** Spherical coordinate system with $dS = r^2 \sin\theta d\theta d\phi = r^2 d\Omega$, where $\Omega = \sin\theta d\theta d\phi$ is the solid angle.

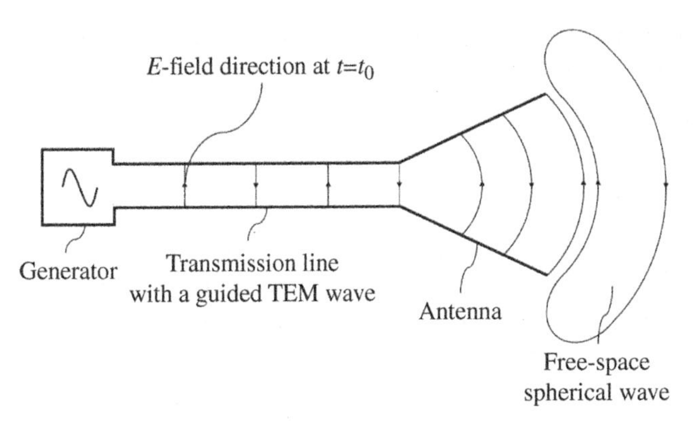


**Figure 5.2** Illustration of the transition of a guided TEM wave along a transmission line to radiate spherical waves in free space using an antenna.

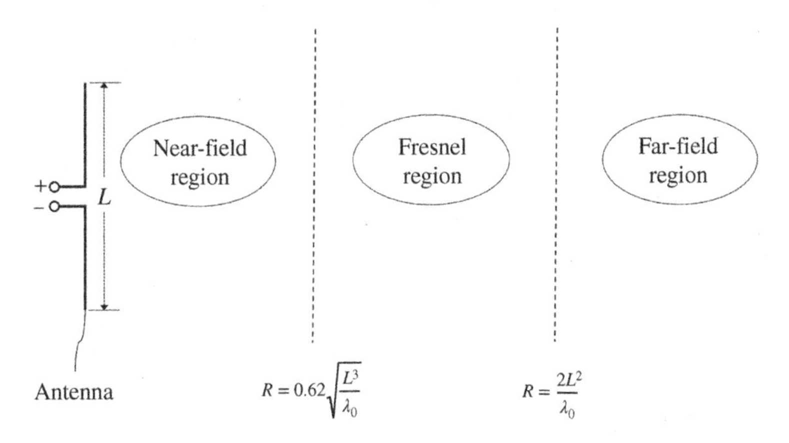


**Figure 5.3** Field regions around an antenna.

region. Figure 5.3 shows a dipole antenna and corresponding field regions. In the far-field region, the electric and magnetic field components are orthogonal to each other and orthogonal with respect to the direction of propagation, forming a so-called *transverse electromagnetic* (TEM) wave, also known as *plane wave*. When the antenna is located in the origin of the coordinate system of Figure 5.1, the transport of electromagnetic energy will be along the $\vec{e}_r$ direction, where $\vec{e}_r$ is the unit vector along the $r$ axis. Although there is not a very sharp transition between the various field regions, the most commonly used criterion to define the far-field region is the so-called *Fraunhofer distance*:

$$R = \frac{2L^2}{\lambda_0}, \tag{5.2}$$

where $R$ is the radial distance from the antenna, $L$ is the largest dimension of the antenna and $\lambda_0 = \frac{c}{f}$ is the free-space wavelength of the radiated electromagnetic wave. Note that $c = 3 \times 10^8$ (m/s) is the speed of light.

### 5.1.3 Radiation Characteristics

The radiation pattern describes the amount of power that is radiated in a certain direction in the far-field of the antenna. It is a two- or three-dimensional graphical representation of the radiation properties of the antenna. In the far-field region the electric field $\vec{E}$ has an $r^{-1}$ dependence. The radiated power per element of solid angle $d\Omega = \sin\theta d\theta d\phi$ can be found from (see also Figure 5.1):

$$P(\vec{r}) = P(\theta,\phi) = |r^2 \vec{S}_p(\vec{r})|, \tag{5.3}$$

where Poynting's vector is found by $\vec{S}_p(\vec{r}) = \frac{1}{2}Z_0^{-1}|\vec{E}(\vec{r})|^2\vec{e}_r$. Note that relation (5.3) is only valid in the far-field region in which the radiated power per element of solid angle is independent of $r$. In general, $P(\theta,\phi)$ will have a maximum value. Let us assume that this maximum occurs at an angle $(\theta,\phi) = (0,0)$. The *normalized* radiation pattern $F(\theta,\phi)$ is now defined as:

$$F(\theta,\phi) = \frac{P(\theta,\phi)}{P(0,0)}. \tag{5.4}$$

Some typical examples are shown in Figures 5.4 and 5.5. The *main beam* or *main lobe* of the antenna is directed towards the direction $\theta = 0$. The other local maxima in the pattern are called *side lobes*.

The main beam of an antenna is characterized by the *half-power beam width* (HPBW) in the principle planes, i.e. the $\theta = 0$ and $\phi = 0$ planes. The *beam width* defines the beam area for which the radiated or received power is larger than half of the maximum power. The beam width in the principle planes is expressed by $\theta_{HP}$ or $\phi_{HP}$.

The *directivity* is a measure that describes the beam-forming capabilities of the antenna, that is, the concentration of radiated power in the main lobe w.r.t. other directions. Let us consider an antenna that radiates a total power of $P_t$ watts. When this antenna is an *isotropic* radiator, the radiated power will be equally distributed over all directions with a corresponding radiated power density of $\frac{P_t}{4\pi}$ per unit of solid angle. The directivity function $D(\theta,\phi)$ of a real antenna is now defined as the power density per unit of solid angle in the direction $(\theta,\phi)$ relative to an isotropic radiator and is given by:

$$D(\theta,\phi) = \frac{P(\theta,\phi)}{P_t/4\pi}. \tag{5.5}$$

The maximum of the directivity function is the directivity $D = \max[D(\theta,\phi)]$. All antennas have some form of losses, including ohmic-losses in metal structures,

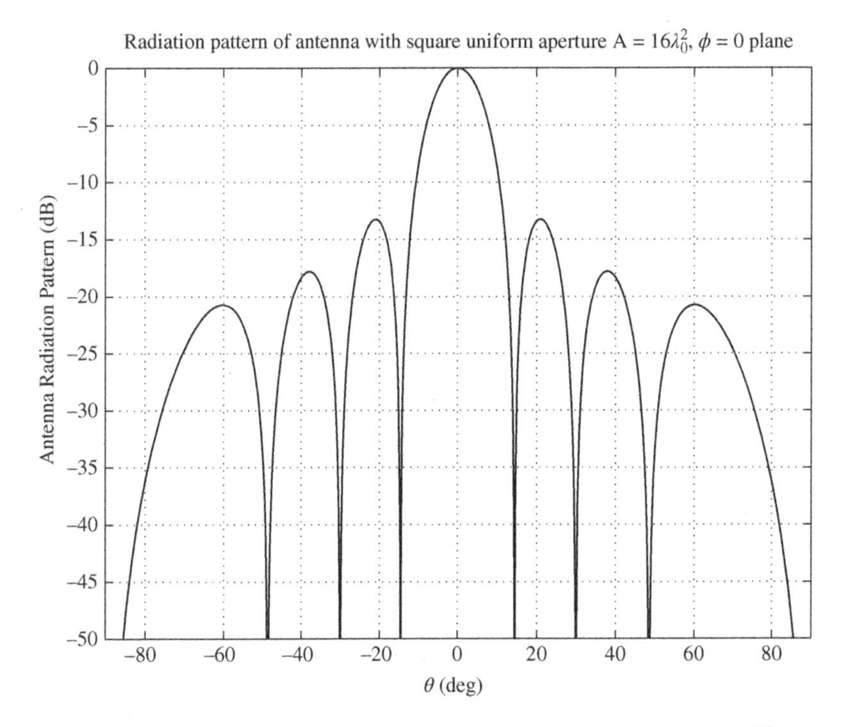


**Figure 5.4** Two-dimensional normalized radiation pattern. A cut in the $\phi = 0°$ plane is shown. The antenna has a uniformly illuminated square-shaped aperture of size $A = (4\lambda_0)^2$. The first side lobe has a maximum level of −13.2 dB.

losses in a supporting dielectric substrate, and impedance-matching losses due to a mismatch between the antenna and the transmission line (see also Figure 5.2). Due to these losses not all the power that is provided by the source will be radiated. When we replace the total radiated power $P_t$ in equation (5.5) by the total input power $P_{in}$ that is provided to the antenna, we obtain the *antenna gain function* $G(\theta, \phi)$, with:

$$G(\theta, \phi) = \frac{P(\theta, \phi)}{P_{in}/4\pi}. \tag{5.6}$$

The maximum of the antenna gain function is called the *antenna gain*, $G = \max[G(\theta, \phi)]$. The efficiency $\eta$ of the antenna is now defined as:

$$\eta = \frac{P_t}{P_{in}}. \tag{5.7}$$

This results in the following relation between the directivity $D$ and antenna gain:

$$G = \eta D. \tag{5.8}$$

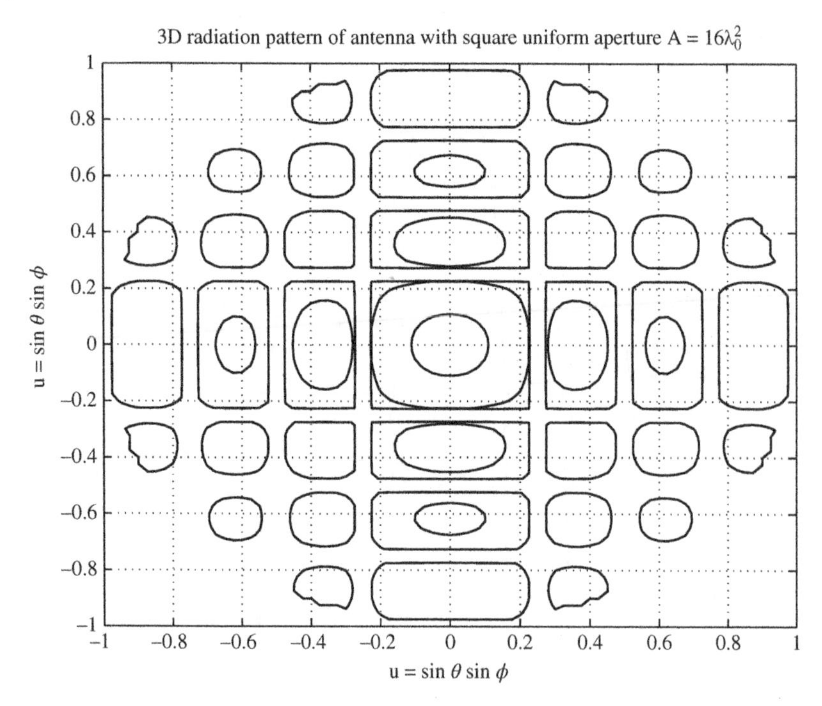


**Figure 5.5** Three-dimensional normalized radiation pattern of the antenna described in Figure 5.4. The indicated isolines correspond to the [−40, −20, −3] dB level relative to the main lobe.

When the antenna is used in receive mode, the *effective antenna aperture* is a more useful measure. The relation between the antenna gain $G$ and the effective antenna aperture $A_e$ is given by:

$$G = \frac{4\pi A_e}{\lambda_0^2}. \tag{5.9}$$

### 5.1.4 Polarization Properties of Antennas

Consider an antenna placed at the origin of our spherical coordinate system (see Figure 5.1). The radiated electric field vector (in the far-field region) from the antenna usually will have two components, $E_\theta$ and $E_\phi$, both perpendicular to the direction of propagation $\vec{e}_r$. Both components (represented in the frequency domain) will usually have a relative phase difference with respect to each other. When the phase difference is equal to 0, the resulting field will be *linearly polarized*, since both components can simply be added vectorially. So the direction of the electric field vector in the time domain is constant. In case of a phase difference between $E_\theta$ and $E_\phi$, the direction of the resulting time-domain electric field $\vec{\mathcal{E}}(\vec{r}, t)$ will vary over time with a period of $T = 2\pi/\omega$. The time-domain electric field vector will now describe an ellipse, therefore this situation is called *elliptical polarization*. Let us investigate this situation in more detail.

At far-field distances from the antenna, we can assume that the fields are locally flat and can be represented by a plane wave. Now assume that we have a plane wave propagating in the $\vec{e}_r$ direction. The time-domain electric field can be written as:

$$\begin{aligned} \mathcal{E}_\theta(\vec{r}, t) &= E_1(\vec{r}) \cos(\omega t - k_0 r), \\ \mathcal{E}_\phi(\vec{r}, t) &= E_2(\vec{r}) \cos(\omega t - k_0 r + \varphi), \end{aligned} \tag{5.10}$$

where $\varphi$ is the phase difference between both electric-field components $E_\theta$ and $E_\phi$ (frequency domain). To reduce the complexity in our notation, we will omit the place-time dependence $(\vec{r}, t)$ in the rest of this section. In addition, we will choose $r = 0$. In order to determine the equation that describes the elliptical trajectory of the resulting electric field vector $\mathcal{E} = (\mathcal{E}_\theta, \mathcal{E}_\phi)$, we need to eliminate the time-dependence. This can be done by writing $\mathcal{E}_\phi$ in the following form:

$$\begin{aligned} \mathcal{E}_\phi &= E_2(\cos \omega t \cos \varphi - \sin \omega t \sin \varphi) \\ &= E_2(\frac{\mathcal{E}_\theta}{E_1} \cos \varphi - \sin \omega t \sin \varphi), \end{aligned} \tag{5.11}$$

resulting in:

$$\begin{aligned} \left(\frac{\mathcal{E}_\phi}{E_2} - \frac{\mathcal{E}_\theta}{E_1} \cos \varphi\right)^2 &= \sin^2 \omega t \sin^2 \varphi = (1 - \cos^2 \omega t) \sin^2 \varphi \\ &= \left(1 - \left[\frac{\mathcal{E}_\theta}{E_1}\right]^2\right) \sin^2 \varphi. \end{aligned} \tag{5.12}$$

This can be rewritten in the following form:

$$\left(\frac{\mathcal{E}_\theta}{E_1}\right)^2 + \left(\frac{\mathcal{E}_\phi}{E_2}\right)^2 - \frac{2\mathcal{E}_\theta \mathcal{E}_\phi}{E_1 E_2} \cos \varphi = \sin^2 \varphi. \tag{5.13}$$

It can be shown that the resulting equation describes an ellipse in the plane perpendicular to the direction of propagation $\vec{e}_r$. The orientations of the major and minor axes of the ellipse usually do not coincide with the unit vectors $\vec{e}_\theta$ and $\vec{e}_\phi$. The ratio between the major and minor axes of the ellipse is the so-called *axial ratio* (AR), with $AR \geq 1$. Figure 5.6 illustrates the polarization ellipse and orientation of the major and minor axes of the ellipse.

The AR is an important quantity to describe the quality of circularly polarized antennas. Now let us investigate this special case in more detail.

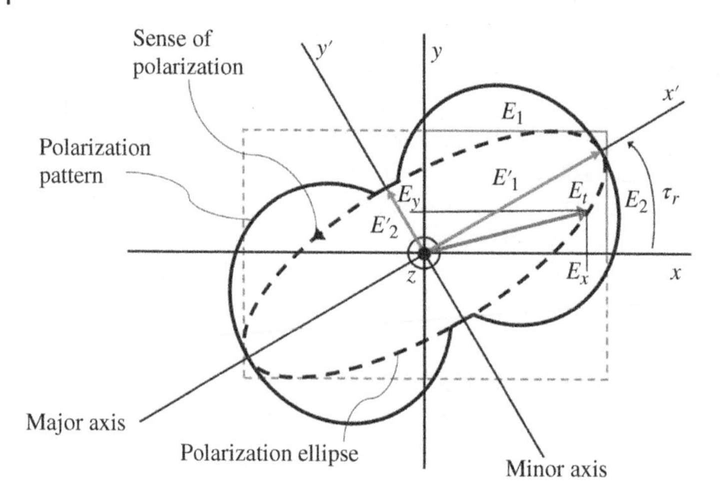


**Figure 5.6** Polarization ellipse (a function of time) and the polarization pattern (a function of space) [1, 2]. Note that the AR is determined by $AR = \frac{|E_1'|}{|E_2'|} \geq 1$. We have assumed in this figure that $\vec{e}_\theta$ coincides with the $x$ direction and $\vec{e}_\phi$ coincides with the $y$ direction [3].

i. *Circular polarization* ($E_1 = E_2 = E$ and $\varphi = \pm\pi/2$). Then (5.13) can be written as:

$$\left(\frac{\mathcal{E}_\theta}{E}\right)^2 + \left(\frac{\mathcal{E}_\phi}{E}\right)^2 = 1. \tag{5.14}$$

This represents the equation of a circle, therefore we now have a *circularly polarized* field. We can distinguish two particular situations, the first one with $\varphi = \pi/2$:

$$\vec{\mathcal{E}}_L = E(\vec{e}_\theta \cos\omega t - \vec{e}_\phi \sin\omega t), \tag{5.15}$$

and in case $\varphi = -\pi/2$, we obtain:

$$\vec{\mathcal{E}}_R = E(\vec{e}_\theta \cos\omega t + \vec{e}_\phi \sin\omega t). \tag{5.16}$$

In the first situation, the $\vec{\mathcal{E}}$ vector rotates anticlockwise w.r.t. the direction of propagation, therefore this case is called *left-hand circular polarization* (LHCP). In the second situation (5.16), the $\vec{\mathcal{E}}$ vector rotates clockwise, corresponding to *right-hand circular polarization* (RHCP). In a practical situation, antennas will never have perfect circular polarization (CP). We will always have an $\vec{\mathcal{E}}_L$ and $\vec{\mathcal{E}}_R$ field component. The AR describes the quality of the CP and can be expressed in term of the frequency-domain field components $\vec{E}_L$ and $\vec{E}_R$ according to:

$$AR = \left| \frac{|E_L| + |E_R|}{|E_L| - |E_R|} \right|. \tag{5.17}$$

where $\vec{\mathcal{E}}_L = Re[\vec{E}_L e^{j\omega t}]$. When $AR = 1$, we would have a perfectly circularly polarized wave.

ii. *Linear polarization* ($E_1 \neq E_2$ en $\varphi = \pm\pi$). Equation (5.13) can now be written as:

$$\frac{\mathcal{E}_\theta}{E_1} + \frac{\mathcal{E}_\phi}{E_2} = 0, \tag{5.18}$$

which mathematically represents a straight line. In this case we obtain linear polarization, where $AR = \infty$.

## 5.2 Impedance Measurements

### 5.2.1 Circuit Representation of Antennas

The input impedance of the antenna is an important measure that determines how much of the power which is transported along the transmission line is actually radiated by the antenna. The input impedance of an antenna is in general complex, consisting of a resistive part $R_a$ and a reactive part $X_a$ according to:

$$\begin{aligned} Z_a &= R_a + jX_a \\ &= R_r + R_L + jX_a, \end{aligned} \tag{5.19}$$

where $R_r$ represents the *radiated* losses and $R_L$ represents the real losses, such as ohmic or dielectric losses. Note that the *radiation resistance* $R_r$ is directly related to the total radiated power by the antenna and is in fact a measure that defines how well the antenna works. The reactance $X_a$ is related to the stored reactive energy in the close vicinity of the antenna. The antenna can now be represented by an one-port microwave network with input impedance $Z_a = V^p/I^p$, where $V^p$ is the voltage along the input ports of the antenna and $I^p$ represents the current through the antenna. Figure 5.7 shows the equivalent circuit of an antenna which is connected to a generator with internal impedance $Z_g = R_g$. In practise, $R_g = 50\,\Omega$ will often be used. However, for optimal noise matching of the antenna to a low-noise amplifier other (complex) values of the source impedance are preferred.

At radio frequencies and microwave frequencies it is not possible to measure voltages and currents directly. As a consequence, we cannot measure the input impedance $Z_a$ directly. However, at these frequencies it is possible to measure the

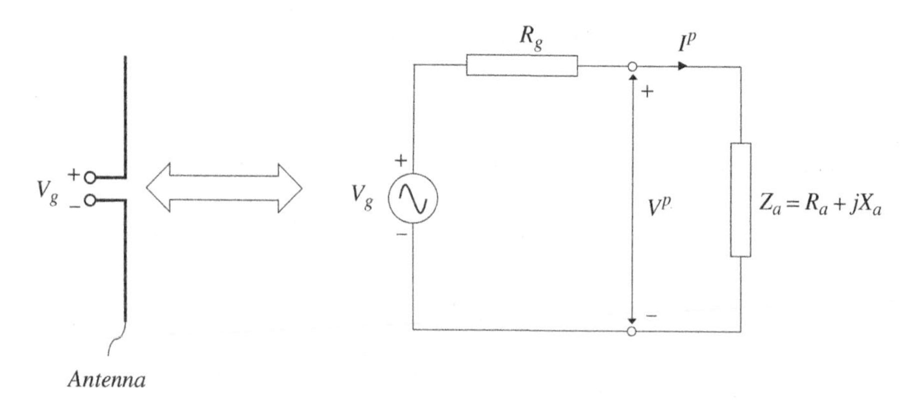


**Figure 5.7** Equivalent circuit representation of an antenna. The antenna is connected to a generator with internal impedance $R_g$.

(complex) amplitude of incident and reflected waves that travel along a transmission line. The input reflection coefficient of the antenna can be experimentally determined with a *vector network analyser* (VNA). Now let us assume that the antenna is connected to a transmission line with a characteristic impedance equal to $Z_0^t$. The relation between the reflection coefficient and the input impedance at the input of the antenna is given by:

$$\Gamma = \frac{Z_a - Z_0^t}{Z_a + Z_0^t}. \tag{5.20}$$

In the case where the antenna is matched to the transmission line with $Z_a = Z_0^t$, we will have zero reflection with $\Gamma = 0$.

For determining the input impedance of an integrated *millimeter-wave* (mmWave) antenna, a generic wafer-probe station can be used, like the Model 9000 probe station from Cascade Microtech, Inc., shown in Figure 5.8. It consists of a metallic chuck on which the AUT can be placed. A built-in vacuum fixation avoids any movement of the AUT such that a mmWave measurement probe can easily be landed using the adjustable probe holders and the microscope, which are also shown in the figure. The measurement probe itself connects the antenna via a coaxial cable to a VNA, as depicted in the sketched measurement setup shown in Figure 5.9. Various probe types with different pin configurations on the antenna side are commercially available. The basic pin structure of typical probes (Picoprobe Models 67A in ground-signal-ground (GSG), ground-signal (GS), and signal-ground (SG) configurations were used, see http://www.ggb.com/67a.html) is illustrated in Figure 5.10. For each of those probes a calibration substrate is available such that the input reflection coefficient, $\Gamma_{\text{in}}$, of the AUT can be directly measured as seen from the probe tip's position. The calibration substrate used

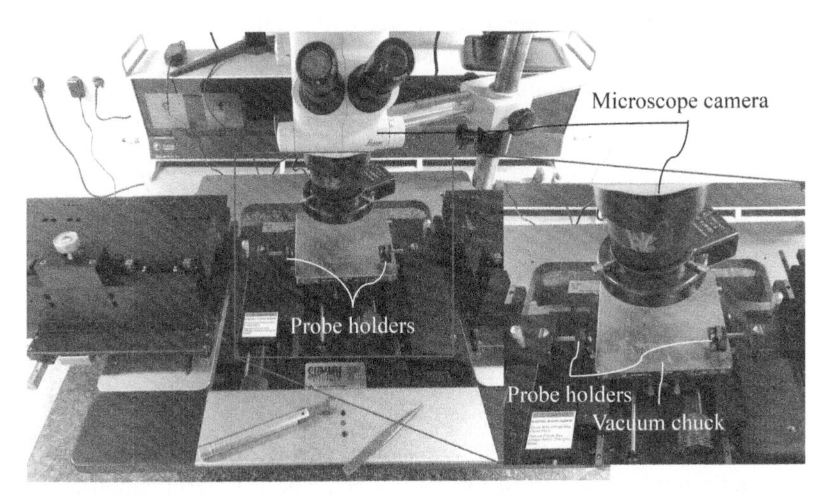


**Figure 5.8** Photograph of a probe station model 9000 from Cascade Microtech, Inc. (see https://www.cascademicrotech.com/products/probes/rfmicrowave/t-wave-probe/t-wave-probe).

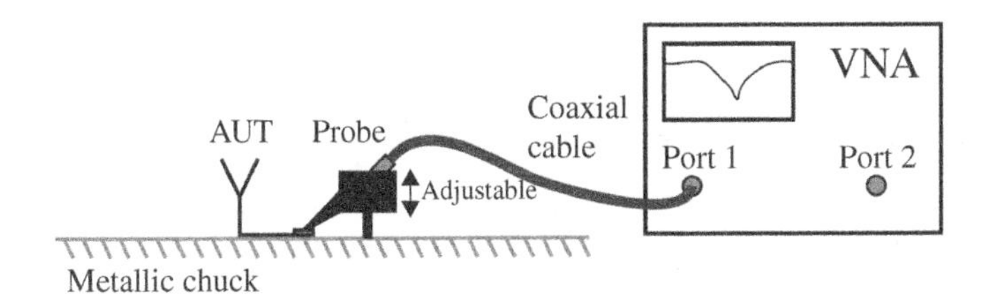


**Figure 5.9** Sketch of the input-impedance measurement setup.

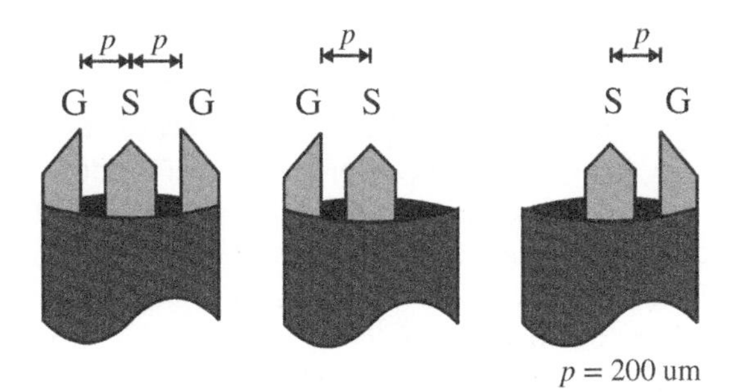


**Figure 5.10** Illustration of probe tips in GSG, GS, and SG configurations. Where $P$ is the distance between the probe tips.

to characterize a particular *antenna-in-package* (AiP) can be made by putting three exact copies of the interconnecting metal pads with short, open, and load termination, respectively, on a similar substrate. Their measured input reflection coefficients are related to the input reflection coefficients of the termination according to

$$\Gamma_{\text{meas.,short}} = S_{11,\text{pads}} + \frac{S_{12,\text{pads}} S_{21,\text{pads}} \Gamma_{\text{short}}}{1 - S_{22,\text{pads}} \Gamma_{\text{short}}}, \tag{5.21}$$

$$\Gamma_{\text{meas.,open}} = S_{11,\text{pads}} + \frac{S_{12,\text{pads}} S_{21,\text{pads}} \Gamma_{\text{open}}}{1 - S_{22,\text{pads}} \Gamma_{\text{open}}}, \tag{5.22}$$

$$\Gamma_{\text{meas.,load}} = S_{11,\text{pads}} + \frac{S_{12,\text{pads}} S_{21,\text{pads}} \Gamma_{\text{load}}}{1 - S_{22,\text{pads}} \Gamma_{\text{load}}}, \tag{5.23}$$

where the left-hand side shows the measured quantities. The right-hand side consists of the reflection coefficients of the short, open, and load terminations, i.e. $\Gamma_{\text{short}} = -1$, $\Gamma_{\text{open}} = 1$, and $\Gamma_{\text{load}} = 0$, respectively, and the two-port $S$ parameters of the bond pads. From the three linearly independent equations (5.21) to (5.23), the parameters $S_{11,\text{pads}}$ and $S_{22,\text{pads}}$ as well as the product $S_{12,\text{pads}} S_{21,\text{pads}}$ can be determined. The input reflection coefficient $\Gamma_{\text{AUT}}$ of the on-chip antenna is then given by

$$\Gamma_{\text{AUT}} = \frac{\Gamma_{\text{AUT+pads}} - S_{11,\text{pads}}}{S_{22,\text{pads}}(\Gamma_{\text{meas}} - S_{11,\text{pads}}) + (S_{12,\text{pads}} S_{21,\text{pads}})}, \tag{5.24}$$

where $\Gamma_{\text{AUT+pads}}$ is the measured input reflection coefficient of the AUT, i.e. including the metal pads. The input impedance of the antenna can now be calculated with the aid of (5.20).

Figure 5.11 shows a photograph of a manufactured AiP prototype together with its design parameters. Here, the metalized top-side of the RO4003C laminate also serves as ground plane and heat sink for the front-end IC. To assure that no parallel plate modes will be excited, via fences connect the upper and lower metal plates around the dipole. Furthermore, the coupled microstrip lines are extended with a transmission line that allows the landing of a mmWave measurement probe with a pitch of 200 μm. A comparison between the measured and simulated input impedance and the input reflection coefficient of a randomly picked sample is depicted in Figures 5.12a and 5.12b, respectively. The measurement was carried out using a single-ended Picoprobe Model 67A-GS-200-DP in a GS configuration. By landing the GS probe on the signal lines, the antenna is excited in the desired differential mode. From Figures 5.12a and 5.12b a good agreement between simulation and measurement can be concluded.

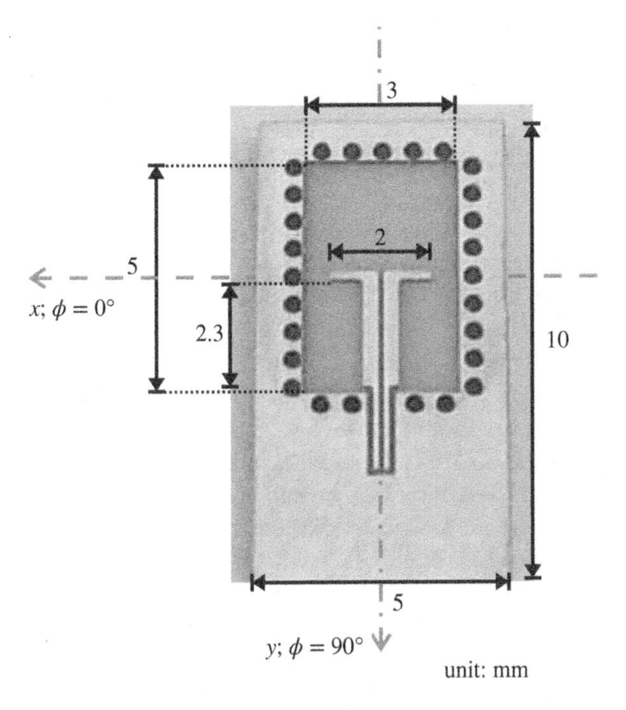


**Figure 5.11** Photograph of a fabricated 60-GHz AiP prototype.

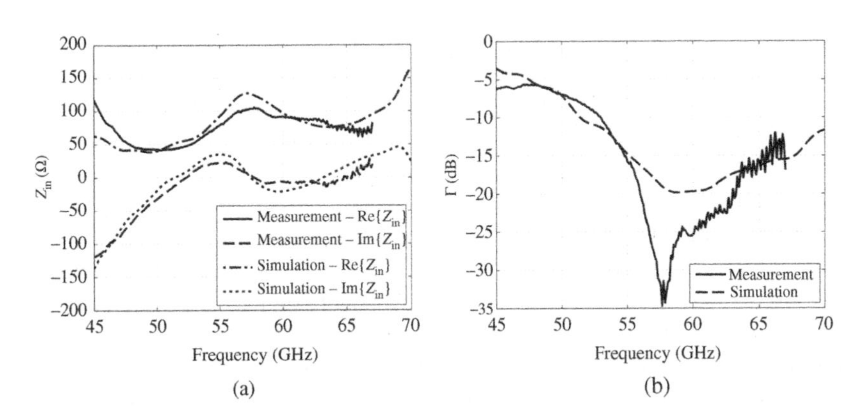


**Figure 5.12** Comparison between measured and simulated antenna parameters of the AiP prototype in Figure 5.11: (a) input impedance and (b) input reflection coefficient ($Z_0 = 100\ \Omega$).

## 5.3 Anechoic Measurement Facility for Characterizing AiPs

Since the beginning of this century there has been a growing interest in the development of new test facilities to characterize mmWave antennas at frequencies beyond 30 GHz. At these frequencies, low-gain antennas have sizes of 1 cm or less, integrated on chip [4], in package, or in a system [5]. As a result, the chamber could be table-top size. Furthermore, because of the small size of the antenna new interface technologies are required [6]. The first mmWave measurement facilities appeared in 2006 and were designed and built by universities [7, 8]. Since 2015 commercial mmWave antenna test facilities have become available [see Micro vision group (http://www.mvg-world.com/en), Orbit (http://www.orbitfr.com/content/solutions-automotive), and Rohde-Schwarz (https://www.rohde-schwarz.com/ae/about/news-press/)]. Most of these measurement systems have all kinds of bulky and metallic moving parts and probe stations inside the anechoic environment, causing a lot of interference when measuring the antenna pattern.

The design of the *mmWave anechoic chamber* (MMWAC) has a confined space where the AUT is positioned in such a way that moving and supporting objects have a negligible effect on the measured radiation characteristics. The motors and mechanical structure supporting the reference antenna are situated outside the anechoic chamber. Through a slit, the reference antenna is guided inside the anechoic environment. In addition, a special construction for the support of the probe needs to be made to separate the probe station from the probe.

### 5.3.1 Design of the mmWave Anechoic Chamber

The measurement system, shown in Figure 5.13a (illustration) and Figure 5.13b (realized model) is an in-house developed compact, moveable MMWAC with dimensions $L = 1.8$ m, $W = 1.1$ m, and $H = 2.0$ m.

The chamber is enclosed with a metal cover and covered with EM absorbers which have been proven to be effective up to 90 GHz [9]. The AUT is positioned in the middle of the room on a translation table. The translation table can move along the $x$ and $y$ axis[1] to align the AUT with respect to the reference antenna. An alignment camera is used to support this alignment procedure, as shown in Figure 5.14. The translation table can also move along the $z$ axis to place the antenna in the phase center for a specific frequency during a gain measurement. The chamber has a probe station incorporated, which is located underneath the translation table. Support structures connect the probe with the probe station such that it can be

1 The illustration shows a circularly polarized antenna where the $\phi$ rotation is also of importance. Both the cross-hair of the camera and the marking on the printed circuit board ensure correct alignment and repeatability.

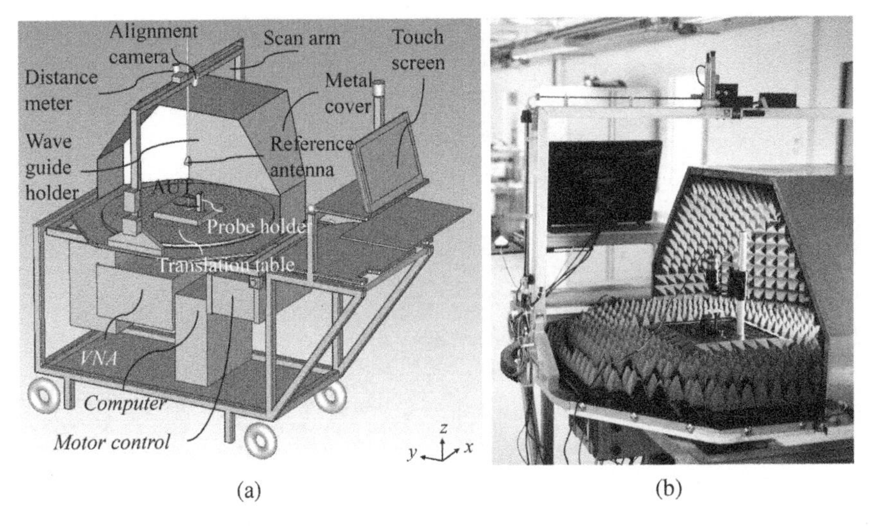


**Figure 5.13** (a) Illustration of a mmWave anechoic chamber and (b) a photograph of the realized mmWave anechoic chamber.

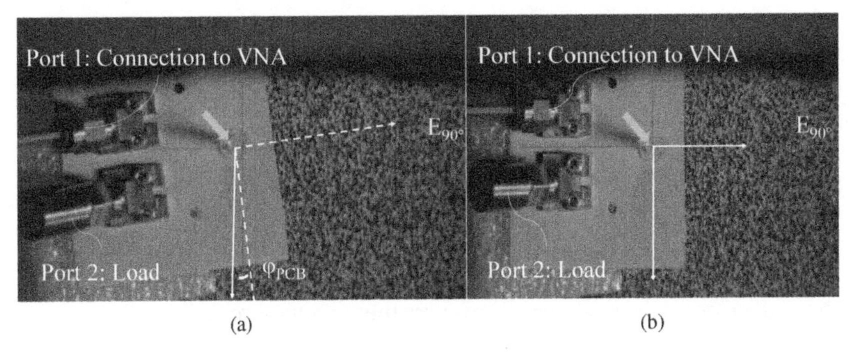


**Figure 5.14** Circularly polarized rod antenna (a) misaligned with rotational insert angle and (b) aligned with respect to the reference antenna. The location of the antenna is indicated by the yellow arrow.

moved with a resolution of 1 μm. As can be observed, an unique probe setup is realized with the help of an existing probe [10], as shown in Figure 5.15. A thorough study of the effect of the probe on the behavior of the antenna has been discussed in [10–12] addressing the interconnection uncertainties.

### 5.3.2 Defining Antenna Measurement Uncertainty

Any analysis, realization, and measurement of an antenna is subject to uncertainties [13]. As discussed in [14–17], quantifying the different types of errors,

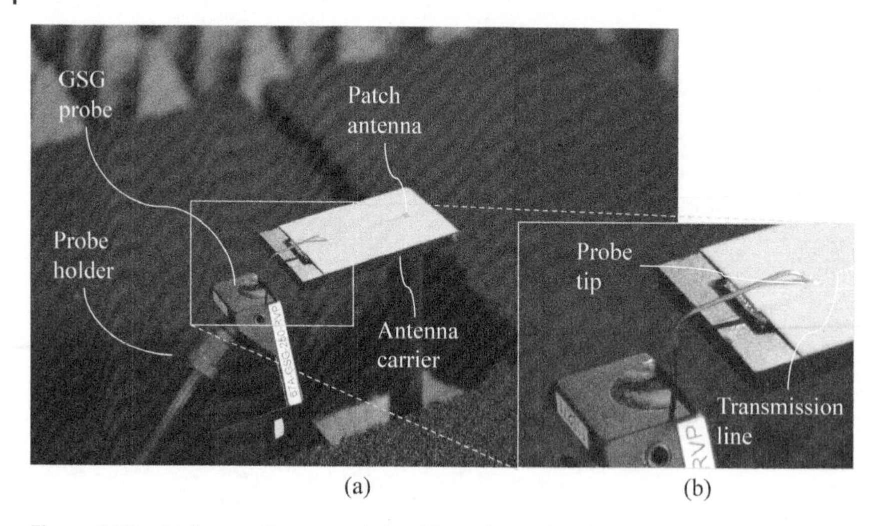


**Figure 5.15** (a) Perspective view of the side probe positioned on a co-planar transmission line connected to a linear polarized patch antenna and (b) detailed view of the transition of the probe and AUT.

which are an estimate of measurement uncertainties, is not always done correctly, therefore we have illustrated the interrelationship in Figure 5.16. The definitions for the errors are expressed as follows:

- the *systematic error* is a "component of measurement error that in replicate measurements remains constant or varies in a predictable manner" [14]
- the *random error* is a "component of measurement error that in replicate measurements varies in an unpredictable manner" [14]
- the *total error* is "the systematic error plus the random error".

For the performance characteristics:

- the *trueness* is "the closeness of agreement between the average of an infinite number of replicate measured quantity values and a reference quantity value" [14]
- the *accuracy* is "the closeness of the agreement between the result of a measurement and a true value of the measurand" [14]
- the *precision* is "the closeness of agreement between indications or measured quantity values obtained by replicate measurements on the same or similar objects under specified conditions" [14].

For their quantitative expression:

- the *bias* is "an estimate of a systematic measurement error" [14]

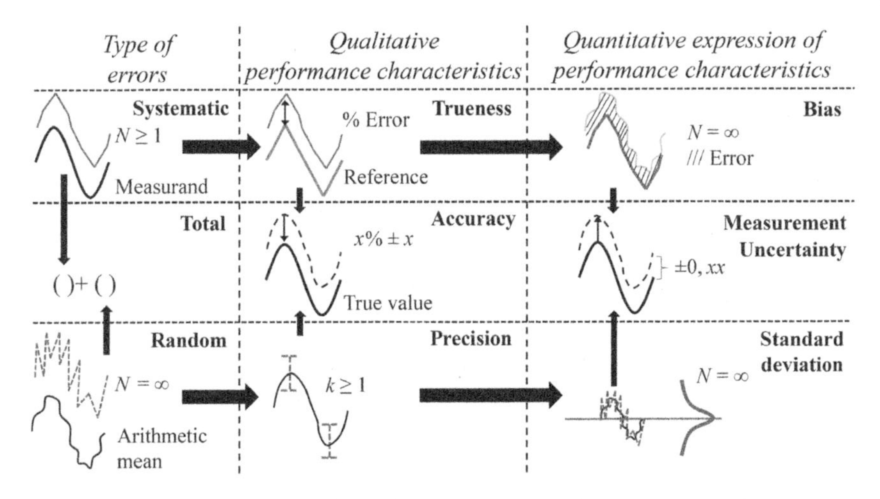


**Figure 5.16** Illustration of the interrelations between various error types, the performance characteristics used to estimate them, and ways to express the estimates quantitatively. $N$ represents the number of measurements and $k$ is the coverage factor.

- the *measurement uncertainty* is "a closeness of agreement between a measured quantity value and a true quantity value of a measurand" [14]
- the *standard deviation* is "a measure that is used to quantify the amount of variation or dispersion of a set of data values" [17].

Both *trueness* and *accuracy* require a reference according to the definition. For this, we could use the *true value*, which is defined as:

- "a quantity value consistent with the definition of a quantity" [14].

However, the true value is unknown and can never be exactly determined, hence we need to use a reference value instead. The absolute *systematic error* is then quantitatively expressed by a *bias* with:

$$bias = \left[1 - \left(\frac{|R_v|}{||R_v - M_v| + |R_v||}\right)\right] \times 100\%, \tag{5.25}$$

where $R_v$ is the reference value and $M_v$ is the measured value. This implies that when $M_v$ is equal to $R_v$, the bias is 0% and, as a result, the trueness equals 100%. The random error is represented by the unbiased standard deviation $s$ and is expressed by [15]:

$$s_{(x)} = \sqrt{\frac{1}{N-1}\sum_{n=1}^{N}(x_n - \overline{x})^2} \tag{5.26}$$

where $x_n$ is the $n$th random variable and $\overline{x}$ is the arithmetic mean or average. The $N - 1$ corresponds to the number of degrees of freedom and $N$ to the total amount of observations $x$, i.e. a sample of the population. To increase the level of confidence, the standard deviation $s_x$ must be multiplied by the coverage factor $k$, identifying which part of the normal distribution is taken into account. According to [14], $k = 1$ corresponds to a level of confidence larger than 68% and $k = 3$ larger than 99%. This interrelation for analyzing any measurement result is widely used in the antenna test community.

### 5.3.3 Uncertainty in the mmWave Antenna Test Facility

To be able to measure specific antenna characteristics, like the radiation pattern, the reference antenna will be, in this case, moved around the AUT. To measure the polarization characteristics the reference antenna needs to be rotated around its own axis. These specific movements cause uncertainties, $\mu_{sys}$, in the obtained measurement data. The motor movements within the MMWAC for various type of measurements are described in Table 5.1 and illustrated in Figure 5.17. The *multiple amplitude phase component method* (MAPCM) is used to measure the radiation properties of circularly polarized antennas (see section 5.3.4 for more details).

The orientations related to the specific movements during antenna measurements are illustrated in Figure 5.17. These movements are related to the spherical coordinate system (see Figure 5.17) where the co- and cross-polarization are defined according to [18]. Figure 5.17 also illustrates the movements (*El* for elevation, *Az* for azimuth) and the maximum angles that can be reached.

The rotation of the scan arm is related to $\phi$, the movement of the scan arm to $\theta$, and the rotation of the reference antenna to $\phi_{pol}$ (see Figure 5.17). The latter movement is used to rotate the reference antenna to measure the co- or cross-polarization for a specific plane of the AUT. With these movements, the half-sphere realized gain of the AUT can be measured. The scan-arm (see Figure 5.17) moves around the AUT. The AUT remains static during measurements to avoid mechanical stress on the RF connection.

### 5.3.4 Case Study AiP: Characterization of a mmWave Circularly Polarized Rod Antenna

Figure 5.18a shows an illustration of a mmWave short-range bi-directional communication system equipped with a circularly polarized rod antenna. The

**Table 5.1** Motor movements during various types of antenna measurements.

| Measurement | $\phi_{pol}$ | $\theta_{scan}$ | Measured duration |
|---|---|---|---|
| Radiation pattern co-polarization | 0° | −90° to 90° | 2 min |
| Radiation pattern cross-polarization | 90° | −90° to 90° | 3 min |
| MAPCM (CP) | 0°, 45°, 90°, 135° | −90° to 90° | 10 min |

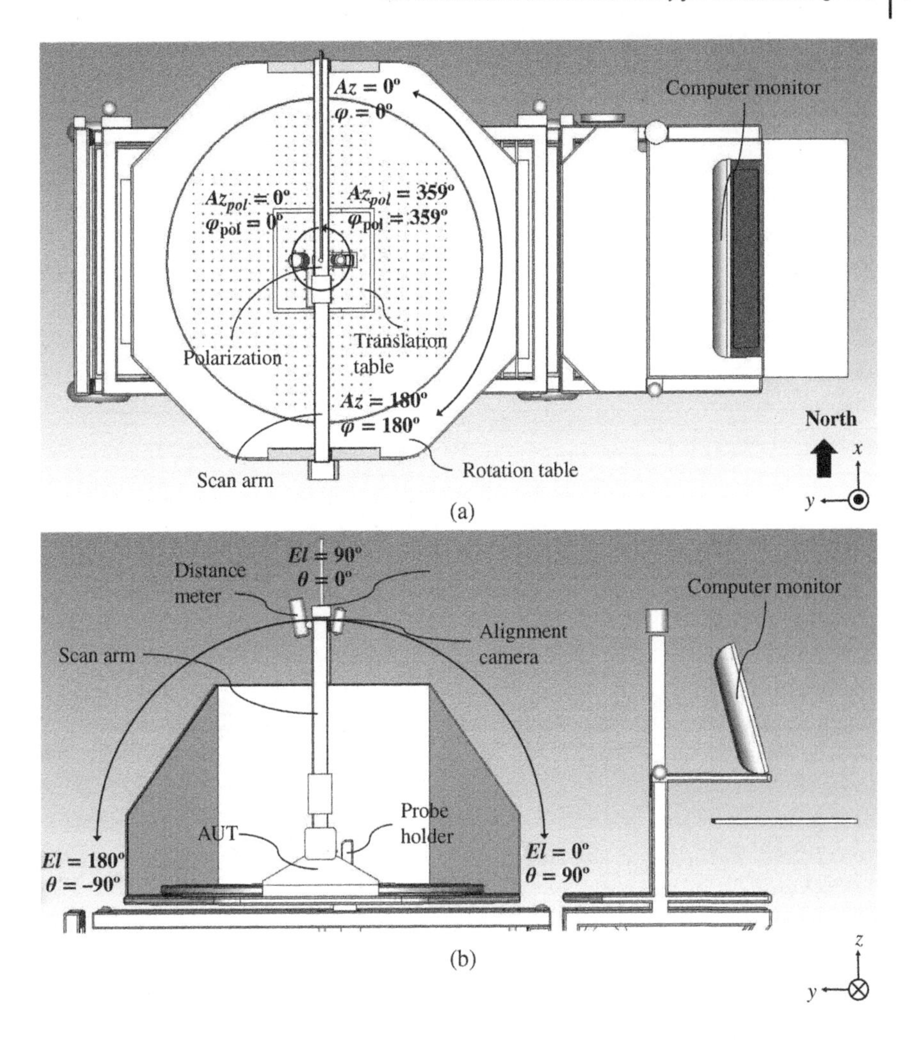


**Figure 5.17** Motion orientation pertaining to rotations and translations in the mmWave anechoic chamber: (a) top view and (b) side view.

motivation for using CP and for using this specific housing is determined by the application, which requires wireless high-speed communication in a highly reflective environment that is oil contaminated.

The AiP consists of a printed circuit board which has the antenna incorporated on one side and the transceiver electronics on the other. The electronics are sealed with a specific housing. The power supply is provided by inductive coupling. The design requirements of the rod antenna are summarized in Table 5.2.

The antenna consists of a dual-fed, microstrip patch antenna (see Figure 5.18b) that generates CP by using a microstrip hybrid to create the required phase shift of 90°. Two electromagnetically coupled microstrip transmission lines feed

**Table 5.2** Antenna design requirements of the circularly polarized rod antenna [19].

| Parameter | Antenna design requirement |
|---|---|
| Center frequency | 61 GHz |
| Bandwidth at −10 dB input reflection | >9 GHz |
| Isolation bandwidth at −20 dB between | >9 dBi |
| Gain | >9 dBi |
| Half-power beam width | <30° |
| Polarization | Dual circular (RHCP and LHCP) |
| Axial ratio at broadside | 0.5 dB |
| Impedance | 50 Ω |

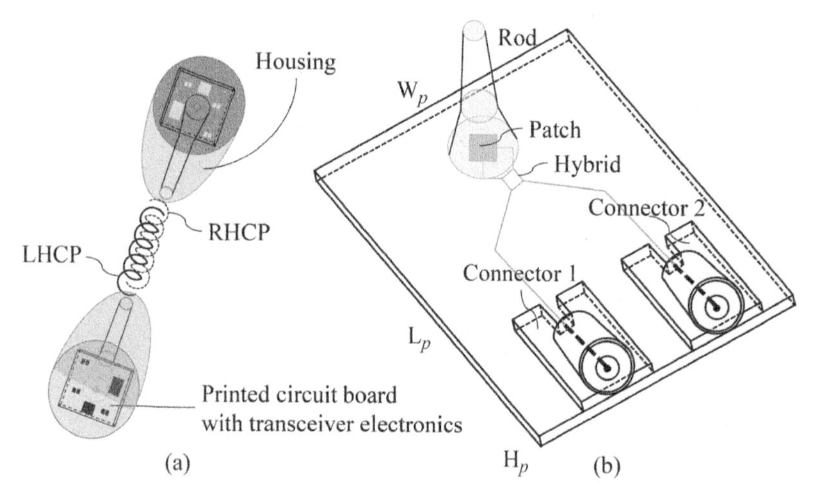


**Figure 5.18** (a) Illustration of the short-range communication application using a circularly polarized rod antenna and (b) the same circularly polarized rod antenna, but now in a measurable configuration to perform an accurate characterization.

the microstrip patch. The hybrid makes it also possible to excite the antenna with both RHCP and LHCP. A dielectric tapered rod is placed on top of the patch antenna, where the shape and material properties are chosen such that it maximizes the antenna gain. In order to characterize the circularly polarized rod antenna, the transceiver is replaced with two connectors of type 08K80A-40ML5 [27]. The substrate is extended to add the connectors as illustrated in Figure 5.18b. Figure 5.19 shows the realized model situated in the mmWave anechoic chamber.

Various polarization pattern methods can be used to characterize circularly polarized antennas, as described in [20] and see Table 5.3. The symbol 'V'

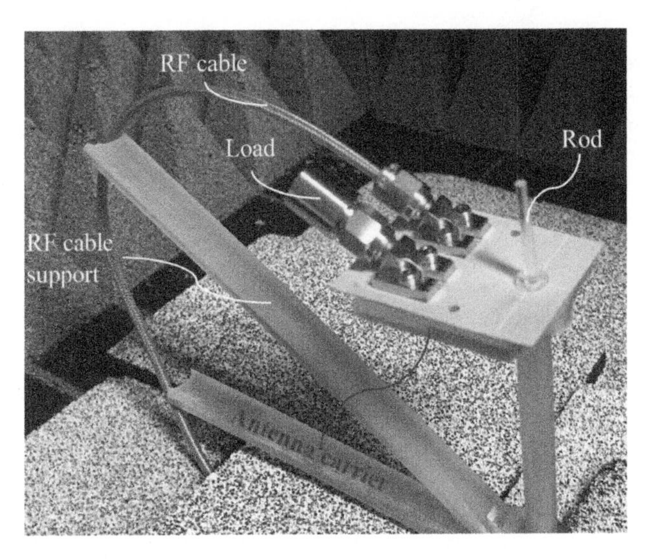


**Figure 5.19** Circularly polarized rod antenna with connector interface and cable support to minimize positioning errors.

**Table 5.3** Measurement methods to determine antenna polarization characteristics [20].

| Measurement method | Acronym | Axial ratio | Tilt angle | Sense of orientation | Radiation pattern |
|---|---|---|---|---|---|
| Polarization pattern method | PPM | V | V | X | X |
| Rotating source method | RSM | V | X | X | X |
| Multiple amplitude component method | MACM | V | V | V | X |
| Phase amplitude method | PAM | V | V | V | V |
| Multiple amplitude phase component method | MAPCM | V | V | V | V |

marks which antenna characteristics are obtained with the specific polarization pattern method and 'X' indicates which are not. The MAPCM is preferred over the *phase amplitude method* (PAM), since the PAM requires a very expensive dual-polarized reference antenna that can only be used for characterizing circularly polarized antennas.

The effects of the measurement system on the antenna characteristics, like the AR and antenna gain, are shown in Table 5.4.

We can see from Table 5.4 that the influence of the measurement system is minimal. The characterization of the circularly polarized rod antenna (see 5.19)

**Table 5.4** Uncertainty caused by the motor movements and positioning of the AUT and the effect on the measured results of the axial ratio (AR) and gain (G).

| # | Orientation | $\mu_{sys}$ | $\mu_{dB}$ $\theta$ and $\phi = 0°$ |
|---|---|---|---|
| 11 | Rotation | ±0.0055° | AR: $\pm 60e^{-4}$ dB<br>G: $\pm 1e^{-4}$ dB |
| 12 | Scan | ±0.0011° | AR: $\pm 12e^{-4}$ dB<br>G: $\pm 3e^{-4}$ dB |
| 13 | Polarization | ±0.0027° | AR: $\pm 30e^{-4}$ dB<br>G: $\pm 6e^{-4}$ dB |
| 14 | Translation ($x$ axis) | ±0.0028 mm | AR: $\pm 3e^{-5}$ dB<br>G: $\pm 5e^{-4}$ dB |
| 15 | Translation ($y$ axis) | ±0.0048 mm | AR: $\pm 5e^{-5}$ dB<br>G: $\pm 8e^{-4}$ dB |
| 16 | Translation ($z$ axis) | ±0.0024 mm | AR: $\pm 3e^{-5}$ dB<br>G: $\pm 4e^{-4}$ dB |
| 17 | Antenna placement | ±0.12 mm | $\tau_r$: ±9°<br>G: ±0.04 dB |
| 18 | Cable position | Not determined | |

is performed with the antenna measurement system described in subsection 5.3.1, a VNA and a linearly polarized reference antenna. The AUT placed in the measurement setup is shown in Figure 5.19. This results in the measured precision of the AR shown in Figure 5.20a (the standard deviation 1$s$ is illustrated by the transparent colored band) and the trueness in Figure 5.20b (pink lines). The measured precision of the realized gain is shown in Figure 5.20c and the trueness in Figure 5.20d. A simulation exercise as discussed in [2] using Computer Simulation Technology MicroWave Studio (CST MWS; https://www.cst.com/products/cstmws) has led to an outcome defined with boundary lines (minimum and maximum) to illustrate the effect of uncertainties of the material properties and manufacturing tolerances of the realized design [21, 22]. To quantify the measured results, both the trueness and precision of the AR and the realized gain have been determined.

The intrinsic settings of the simulation tool for the material properties and manufacturing tolerances are:

- software settings: *waveguide port* (WGP), finite integration technique (time-domain simulation)
- material properties: $\epsilon_r = 3.16$, $\tan\delta = 0.004$
- manufacturing tolerances.

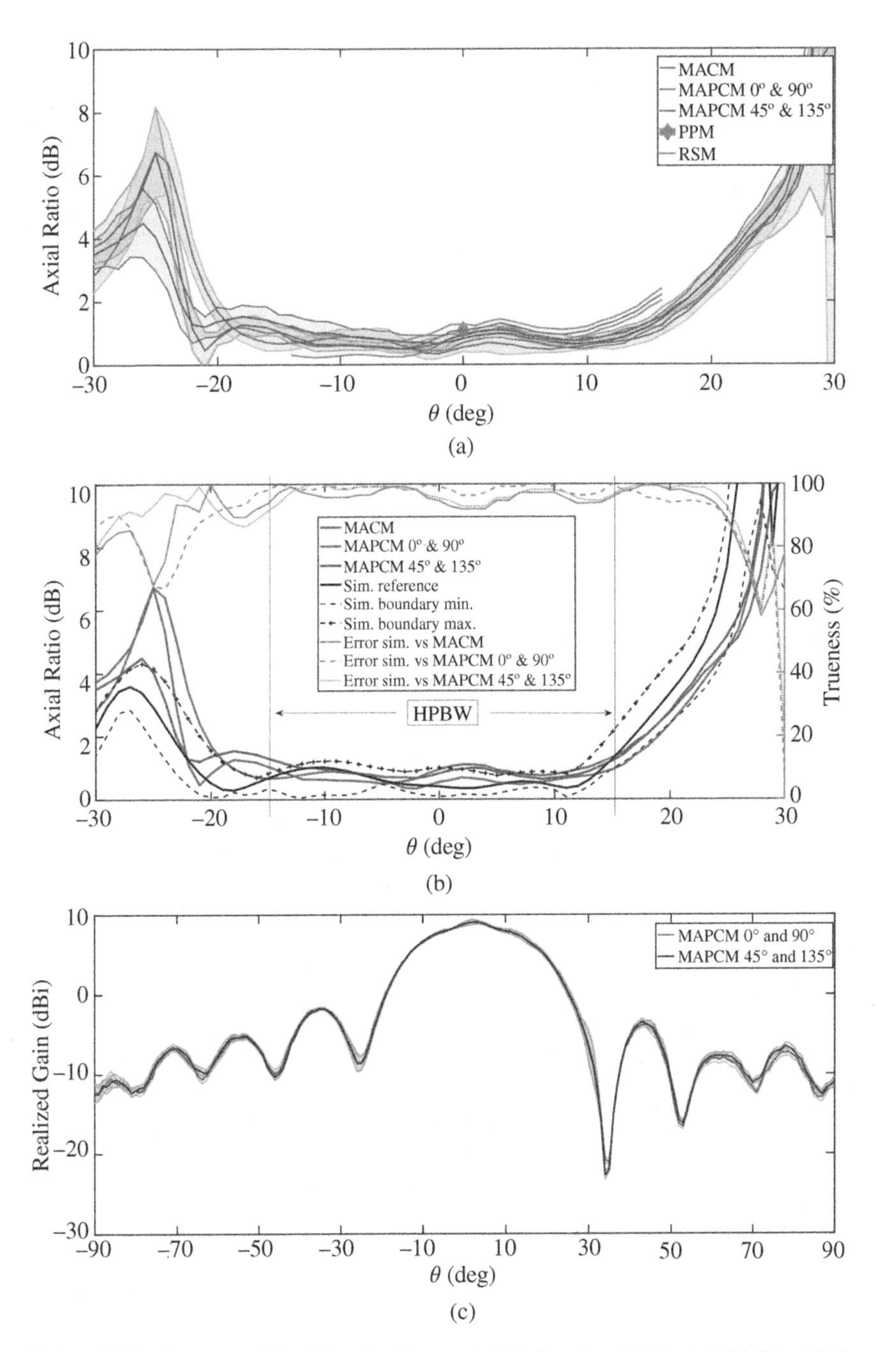


**Figure 5.20** Accuracy of the AR and gain versus theta angles obtained at 61 GHz: (a) the precision for 1*s* is indicated with the transparent colored band around the mean, (b) the trueness with the axis on the right side in percentage of the measured AR, (c) the precision, and (d) the trueness of the radiation pattern.

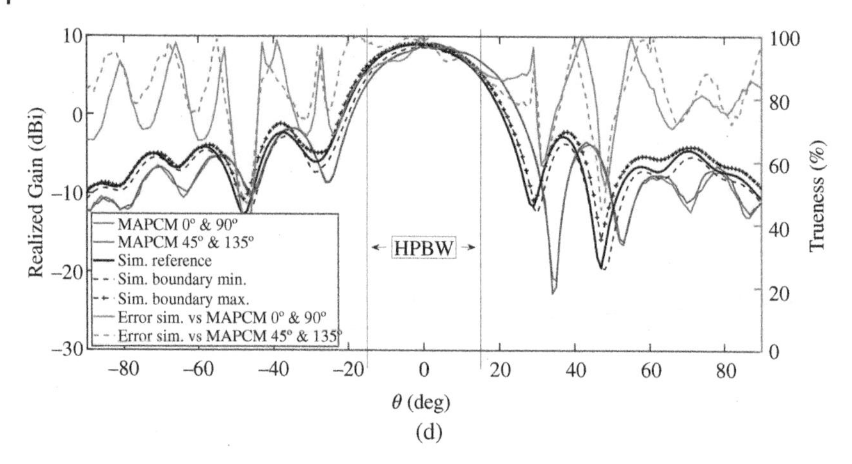


**Figure 5.20** *(Continued)*

The various polarization method measurements that have been conducted are characterized by:

- 20 repeats [14] for the MACM and PPM method and 10 repeats [14] for the RSM
- performed by two different people
- performed over a period of several weeks.

The gain and polarization loss are determined as described in [23–25]. A transmission measurement between two identical reference antennas is performed to determine the $S_{21}$ scattering coefficient. After the $S_{21}$ measurement, one of the reference antennas is replaced by the AUT. The $S_{21}$ measurement is repeated and the obtained value is compared to the gain of the reference antenna derived from the data sheet. The uncertainty of the misalignment between the reference antenna and the AUT related to the phase center is 0.25 dB. The precision of the AR and the radiation pattern is shown in Figures 5.20a and 5.20c, respectively. The corresponding trueness is shown in Figures 5.20b and 5.20d. In Figures 5.20a and 5.20c, the standard deviation is indicated by the transparent colored area. It is observed that the repeatability of the measurements is less than 0.1 dB for both AR and realized gain. In Figures 5.20b and 5.20d the trueness is expressed as a percentage, and the trueness shown in Figure 5.20b for the AR and in Figure 5.20d for the gain is determined with the simulation results of the AR and gain as a reference. The trueness of the AR is more than 90% and more than 85% for the realized gain within the HPBW. With increasing angle, larger than the HPBW, the trueness decreases by more than 20%. The deviation observed in Figure 5.20d, especially for $\theta$ angles (see Figure 5.17) larger than 30°, is caused by the load attached to port 2 (see Figure 5.19). The AR obtained by measuring $\phi_{pol} = 0°$ (E1) and $\phi_{pol} = 90°$ (E2) is different compared to the AR obtained by measuring $\phi_{pol} = 45°$ (E3) and

$\phi_{pol} = 135°$ (E4). These differences, as can be observed in Figure 5.20, are caused by interference related to the orientation of the connectors (see Figure 5.17). The effect of the uncertainties on the measured AR, tilt angle (TA) and gain (G), as shown in Table 5.4, are separately expressed as $\mu_{dB}$ for $\theta$ and $\phi$ angles of 0°. Furthermore, #11 to #16 will affect the repeatability or precision (see Figure 5.16) and the reproducibility of the measurement. Items #17 and #18 will only affect the reproducibility.

## 5.4 Over-the-air System-level Testing

The performance of a wireless system is traditionally determined by measuring the conducted characteristics of a transceiver and the antenna performance in an anechoic facility. By combining these results, the overall system performance can be estimated. However, there is a growing number of systems in which such a traditional concept cannot be used anymore:

i. *MIMO and beamforming*: *Multiple input, multiple output* (MIMO) wireless systems use spatial multiplexing and beamforming through time/space coding resulting in a much higher system capacity as compared to traditional single input, single output systems. MIMO systems adapt their settings depending on the propagation channel. Also, in phased-array radar systems *over-the-air* (OTA) testing is required to determine the overall antenna system performance and for calibration purposes [26, 27], therefore traditional experimental validation concepts do not apply anymore [28].
ii. *Integrated antennas*: In mmWave wireless systems a high level of integration is required in order to optimize overall performance and reduce cost. As a result, separate "conducted" measurements of the electronics and antenna are not possible anymore. In addition, integrated antenna systems might provide a frequency synthesizer on the chip, which could be used for test purposes as well.

In these situations OTA testing [29, 30] will provide a much better prediction of the performance of antenna systems. In OTA testing, a communication link is set up and some form of information is exchanged between the antenna system under test and the tester, for example by means of transferring data packets in order to determine the bit error rate and the packet error rate. In this case a communication tester can be used instead of a vector network analyzer. An example of OTA testing of a complete integrated antenna system is shown in Figure 5.21 (see [31]).

The total package includes a complete 60-GHz radar system implemented in a single 200 μm thick BiCMOS-chip. It consists of a transmitter and two independent receivers. The on-chip monopole antennas excite in-package antenna structures. In this way, the antennas are directly connected to the amplifiers, with the on-chip

**Figure 5.21** Complete frequency modulated continuous wave (FMCW) radar operating in the 60-GHz band [31].

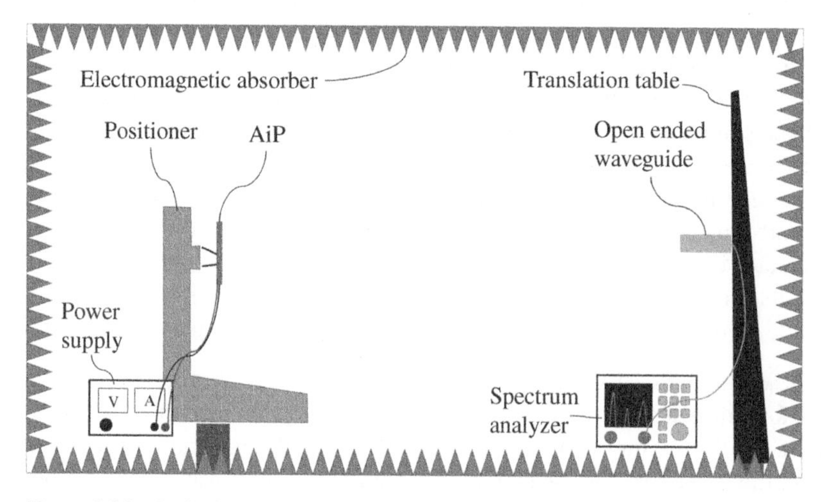


**Figure 5.22** Radiation pattern measurement setup using OTA testing. The RF signal is generated from the on-chip frequency synthesizer and radiated through the on-chip/in-package transmit antenna. A standard-gain horn antenna receives the signal, which is down-converted for further analysis in a spectrum analyzer [31].

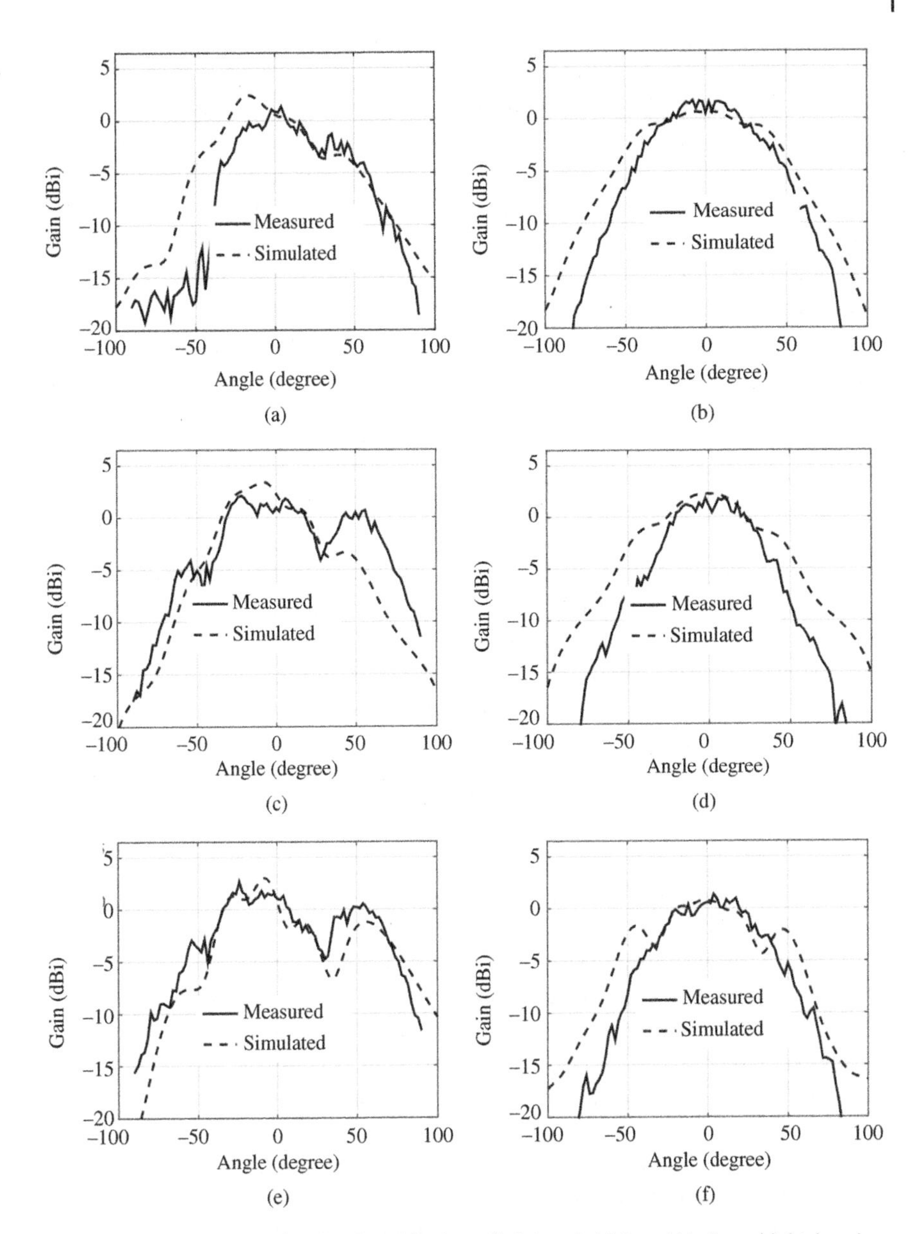


**Figure 5.23** Measured and simulated E-plane (left-hand side) and H-plane (right-hand side) radiation patterns of the transmit antenna of the single-chip FMCW radar: (a) and (b) at 57 GHz, (c) and (d) at 60 GHz, and (e) and (f) at 64 GHz [31].

electronics acting as a ground plane for the monopoles. The whole single-chip radar module is overmolded with a plastic compound, resulting in the packaged modules shown in Figure 5.21. The performance of the integrated antennas was validated by measuring the gain and radiation pattern of the transmit antenna with OTA testing in a mmWave anechoic measurement setup, as illustrated in Figure 5.22. The 60-GHz signal is generated from the on-chip frequency generator and radiated through the on-chip transmit antenna. A standard-gain horn antenna receives the signal, which is down-converted to the base band by an external mixer for further analysis with a spectrum analyzer. The measured E- and H-plane radiation patterns are shown in Figure 5.23. The output power of the power amplifier was measured from a special test chip without the on-chip transmit antenna.

## 5.5 Summary and Conclusions

In this chapter we introduced the antenna parameters which need to be validated in an antenna test facility. The design of an anechoic test facility which is optimized for testing mmWave integrated antenna concepts was presented. This facility can be used for both probe-based and connector-based measurements. Based on a test case, we discussed the accuracy and precision of AiP measurements in such a facility. Finally, OTA system-level measurements were discussed based on a test case of a single-chip FMCW radar.

## References


1 G.E. Evans, *Antenna Measurement Techniques*, Artech House, Inc., Norwood, MA, 1990.
2 A. Reniers, Q. Liu, M. Rousstia et al., Statistical analysis applied to simulating and measuring circularly-polarized millimeter-wave antennas [measurements corner]. *IEEE Antennas and Propagation Magazine*, vol. 61, no. 1, pp. 98–138, February 2019.
3 J. Kraus, *Antennas*, 2nd ed. McGraw-Hill, New-York, 1988.
4 U. Johannsen, A.B. Smolders, and A.C.F. Reniers, Measurement package for mm-wave antennas-on-chip. In *20th International Conference on Applied Electromagnetics and Communications*, pp. 1–4, September 2010.
5 U. Johannsen, A.B. Smolders, A.C.F. Reniers et al., Integrated antenna concept for millimeter-wave front-end modules in proven technologies. In *6th European Conference on Antennas and Propagation*, pp. 2560–2563, March 2012.

6 T. Zwick, C. Baks, U.R. Pfeiffer et al., Probe based MW antenna measurement setup. In *IEEE Antennas and Propagation Society Symposium*, vol. 1. pp. 747–750, June 2004.
7 J.A.G. Akkermans, R. van Dijk, and M.H.A.J. Herben, Millimeter-wave antenna measurement. In *2007 European Microwave Conference*, pp. 83–86, October 2007.
8 D. Titz, F. Ferrero, and C. Luxey, Development of a millimeter-wave measurement setup and dedicated techniques to characterize the matching and radiation performance of probe-fed antennas [measurements corner]. *IEEE Antennas and Propagation Magazine*, vol. 54, no. 4, pp. 188–203, August 2012.
9 A.C.F. Reniers and A.B. Smolders, Guidelines for millimeter-wave antenna measurements. In *IEEE Conference on Antenna Measurements Applications*, pp. 1–4, September 2018.
10 A.C.F. Reniers, A.R. van Dommele, M.D. Huang et al., Disturbing effects of microwave probe on mm-wave antenna pattern measurements. In *8th European Conference on Antennas and Propagation*, pp. 161–164, April 2014.
11 A.C.F. Reniers, A.R. van Dommele, A.B. Smolders et al., The influence of the probe connection on mm-wave antenna measurements. *IEEE Transactions on Antennas and Propagation*, vol. 63, no. 9, pp. 3819–3825, September 2015.
12 Q. Liu, A.C.F. Reniers, U. Johannsen et al., Influence of on-wafer probes in mm-wave antenna measurements. In *12th European Conference on Antennas and Propagation*, pp. 1–5, April 2018.
13 A.C.F. Reniers, Q. Liu, M.H.A.J. Herben, et al., "Review of the accuracy and precision of mm-wave antenna simulations and measurements. In *10th European Conference on Antennas and Propagation*. pp. 1–5, April 2016.
14 J. Csavina, J.A. Roberti, J.R. Taylor et al., Traceable measurements and calibration: a primer on uncertainty analysis. *Ecosphere*, vol. 8, no. 2, pp. e01 683 2017. Available: http://dx.doi.org/10.1002/ecs2.1683.
15 A. Menditto, M. Patriarca, and B. Magnusson, Understanding the meaning of accuracy, trueness and precision. *Accreditation and Quality Assurance*, vol. 12, no. 1, pp. 45–47, January 2007. Available: https://doi.org/10.1007/s00769-006-0191-z.
16 J.M. Bland and D.G. Altman, Statistics notes: Measurement error. *British Medical Journal*, vol. 313, no. 7059, p. 744, 1996. Available: https://www.bmj.com/content/313/7059/744.1.
17 A. Ludwig, The definition of cross polarization. *IEEE Transactions on Antennas and Propagation*, vol. 21, no. 1, pp. 116–119, January 1973.
18 M.W. Rousstia and M.H.A.J. Herben, High performance 60-GHz dielectric rod antenna with dual circular polarization. In *2013 European Microwave Conference*, pp. 1671–1674, October 2013.

**19** Ieee standard test procedures for antennas. ANSI/IEEE Std 149-1979, pp. 0–1, 1979.

**20** Joint Committee for Guides in Metrology, International vocabulary of metrology - basic and general concepts and associated terms (VIM). Available: https://www.bipm.org/en/publications/guides/vim.html.

**21** G.A.E. Vandenbosch, State-of-the-art in antenna software benchmarking: "Are we there yet?". *IEEE Antennas and Propagation Magazine*, vol. 56, no. 4, pp. 300–308, August 2014.

**22** B.E. Fischer, I.J. LaHaie, M.D. Huang et al., Causes of discrepancies between measurements and em simulations of millimeter-wave antennas [measurements corner]. *IEEE Antennas and Propagation Magazine*, vol. 55, no. 6, pp. 139–149, December 2013.

**23** K. Tian, Y. Zhang, and L. You, A simple method for an ultra-large electrical-size circular polarization antenna measurement. In *Proceedings of 2014 3rd Asia-Pacific Conference on Antennas and Propagation*, pp. 857–860, July 2014.

**24** H.-C. Lu and T.-H. Chu, Antenna gain and scattering measurement using reflective three-antenna method. In *IEEE Antennas and Propagation Society International Symposium. 1999 Digest.* Held in conjunction with *USNC/URSI National Radio Science Meeting* (Cat. No.99CH37010), vol. 1, pp. 374–377, July 1999.

**25** T. Milligan, Polarization loss in a link budget when using measured circular-polarization gains of antennas. *IEEE Antennas and Propagation Magazine*, vol. 38, no. 1, pp. 56–58, February 1996.

**26** G.A. Hampson and A.B. Smolders, A fast and accurate scheme for calibration of active phased-array antennas. In *IEEE Antennas and Propagation Society International Symposium Digest.* Held in conjunction with *USNC/URSI National Radio Science Meeting* (Cat. No.99CH37010), vol. 2, pp. 1040–1043, July 1999.

**27** A.B. Smolders, Design and construction of a broadband wide-scan angle phased-array antenna with 4096 radiating elements. In *Proceedings of International Symposium on Phased Array Systems and Technology*, pp. 87–92, October 1996.

**28** M.D. Foegelle, The future of mimo over-the-air testing. *IEEE Communications Magazine*, vol. 52, no. 9, pp. 134–142, September 2014.

**29** J. Mroczkowski and D. Campion (2019), Production test interface solutions for mmwave and antenna in package (AiP). Available: https://xcerra.com/production-test-interface-solutions-for-mmwave-and-aip.

**30** D. Bock and J. Damm (2019), New test methodologies for 5G wafer high-volume production. Available: https://www.formfactor.com/download/new-test-methodologies-for-5g/?wpdmdl=15616&refresh=5c9df958e90231553856856.

**31** B.B. Adela, P.T.M. van Zeijl, U. Johannsen et al., On-chip antenna integration for millimeter-wave single-chip fmcw radar, providing high efficiency and isolation. *IEEE Transactions on Antennas and Propagation*, vol. 64, no. 8, pp. 3281–3291, August 2016.

# 6

# Antenna-in-package Designs in Multilayered Low-temperature Co-fired Ceramic Platforms

*Atif Shamim and Haoran Zhang*

*Computer, Electrical and Mathematical Sciences and Engineering Division, King Abdullah University of Science & Technology (KAUST), KSA*

## 6.1 Introduction

Over the past few decades, the evolution of radio frequency (RF) wireless systems has brought convenience to our daily lives in the form of communication through smart phones, navigation by GPS, information sharing through wireless sensor networks, and safety by automotive radar. Electronics in general, and wireless systems in particular, have been shrinking in size without losing functionality. In fact, more and more functionalities have been added in later versions. Traditionally, more focus has been on the miniaturization of transistors in the integrated circuit (IC) environment and this miniaturization has been well-governed by the famous Moore's law. However, ICs (commonly known as chips) form a small part of the system. The bulk of the system is formed by other components such as interconnects and discrete components such as resistors, capacitors, inductors, etc. To miniaturize the bulk of the system, a concept known as system-on-package (SoP) was introduced a couple of decades ago [1]. It is well-known that all chips require packages for mechanical protection. Despite the fact that these packages take considerable space and cost money, they do not provide any functionality except protection from the environment. The idea behind SoP was to make the package functional by utilizing it for the implementation of system components. This can help in overall system-level miniaturization as well as in cost savings. Figure 6.1 illustrates the SoP concept where a radio frequency integrated circuit (RFIC) chip is placed inside a package cavity and connected to other passive components, such as filters, inductors, capacitors, and antennas, through bond wires, vias, and metal traces on each layer. It is also noted from Figure 6.1 that the antennas can be

*Antenna-in-Package Technology and Applications,* First Edition.
Edited by Duixian Liu and Yueping Zhang.

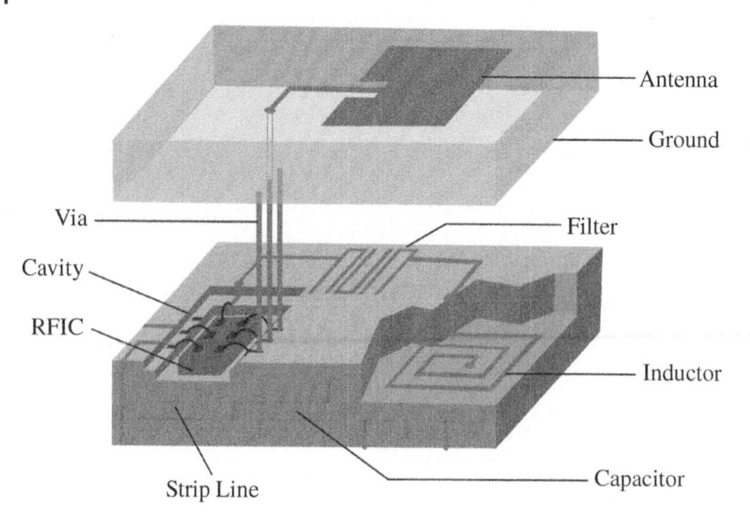


**Figure 6.1** An enlarged, 3D conceptual illustration of RF SoP to demonstrate the embedded components and high level of integration possible in the SoP concept.

implemented on the package, leading to the antenna-in-package (AiP) concept, which is the focus of this chapter.

Consistent with the SoP concept, AiP is an antenna that is realized on the package of the driving circuit [2]. There are many advantages of AiP that are beneficial for RF systems, the most prominent being the ability to realize multilayer (vertically integrated) antennas that reduce the overall size of the system. Furthermore, no separate printed circuit boards (PCBs) for antenna realization or connections to an external antenna board are required, increasing compactness and cost savings. Finally, efficient antennas can be realized in low-loss packaging technologies, particularly at millimeter-wave (mmWave) bands. These aspects are discussed in detail in the following sections.

Low-temperature co-fired ceramic (LTCC) is one of the mainstream technologies for AiP designs. This chapter focuses on AiP designs in LTCC technology. Before moving on to details of AiP design, LTCC technology is introduced in the next section.

## 6.2 LTCC Technology

LTCC is a multilayer ceramic tape system that has recently gained popularity not only as a packaging technology but also as an efficient medium to realize passive components. In this section, we first briefly introduce LTCC technology and then discuss the advantages and issues related to it. Later sections focus on the LTCC process flow in detail. Finally, we list the major LTCC material suppliers and worldwide service foundries as a quick reference guide for readers.

### 6.2.1 Introduction

LTCC has been the technology of choice for AiP designs due to its excellent electrical properties in addition to its suitable packaging properties for driving circuits. For these reasons, LTCC technology has found uses in many key application areas such as wireless communication and automotive radars. LTCC technology has also been equally popular for military and space applications. Although the majority of LTCC work has been in the research domain, commercial products, such as wireless personal area networks (WLANs) and GPS LTCC antennas from Digi-Key [3], are already available in the market, verifying the maturity of this technology. The core element behind LTCC technology is ceramic/glass dielectric tape (also known as GreenTape). Multiple GreenTapes can be stacked together and then co-fired to realize vertically integrated passive components in a compact fashion.

This technology is an attractive solution for realizing passive components for the following reasons. First, its multilayered fabrication and metallization enables the integration of RFICs and other passive components, including antennas, as a three-dimensional (3D) integrated system. With LTCC technology, up to 100 layers can be stacked up together [4], thus a great deal of compactness can be achieved with this high level of vertical integration. LTCC-based AiP designs can therefore be considerably more compact compared to antennas realized on typical planar platforms such as single-layer PCBs. Second, there is a variety of low-loss and varying dielectric constant LTCC tape systems that are beneficial for passive component implementation. Most LTCC tapes have very-low-loss tangents (in the range of $10^{-3}$), and many tapes maintain this low loss even at very high frequencies, such as Ferro A6M tape with tan $\delta < 0.002$ at 100 GHz [5]. LTCC tape dielectric constants can vary from ~4 to ~70. Some of the LTCC tape systems with high dielectric constants, such as CT765 LTCC ($\varepsilon_r = 68.7$), can greatly assist in the miniaturization of passive components. For an antenna designer, the capability of co-firing ceramics with different dielectric constants brings flexibility. In addition, LTCC technology can use low-loss conductors for the metallization steps because the multilayered LTCC tapes are laminated and co-fired at relatively low temperatures (850–900 °C), which is lower than the melting temperatures of the noble metals with excellent conductivities, such as Ag and Au. These low-loss conductors can further improve the radiation performance of the AiP. Third, LTCC has good packaging properties, such as high thermal conductivity (~3 W/mK), low thermal expansion (~4 to ~7 ppm/K), and decent hermetic sealing capability (low water absorption). These properties are important for protecting driving circuits and antennas when operating in high-power conditions or harsh environments.

### 6.2.2 LTCC Fabrication Process

In general, the LTCC fabrication process comprises six main steps: tape preparation, formation of metal traces and vias, stacking, lamination, co-firing, and post-processing. Figure 6.2 shows the generalized LTCC fabrication process. In the first step (highlighted in pink), the LTCC tape is prepared. Raw materials composed of organic binders and inorganic recrystallized glass, ceramic powder and other solid or liquid additives are mixed and then ball-milled for more than 48 h to produce the LTCC slurry [6]. The slurry is then cast and cut into LTCC tape with the required thickness. Commercial LTCC tapes are mainly fabricated by LTCC material suppliers who supply them to LTCC foundries.

The second step (highlighted in gray) is the formation of metal traces and vias. It starts with tape blanking, which is just cutting the LTCC tape into the required tape size for each layer. After tape blanking, via punching can be done by either a laser source or using a mechanical method, such as punching through a computer numerical controlled (CNC) punching machine. Laser punching enables more precise punching locations and better via quality, while the mechanical method has higher efficiency by punching multiple vias simultaneously with a high-speed multihead CNC punching machine [6]. After punching, the vias are filled with conductor paste through a screen printer. The mask of the screen printer has

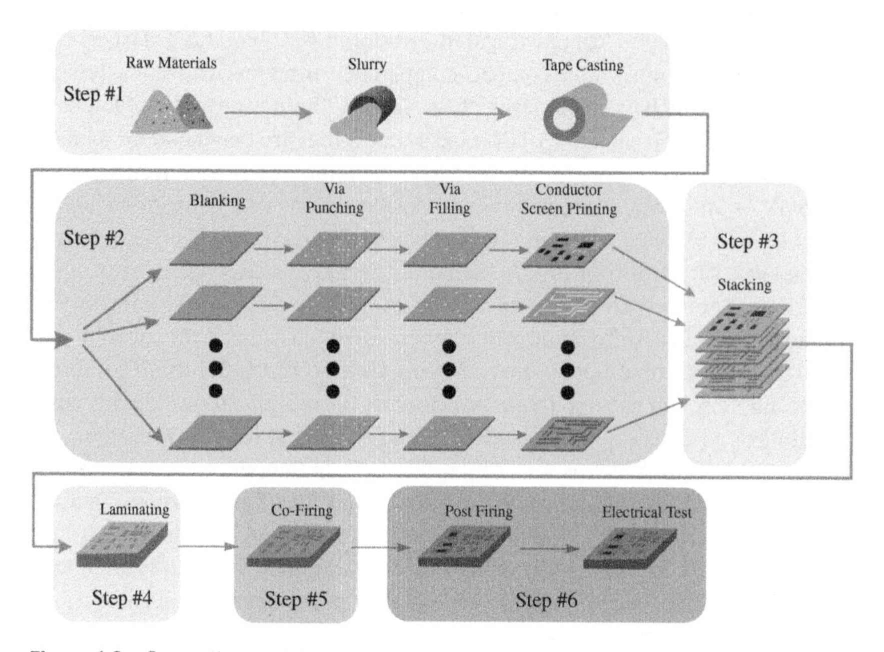


**Figure 6.2** Generalized LTCC fabrication process.

a precise map of all via holes, which is also aligned above the tape. In the screen-printing process, the conductor paste is first placed on top of the mask (screen), then the paste is pushed through the mask openings by a squeegee. After via filling, the solid conducting vias are ready to be connected to the various metal traces between the different layers. The metal traces on each layer are also made by screen printing. Several non-contact masks are used to transfer the metal patterns to all tape layers.

After the formation of metal traces and vias, all the tape layers are aligned and stacked together in order. The alignment of the stacking is critical because a small alignment error can lead to disconnection of the whole system. The stacked tapes are then laminated according to the tape specifications from the manufacturer. For example, in the lamination step, Ferro A6M tapes are pressed between two heated plates at 70 °C, 200 bar for 10 min.

After lamination, the tapes are co-fired according to a specific sintering profile (highlighted in blue in Figure 6.2). A typical sintering profile starts with temperatures slowly rising up to 450 °C at a rate of 2–5 °C per minute. The temperature then rises up to 850–900 °C with a dwell time of about 15–30 min. Once the temperature cools down to room temperature, the sintered LTCC packages are ready for post-firing and electrical testing. In post-firing, steps such as chip attachment, wire bonding, flip-chip assembly, or other back-end processes can be performed. After the post-firing step, the samples are tested to characterize their physical and electrical properties.

Despite the advanced fabrication processes, there are still some technological issues in LTCC technology. The major issue is that of tape shrinkage during the co-firing step. For example, DuPont 951 tape shows 12.7% shrinkage along the $x$–$y$ axis and 15% shrinkage along the $z$ axis after co-firing [4]. Although foundries provide shrinkage information in advance, the precision of this information depends on a particular fabrication run and its conditions. Another issue is the alignment between multiple layers. This includes the layer-to-layer alignment and conductor-to-via alignment. Good alignment ensures a robust connection between each layer, which is critical for a reliable AiP design. However, the alignment accuracy also varies from foundry to foundry and one LTCC process to another. Having said that, LTCC fabrication processes have improved considerably since their inception and the above-mentioned issues have become less severe compared to the previous decades.

### 6.2.3 LTCC Material Suppliers and Manufacturing Foundries

There are several LTCC material suppliers in the market, for example DuPont, Heraeus, Ferro, ESL, CeramTec, Kyocera, and KOA. They provide many commercial LTCC tape products. Some of the most commonly used LTCC tape systems from

**Table 6.1** Characteristics of LTCC material systems.

| Property | DuPont DP 951 | DuPont DP 943 | Heraeus CT 2000 | Heraeus HL 2000 | Ferro A6M | FSL 41110-70C |
|---|---|---|---|---|---|---|
| Dielectric constant | 7.85 | 7.5 | 9.1 | 7.3 | 5.9 | 4.3–4.7 |
| tan $\delta$ | 0.0045 | 0.001 | 0.002 | 0.0026 | 0.002 | 0.004 |
| Breakdown voltage (V/25 μm) | >1000 | >1000 | >1000 | >800 | >1000 | >1500 |
| Insulation resistance (Ω/cm) | $>10^{12}$ | $>10^{12}$ | $>10^{13}$ | $>10^{13}$ | $>10^{12}$ | $>10^{12}$ |
| Thickness-green (μm) | 50, 112, 250 | 125 | 25, 98, 250 | 131 | 125 250 | 125 |
| Thickness-fired (μm) | 42, 95, 212 | 112 | 20, 77, 200 | 87–94 | 92 185 | 105 |
| Shrinkage $x$–$y$ axis (%) | 12.7 ± 0.3 | 9.5 ± 0.3 | 10.6 ± 0.3 | 0.16–0.24 | 14.8 ± 0.2 | 13 ± 0.5 |
| Shrinkage $z$ axis (%) | 15.0 ± 0.5 | 10.3 ± 0.3 | 16.0 ± 1.5 | 32 | 27 ± 0.5 | 16 ± 1 |
| Thermal conductivity (W/mK) | 3 | 4.4 | 3 | 3 | 2 | 2.5–3 |
| CTE (ppm/K) | 5.8 | 4.5 | 5.6 | 6.1 | 7 | 6.4 |
| Flexural strength (MPa) | 320 | 230 | 210 | >200 | >170 | NA |

these manufacturers are DP 951 and DP 943 from DuPont, microwave tape A6M from Ferro, CT 2000 from Heraeus, and 41110 from ESL. The characteristics of these LTCC tapes are listed in Table 6.1. It can be seen that there are many options for low-loss LTCC tapes with different dielectric constants. In addition, most of the LTCC tapes shown in Table 6.1 have good thermal conductivity and low thermal expansion coefficient. The tape shrinkage issue can also be observed. For example, HL 2000 from Heraeus shows a $z$-axis shrinkage of 32%.

LTCC foundries are located all around the world, supplying manufacturing services that employ commercial tape systems. Some foundries have also developed their own LTCC tapes to meet their own needs regarding fabrication. As a quick reference for readers, Table 6.2 lists some of the major LTCC foundries around the world. All the foundries have their own design rules, which are available from the foundries. The design rules define the fabrication boundaries, such as minimum conductor gaps, maximum package size, maximum layer number, and so on.

Table 6.2 Worldwide LTCC foundries.

| North America | Europe | Asia |
|---|---|---|
| VisPro, ATC, C-MAC, CoorsTek, CTS, Johanson Technology | VTT, VIA Electronic, Selmic, IMST, MSE, Micro-Hybrid Electronic, EPCOS | ACX, Hitachi, Murata, Nikko, Taiyo Yuden, Yokowo |

These rules must be taken into consideration by antenna or RF system designers at the design and simulation stages.

## 6.3 LTCC-based AiP

It can be seen from section 6.2 that LTCC technology is highly suitable for AiP design due to its numerous advantages, such as vertical integration in multilayer substrates, low-loss material especially at higher frequencies, and good thermal stability. For these reasons, a number of LTCC-based AiP designs have been developed in the last two decades covering various application spectrums. LTCC technology has peculiar advantages for certain types of AiP designs. For example, LTCC technology is highly suitable for substrate-integrated waveguide (SIW)-based antennas. This is because LTCC is a multilayered technology in which conductive vias are an integral part of the fabrication process. Because SIW technology requires many via holes in the design process and can benefit from the multiple layers, LTCC technology has become a natural choice for SIW-based designs. Similarly, due to low loss at higher frequencies, LTCC is a suitable medium for mmWave antennas. Some AiP applications for harsh environments can also benefit from the good mechanical stability and thermal conductivity of LTCC tapes. Finally, LTCC, due to its excellent packaging capabilities, is highly suitable for active antennas where the antenna is realized on the package of the driving circuits. These AiPs, due to their suitability for LTCC technology, are a special focus of this section. In addition, this section reviews various gain enhancement techniques for AiP designs. Towards the end of the section, Ferrite LTCC base reconfigurable AiPs – an emerging trend in this field – are discussed.

### 6.3.1 SIW AiP

The traditional planar transmission lines, such as microstrip lines and coplanar waveguides, have the disadvantages of radiation losses and coupling to other RF components. It is well-known that metallic waveguides have the best transmission performance among most transmission lines, however they cannot be integrated in planar substrates due their bulky sizes and the fabrication complexity of the vertical metallic walls. The SIW concept was developed to overcome this problem [7]. SIW mainly consists of a dielectric with top and bottom metalized surfaces, while the two narrow walls on the sides are metallized through metallic via holes, thus realizing a waveguide in a substrate. The vias are arranged in a fashion such that electromagnetic waves are confined inside the waveguide with minimum leakage from the side walls, thus there is mutual coupling and radiation losses are minimized [8]. SIW can be integrated into antenna packages while preserving the advantages of low insertion loss and high quality associated with conventional waveguides. More importantly, the combination of multilayer integration, low dielectric loss, and easy via hole fabrication in LTCC technology makes it a perfect candidate for SIW-based AiP designs. In addition, the high permittivity of LTCC tapes can further miniaturize SIW sizes for compact AiP designs.

Recently, LTCC-based SIW structures have been widely used in AiP and feed network designs. However, most antenna designs are limited to higher frequencies (such as mmWave and above) to facilitate substrate integration. The slotted waveguide antenna array has been a popular choice, particularly for applications such as radar, navigation systems, and high power wireless systems, due to its small volume, high efficiency, and high power handling capability. Several slotted SIW antenna arrays in LTCC technology have been described in the literature [8–10]. In [9], an SIW-based slotted antenna array was demonstrated, as shown in Figure 6.3. Five layers of LTCC tapes have been used to build the SIW structure and the radiating slots have been realized on the top metalized surface. Instead of a conventional straight slot, a bent slot shape in this example provides more degrees of freedom for the antenna design.

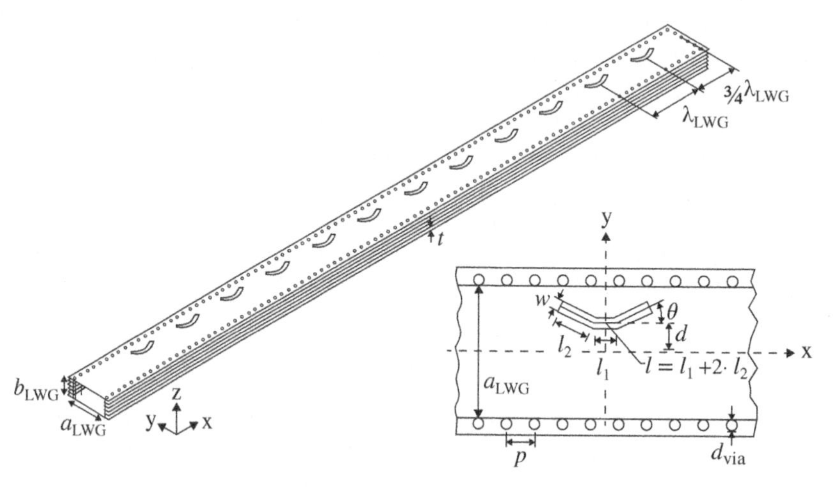


**Figure 6.3** 3D view and slot geometry of the 79-GHz SIW 12-slot array antenna in LTCC [9].

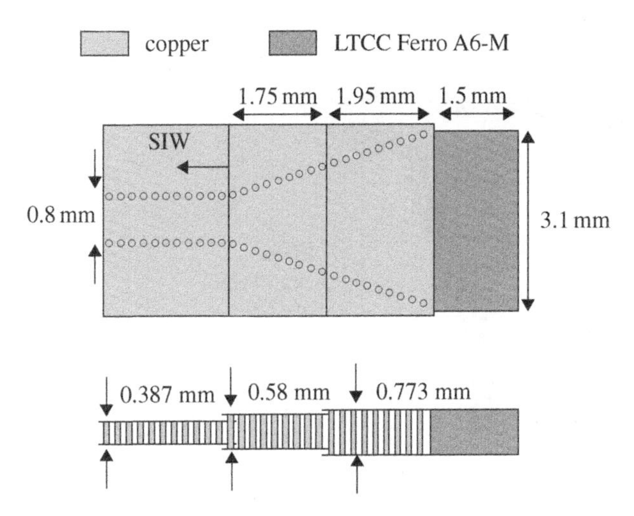


**Figure 6.4** Top and side view of the 140-GHz SIW horn antenna in LTCC [11].

Waveguide horn is another antenna type that can be realized in SIW technology. However, for ease in integration with the substrates, mostly H-plane horns have been realized. The first implementation of a horn antenna in LTCC-based SIW technology was reported in [11]. In this example, a 140 GHz H-plane horn antenna was demonstrated. This SIW horn antenna incorporates steps in the top and bottom flared walls while the side walls are realized with via (as shown in Figure 6.4). The design achieved a decent gain of 13.3 dBi in the end-fire direction. In subsequent work in this area, a vertical E-plane SIW horn antenna was demonstrated for the first time in [12], as shown in Figure 6.5. The E-plane horn became feasible due to a very high operating frequency of 300 GHz. The rectangular horn antenna was built in the LTCC substrate with an air cavity in the center, while the outer surface was formed by metal layers and via fences. It achieved 100 GHz of impedance bandwidth and 18 dBi of peak gain at 300 GHz. Despite the high operating frequency, it required 27 LTCC layers, which is not very practical due to the high costs associated with an increasing number of LTCC layers.

Other than the realization of antennas, SIW-based LTCC technology can also be used in feeding networks. Low-loss feeding networks are highly desirable for antenna arrays, especially large arrays. SIW, as a waveguide-like structure, has low radiation loss, which is a large issue with typical microstrip or coplanar waveguide (CPW)-based feed networks. By realizing SIW-based feed networks in LTCC technology, the dielectric and conductor losses are also minimized due to excellent dielectric and conductive properties inherent in the technology. In addition, the multilayered integration possibility of LTCC technology to make 3D feeding networks results in extremely compact sizes, which resolves another major issue for

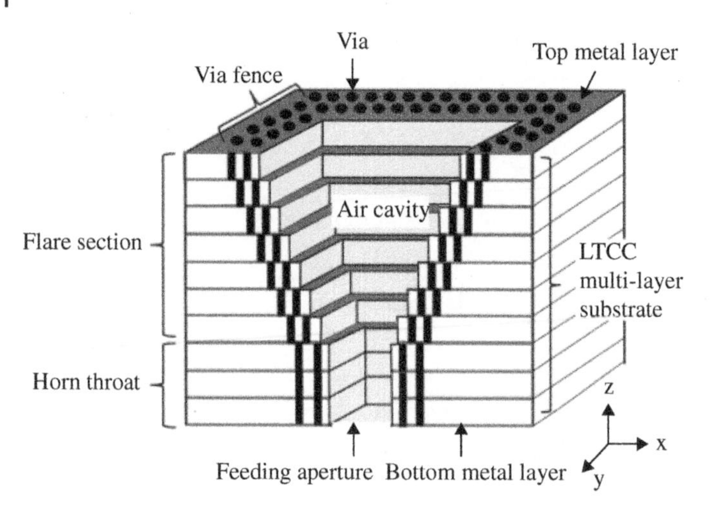


**Figure 6.5** Cross-section view of the 300-GHz SIW horn antenna in LTCC [12].

typical large arrays. Many examples of SIW-based feeding networks in LTCC have been reported in the literature [13–15]. For example, in [13], a 4 × 4 slot antenna array at 60 GHz is fed by a low-loss SIW-based feeding network. The feeding network occupies the bottom seven layers as a folded butler matrix for beamforming, and layers 8 to 15 as a power divider, while the antenna array is realized on the top five layers, as shown in Figure 6.6. This antenna achieves a high gain of 15.3 dBi at 60 GHz and a wide bandwidth of 4.2 GHz. This LTCC antenna array demonstrates a size reduction of approximately 80% compared to a similar 4 × 4 planar design reported in [16].

### 6.3.2 mmWave AiP

With the rising interest in mmWave wireless communication systems, such as 5G communication systems, automotive radar, and imaging systems, high gain and high radiation efficiency mmWave antennas have become very important [17]. The mmWave is the frequency band from 30 to 300 GHz and corresponds to wavelengths from 10 to 1 mm. In the mmWave band, the antenna is small enough to be efficiently integrated into the package. Hence, considerable AiP work has been done for mmWave applications. LTCC technology is considered a highly suitable platform for mmWave AiP designs due to the benefits mentioned in section 6.2, particularly its low-loss aspect at these frequencies. In this subsection, LTCC-based mmWave AiPs are discussed and examples of different mmWave frequency bands and applications are demonstrated.

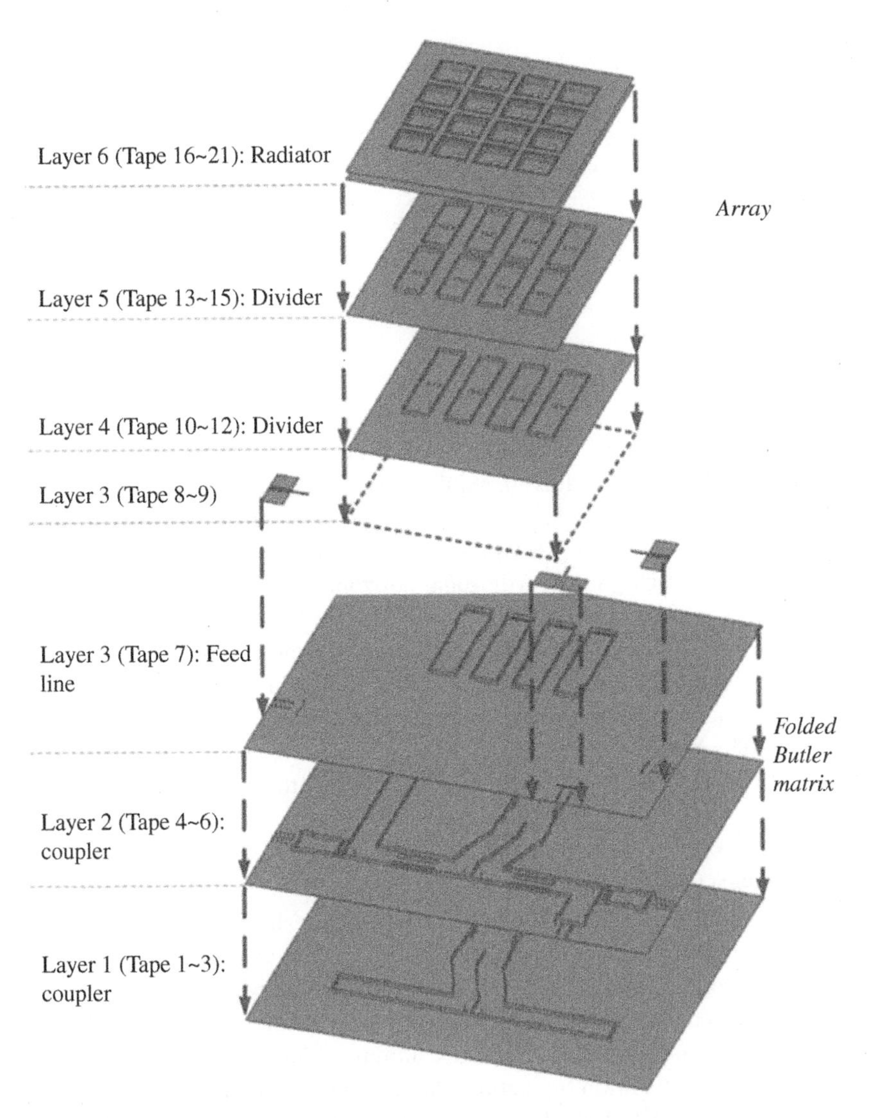


**Figure 6.6** A 60-GHz 4 × 4 antenna array with multilayered SIW-based feeding network [13].

#### 6.3.2.1 5G AiP

The latest 5G mobile communication systems have attracted equal interest from industry and the research community. According to the 3rd Generation Partnership Project (3GPP), there are several mmWave frequency bands that have been assigned for 5G mobile communications, such as 28, 37, 39, and 47 GHz bands [18].

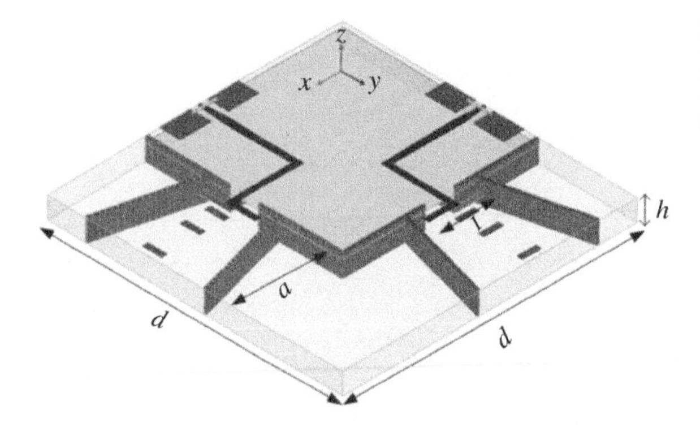


**Figure 6.7** Multilayered 3D structure of the 38-GHz LTCC quasi-Yagi antenna [20].

The use of these mmWave bands for 5G systems can enable high data rates, and antennas with large bandwidths are required to support these high data rates. Several LTCC-based 5G AiP designs have been demonstrated in the literature [19–23]. For example, in [20], two 38-GHz quasi-Yagi antennas were implemented in one LTCC substrate to achieve two orthogonal horizontal polarizations, as shown in Figure 6.7. In order to improve the gain and isolation between the two antennas, the vertical via walls were smartly designed around the radiating elements by making good use of the LTCC multilayered integration capability. This LTCC-based 5G quasi-Yagi antenna achieves a maximum gain of 6 dBi with over 80% radiation efficiency. The horizontally polarized antenna demonstrates a wide impedance bandwidth of 11.5 GHz (30.5–42 GHz), which covers the entire 37 and 39 GHz bands.

#### 6.3.2.2 WPAN (60-GHz) AiP

The unlicensed 60-GHz frequency band can be used for short-range wireless communication systems with multi-gigabit data rates. One of the most anticipated applications for this band is the high data rate wireless personal area network (WPAN) due to the large bandwidth (~7 GHz) dedicated for this band [24]. Most of these applications are for short-range communication due to the large loss of approximately15–30 dB/km from atmospheric absorption in this band [25]. Considerable work on AiP designs in the 60-GHz band has been presented in the literature [26–30]. In [27], a 60-GHz $4 \times 6$ stacked circular patch antenna array was demonstrated. As shown in Figure 6.8a,b, this AiP design consists of ten LTCC layers, where the antenna array is realized on the top three layers and the other layers are used for the array feeding network. In addition, an RFIC chip is attached to the LTCC substrate through flip-chip technology to feed the antenna array with different phase distribution, thus the beam-steering ranges of 45° in the both E-plane and the H-plane are achieved. A maximum gain of more than 14.5 dBi has been demonstrated at the boresight. Moreover, as shown in Figure 6.8c, this design

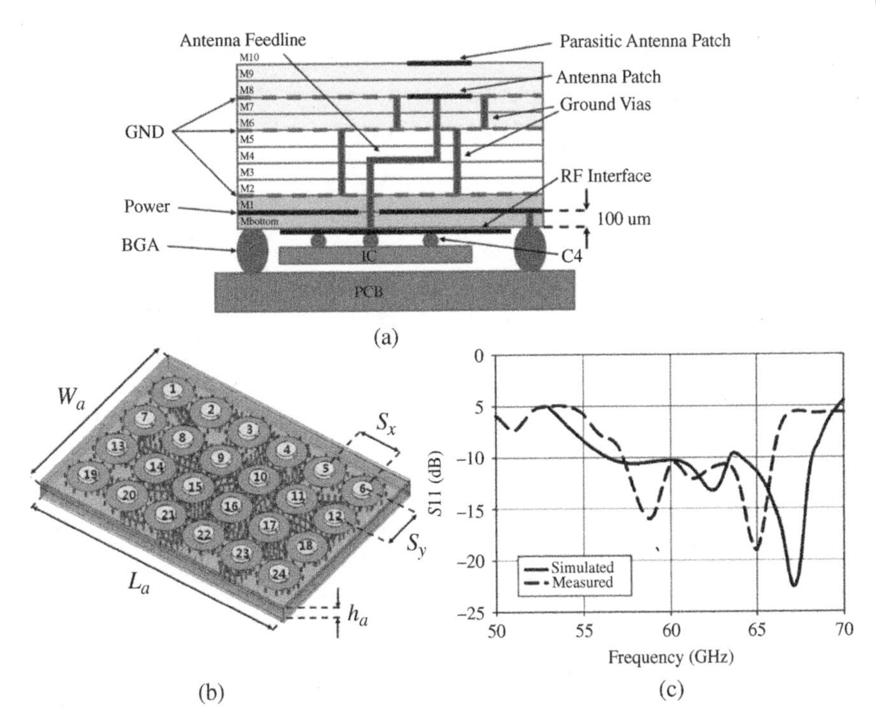


**Figure 6.8** A 60-GHz 4 × 6 LTCC-based patch antenna array: (a) side view of the AiP stack-up, (b) simulation model, and (c) antenna array reflection coefficient [27].

realizes a wide bandwidth of 9 GHz from 57 to 66 GHz, which makes it suitable for high data rate WPAN applications.

#### 6.3.2.3 Automotive Radar (79-GHz) AiP

Another popular frequency band for AiP work is the 79-GHz band (76–81 GHz), which has been allocated for automotive radar applications [31]. Antenna designs play an important role in functions such as collision avoidance and blind spot detection to improve driving efficiency and safety. LTCC technology has been widely used in the design of high-resolution automotive radar systems [9, 32–35] because LTCC materials have excellent thermal and packaging properties in addition to good electrical properties, thus it is a good medium for high-performance AiP designs, particularly for harsh environments. In [32], ceramic-filled cavity resonators were designed using LTCC technology as an antenna array for 79-GHz automotive applications. As shown in Figure 6.9, this antenna consists of eight ceramic-filled cavities which are excited by transverse slots on top of the laminated waveguide (between the B–B′ and C–C′ layers). The laminated waveguide is fed by the WR-12 waveguide at the bottom and divides the RF power into

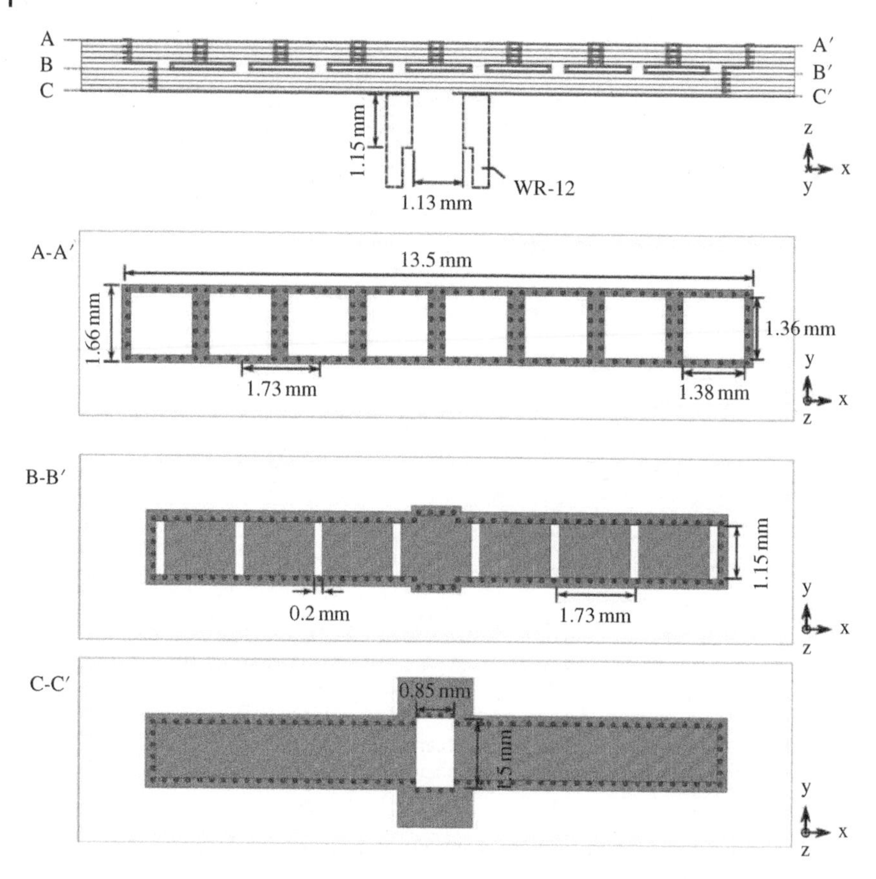


**Figure 6.9** Configuration of the 79-GHz cavity resonator antenna array in LTCC [32].

eight parts to feed each cavity resonator. This prototype shows a decent radiation performance with a gain of 13.1 dBi and a radiation efficiency of 65% at 79 GHz. More importantly, this antenna presents a stable E-plane radiation pattern with a 13° 3-dB beamwidth over the whole operating bandwidth from 76 to 81 GHz, while the side-lobe level stays 11 dB below the main radiation for the entire band. The features of high gain, sharp beam, and low-level side lobe make this antenna a perfect candidate for automotive radar applications. This compact design of ceramic-filled resonators has made good use of high permittivity LTCC tapes and their capacity for multilayer integration.

#### 6.3.2.4 Imaging and Radar (94-GHz) AiP

The 94-GHz band has also attracted research interest in the past decade because of the small wavelength and relatively low atmosphere absorption in this

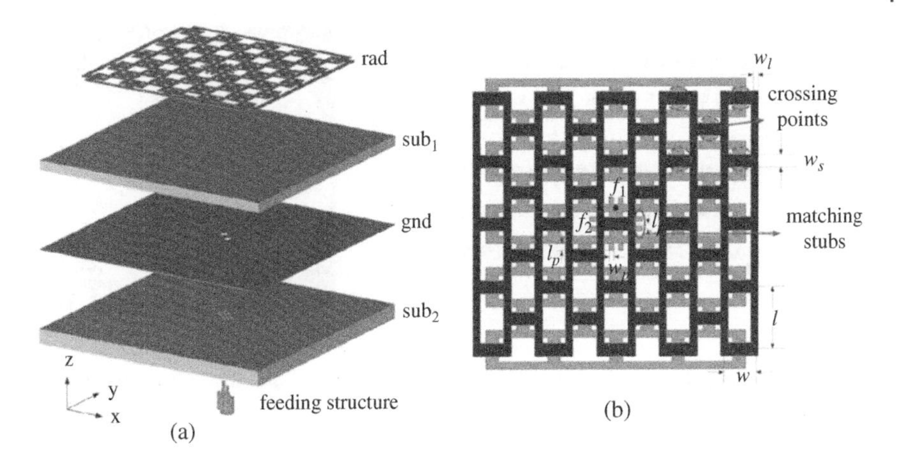


**Figure 6.10** A 94-GHz microstrip mesh array antenna: (a) 3D exploded view and (b) radiator (top view) [38].

frequency band. According to the Federal Communications Commission (FCC), the 94-GHz band (92–95 GHz) has been allocated for imaging and military radar and point-to-point communication systems [36]. For all these applications, high gain and compact designs are desirable features from the antenna design perspective. Several examples of 94-GHz LTCC AiPs have been demonstrated in the literature [15, 37, 38]. For example, in [38], a dual-polarized mesh array antenna was designed in LTCC technology for 94-GHz band applications. As shown in Figure 6.10, this antenna consists of two grid arrays in two orthogonal directions with specially designed matching stubs to compensate the input impedance alteration, which is caused by the direct coupling of the complementary array of another polarization [38]. This mesh array antenna achieves a high gain of 13.3 dBi at 93.6 GHz, a 10 dB impedance bandwidth of 6.5 GHz, and a 3-dB gain bandwidth of 3.0 GHz, which can cover the whole 94-GHz band.

#### 6.3.2.5 Sub-THz (Above-100-GHz) AiP

The demand for massive data transfer has driven the rapid development of high data rate wireless communication systems in the above-100-GHz mmWave bands or sub-terahertz ranges, such as the D band (110–170 GHz), 270-GHz band, and 300-GHz band, in which large bandwidth can be easily achieved. Recently, researchers have shown many high data rate wireless systems above 100 GHz employing LTCC AiP designs [12, 39–41]. For example, a 300-GHz step-profiled horn antenna realized in 27 LTCC layers achieved a high gain of around 18 dBi over 72-GHz bandwidth [12]. Other than high data rate wireless communications, the sub-terahertz frequency band is also attractive for imaging in which a sharp radiation beam is required for better resolution. In [40], a radial waveguide is

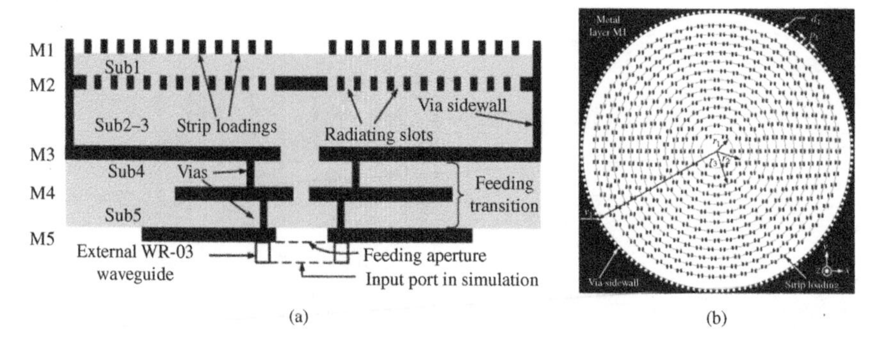


**Figure 6.11** Radial line slotted antenna array in LTCC: (a) side view and (b) top view [40].

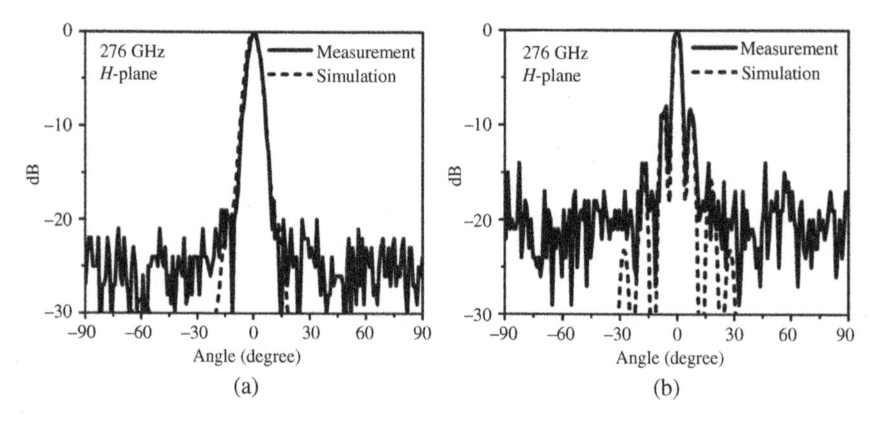


**Figure 6.12** Measured and simulated radiation patterns at 276 GHz: (a) H-plane and (b) E-plane [40].

formed by utilizing vias as the side wall in LTCC layers 2 and 3, as shown in Figure 6.11. The radiating slots are implemented on metal layer 2, loaded by metal strips on metal layer 1. Through this design of a radial line slotted antenna array, a high gain of 27.6 dBi is achieved at 275.2 GHz with a wide 3-dB gain bandwidth of 9.8 GHz. In addition, as shown in Figure 6.12, the antenna demonstrates a half-power beam width of around 7° in the H-plane and around 4° in the E-plane.

### 6.3.3 Active Antenna in LTCC

In the previous sections, the emphasis has been just on the AiP. However, another important aspect for AiP is its interface or integration with the circuits under the package. The true benefit of AiP is that the packaged circuits can be integrated with antennas in the same package for a compact system. Typically, the antennas that are integrated and characterized with the driving circuits are known as active antennas. Active antennas require additional routings for DC bias, which can be efficiently realized in multilayered LTCC technology. LTCC technology is highly suitable for active antennas as it can provide excellent packaging characteristics (such as hermetic sealing, good thermal stability, and low thermal expansion) for RF chips as well as an efficient platform for high-performance integrated antennas.

In terms of chip placement methods, there are two main approaches in LTCC technology. The chips can be mounted either on the top or the bottom of the LTCC substrate (surface mounted) or they can be completely buried in the LTCC cavities. Examples of both these cases are described and discussed later in this section. In addition to chip placement methods, there are two main chip attachment techniques that have been widely adapted in LTCC technology. First is the traditional wire bonding approach, in which the bond pads on the chip are connected to the bond pads on the package through bond wires. Second is the flip-chip approach, in which the bond pads on the chip are placed on specialized solder balls or bumps, so no wires are required for connection.

For the surface mounting method in LTCC, the flip-chip technique has mostly been used to attach the RFIC chips to the surface of the LTCC substrate [28, 42, 43]. For example, in [42], a 2 × 2 circularly polarized patch antenna array was integrated with a 60-GHz Doppler complementary metal oxide semiconductor (CMOS) radar chip in a LTCC substrate, as shown in Figure 6.13a. The CMOS radar chip is mounted on the bottom of the LTCC substrate through flip-chip technology. In this LTCC substrate, a through-substrate via is smartly built as a coaxial cable with the help of six grounded multilayered vias to realize the vertical connection between the chip and the radiating elements on the top side of the

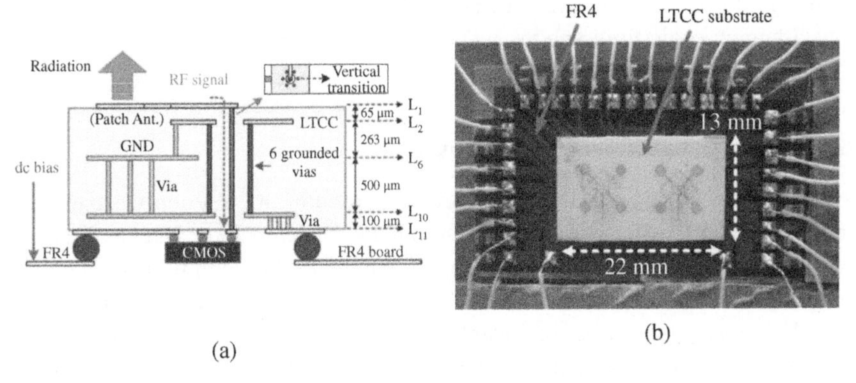


**Figure 6.13** A 60-GHz active antenna in an LTCC package: (a) layer profile and (b) fabricated sample [42].

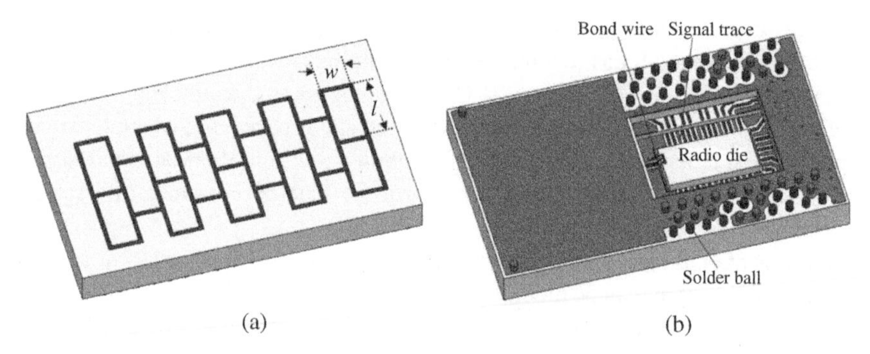


**Figure 6.14** Grid array antenna in LTCC: (a) top view and (b) bottom view [45].

substrate. Figure 6.13b shows the LTCC active antenna array package mounted on a regular FR4 board. In the case of surface-mounted chips, the complete package is provided by a combination of LTCC and PCB technology.

In the other approach, cavity structures have been utilized in LTCC for chip placement [26, 44–49]. By cutting the desired size of hole in each layer and then aligning these holes in multiple layers, cavity structures can be easily realized in LTCC technology. More importantly, the cavity structures can be designed to have a step shape for different layers, thus providing more space for bond wire connections to different layers, as shown in the multistep cavity in Figure 6.14. Placing the chips in buried cavities, instead of surface mounting, provides complete packaging to the chip. In [45], a 60-GHz grid array antenna was integrated with a receiver chip in LTCC. As shown in Figure 6.14, the chip is placed in a multitier cavity and connected to the metal traces in LTCC through bond wires. One of the important aspects of this example is that the antenna array and the receiver chip are designed on the opposite sides of the LTCC substrate and separated by a ground plane to minimize the interference between the antenna and the chip, which is a well-known co-design issue for active antenna designs. This design achieves a high gain of 13.5 dBi at 60 GHz.

### 6.3.4 Gain Enhancement Techniques in LTCC

With unprecedented proliferation in wireless communication systems, expanding to mmWave and terahertz bands, there is increasing demand for antenna designs with high gain and high radiation efficiency to enhance the communication range or compensate for high path loss in higher frequency bands. In addition, the signal-to-noise ratio can be significantly improved by the use of high gain antennas in wireless communication systems. Although LTCC is a low-loss medium that is suitable for efficient antenna design, gain improvements are still required in many cases, particularly in designs with relatively higher dielectric constants.

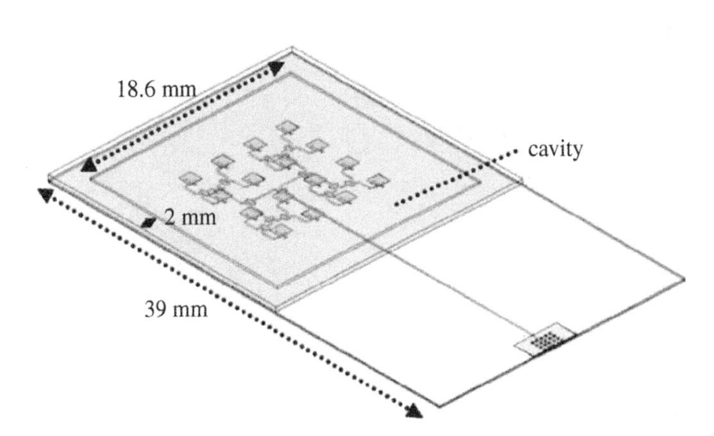


**Figure 6.15** A 4 × 4 patch antenna array in LTCC with embedded cavity [53].

Several gain enhancement techniques have been investigated to achieve high gain antennas in LTCC technology. One of the most common methods is to introduce open cavities around the radiating elements [50]. Compatible with the LTCC process, the open cavities can reduce losses caused by dielectric substrates to achieve better gains. In addition, this also helps with issues related to excitation of surface waves, which are common in typical high permittivity and thick LTCC substrates. Other than the open cavities, embedded cavities have also been demonstrated to improve gain performance [51–53]. For example, in [53], a 60-GHz antenna array is designed on a cavity-embedded substrate, which achieves 2.8-dB gain enhancement without changing the antenna array structure, as shown in Figure 6.15. However, the embedded cavities require an additional fabrication process in which, typically, a sacrificial volume material is inserted in the cavity before lamination and requires removal from a special escaping orifice during co-firing [54]. The additional process can increase the complexity and cost of fabrication.

Another approach for gain enhancement is to focus the antenna radiation beam by using 3D structures in a multilayered LTCC, such as a superstrate layer or a lens structure [55, 56]. For example, an LTCC-based Fresnel lens structure was demonstrated to enhance the antenna gain in [56]. As shown in Figure 6.16, the antenna array was realized in a standard LTCC tape system (CT 707, $\varepsilon_r = 6.4$), whereas to make a compact Fresnel lens, a very high permittivity LTCC substrate (CT765, $\varepsilon_r = 68.7$) was used. The 1 × 4 fractal patch array is fed through the apertures in the ground plane. The fractal patch is used for bandwidth improvement. The Fresnel lens is realized through four layers of planar grooved plates and provides 6 dB of gain enhancement. This design eventually achieves a gain of 15 dBi at 24 GHz, while maintaining a high level of compactness compared to similar designs [57, 58].

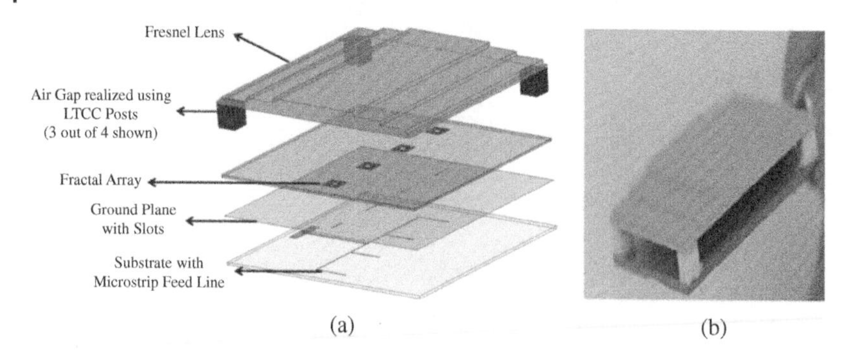


**Figure 6.16** Geometry of a 1 × 4 fractal antenna array with integrated Fresnel lens: (a) schematic view and (b) fabricated sample [56].

In addition to gain enhancement techniques, there has been work focusing on surface wave suppression in LTCC. As mentioned above, this issue is relevant to LTCC AiP designs because LTCC substrates have relatively high dielectric constants and can be quite thick in some designs. One effective surface wave suppression method is to use the electromagnetic band gap (EBG) structure, which is perfectly compatible with multilayered LTCC fabrication [59–61]. The soft surface structure is one class of EBG structures that behaves as a perfect electrical conductor for the transverse electric (TE) mode and as a perfect magnetic conductor for the transverse magnetic (TM) mode, therefore the soft surface has stopband characteristics along the soft direction, as shown in Figure 6.17a [61]. As can be seen in Figure 6.17b, the soft surfaces are implemented between the LTCC-based patch antenna array to realize a stopband for suppressing the surface waves. Although

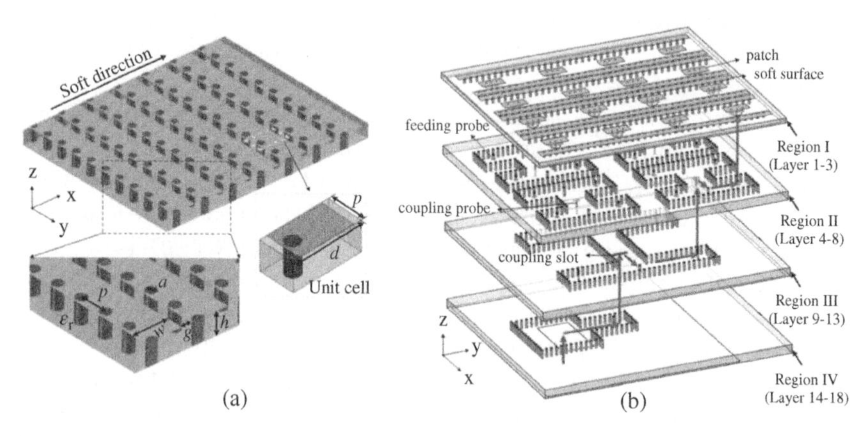


**Figure 6.17** (a) Geometry of the via-loaded strip soft surface structure and (b) 4 × 4 patch antenna array with soft surfaces [61].

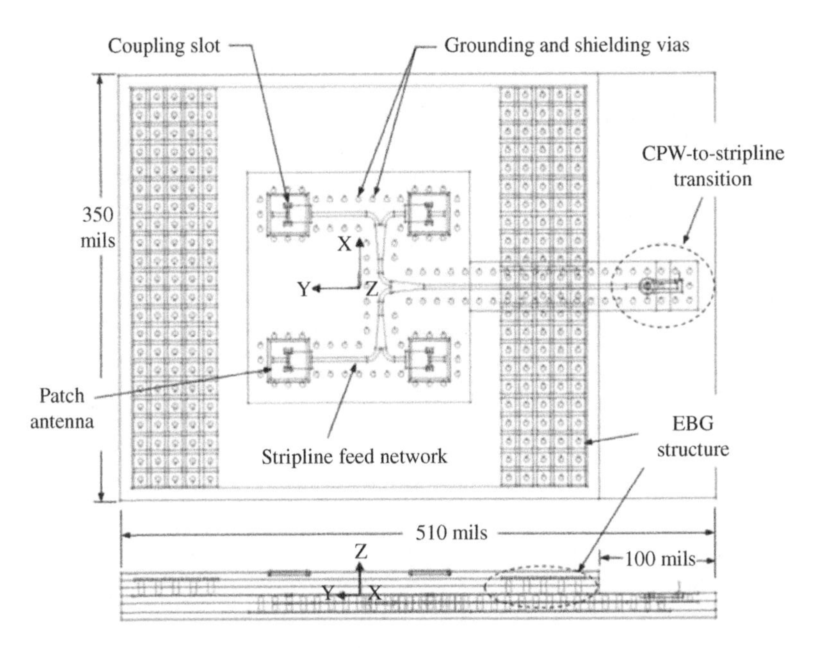


**Figure 6.18** Sievenpiper EBG embedded 60-GHz 2 × 2 antenna array [60].

the soft surface structures have been proved to be useful for surface wave suppression, the soft surface, as an anisotropic structure, only works along the soft direction. In contrast, the Sievenpiper EBG structure, as an isotropic structure, is able to suppress surface waves in all directions. Moreover, periodic and interlayered EBG structures are easily implemented in LTCC technology. In [60], a 60-GHz 2 × 2 patch antenna array is designed and fabricated in the LTCC substrate, where two five-row Sievenpiper EBG structures have been embedded on the two sides to suppress the surface waves. A 4-dB gain enhancement and 8-dB side lobe improvement are achieved by the implementation of Sievenpiper EBG structures, as shown in Figure 6.18.

### 6.3.5 Ferrite LTCC-based Antenna

In the above sections, many examples of functional packages have been shown where the LTCC package acts as a substrate for the antenna and provides protection to the circuits. The next question is whether the components realized in the package, such as AiP, can be controlled. In other words, in addition to adding functionality to the package, can we also add control to it? The answer is yes, by mixing controllable materials with the ceramic powder control can be added to the LTCC package. One example is mixing iron oxide with the ceramic powder,

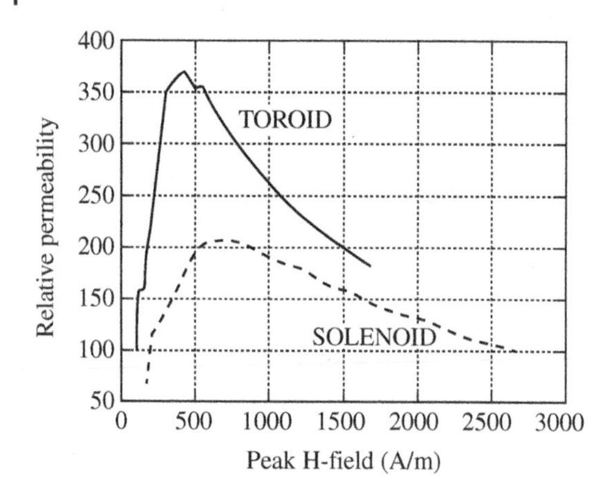


**Figure 6.19** Measured relative linear permeability of ESL 40012 [62].

which has resulted in ferrite LTCC tape systems. Ferrite materials are magnetic dielectrics whose permeability can be altered dynamically by applying magnetic fields across them. For example, according to [62], the commercially available ferrite ESL 40012 shows a relative permeability tuning range from 100 to 370 when magnetic fields ranging from 100 to 2600 A/m are applied across it, as shown in Figure 6.19. The capability of tuning the permeability tensor through an applied magnetic field means that any RF component realized on this tape system can be tuned or reconfigured. For AiP, this reconfigurability can be in resonant frequency [63], radiation pattern [10], or even in polarization [64]. Such tunable and reconfigurable antennas are highly desirable for modern wireless communication systems as they can provide diversity and flexibility in the use of available spectra as well as radiation space.

Although the ferrite LTCC is a relatively new concept for AiP designs, some examples of tunable and reconfigurable AiP designs in this tape system can be found in the literature. In one of the first demonstrations of a tunable and reconfigurable AiP in ferrite LTCC technology [63], a 12-GHz microstrip patch antenna has been shown to vary its resonant frequency by 600 MHz when biased through an embedded winding, as shown in Figure 6.20. Two kinds of embedded windings, a solenoid and a toroid, were explored in that work. Although the antenna and

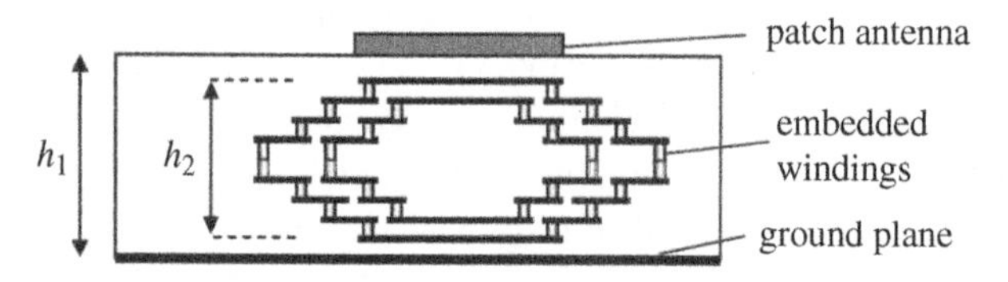


**Figure 6.20** Cross sectional view of the tunable antenna module [63].

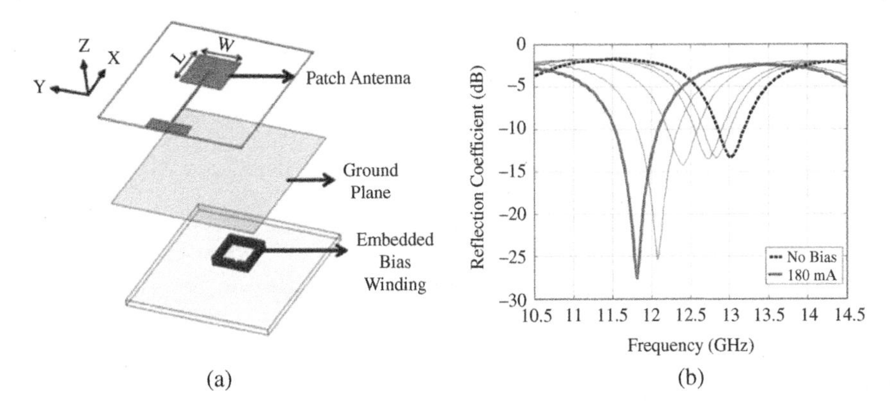


**Figure 6.21** Tunable antenna on a ferrite LTCC substrate with embedded windings: (a) schematic view and (b) measured frequency response with bias from 0 to 180 mA (from right to left) [65].

the winding have both been implemented monolithically in the multilayer ferrite LTCC process, one issue was the interference of the RF fields with the embedded windings. In subsequent work [65], this problem has been resolved by moving the antenna ground plane above the embedded windings to restrict the RF fields to the ground plane while the DC magnetic fields from the windings can still influence the antenna substrate layers to achieve the required tuning. This is conceptually shown in Figure 6.21a. This antenna demonstrates a tunability of 1.25 GHz in resonance frequency, varying from 11.8 to 13.05 GHz when the bias current changes from 180 to 0 mA, as shown in Figure 6.21b. In another interesting design, the bias windings have been optimized to act as a helical antenna as well [66]. This means that the helical antenna/winding is not only providing the bias for tunability but is also the radiating element. Other than tunable AiP design, ferrite LTCC technology has also been widely used for other RF components, such as filter and inductor designs [67–69].

Ferrite LTCC has also been used for antenna designs with radiation pattern reconfigurability, in which the most important component is the embedded tunable phase shifter [70, 71]. Only a handful of radiation pattern reconfigurable antennas with an embedded phase shifter in ferrite LTCC have been demonstrated in the literature [10, 72]. The concept of a phase shifter, which comprises an SIW section and embedded windings, feeding a slotted SIW 4 × 4 phase array antenna has been demonstrated in [10] and is shown in Figure 6.22. The insertion phase of the SIW can be tuned according to the strength of the magnetic field applied to the ferrite substrate. By applying different DC current values to the embedded windings, magnetic fields with tunable strengths are generated, leading to the change of SIW substrate permeability. As a result, signals with different phases

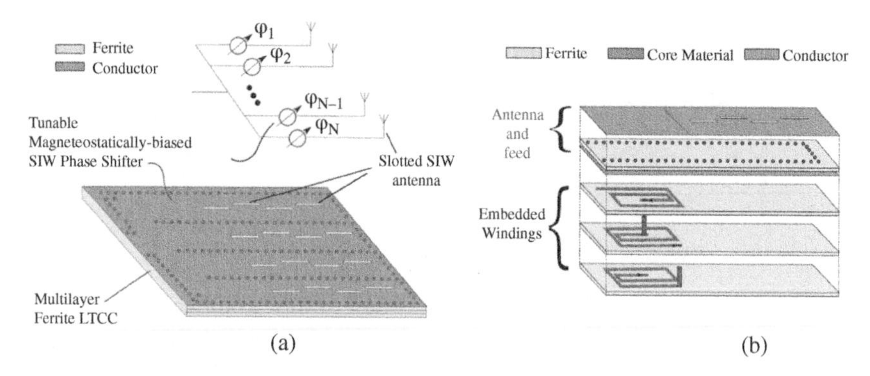


**Figure 6.22** (a) Conceptual sketch of a monolithic slotted SIW phased antenna array in ferrite LTCC and (b) tunable SIW phase shifter with embedded windings [10].

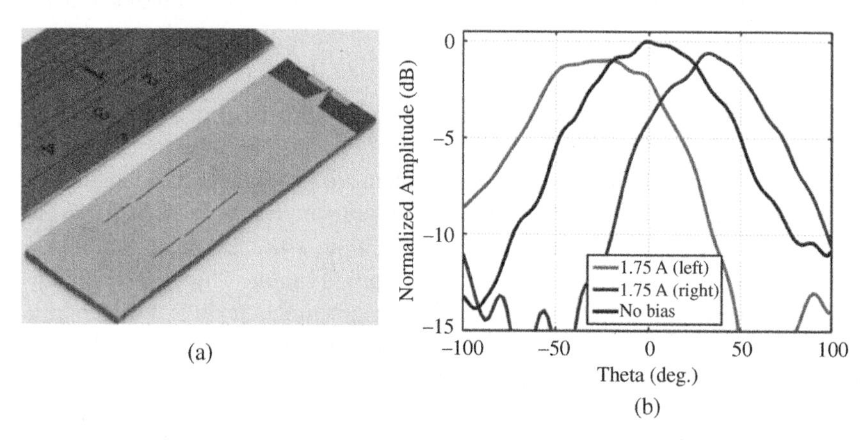


**Figure 6.23** (a) Fabricated 2 × 3 antenna array prototype and (b) measured radiation pattern scanning range [10].

are fed to each antenna subarray. Eventually, the proposed antenna array achieves the beam-scanning function. Figure 6.23b shows the experimental verification of a 2 × 3 Ku-band slotted SIW antenna array. It shows a beam scanning range of ±19° with embedded winding and an enhanced beam scanning of ±28° with an external winding. This design verifies that monolithic phased arrays can be realized in ferrite LTCC where the phase can be controlled through the applied magnetic field. In a subsequent work, a half-mode shifter has been demonstrated for antenna radiation pattern reconfigurability, which does not need two solenoids for its operation [72].

Although ferrite LTCCs show many good features for tunable antennas, there are some issues which need to be resolved for future advancement in this area. One of the main drawbacks is that ferrite LTCC tapes have relatively higher loss (tan $\delta > 0.001$) compared to conventional low-loss LTCC tape. In addition, the existing ferrite LTCC tapes have decent operation up to ~20 GHz of frequency, as beyond that substantial loss restricts their use. One of the suggested solutions is to design the antennas on conventional low-loss LTCC tapes and design tunable switches, such as a phase shifter, on the ferrite LTCC tapes. This solution ensures the high radiation performance of the antenna and good tunability of switches, but it requires the technique of LTCC fabrication with multiple tape types. In addition, the resistance of the embedded windings needs to be reduced to mitigate its self-heating effect if the windings are integrated for biasing [10].

## 6.4 Challenges and Upcoming Trends in LTCC AiP

For the foreseeable future, LTCC technology will continue to be one of the most attractive approaches to realize AiP designs due to all its above-mentioned advantages. However, several challenges still remain for widespread LTCC-based AiPs use. One of the main challenges is the improvement of LTCC fabrication resolution and repeatability. The AiP designs, especially for mmWave frequencies and beyond, have very small feature sizes because of the small wavelength, so fabrication accuracy becomes critical for high-frequency AiP designs. However, LTCC technology has limited fabrication resolution. For example, according to VTT foundry design guidelines, it has minimum conductor line width of 100 μm and spacing of 150 μm [73]. However, resolution on the order of tens of microns (or even lower) is required for AiP designs at sub-terahertz frequencies and beyond, so the LTCC fabrication resolution needs to be improved. There are other issues in the fabrication process that can affect fabrication accuracy. The tape shrinkage issue in LTCC package fabrication is a major concern for fabrication accuracy. Although this aspect has been improved over time, some unpredictability remains. Tape shrinkage occurs during sintering, which causes variation in design dimensions, therefore the tape shrinkage issue should be taken into account during the design stage to obtain the desired design dimensions. As well as tape shrinkage, layer-to-layer alignment is another important issue related to fabrication accuracy. Good alignment ensures a robust connection between each layer and is critical for a reliable AiP design. Researchers are currently exploring methods to realize fine line printing for more accurate fabrication, such as high-precision screen printing, inkjet printing, and gravure offset printing.

With the rapid growth in LTCC-based AiP designs, there is a need for LTCC tape systems with different functionalities. One development trend is the implementation of multi-type LTCC tape systems, referring to integrating different types of LTCC tapes, including regular, ferrite, ferroelectric, and high-permittivity LTCC tapes, in one single package to achieve some special functions, as shown in

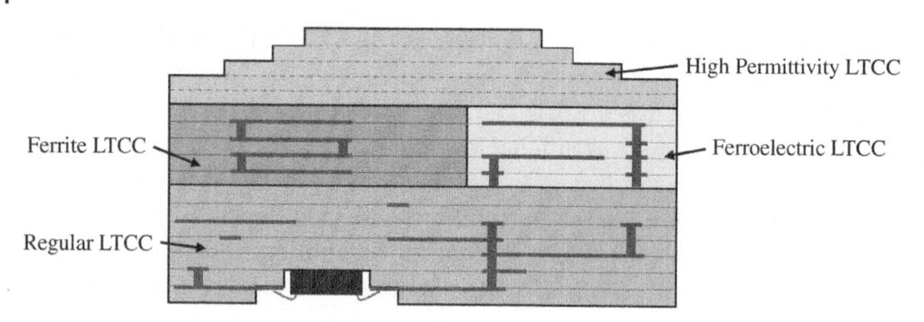


**Figure 6.24** Conceptual drawing of multi-type LTCC tape system.

Figure 6.24. In addition to ferrite LTCC, ferroelectric LTCC is another new LTCC tape system [74] that features tunable permittivity. Both ferrite and ferroelectric LTCC can therefore be used to build tunable components, such as ferrite-based tunable inductors, ferroelectric-based tunable capacitors, and tunable phase shifters, which are helpful for the realization of reconfigurable antenna designs. In addition, high-permittivity LTCC tapes can be useful for compact AiP designs and the on-package realization of some special structures, such as a lens for antenna radiation beam shaping. However, stacking different LTCC tape systems on each other has compatibility and adhesion issues because each LTCC tape system has its own laminating and co-firing conditions, which make the realization of multi-type LTCC tape systems challenging.

## References

1 R.T. Rao, Introduction to system-on-package (SOP) technology. *System on Package: Miniaturization of the Entire System*, McGraw Hill Professional, Access Engineering, 2008.

2 Y.P. Zhang and D. Liu, Antenna-on-chip and antenna-in-package solutions to highly integrated millimeter-wave devices for wireless communications. *IEEE Transactions on Antennas and Propagation*, vol. 57, no. 10, pp. 2830–2841, 2009.

3 Digi-Key, LTCC Antennas – TDK's tiny multi-layer chip antennas for connectivity applications. 9 August 2013. Available: https://www.digikey.com/en/product-highlight/t/tdk/ltcc-antennas.

4 DuPont, DuPont™ GreenTape™ low temperature co-fired ceramic system – Design and layout guidelines. 11 September 2009. Available: https://www.dupont.com/content/dam/dupont/products-and-services/electronic-and-electrical-materials/documents/prodlib/GreenTape_Design_Layout_Guidelines.pdf.

**5** Ferro, Low temperature co-fired ceramic systems A6M/A6M-E high frequency LTCC tape system. November 2015. Available: https://www.etsmtl.ca/Unites-de-recherche/LTCC/Services-offerts/Ferro_A6M.pdf.

**6** R. Sturdivant, Fundamentals of packaging at microwave and millimeter-wave frequencies. In *RF and Microwave Microelectronics Packaging*, K. Kuang, F. Kim, and S. S. Cahill (eds), Springer US, Boston, MA, pp. 1–23, 2010.

**7** D. Deslandes and K. Wu, Integrated microstrip and rectangular waveguide in planar form. *IEEE Microwave and Wireless Components Letters*, vol. 11, no. 2, pp. 68–70, 2001.

**8** J. Xu, Z.N. Chen, X. Qing et al., 140-GHz $TE_{20}$-mode dielectric-loaded SIW slot antenna array in LTCC. *IEEE Transactions on Antennas and Propagation*, vol. 61, no. 4, pp. 1784–1793, 2013.

**9** F. Bauer and W. Menzel, A 79-GHz resonant laminated waveguide slotted array antenna using novel shaped slots in LTCC. *IEEE Antennas and Wireless Propagation Letters*, vol. 12, pp. 296–299, 2013.

**10** A. Nafe, F.A. Ghaffar, M.F. Farooqui et al., A ferrite LTCC-based monolithic SIW phased antenna array. *IEEE Transactions on Antennas and Propagation*, vol. 65, no. 1, pp. 196–205, 2017.

**11** S.B. Yeap, X. Qing, M. Sun et al., 140-GHz 2 × 2 SIW horn array on LTCC. *IEEE Asia-Pacific Conference on Antennas and Propagation*, pp. 279–280, 27–29 August 2012.

**12** T. Tajima, H. Song, K. Ajito et al., 300-GHz step-profiled corrugated horn antennas integrated in LTCC. *IEEE Transactions on Antennas and Propagation*, vol. 62, no. 11, pp. 5437–5444, 2014.

**13** Y.J. Cheng, X.Y. Bao, and Y.X. Guo, 60-GHz LTCC miniaturized substrate integrated multibeam array antenna with multiple polarizations. *IEEE Transactions on Antennas and Propagation*, vol. 61, no. 12, pp. 5958–5967, 2013.

**14** B. Cao, H. Wang, Y. Huang et al., High-gain L-probe excited substrate integrated cavity antenna array with LTCC-based gap waveguide feeding network for W-band application. *IEEE Transactions on Antennas and Propagation*, vol. 63, no. 12, pp. 5465–5474, 2015.

**15** W. Yang, Y. Yang, W. Che et al., 94-GHz compact 2-D multibeam LTCC antenna based on multifolded SIW beam-forming network. *IEEE Transactions on Antennas and Propagation*, vol. 65, no. 8, pp. 4328–4333, 2017.

**16** Y.J. Cheng, W. Hong, and K. Wu, Millimeter-wave multibeam antenna based on eight-port hybrid. *IEEE Microwave and Wireless Components Letters*, vol. 19, no. 4, pp. 212–214, 2009.

**17** W.C. Yang, H. Wang, W.Q. Che et al., High-gain and low-loss millimeter-wave LTCC antenna array using artificial magnetic conductor structure. *IEEE Transactions on Antennas and Propagation*, vol. 63, no. 1, pp. 390–395, 2015.

**18** 3GPP, Release 15: TR 21.915. 26 April 2019. Available: https://www.3gpp.org/release-15.

**19** Y. Hsu and Y. Lin, A cavity-backed aperture antenna on LTCC for 5G mobile communications. *IEEE International Symposium on Radio-Frequency Integration Technology*, pp. 1–3, 24–26 August 2016.

**20** T. Huang, Y. Hsu, and Y. Lin, End-fire quasi-Yagi antennas with pattern diversity on LTCC technology for 5G mobile communications. *IEEE International Symposium on Radio-Frequency Integration Technology*, pp. 1–3, 24–26 August 2016.

**21** J. Park, D. Choi, and W. Hong, 37–39 GHz vertically-polarized end-fire 5G antenna array featuring electrically small profile. *IEEE International Symposium on Antennas and Propagation & USNC/URSI National Radio Science Meeting*, pp. 637–638, 8–13 July 2018.

**22** J. Park, D. Choi, and W. Hong, 28 GHz 5G dual-polarized end-fire antenna with electrically-small profile. *12th European Conference on Antennas and Propagation*, pp. 1–4, 9–13 April 2018.

**23** M. Peng and A. Zhao, LTCC-based phased array antenna for 5G millimeter-wave application in mobile device. *IEEE International Symposium on Antennas and Propagation & USNC/URSI National Radio Science Meeting*, pp. 245–246, 8–13 July 2018.

**24** R. Fisher, 60-GHz WPAN standardization within IEEE 802.15.3c. *International Symposium on Signals, Systems and Electronics*, pp. 103–105, 30 July–2 August 2007.

**25** R.C. Daniels and R.W. Heath, 60-GHz wireless communications: Emerging requirements and design recommendations. *IEEE Vehicular Technology Magazine*, vol. 2, no. 3, pp. 41–50, 2007.

**26** Y.P. Zhang, M. Sun, K.M. Chua et al., Antenna-in-package design for wirebond interconnection to highly integrated 60-GHz radios. *IEEE Transactions on Antennas and Propagation*, vol. 57, no. 10, pp. 2842–2852, 2009.

**27** W. Hong, A. Goudelev, K. Baek et al., 24-element antenna-in-package for stationary 60-GHz communication scenarios. *IEEE Antennas and Wireless Propagation Letters*, vol. 10, pp. 738–741, 2011.

**28** D.G. Kam, D. Liu, A. Natarajan et al., LTCC packages with embedded phased-array antennas for 60-GHz communications. *IEEE Microwave and Wireless Components Letters*, vol. 21, no. 3, pp. 142–144, 2011.

**29** Y.P. Zhang, M. Sun, D. Liu et al., Dual grid array antennas in a thin-profile package for flip-chip interconnection to highly integrated 60-GHz radios. *IEEE Transactions on Antennas and Propagation*, vol. 59, no. 4, pp. 1191–1199, 2011.

**30** E. Cohen, M. Ruberto, M. Cohen et al., A CMOS bidirectional 32-element phased-array transceiver at 60 GHz with LTCC antenna. *IEEE Transactions on Microwave Theory and Techniques*, vol. 61, no. 3, pp. 1359–1375, 2013.

**31** Federal Communications Commission, Radar services in the 76-81 GHz band, 22 June 2017. Available: https://www.fcc.gov/document/radar-services-76-81-ghz-band.

**32** F. Bauer and W. Menzel, A 79-GHz planar antenna array using ceramic-filled cavity resonators in LTCC. *IEEE Antennas and Wireless Propagation Letters*, vol. 12, pp. 910–913, 2013.

**33** F. Bauer, X. Wang, W. Menzel et al., A 79-GHz radar sensor in LTCC technology using grid array antennas. *IEEE Transactions on Microwave Theory and Techniques*, vol. 61, no. 6, pp. 2514–2521, 2013.

**34** C. Rusch, S. Beer, and T. Zwick, LTCC endfire antenna with housing for 77-GHz short-distance radar sensors. *IEEE Antennas and Wireless Propagation Letters*, vol. 11, pp. 998–1001, 2012.

**35** X. Wang and A. Stelzer, A 79-GHz LTCC patch array antenna using a laminated waveguide-based vertical parallel feed. *IEEE Antennas and Wireless Propagation Letters*, vol. 12, pp. 987–990, 2013.

**36** Federal Communications Commission, Notice of Proposed Rule Making (FCC 02-180), Allocations and Service Rules for the 71–76 GHz, 81–86 GHz and 92–95 GHz Bands. 28 June 2002. Available: https://www.fcc.gov/wireless/bureau-divisions/broadband-division/millimeter-wave-708090-ghz-service/millimeter-wave-70#block-menu-block-4.

**37** Z. Chen, Y.P. Zhang, A. Bisognin et al., An LTCC microstrip grid array antenna for 94-GHz applications. *IEEE Antennas and Wireless Propagation Letters*, vol. 14, pp. 1279–1281, 2015.

**38** Z. Chen, Y.P. Zhang, A. Bisognin et al., A 94-GHz dual-polarized microstrip mesh array antenna in LTCC technology. *IEEE Antennas and Wireless Propagation Letters*, vol. 15, pp. 634–637, 2016.

**39** J. Xu, Z.N. Chen, and X. Qing, 270-GHz LTCC-integrated high gain cavity-backed Fresnel zone plate lens antenna. *IEEE Transactions on Antennas and Propagation*, vol. 61, no. 4, pp. 1679–1687, 2013.

**40** J. Xu, Z.N. Chen, and X. Qing, 270-GHz LTCC-integrated strip-loaded linearly polarized radial line slot array antenna. *IEEE Transactions on Antennas and Propagation*, vol. 61, no. 4, pp. 1794–1801, 2013.

**41** B. Zhang H. Gulan, T. Zwick et al., Integration of a 140 GHz packaged LTCC grid array antenna with an InP detector. *IEEE Transactions on Components, Packaging and Manufacturing Technology*, vol. 5, no. 8, pp. 1060–1068, 2015.

**42** T. Shen, T.J. Kao, T. Huang et al., Antenna design of 60-GHz micro-radar system-in-package for noncontact vital sign detection. *IEEE Antennas and Wireless Propagation Letters*, vol. 11, pp. 1702–1705, 2012.

**43** Y. Lu and Y. Lin, Electromagnetic band-gap based corrugated structures for reducing mutual coupling of compact 60-GHz cavity-backed antenna arrays in low temperature co-fired ceramics. *IET Microwaves, Antennas & Propagation*, vol. 7, no. 9, pp. 754–759, 2013.

**44** M. Sun, Y.P. Zhang, K.M. Chua et al., Integration of Yagi antenna in LTCC package for differential 60-GHz radio. *IEEE Transactions on Antennas and Propagation*, vol. 56, no. 8, pp. 2780–2783, 2008.

**45** M. Sun, Y.P. Zhang, D. Liu et al., A ball grid array package with a microstrip grid array antenna for a single-chip 60-GHz receiver. *IEEE Transactions on Antennas and Propagation*, vol. 59, no. 6, pp. 2134–2140, 2011.

**46** Y.P. Zhang, M. Sun, and W. Lin, Novel antenna-in-package design in LTCC for single-chip RF transceivers. *IEEE Transactions on Antennas and Propagation*, vol. 56, no. 7, pp. 2079–2088, 2008.

**47** S. Wi, J. Kim, N. Kang et al., Package-level integrated LTCC antenna for RF package application. *IEEE Transactions on Advanced Packaging*, vol. 30, no. 1, pp. 132–141, 2007.

**48** W. Sang-Hyuk, Y.P. Zhang, L. Yongshik et al., Co-design of antenna and feeding network in LTCC package for UWB single-chip radios. *IEEE Antennas and Propagation Society International Symposium*, pp. 329–332, 9–15 June 2007.

**49** G. Brzezina, L. Roy, and L. MacEachern, Planar antennas in LTCC technology with transceiver integration capability for ultra-wideband applications. *IEEE Transactions on Microwave Theory and Techniques*, vol. 54, no. 6, pp. 2830–2839, 2006.

**50** S.B. Yeap, Z.N. Chen, and X. Qing, Gain-enhanced 60-GHz LTCC antenna array with open air cavities. *IEEE Transactions on Antennas and Propagation*, vol. 59, no. 9, pp. 3470–3473, 2011.

**51** A. Panther, A. Petosa, M.G. Stubbs et al., A wideband array of stacked patch antennas using embedded air cavities in LTCC. *IEEE Microwave and Wireless Components Letters*, vol. 15, no. 12, pp. 916–918, 2005.

**52** I.K. Kim, N. Kidera, S. Pinel et al., Linear tapered cavity-backed slot antenna for millimeter-wave LTCC modules. *IEEE Antennas and Wireless Propagation Letters*, vol. 5, pp. 175–178, 2006.

**53** A.E.I. Lamminen, J. Saily, and A.R. Vimpari, 60-GHz patch antennas and arrays on LTCC with embedded-cavity substrates. *IEEE Transactions on Antennas and Propagation*, vol. 56, no. 9, pp. 2865–2874, 2008.

**54** K. Malecha, Fabrication of cavities in low loss LTCC materials for microwave applications. *Journal of Micromechanics and Microengineering*, vol. 22, no. 12, p. 125004, 2012.

**55** F.A. Ghaffar, M.U. Khalid, A. Shamim et al., Gain-enhanced LTCC system-on-package for automotive UMRR applications. *53rd IEEE International Midwest Symposium on Circuits and Systems*, pp. 934–937, 1–4 August 2010.

**56** F.A. Ghaffar, M.U. Khalid, K.N. Salama et al., 24-GHz LTCC fractal antenna array SoP with integrated Fresnel lens. *IEEE Antennas and Wireless Propagation Letters*, vol. 10, pp. 705–708, 2011.

**57** A. Petosa and A. Ittipiboon, Design and performance of a perforated dielectric Fresnel lens. *IEE Proceedings – Microwaves, Antennas and Propagation*, vol. 141, no. 5, p. 309, 1994.

**58** I. Kadri, A. Petosa, and L. Roy, Ka-band Fresnel lens antenna fed with an active linear microstrip patch array. *IEEE Transactions on Antennas and Propagation*, vol. 53, no. 12, pp. 4175–4178, 2005.

**59** A.E.I. Lamminen, A.R. Vimpari, and J. Saily, UC-EBG on LTCC for 60-GHz frequency band antenna applications. *IEEE Transactions on Antennas and Propagation*, vol. 57, no. 10, pp. 2904–2912, 2009.

**60** W.E. McKinzie, D.M. Nair, B.A. Thrasher et al., 60-GHz 2 × 2 LTCC patch antenna array with an integrated EBG structure for gain enhancement. *IEEE Antennas and Wireless Propagation Letters*, vol. 15, pp. 1522–1525, 2016.

**61** H. Jin, W. Che, K. Chin et al., 60-GHz LTCC differential-fed patch antenna array with high gain by using soft-surface structures. *IEEE Transactions on Antennas and Propagation*, vol. 65, no. 1, pp. 206–216, 2017.

**62** A. Shamim, J. Bray, L. Roy et al., Microwave and magnetostatic characterization of ferrite LTCC for tunable and reconfigurable SiP applications. *IEEE/MTT-S International Microwave Symposium*, pp. 691–694, 3–8 June 2007.

**63** A. Shamim, J.R. Bray, N. Hojjat et al., Ferrite LTCC-based antennas for tunable SoP applications. *IEEE Transactions on Components, Packaging and Manufacturing Technology*, vol. 1, no. 7, pp. 999–1006, 2011.

**64** F.A. Ghaffar, M. Vaseem, L. Roy et al., Design and fabrication of a frequency and polarization reconfigurable microwave antenna on a printed partially magnetized ferrite substrate. *IEEE Transactions on Antennas and Propagation*, vol. 66, no. 9, pp. 4866–4871, 2018.

**65** F.A. Ghaffar, J.R. Bray, and A. Shamim, Theory and design of a tunable antenna on a partially magnetized ferrite LTCC substrate. *IEEE Transactions on Antennas and Propagation*, vol. 62, no. 3, pp. 1238–1245, 2014.

**66** F.A. Ghaffar and A. Shamim, A ferrite LTCC-based dual purpose helical antenna providing bias for tunability. *IEEE Antennas and Wireless Propagation Letters*, vol. 14, pp. 831–834, 2015.

**67** R. Hahn, S. Krumbholz, and H. Reichl, Low profile power inductors based on ferromagnetic LTCC technology. *56th Electronic Components and Technology Conference 2006*, p. 6, 30 May–2 June 2006.

**68** E. Arabi, M. Lahti, T. Vähä-Heikkilä et al., A 3-D miniaturized high selectivity bandpass filter in LTCC technology. *IEEE Microwave and Wireless Components Letters*, vol. 24, no. 1, pp. 8–10, 2014.

**69** E. Arabi, F.A. Ghaffar, and A. Shamim, Tunable bandpass filter based on partially magnetized ferrite LTCC with embedded windings for SoP applications. *IEEE Microwave and Wireless Components Letters*, vol. 25, no. 1, pp. 16–18, 2015.

**70** J.R. Bray and L. Roy, Development of a millimeter-wave ferrite-filled antisymmetrically biased rectangular waveguide phase shifter embedded in low-temperature cofired ceramic. *IEEE Transactions on Microwave Theory and Techniques*, vol. 52, no. 7, pp. 1732–1739, 2004.

**71** A. Nafe and A. Shamim, An integrable SIW phase shifter in a partially magnetized ferrite LTCC package. *IEEE Transactions on Microwave Theory and Techniques*, vol. 63, no. 7, pp. 2264–2274, 2015.

**72** F.A. Ghaffar and A. Shamim, A partially magnetized ferrite LTCC-based SIW phase shifter for phased array applications. *IEEE Transactions on Magnetics,* vol. 51, no. 6, pp. 1–8, 2015.

**73** VTT, Design guidelines: Low-temperature co-fired ceramic modules. 15 January 2014. Available: https://www.vtt.fi/files/research/mel/ltcc_design_rules.pdf.

**74** H. Jantunen, T. Hu, A. Uusimäki et al., Ferroelectric LTCC for multilayer devices. *Journal of the Ceramic Society of Japan*, vol. 112, pp. S1552–S1556, 2004.

# 7

# Antenna Integration in Packaging Technology operating from 60 GHz up to 300 GHz (HDI-based AiP)

*Frédéric Gianesello[1], Diane Titz[2], and Cyril Luxey[2]*

[1] *ST Microelectronics, Technology R&D, Silicon Technology Development, 850 Rue Jean Monnet, 38920 Crolles, France*
[2] *Université Nice Sophia Antipolis, Polytech'Lab, 930 Route des Colles, 06410 Biot, France*

## 7.1 Organic Packaging Technology for AiP

With the rise of consumer applications at millimeter-wave (mmWave) frequencies [e.g. wireless gigabit alliance (WiGig) communications or automotive radars] during the past ten years, we have witnessed strong research activity dealing with the integration of mmWave antennas in the package of electronic sub-systems. The objectives were both to ease the integration and lower the cost of the mmWave system by integrating the antenna in a system-in-package approach (SiP). First investigations have considered mature mmWave laminate technologies such as ceramic (either high-temperature co-fired ceramics or low-temperature co-fired ceramics) or high-end glass reinforced material (e.g. Rogers). Unfortunately, these technologies are high-end and consequently quite expensive. In order to achieve a cost structure more in line with standard consumer products, SiP using conventional "organic" microelectronic packaging technology haS been considered in order to develop a high-performance and low-cost antenna-in-package (AiP) module.

This chapter provides an overview of the work performed by STMicroelectronics and the University of Nice (Polytech'Lab) to design innovative AiP solutions integrated on "organic" microelectronic packaging technology and assess their achievable performance up to 300 GHz.

### 7.1.1 Organic Package Overview

Chip density continuous improvement (following Moore's law) has driven the development of packaging technology in order to achieve higher pin counts, and smaller and thinner packages. Dual in-line package (using pin insertion) was

*Antenna-in-Package Technology and Applications*, First Edition.
Edited by Duixian Liu and Yueping Zhang.

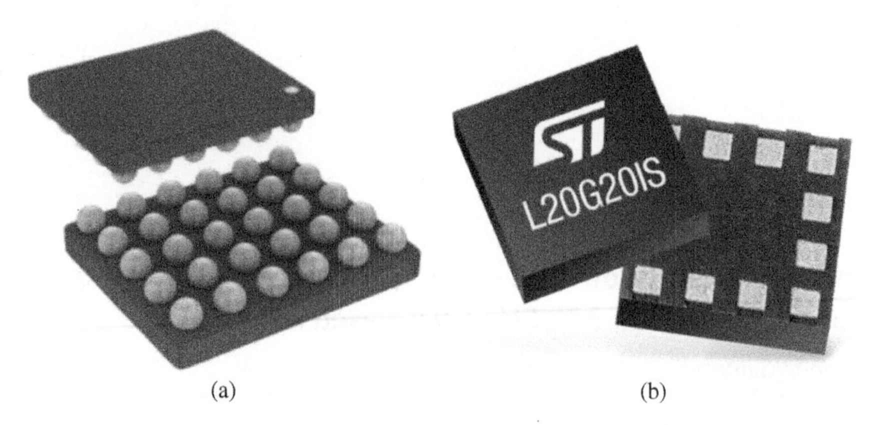


**Figure 7.1** Example of (a) BGA package (STMicroelectronics ST25R3912-AWLT) and (b) LGA package (STMicroelectronics L20G20IS) using organic substrates.

first used in the first half of the 1970s and subsequently quad flat package (using solder reflow) was introduced to meet the requirement of the narrower pin's pitch. The introduction of high-density logic integrated circuits (ICs) has been supported by the development of multilayer substrates embedding some routing in the package. This has led to the development of land grid array (LGA) and ball grid array (BGA) packages as illustrated in Figure 7.1.

During the early years of LGA and BGA packages, ceramics was the main logic semiconductor packaging substrate. Lower-cost copper-clad laminates and associated buildup technology were introduced around 1993, making organic packaging substrates the chosen technology to support the development of fine pitch wiring technology. In order to connect the IC to the substrate, two approaches can be considered, as illustrated in Figure 7.2: wire bonding (which has a cost advantage) and flip chip (which enables better electrical performance to be achieved).

### 7.1.2 Buildup Architecture

Standard packaging substrates use a symmetrical buildup (also called stackup) made of a central core (substrate) sandwiched between two sheets of copper foil laminated (substrates) on the outside with the bonding provided by a layer of glass cloth pre-impregnated with uncured resin (usually called prepreg, more details provided in the next section). As illustrated in Figure 7.3, by taking the panel multiple times through production process, additional layers can be laminated (using additional prepreg) in order to achieve higher complexity substrate. The main difference between the core and the prepreg concerns via formation: the core is generally embedding drilled plated-through-hole (PTH) while prepreg are

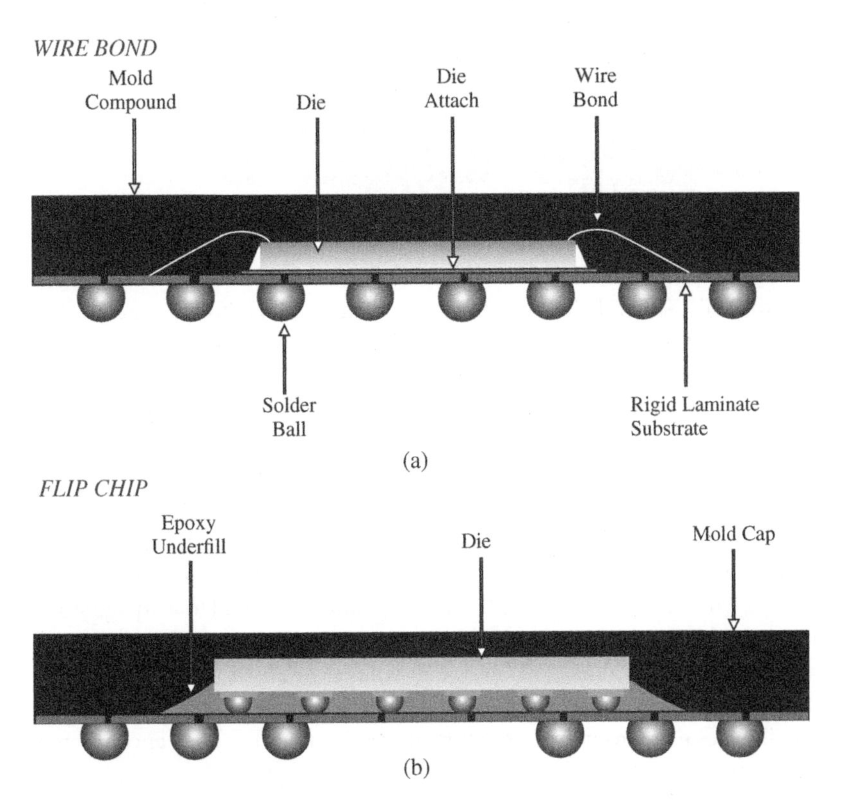


**Figure 7.2** Illustration of (a) wire bonding and (b) flip-chip assembly of a BGA substrate over a PCB application board.

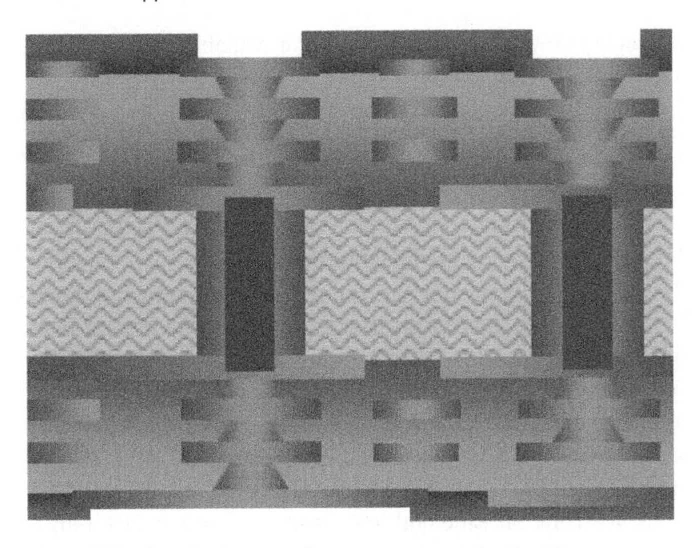

**Figure 7.3** Standard core and prepreg symmetrical buildup.

**Figure 7.4** Four-layer coreless buildup example.

incorporating filled-laser drilled-via. This technology is very similar to the printed circuit board technology (PCB) but slightly differs in terms of substrates' materials and design rules that can be used.

In order to satisfy the increasing demand for higher complexity, superior electrical performance and lower cost, substrate packaging technology has evolved toward coreless substrates (see Figure 7.4). The main concept consists of removing the core layer and using laser-drilled micro-vias (μvias) everywhere in the buildup. By doing this, better electrical performance (since μvias are smaller than drilled PTH) and a thinner substrate can be achieved (the core layer being thicker than the prepreg). The main challenge with coreless buildup is related to warpage since the stiffness of the core layer cannot be relied on. Coreless substrate technology has been pioneered by Sony since 2004 and high-volume production started in 2007. Today, the most advanced packaging technology uses a coreless substrate approach with possible design rules which are more aggressive than the ones encountered in PCB technology.

### 7.1.3 Industrial Material

In order to build the stackup, several material options are available on the market. We can divide the available materials into two types:

- *Glass clothes copper-clad laminate (CCL)*: To reinforce the material, a glass cloth filled with a polymer resin (epoxy, bismaleimide triazine, polyphenylene, cyanate ester or acrylates) is used. The CCL core and the prepreg are very similar to what is used in traditional PCBs (e.g. FR4, which is an epoxy-based glass-reinforced material) and some manufacturers provide the same material for both PCBs and the packaging substrate (Megtron6 from Panasonic is a good example). However, high-end material using thermoplastic such as polytetrafluoroethylene (e.g. Rogers RO4003) does not exist for packaging substrates. Since CCL prepreg needs to be easily laser-drillable, the main development in recent years has concerned woven glass cloth-style optimization to improve its uniformity (since fiberglass and the surrounding resin with no glass exhibit different ablation rates). Table 7.1 summarizes some examples of standard industrial materials, focusing on Bluetooth (BT) and epoxy-filled CCL (since

**Table 7.1** Examples of industrial CCL material

| Item/test condition | Unit | HL832NX | HL832NSR | HL832NSA | HL972LF |
|---|---|---|---|---|---|
| CTE $x, y$ | ppm\°C | 14 | 8 | 1 | 10 |
| Tg TMA | °C | 230 | 230 | >350 | 270 |
| Tg DMA | °C | 200 | 210 | 270 | 240 |
| Young modulus | GPa | 28 | 30 | 31 | 25 |
| Tensile strength | MPa | 270 | 280 | 350 | 220 |
| Dielectric constant @ 1 GHz | – | 4.9 | 4.5 | 4.0 | 3.5 |
| Loss tangent @ 1 GHz | – | 0.011 | 0.008 | 0.005 | 0.003 |
| Flame resistancy (UL94) | – | V-0 | V-0 | V-0 | V-0 |

CTE, coefficient of thermal expansion; Tg: glass transition temperature; TMA, thermo mechanical analyzer; DMA, dynamic mechanical analyzer.

they currently dominate the market due to their thermal stability and low cost), along with their main properties.

- *Non-reinforced resin-coated copper (RCC) foil*: Since glass-reinforced CCL exhibits some difficulty in performing laser drilling and achieving thin thickness, the industry has worked on some alternative solutions. One solution consists of using the copper foil as a carrier for the dielectric (RCC) so it can be incorporated in the buildup during the lamination process. Of course, the lack of reinforcement causes some challenges about dimensional stability and thickness control. Moreover, RCC foils are generally more expensive than the equivalent prepreg but when considering laser drilling time, the cost of substrate manufacturing is lower. RCC material is today used extensively for coreless substrate and could be used in standard buildup in order to provide access to more aggressive design rules (as will be discussed later). The main provider of RCC material for packaging substrate is Ajinomoto (https://www.aft-website.com/en/electron/abf#block8). Table 7.2 summarizes the properties of the different materials currently provided by Ajinomoto to the market.

### 7.1.4 HDI Design Rules

Moving to advanced packaging substrate technology, more aggressive design rules are necessary, which has led to the rise of high-density integration (HDI) technology. HDI is mainly related to laser-drilled μvias introduced in thin prepreg material (since via diameter is correlated to prepreg thickness due to a maximum optical form factor). As has been mentioned in the previous section, μvias are smaller in diameter and consequently provide improved electrical performance and better integration capability than conventional mechanically drilled PTH. Table 7.3

**Table 7.2** Examples of industrial RCC material provided by Ajinomoto (https://www.aft-website.com/en/electron/abf#block8)

| Item / test condition | Unit | GX13 | GX92 | GX-T31 | GZ41 |
|---|---|---|---|---|---|
| CTE x,y | (tensile TMA) | 46 | 39 | 23 | 20 |
| Tg TMA | °C | 156 | 153 | 154 | 171 |
| Tg DMA | °C | 177 | 172 | 172 | 189 |
| Young modulus | GPa | 4 | 5 | 7.5 | 9 |
| Tensile strength | MPa | 93 | 98 | 104 | 120 |
| Dielectric constant @ 1 GHz | – | 3.35 | 3.3 | 3.47 | 3.3 |
| Loss tangent @ 1 GHz | – | 0.012 | 0.014 | 0.0114 | 0.0058 |
| Flame resistancy (UL94) | – | V0 | V0 | V0 | V0 |

CTE, coefficient of thermal expansion; Tg: glass transition temperature; TMA, thermo mechanical analyzer; DMA, dynamic mechanical analyzer.

**Table 7.3** Current PTH and μvia diameters supported in the packaging industry

| Item | Unit | PTH | μvia |
|---|---|---|---|
| Via size (drill/land) | μm | From 100/190 down to 75/130 | 60/110 |

summarizes the current PTH and μvia diameters that are supported in the packaging industry.

Beyond the design rules of the vias, achievable traces (width and neighbor separation) are also crucial parameters which have been downsized during the years in order to support the design of highly ICs. To do this, four main technologies offer the best compromise between achievable minimum features and manufacturing cost:

- *Subtractive process (SUB)*: The subtractive process is similar to what is used in PCB technology. Thin lines are formed by coating the copper layer with an etch resist, applying photolithography to image the areas where the copper should be retained and etching away the unimaged material. The main drawback of this approach is that vertical chemical etching will also dissolve the copper in the horizontal direction along the trace walls, leading to a trapezoidal shape and consequently the dimensional control may limit the achievable minimum features.
- *Semi-additive process (SAP)*: A thin layer of copper is first coated on the substrate, then a resist is applied followed by a negative pattern design. Copper is electroplated to the desired thickness in the areas where the resist is not present

**Table 7.4** Design rules for SUB, SAP, mSAP and ETS processes

| Item | SUB | MSAP | SAP | SAP + ABF | ETS |
|---|---|---|---|---|---|
| Trace W/S (μm) | 25/25 | 20/20 | 15/15 | 12/12 | 10/10 |

(leading to the additive nature of the process). Finally, the seed copper layer is removed.

- *Modified semi-additive process (mSAP)*: The main difference between the mSAP and SAP processes is the thickness of the seed copper layer. Generally, SAP processing begins with a thin electroless copper coating (<1.5 μm) and mSAP begins with a thin laminated copper foil (>1.5 μm). Consequently, SAP can achieve a more aggressive trace width and separation at the expense of a cost adder.
- *Embedded traces (ETS)*: Moving to a coreless substrate and more aggressive design rules, the removal of the seed layers used in the SAP process has proved to be a challenge. To overcome this limitation, embedded copper traces have emerged as a promising solution since they are buried in the buildup dielectric layer and consequently there is no copper seed layer to etch. Two different process flows are used to embed the traces: copper trace transfer and photo-trench embedding [1].

The current industrial design rules for the SUB, SAP, mSAP, and ETS processes are summarized in Table 7.4.

### 7.1.5 Assembly Constraints and Body Size

Once the sequential lamination process is finished, it is necessary to assemble the IC on the substrate and perform some finishing (over-molding, ball attach etc.). For the IC connection to the substrate, wire bonding and flip-chip solutions can be considered. However, if we intend to operate in higher frequency band, the flip-chip approach is a preferable solution since parasitics effects will be reduced (a flip-chip bump will show a resistance of tens of milliOhm in series with an inductor of tens of picoHenrys). The general process flow used to perform the assembly of the IC on the substrate is summarized in Figure 7.5.

The assembly process uses automated equipment. In order to ease the handling of the samples, the industry delivers strips rather than full panels. A strip is basically a smaller panel with a dedicated format and peripheral holes to ease its handling during the assembly process. There is no general strip format, but Figure 7.6 shows an example of a classical strip used at STMicroelectronics for BGA assembly.

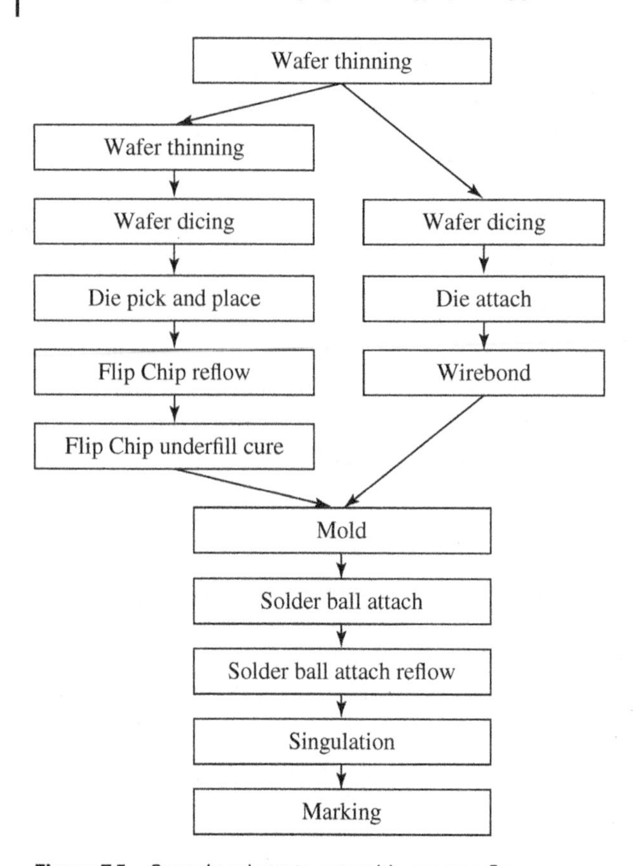


**Figure 7.5** Organic substrate assembly process flow.

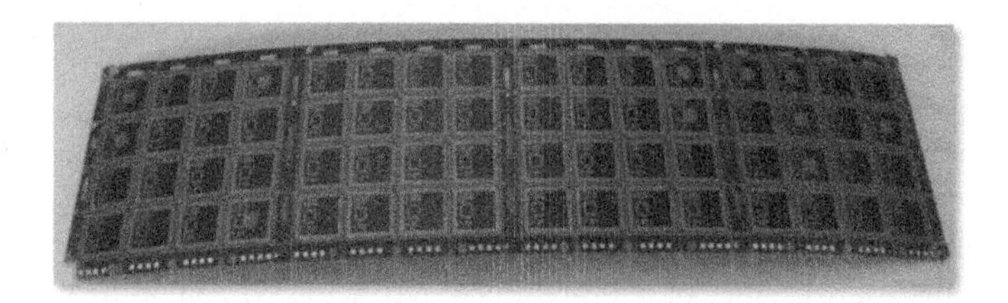

**Figure 7.6** Example of an organic strip comprising 64 BGA substrates before assembly (from [7]).

Since strips have to be used instead of full panels, one cannot assess the cost of a given substrate directly by dividing its body size from a full panel. Different body size and strip formats will lead to different numbers of packages per strip. To illustrate this point, Table 7.5 summarizes the number of substrates per strip for different body sizes for the classical strips used at STMicroelectronics.

**Table 7.5** Number of parts per strip in function of substrate body size

| **Body size (mm × mm)** | 3 × 3 | 4 × 4 | 5 × 5 | 6 × 6 | 8 × 8 | 12 × 12 | 17 × 17 |
|---|---|---|---|---|---|---|---|
| **Number of parts-per-strip** | ~900 | ~484 | ~324 | ~196 | ~144 | ~64 | ~16 |

## 7.2 Integration of AiP in Organic Packaging Technology Below 100 GHz

### 7.2.1 Integration Strategy of the Antenna

Considering the previous sections, one assembly strategy is proposed in (Figure 7.7) where the die is attached on the bottom side of the BGA module and the antenna radiates from the other side. Hence, the die is naturally protected by its package and the final module has a low profile (less than 1 mm), which is a current target for mobile device integration. Moreover, if the antenna topology allows broadside or end fire radiation with large front-to-back ratios, the possible electromagnetic interference originating from the microelectronic circuits of the IC toward the antenna is minimized.

To ensure a broadside radiation on the opposite side of the IC with a high front-to-back ratio, an aperture-coupled patch (ACP) antenna [2] architecture is often selected. This antenna presents wideband performance in terms of return loss and gain. A patch antenna is excited by a slot etched in the ground plane of a microstrip line (Figure 7.8).

By tuning the length of the slot ($L_f$) and the size of the patch ($L_p$), we can produce two close resonances sufficiently coupled to increase the operating bandwidth. However, the bandwidth relies also on the thickness of the substrate. For example, to reach a 15–20% bandwidth at 60 GHz, the thickness of the substrate between the slot and the patch must be at least 400 μm. Then, the length of the stub ($L_s$) is varied to optimize the matching of the antenna. If we choose a simple buildup, one

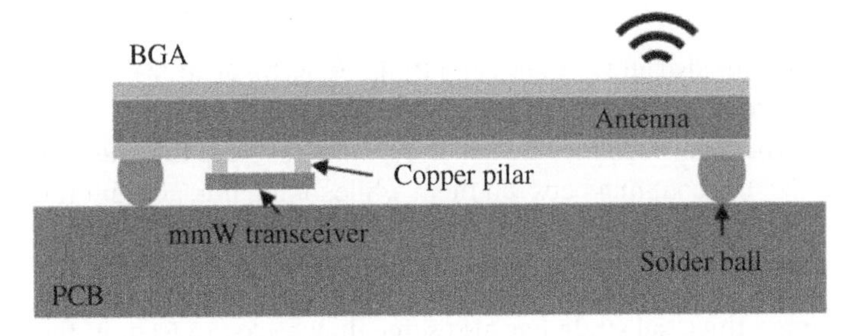


**Figure 7.7** AiP integration strategies: the die is on the opposite side of the module.

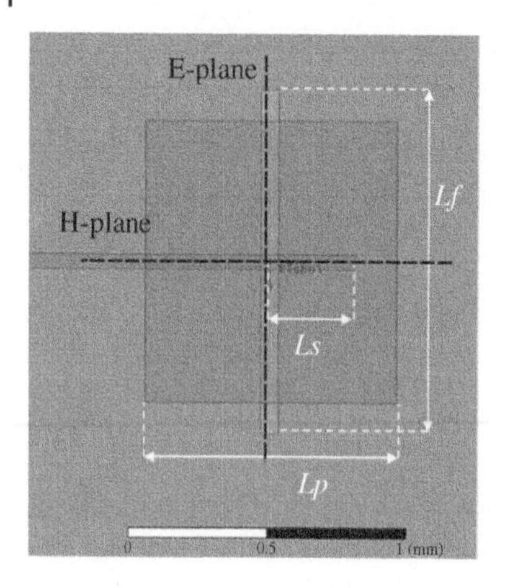


**Figure 7.8** Detailed top view of the patch, slot, and feeding microstrip line comprising an ACP antenna with dimensions to be further optimized.

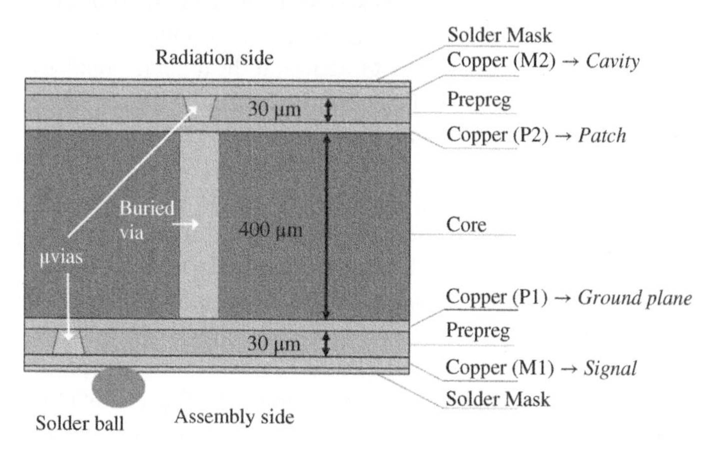


**Figure 7.9** Cross-section of the selected 1-2-1 HDI buildup to embed ACP antenna (from [7]).

core and two prepregs with four metal layers (Figure 7.9), the antenna integration can be the following:

- The microstrip line, but also all DC tracks and RF lines, is drawn at M1 level.
- The ground is at P1 level.
- The substrate thickness between the line and the ground (prepreg thickness) is chosen to maximize the coupling between them while taking into account the minimum allowed resolution.
- The patch antenna is set at P2 level.

However, choosing a thick substrate has also some drawbacks. Indeed, in the ACP design, for instance, the slot couples to the transverse magnetic 0 ($TM_0$)

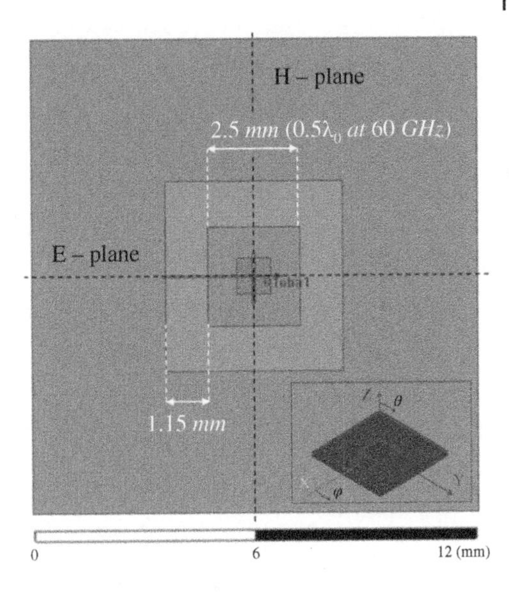


**Figure 7.10** Transparent top view of an ACP with the cavity where the patch is centered in the middle of the BGA module.

surface-wave mode, which propagates in the core substrate in the longitudinal direction of the microstrip line (E-plane) to the edges of the BGA module (Figure 7.10). This $TM_0$ surface-wave mode escapes and diffracts at those edges but is also partly reflected back, which causes constructive and destructive interference in the whole radiation mechanism in the far-field of the antenna. The radiation pattern of the antenna is then distorted in the E-plane (Figure 7.11) and side lobes are created, lowering the gain in the main direction of radiation. This phenomenon was also outlined in [3] where a dipole antenna using a reflecting ground plane was chosen. In order to avoid those effects, a grounded metallic cavity enclosing the antenna can be added to vanish the surface waves on the walls of the cavity. The walls of this cavity, which can be modeled by means of solid copper connecting ground plane P2 to M2 level in simulation, are made of buried vias in the HDI buildup. As expected, the radiation pattern in the H-plane is not largely affected by the addition of this cavity (Figure 7.11) because the $TM_0$ mode propagates in the E-plane direction (apart from a beneficial small increase of the broadside gain and a slight reduction in the level of the secondary lobes). Knowing all the above-mentioned considerations when designing an antenna on an HDI buildup at mmWave frequencies, we will now present some integration examples at 60 and 94 GHz.

### 7.2.2 60-GHz AiP Modules

Using the integration strategy and the antenna choice presented in the previous sections, one of the first modules using HDI mmWave organic technology was

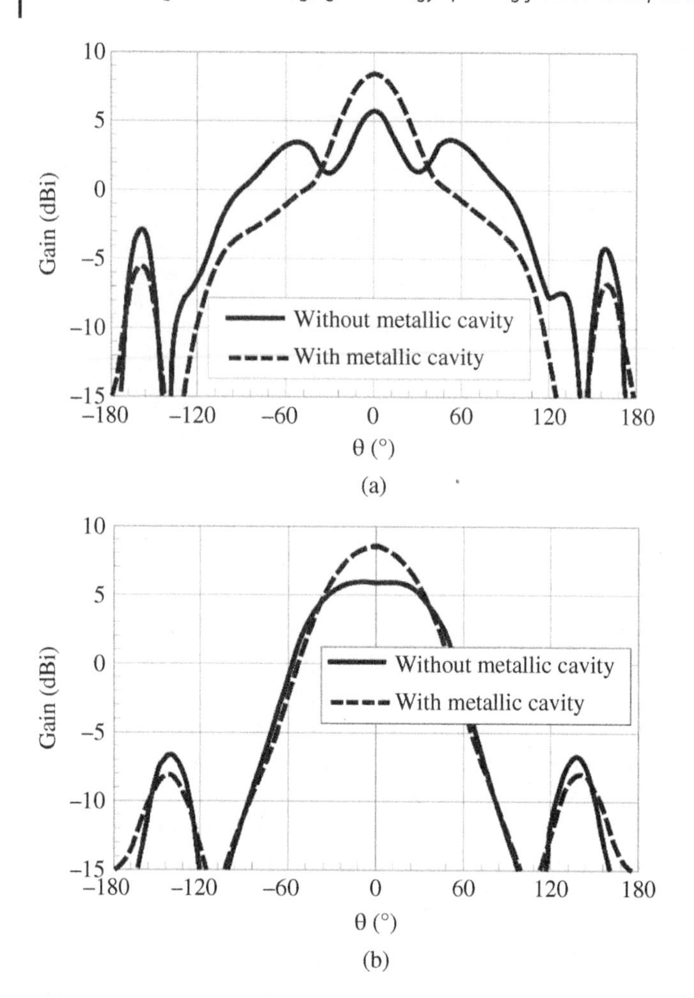


**Figure 7.11** Simulated realized gain radiation patterns for (a) the E-plane ($\phi = 90°$) and (b) the H-plane ($\phi = 0°$) of an ACP with and without a metallic cavity at 66 GHz.

presented by our group in [4]. Two linearly polarized ACP antennas are integrated in this module (Figure 7.12), one for transmission (Tx) and the other for reception (Rx). The material selection was as that time (2010) quite limited since very few materials had been characterized up to mmWave frequencies. To secure this point, the module used traditional PCB mmWave material RO4003C™ from Rogers Corp. (http://www.rogerscorp.com) for the core material, Mercurywave™ 9350 from Park Electrochemical Corp. (http://www.parkelectro.com) for the prepreg, and PFR-800 AUS410™ material from Taiyo America Inc. for the solder mask. The buildup is presented in Figure 7.9 and pictures of the realized module are presented in Figure 7.13.

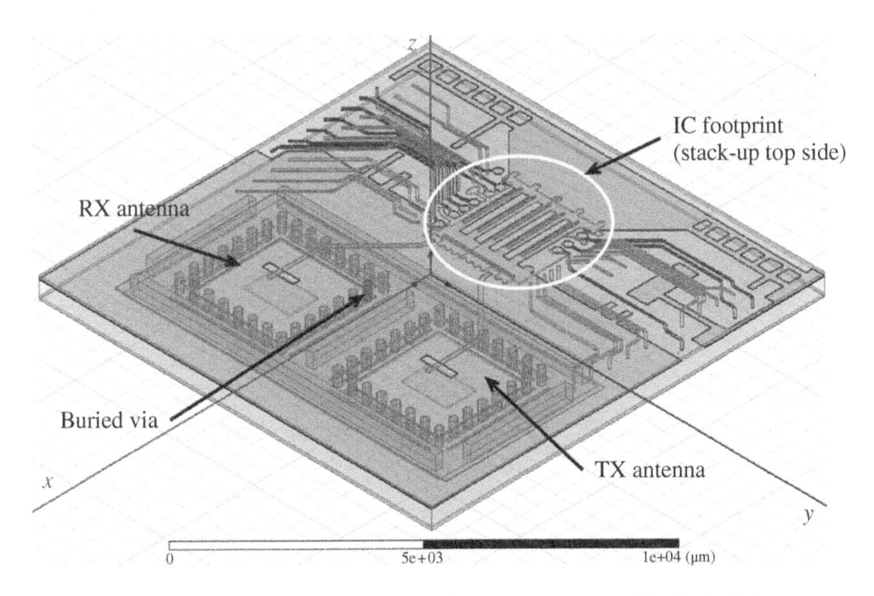


**Figure 7.12** 60-GHz HDI module with ACP antennas surrounded by buried vias.

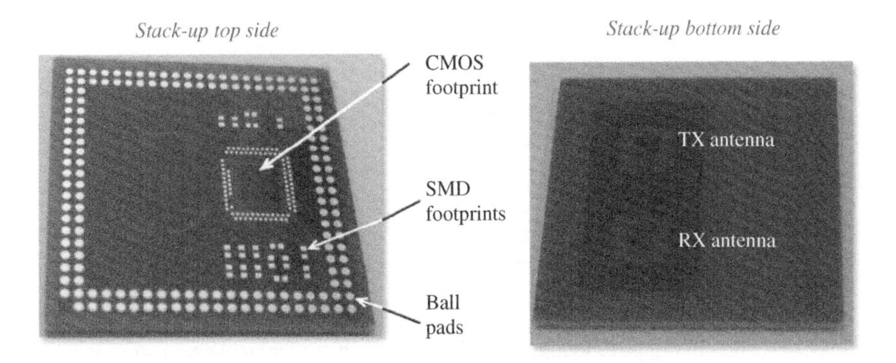


**Figure 7.13** Pictures of the first manufactured HDI mmWave organic 60-GHz module (from [4]).

This module was the first to be manufactured by our group and consequently some discrepancies between the simulation model and realization were found, mainly due to slight deviations for material thicknesses and dielectric characteristics at 60 GHz. The measurement results (obtained from the probe-fed measurement setup developed at the University of Nice Sophia Antipolis [5]) compared with retro-simulations made taking into account realistic values of the manufactured module are presented in Figure 7.14. The obtained matching was not acceptable, but the broadside realized gain reached 5 dBi at 62 GHz. which

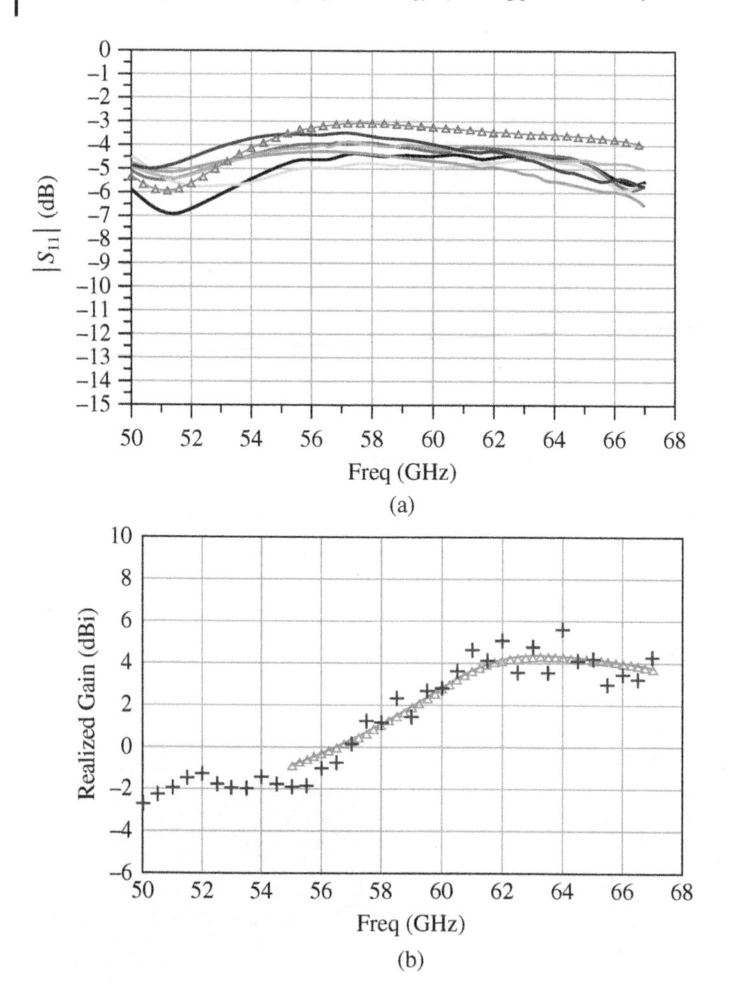


**Figure 7.14** Measurement results: (a) reflection coefficient and (b) realized gain of the first realized module. Retro-simulation results are in small triangles.

was a state-of-the art result at that time and necessary for a 1-m transmission distance. The measurement results and retro-simulations were found to fit with each other.

The 3D radiation patterns at 60 GHz are presented in Figure 7.15. The peak realized gain reaches 3.5 dBi at 60 GHz. The total efficiency computed from the measurement was found to be 35%. This module was covered with a 3D-printed plastic lens [6] to increase the realized broadside gain and achieve a longer WiGig Tx/Rx transmission (80 cm without the lens, 1.5 m with the lens).

Learning from this first prototype, a second one was realized and presented in [7] and [8]. In this module, the Tx antenna was chosen to be linearly polarized, while

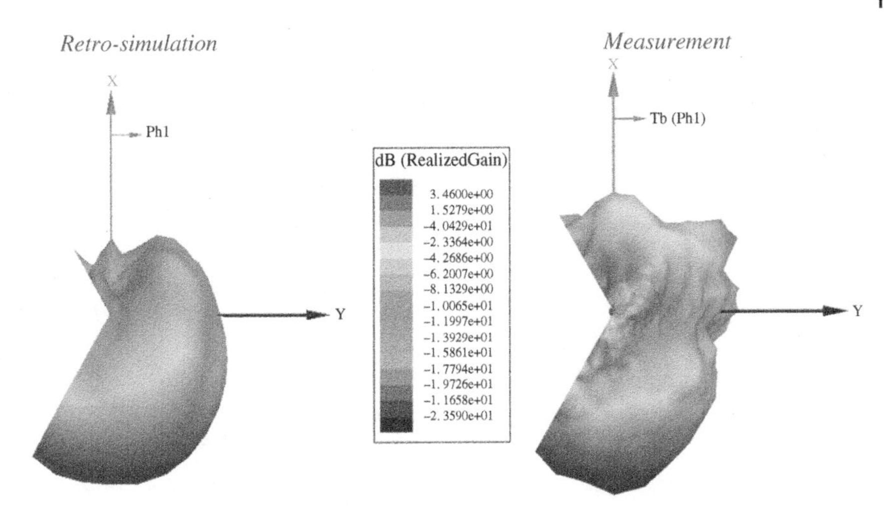


**Figure 7.15** 3D realized gain patterns at 60 GHz of the first realized module (from [5]).

the Rx antenna was designed to operate in circularly polarized mode. Each of the antennas fitted in a $5 \times 5\,mm^2$ area and the circularly polarized (CP) antenna was single-fed by specification. The chosen module buildup was the one presented in Figure 7.9 (using conventional packaging material: Mitsubishi CCL-HL972 for the core and GHPL-970 for the prepreg). The methodology to design the LP is given in section 7.2.1. The innovative CP antenna was made of a rectangular patch (P2) coupled to a double-crossed rectangular slot (made up of slots S1 and S2) with symmetrical circular slots (D1 and D2) as terminations. The cross-slot was excited in its center by means of a microstrip line placed in M1, as in the previous antenna design. A transparent "large view" of the Rx antenna centered on a $12 \times 12\,mm^2$ BGA module is shown in Figure 7.16a.

The three circled stubs in Figure 7.16a on M1 were employed to maximize the matching bandwidth (BW). The parameters of the two slots (S1, S2, D1, and D2 in Figure 7.11b) on P1 were the inputs to the optimization process in order to maximize the axial ratio (AR) BW. The final dimensions for both Tx and Rx antennas are summarized in Table 7.6.

A view of the complete module model is presented in Figure 7.17 and pictures of one of the realized modules in Figure 7.18. In order to simultaneously validate the design and study the manufacturing process variability, several BGA modules from two different manufacturers (called manufacturers #1 and #2) were characterized using the same measurement setup from University Nice Sophia Antipolis [5].

Measurement results for the LP antenna are presented in Figures 7.19 and 7.20. In order to have an idea of the process variability, the reflection coefficient was measured for six different Tx antennas from manufacturer #1 (Figure 7.19a). The

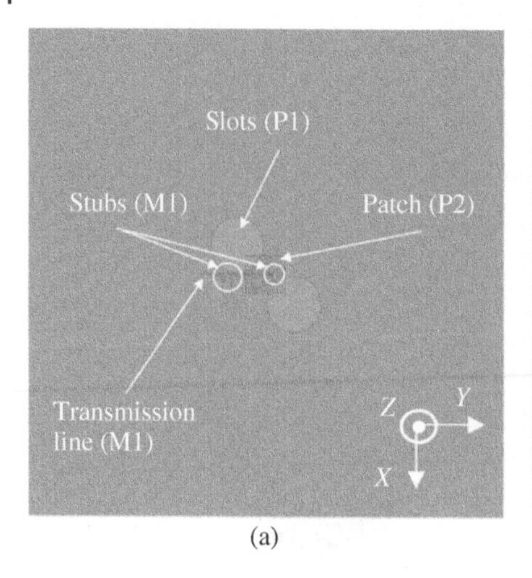


(a)

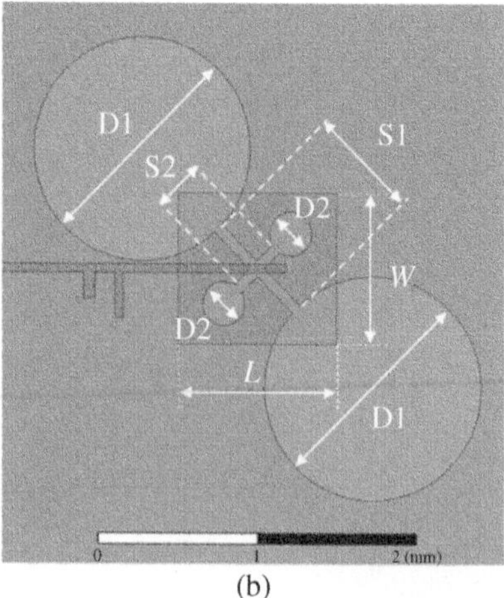


(b)

**Figure 7.16** (a) Transparent top "large" view of the Rx antenna centered in a 12 × 12 × 0.620 mm³ BGA and (b) detailed view of the studied dimensions for optimization once the patch dimensions are fixed (length and width).

**Table 7.6** Antenna final dimensions

| | LP antenna | | CP antenna | |
|---|---|---|---|---|
| Slot | $L_f \times W_s$ (slot width) | $1.1 \times 0.05\,\text{mm}^2$ | D1 × D2 and | $0.122 \times 0.68\,\text{mm}^2$ |
| | | | S1 × S2 | $0.34 \times 0.7\,\text{mm}^2$ |
| Patch | $L_p$ | 0.9 mm | W × L | $0.92 \times 1\,\text{mm}^2$ |
| Stub | $L_s$ | 0.3 mm | | |

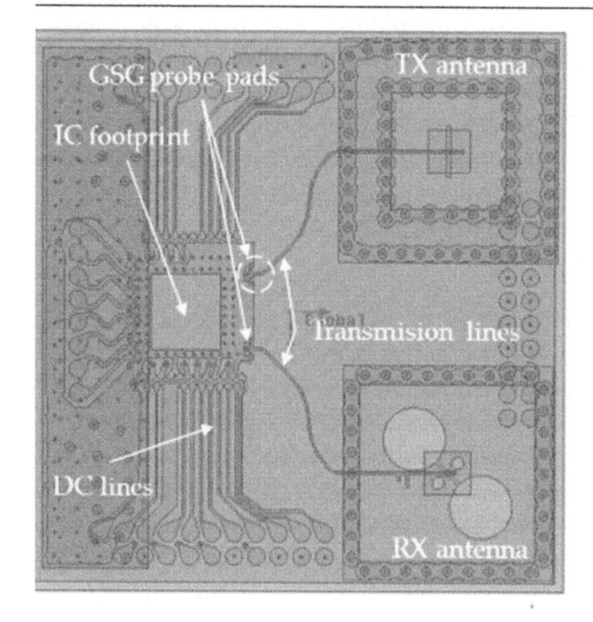


**Figure 7.17** Transparent top view of the complete BGA module (from [7]).

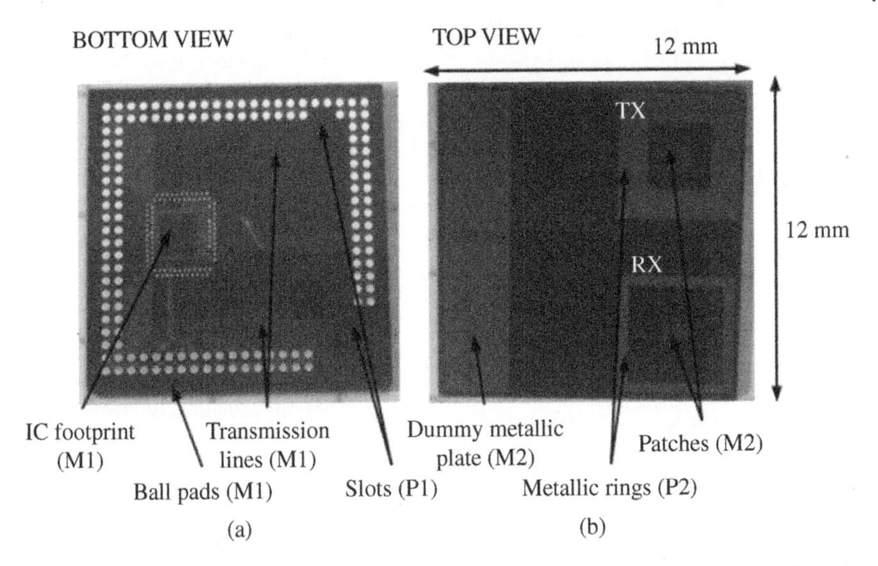


**Figure 7.18** Picture of the BGA module. (a) Bottom view with the chip footprint, the ball pads, and the microstrip feeding lines in transparency (the slots are also visible). (b) Top view with the Tx and Rx patches and a dummy metallic plate at the opposite edge for metal density requirements of the HDI process.

good repeatability of achieved measurements allows both the good repeatability of the fabrication process and the calibration of the measurement system and the measurement itself to be validated. The realized gain was measured from 50 to 67 GHz in the broadside direction of one BGA from manufacturer #1 and is shown in Figure 7.19b for the copolar and crosspolar field components. Considering the $\pm 1.2$ dB error bars representing the uncertainty of the measurement system at 60 GHz, the obtained results are in very good agreement with simulation results. The established gain specification above 5 dBi is fulfilled within the whole WiGig band.

In addition, the discrimination level between the co- and crosspolar components is larger than 25 dB in this frequency band. Finally, the realized gain copolar radiation patterns in the H-plane are shown in Figure 7.20a for the Tx antennas of two different modules from manufacturers #1 and #2 at 60 GHz. The radiation patterns exhibit low variability when the frequency varies from 57 to 66 GHz as can be seen in Figure 7.20b, which shows the radiation patterns of one Tx antenna of a BGA module at five different frequencies.

The simulated and measured magnitudes of the reflection coefficient versus frequency are shown in Figure 7.21a for the Rx antenna integrated in the BGA for two different modules from manufacturers #1 and #2. The antennas from both

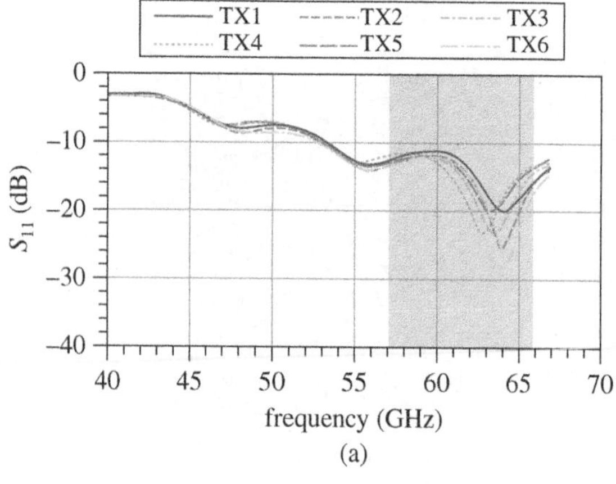


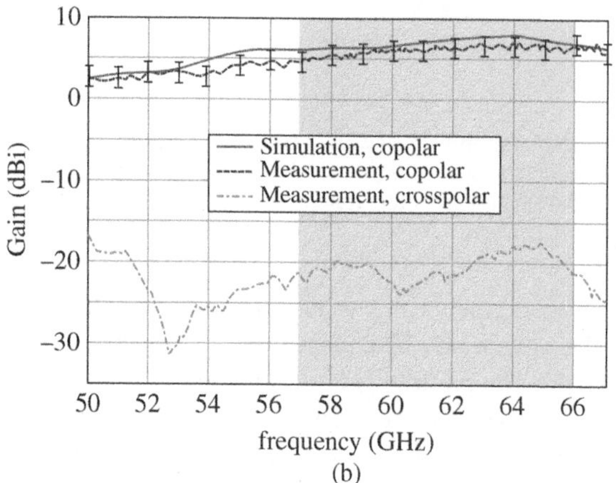


**Figure 7.19** (a) Reflection coefficient of six Tx antenna from six different BGA modules versus frequency from the same manufacturer #1. (b) Simulation and measurement of the broadside realized gain versus frequency of the Tx antenna for the copolar (with ±1.2 dB error bars) and the crosspolar field components from a BGA from manufacturer #1.

manufacturers exhibit a good matching level ($S_{11} < -10$ dB) from 47 to 67 GHz (33% BW). The broadside realized gain measured for $\phi = \{0°, \pm 45°, 90°\}$ is depicted in Figure 7.21b together with the total realized gain, computed as the sum (in linear units) of the 0° and 90° and the +45° and −45° polarizations. The comparison of both total gains shows a less than 0.2 dB difference in the 51–67 GHz band.

The AR of the antennas from manufacturers #1 and #2 is computed from the method described in [5] and plotted in Figure 7.22a with respect to frequency. The agreement between simulation and measurement is fair. The absolute frequency band with values under 3 dB is 7.5 GHz for the antenna from manufacturer #1 and 8.5 GHz for the antenna from manufacturer #2, yielding a relative BW of 14% with respect to 60 GHz. Finally, the measured realized gain radiation patterns for the two modules from both manufacturers are compared to the simulation results at 60 GHz for the H-plane in Figure 7.22b with good agreement with the simulation.

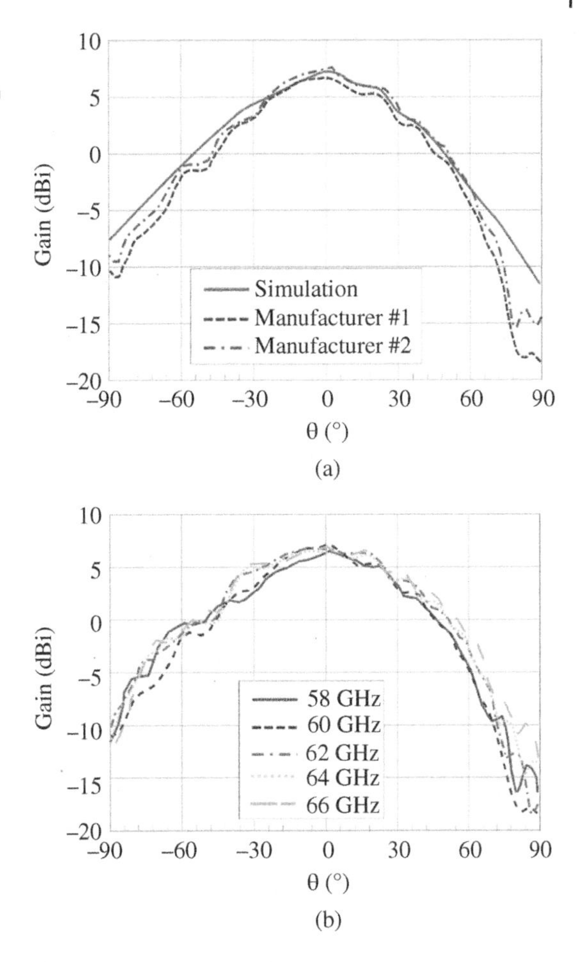


**Figure 7.20** (a) Simulation and measurement of the copolar realized gain radiation patterns of the Tx antenna from two BGA modules from two different manufacturers at 60 GHz in H-plane. (b) Measured realized gain radiation patterns of the copolar component of the Tx antenna of a BGA module from manufacturer #1 at five different frequencies within the WiGig band.

Leveraging the previous development, an array of no more than four ACP antennas was implemented and reported in [9, 10]. A picture of the fabricated module is presented in Figure 7.23.

The goal of this prototype was to study quadrature spatial combining: in-phase and quadrature signals are not combined at the circuit level but rather spatially in free space. To cope with surface waves a large single cavity surrounding each array was inserted. We found measurement results with antennas matched on the whole WiGig band and radiating around a 5 dBi realized gain.

### 7.2.3 94-GHz AiP Module

The previous example demonstrated the suitability of HDI organic technology to integrate 60 GHz antenna in the package along with an IC. We evaluated the achievable performance of the technology higher in frequency (94 GHz). This

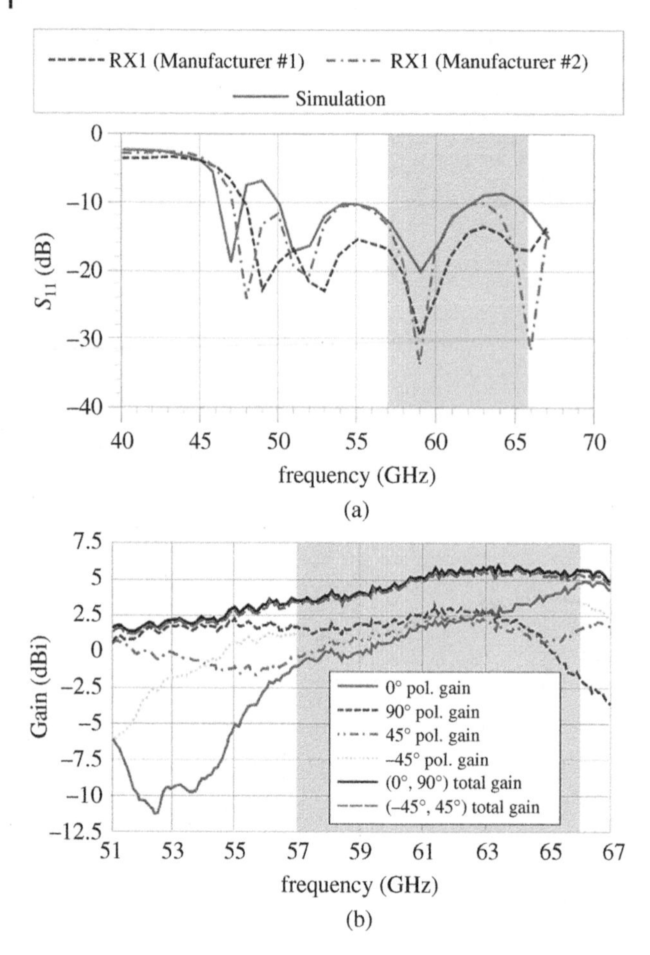


**Figure 7.21** (a) Simulation and measurement of the reflection coefficient versus frequency of the Rx antenna of two BGA modules from two different manufacturers. (b) Measured broadside realized gain of the Rx antenna versus frequency for the 0°, 90°, 45°, and −45° polarizations from a BGA module from manufacturer #1.

work has been reported in [11, 12]. The goal was to achieve a 94-GHz phased array transceiver with four transmitters and four receivers for frequency modulated continuous wave (FMCW) radars. In order to integrate 94-GHz antennas into an HDI module, the buildup had to be modified, especially because of the generation of surface waves. As stated previously, this generation depends on the thickness of the substrate. With the ACP antenna design, we have seen that a grounded cavity could manage the $TM_0$ mode. However, if the thickness of the substrate (mainly the core) is too large, other higher-order modes appear as well. We can calculate and plot (Figure 7.24) the propagation constant versus the substrate

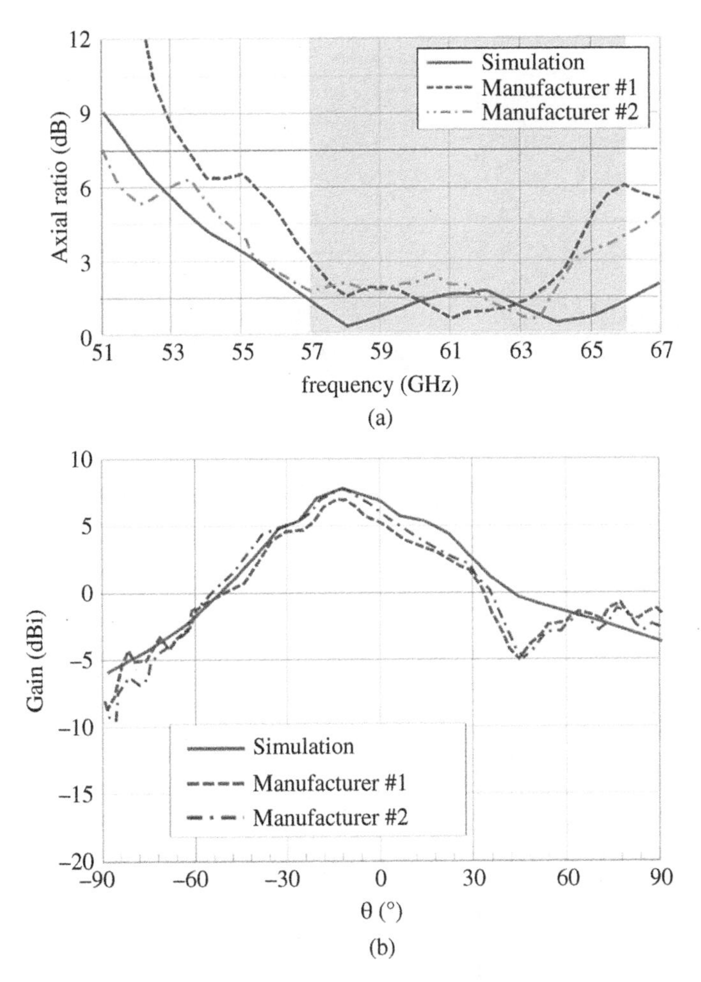


**Figure 7.22** (a) Simulated and computed (from measurement) AR of the Rx antenna versus frequency in the broadside direction of two BGA modules from manufacturers #1 and #2. (b) Simulation and measurement of the copular realized gain radiation patterns of the Rx antenna from two BGA modules from two different manufacturers at 60 GHz in the H-plane ($\phi = 0°$) at 60 GHz.

thickness for these different modes. Three thicknesses of substrate are highlighted in Figure 7.24 for a 94-GHz ACP antenna. It can be seen that the 300 and 400 μm substrates generate only the $TM_0$ mode, but with a 700 μm substrate the $TE_1$ mode is also generated. The chosen core thickness was 300 μm for our design.

A picture of the fabricated module is presented in Figure 7.25. As at 60 GHz, we have here an array of four ACP antennas surrounded by a cavity made of buried vias. For the radar application, the isolation between arrays and inter-elements is

**Figure 7.23** Fabricated module with two antenna arrays (Rx and Tx).

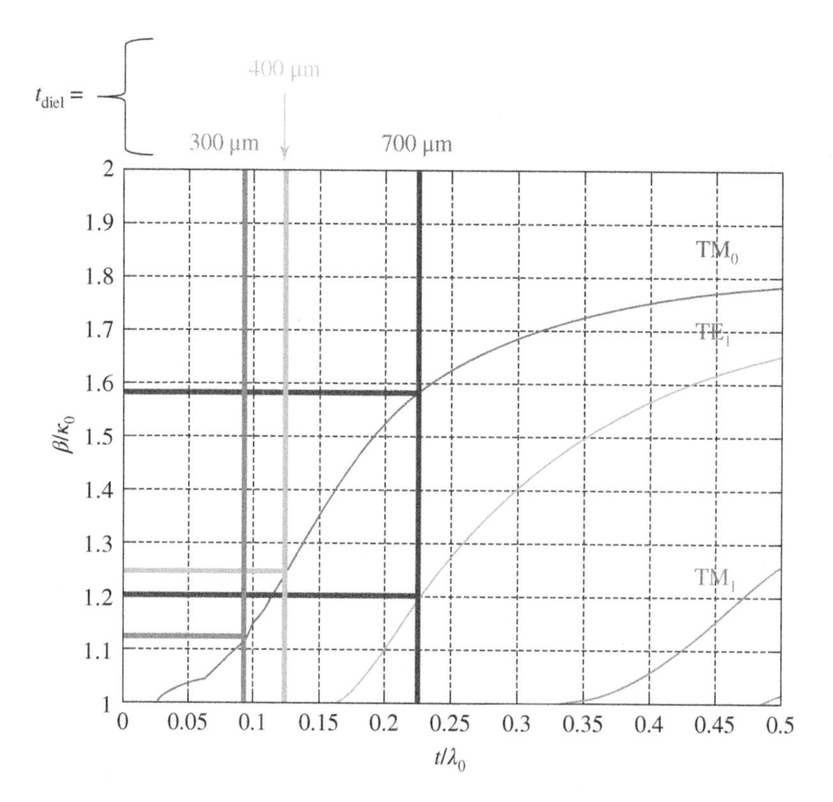


**Figure 7.24** Surface wave propagation constant β normalized to the free-space propagation constant $\kappa_0$ versus substrate thickness $t$ normalized to the space wavelength $\lambda_0$. Three thicknesses of the substrate are illustrated at 94 GHz.

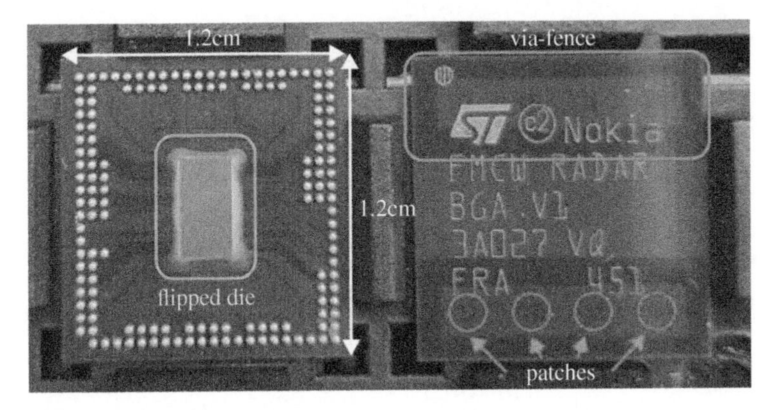


**Figure 7.25** HDI module at 94 GHz with ACP antenna arrays with a flip-chip die for FMCW radar application (from [11, 12]).

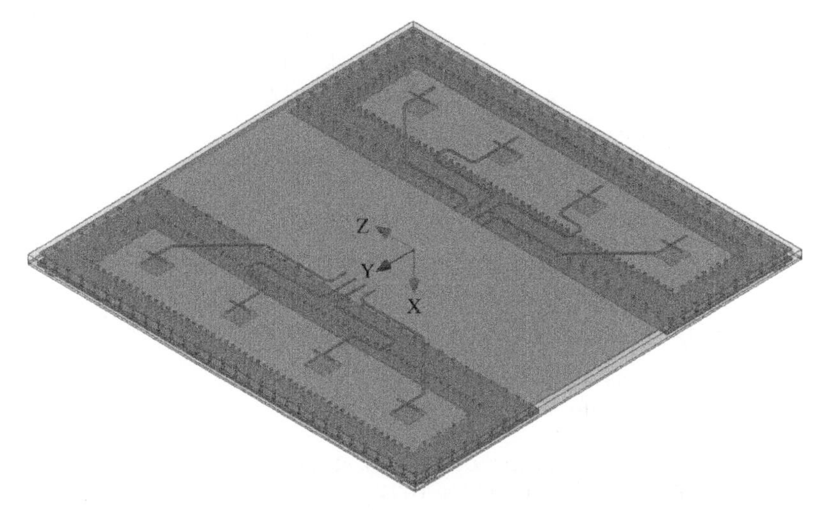


**Figure 7.26** Simulation model of the module at 94 GHz.

extremely important. Thus, the Tx and Rx antenna arrays are located on opposite edges of the module.

The simulation model is presented in Figure 7.26. To reduce element-to-element coupling, the antenna spacing within each array was set to $0.8\lambda$ at 94 GHz. Simulations showed a coupling level lower than −25 dB. The array can only be fed from the die itself, but the reflection coefficient and realized gain of each antenna were measured using the same setup from the University of Nice upgraded at 94 GHz [5].

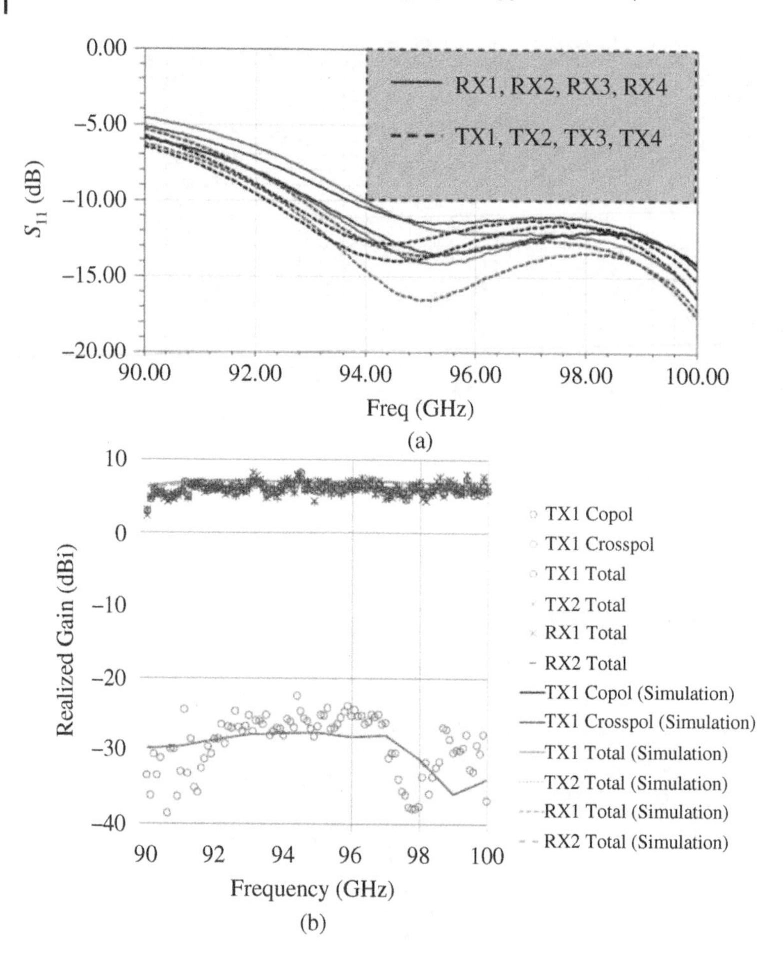


**Figure 7.27** (a) Measured reflection coefficient of each element of the two arrays. (b) Measured and simulated gain of inner (Rx2 and Tx2) and outer (Rx1 and Tx1) elements of each array.

The results are presented in Figure 7.27. The reflection coefficient bandwidth extends from 94 GHz to higher than 100 GHz (the limit of our measurement). The realized gain of all elements is higher than 6 dBi at 94 GHz, which would give a gain around 12 dBi for a whole array. A complete characterization of the module with the flip-chipped die was also carried-out. Beamforming measurements showed a steering range of about ±20° for the main radiated beam while maintaining grating lobes at least 3 dB below this main beam.

Using organic packaging technology, equivalent antenna performance has been achieved at 94 GHz by IBM as described in [13]. Consequently, HDI organic packaging technology has proven its suitability to support the integration of a high-performance and cost-effective AiP module up to 100 GHz.

## 7.3 Integration of AiP in Organic Packaging Technology in the 120–140-GHz Band

AiPs in HDI organic substrate packaging technology having been successfully validated up to 94 GHz, the next step was to assess if such a technology could operate beyond 100 GHz in order to support advanced wireless systems targeting data rate higher than 10 Gb/s. To do this our group developed a 120–140-GHz AiP module [14] integrated in standard HDI organic substrate technology previously operating at 60 and 94 GHz.

### 7.3.1 120–140-GHz AiP Module

Here again, the thickness of the core substrate had to be carefully chosen to get enough bandwidth behavior without generating too many surface waves. We chose a 200-μm thick core substrate (Figure 7.28).

The dimensions of the BGA module were set to $7 \times 7 \times 0.362\ \text{mm}^3$. Figure 7.29 shows a transparent top view of the BGA (left) and a detailed view of the antenna array (right). Previous linearly polarized antenna designs at 60 and 94 GHz have been leveraged to achieve a wide-band ACP antenna at 120 GHz. The BGA module

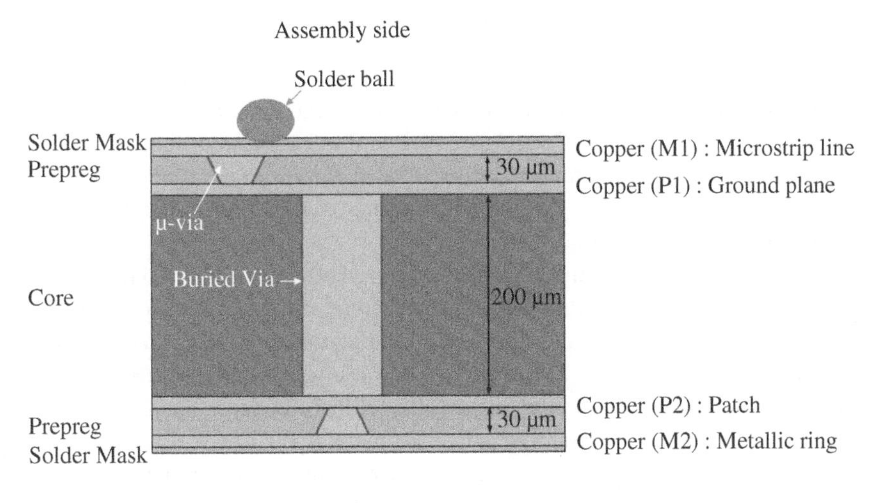


**Figure 7.28** Chosen HDI technology buildup for 120 GHz AiP (from [14]).

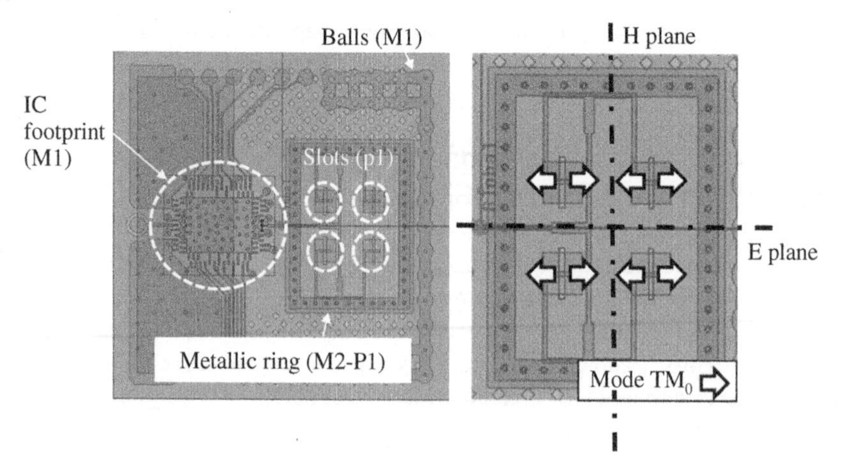


**Figure 7.29** Transparent top view of all the levels of the BGA module. Bottom view of the module with IC footprint and coupling slots fed by microstrip lines (left side). Top view of the module with 2 × 2 array of slot-fed patches (right side) with indications of the direction of propagation of the $TM_0$ mode generated by the slots (from [14]).

integrates a specially designed 2 × 2 array of four ACP antennas. The spacing between the patches was adjusted to obtain optimal illuminating beam width for an elliptical lens: the goal was to obtain a Gaussian radiation pattern in both the E- and H-planes of the BGA module with a power level 10 dB below the peak broadside radiation within an 80–100° angular region. This particular element arrangement was chosen because it partly cancels the $TM_0$ surface-wave mode generated by the slots within the core substrate from out-of-phase natural recombination.

Top and bottom views of the manufactured BGA can be seen in Figure 7.30. The footprint of a dedicated IC [designed by Stanford University using STMicroelectronics 55 nm bipolar complementary metal oxide semiconductor (BiCMOS) technology] has been outlined along with all the necessary DC, digital, and of course mmWave trace routing.

To validate the design within HDI technology, the BGA module radiating into the air was probe-fed and measured with the customized 3D measurement range presented in [15]. As shown in Figure 7.31a, the measured reflection coefficient was consistent with simulated results: $|S_{11}|$ was found to be below −10 dB from 96 to 140 GHz. The wide matching bandwidth achieved is attributed to the complementary behavior of the slot, the patch and the metallic ring surrounding the array antenna. Figure 7.31b shows the simulated and measured realized gain in the broadside direction of the top face of the BGA module. The simulated realized gain varied between 9 and 10 dBi from 116 to 140 GHz, and the measured realized gain values were very close (above 7.8 dBi in the working band at minimum). Note

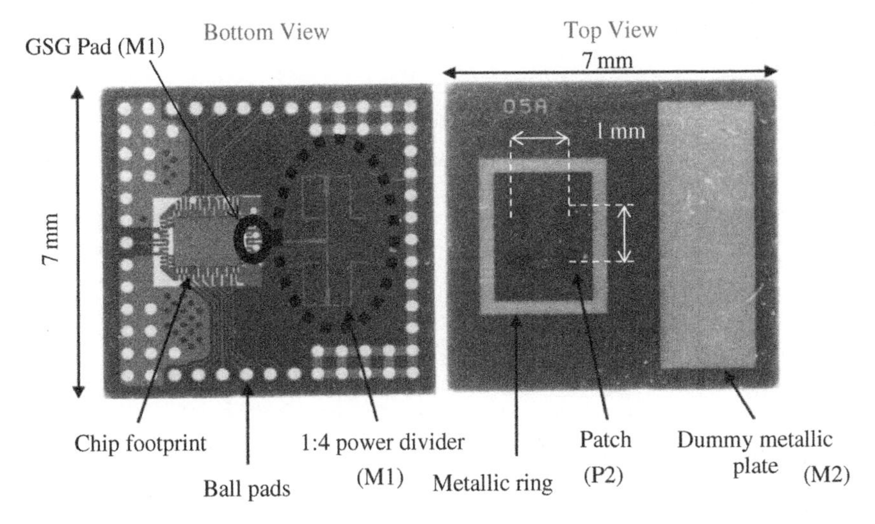


**Figure 7.30** Photographs of bottom and top views of the BGA module integrating the 2 × 2 antenna array (from [14]).

that the uncertainty of the measurement setup [15] is ±1.2 dB at 140 GHz. It is also interesting to note that the total simulated efficiency is greater than 66% within the working band, defined as 120–140 GHz.

A comparison between the simulated and measured realized gain of the E- and H-plane radiation patterns at 130 GHz is presented in Figure 7.32a,b. Very good agreement was observed in the H-plane. Except for the masking effect of the probing system that occurred in the E-plane and prevented angle measurements from −150° to 40°, the agreement between simulation and measurements remains acceptable in this plane.

Targeting a cost-effective antenna system for mass-market production, able to communicate over a few meters, we used 3D printing technology and acrylonitrile butadiene styrene-M30 (ABS-M30) plastic material to fabricate an elliptical lens and achieve higher antenna gain (using the BGA module as an antenna source). A simulated 3D model and the achieved elliptical lens are presented in Figure 7.33 (design details can be found in [14]).

A comparison between the lens antenna patterns obtained from simulations and from near-field (NF) measurements followed by a NF–far-field (FF) transformation is plotted in Figure 7.34 at 140 GHz. The agreement between the FF simulation and the NF–FF transformation obtained from the measurements was good for both planes of the main lobe and the first side lobes within the valid margin of the NF–FF transformation. The directivity was calculated by integrating the radiation pattern, which yielded 35.7 dB for simulation results and 34.6 dB in the case of NF measurements followed by a NF–FF transformation. The lens antenna

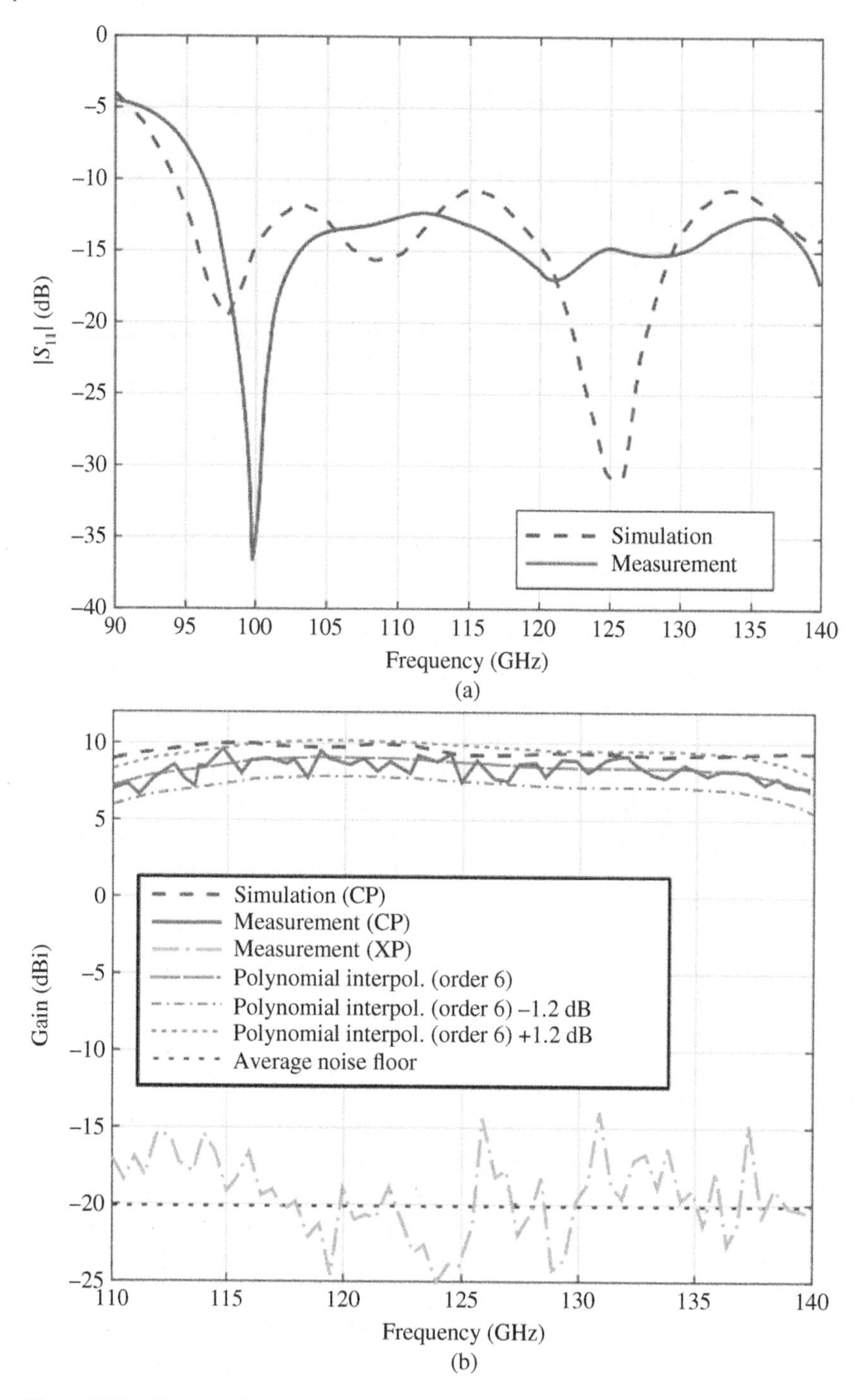


**Figure 7.31** Simulated and measured (a) $|S_{11}|$ of the array antenna of the BGA module radiating in air and (b) broadside realized gain (co- and cross-polarizations) versus frequency of the BGA array antenna (from [14]).

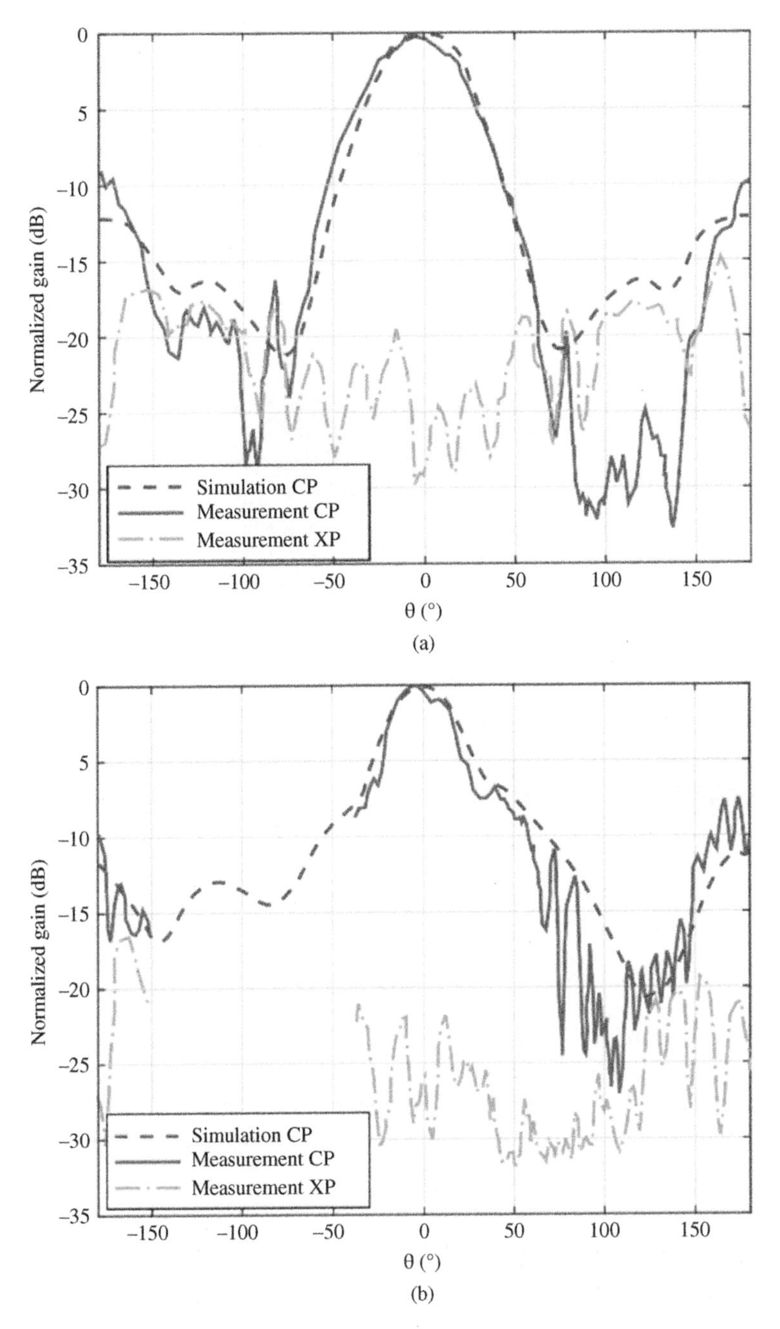


**Figure 7.32** Simulated and measured realized gain of the (a) H- and (b) E-plane radiation patterns at 130 GHz (from [14]).

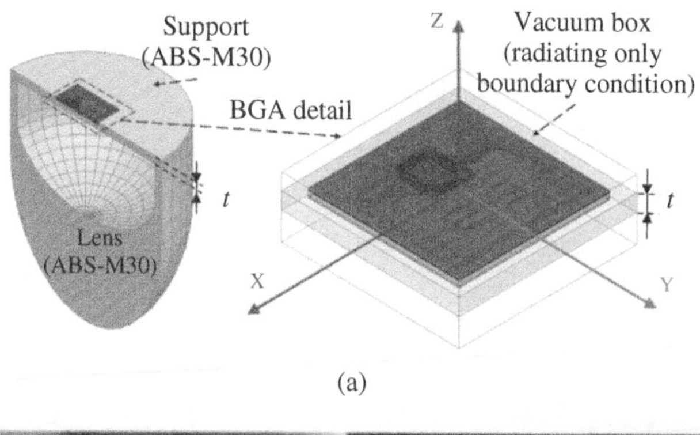


(a)

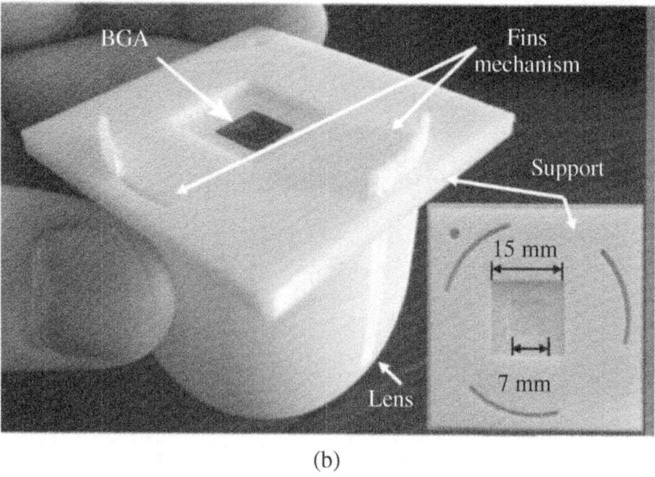


(b)

**Figure 7.33** (a) 3D view of the dome lens antenna (left). HFSS model (right) of the BGA source. (b) Photograph of the assembly of the elliptical dome lens with the BGA source (from [14]).

gain was obtained by means of the well-known intercomparison technique [16] with a standard gain horn in the F band. As the lens antenna gain is measured in the NF region, NF–FF compensation must be applied [14]. Finally, we extracted at 140 GHz a gain of 28.5 dBi, just 0.5 dB lower than the value obtained from FF simulations.

### 7.3.2 Link Demonstration Using a BiCMOS Chip with the 120-GHz BGA Module

In order to use the excellent antenna performance achieved at 140 GHz using industrial organic packaging and plastic 3D printing technologies, the previous

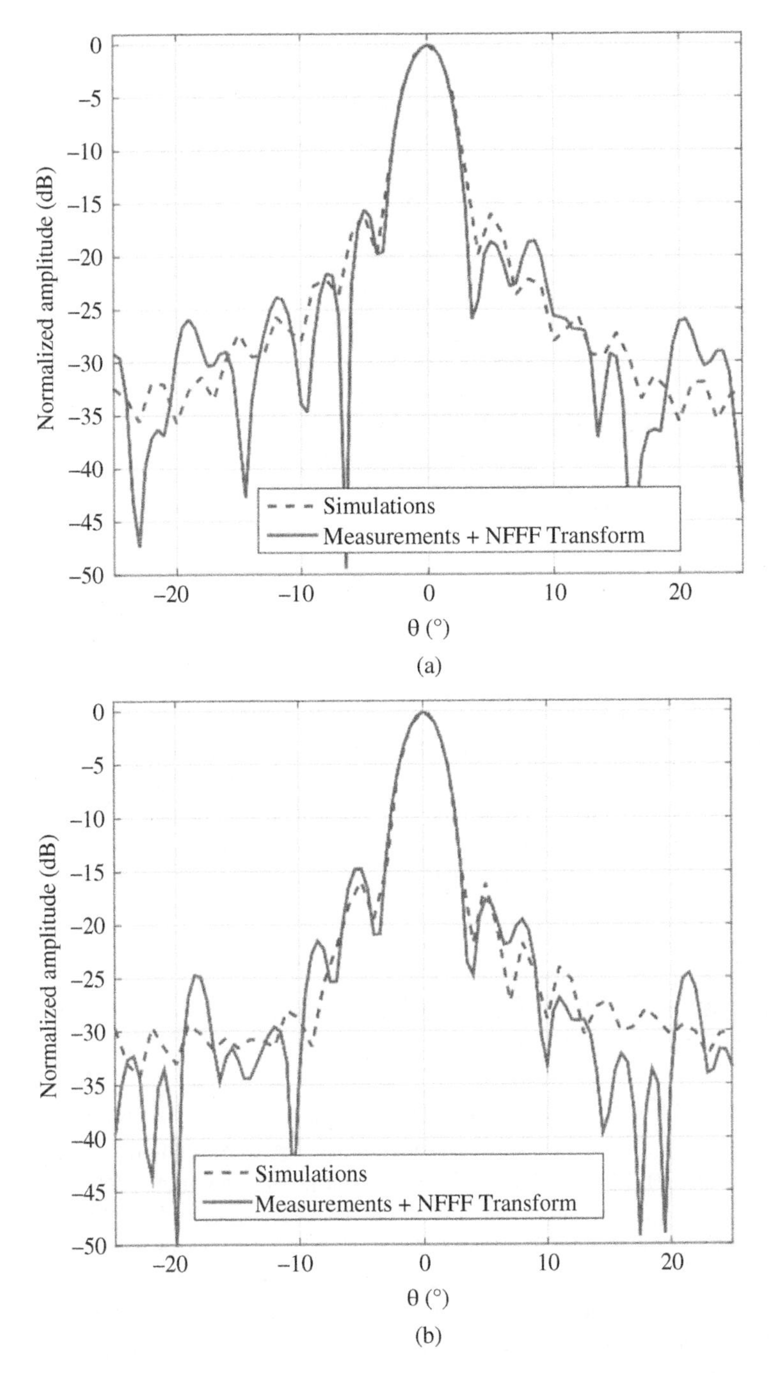


**Figure 7.34** Full antenna system radiation patterns at $f = 140$ GHz. Comparison between FF simulations and measurement results: (a) H-plane ($\phi = 0°$) and (b) E-plane ($\phi = 90°$) (from [14]).

**Figure 7.35** Complete antenna system with active chip and PC board for 5G backhaul links (from [14, 17]).

AiP module was assembled with a 55-nm BiCMOS IC developed by Stanford University [17]. The fabricated antenna system is presented Figure 7.35.

To perform measurements using a single bit error rate tester (BERT) and also observe the performance in complex mmWave channels, a metal reflector was used to close the Tx–Rx loop when transceivers are placed side by side with an absorber in between to avoid direct feedthrough paths (Figure 7.36). Even in the complex channel formed by the reflector, 12.5 Gb/s data transmission with $10^{-6}$ BER was measured at 5 m distance.

## 7.4 Integration of AiP in Organic Packaging Technology Beyond 200 GHz

Previous results achieved up to 140 GHz have demonstrated the suitability of industrial organic packaging technology to support the development of the AiP system in the 120–140 GHz band. Moreover, the good agreement achieved between simulation and measurements also indicates that we should be able to push the technology higher in frequency. Consequently, we have designed an AiP prototype in the 200–280 GHz band using previous antenna topology and integration strategy. The objective was to understand if organic packaging technology could be a viable solution to support an emerging wireless system beyond 200 GHz (the IEEE802.15.3d standard recently introduced).

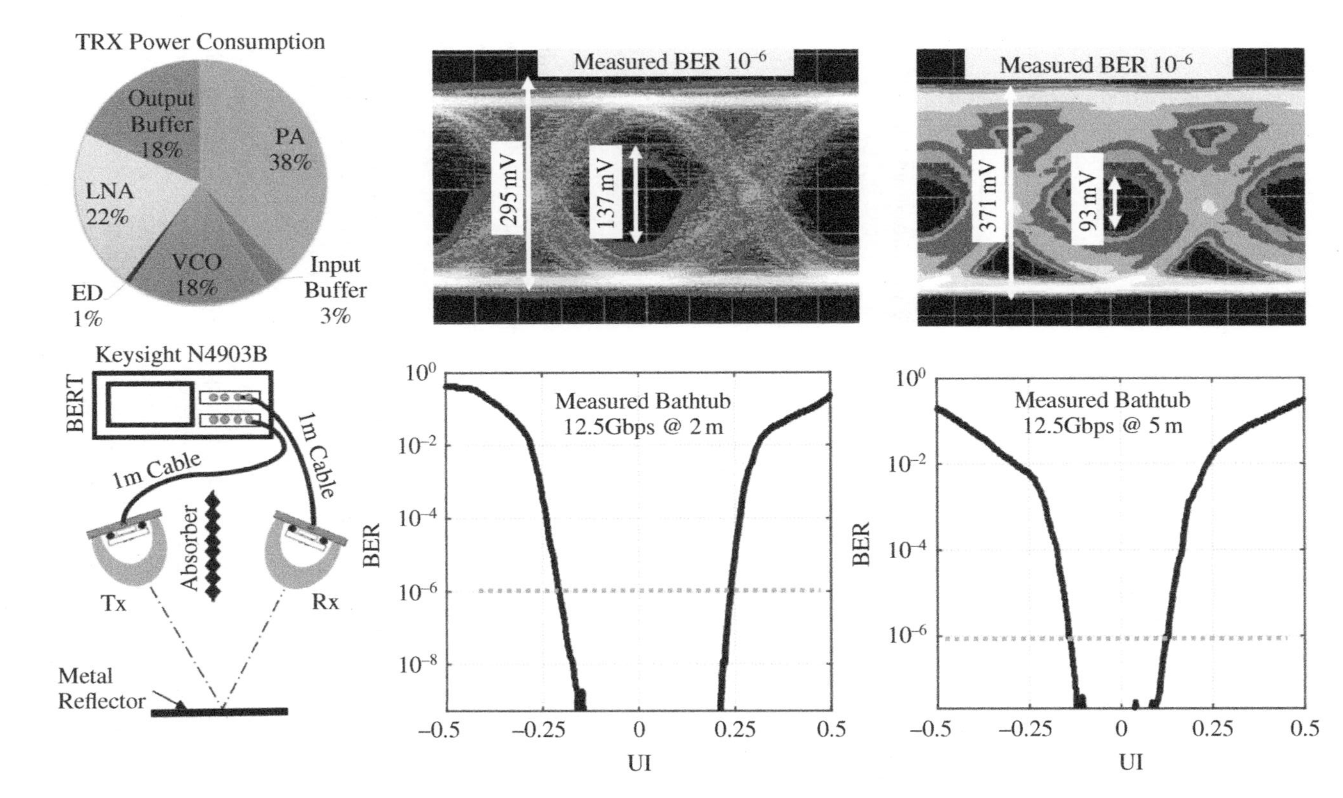


**Figure 7.36** Transceiver power consumption breakdown, TRX end-to-end measurement setup with the reflector, eye diagram and bathtub curves for data transmission at 2 and 5 m (from [17]).

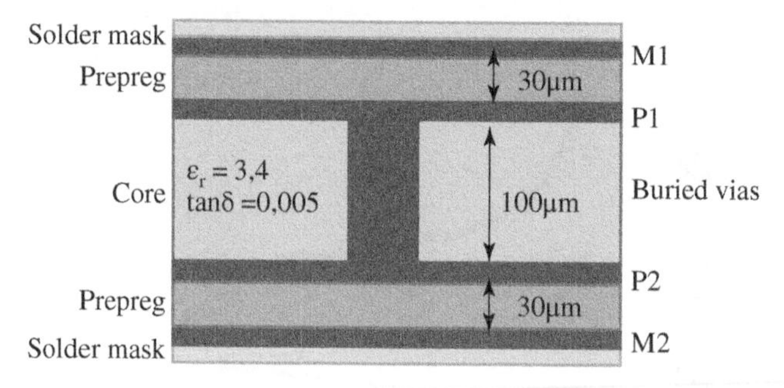


**Figure 7.37** Chosen HDI technology buildup for 240 GHz AiP (from [19]).

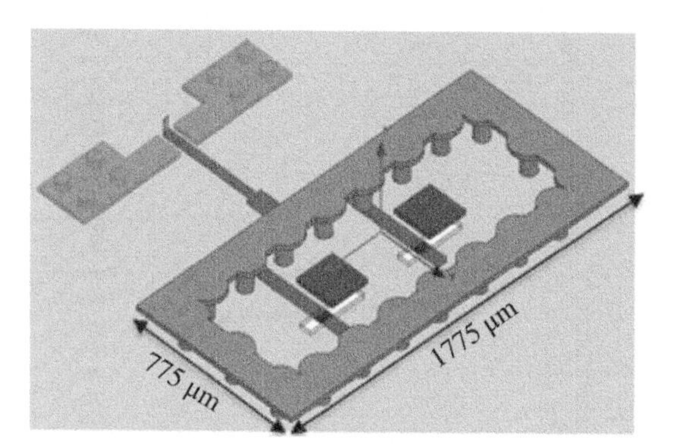


**Figure 7.38** 3D view of the HFSS model of the 240-GHz BGA source (from [19]).

In terms of antenna topology and AiP integration strategy, we used what we did at 140 GHz and carefully scaled the APC design. Here again, the thickness of the core substrate had to be carefully chosen to get enough bandwidth behavior without generating too many surface waves. We chose a 100-μm thick core substrate (Figure 7.37).

The AiP is integrated on a 5 × 5 mm$^2$ HDI organic substrate. It consists of a 2 × 1 array of ACPs (Figure 7.38) with a grounded metal cavity surrounding the two patches. The antenna design was optimized in order to operate in the 200–280-GHz band.

The antenna was measured using a probe-based measurement setup [18] operating in the 200–325-GHz band. The copolarization realized gain in the broadside direction and the reflection coefficient $|S_{11}|$ are first measured between 200 and 300 GHz (Figure 7.39). We noticed a strong gain notch at ~270 GHz, with a corresponding $|S_{11}|$ mismatch. This notch divides the usable expected huge frequency band into two dual bands: 215–240 GHz where the gain is over 5 dBi (7 dBi max. at 235 GHz) and 280–300 GHz where the gain is over 2 dBi. The result is a poor agreement between simulations and measurements.

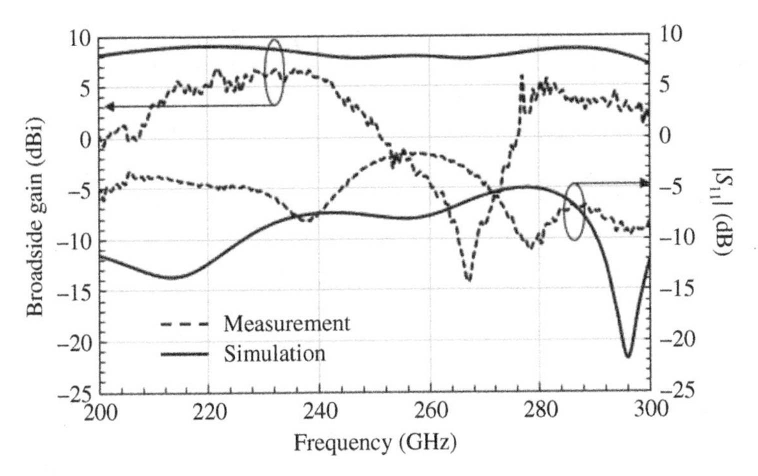


**Figure 7.39** Simulated and measured broadside realized copolarization gain and reflection coefficient versus the frequency of the proposed 240-GHz AiP (from [19]).

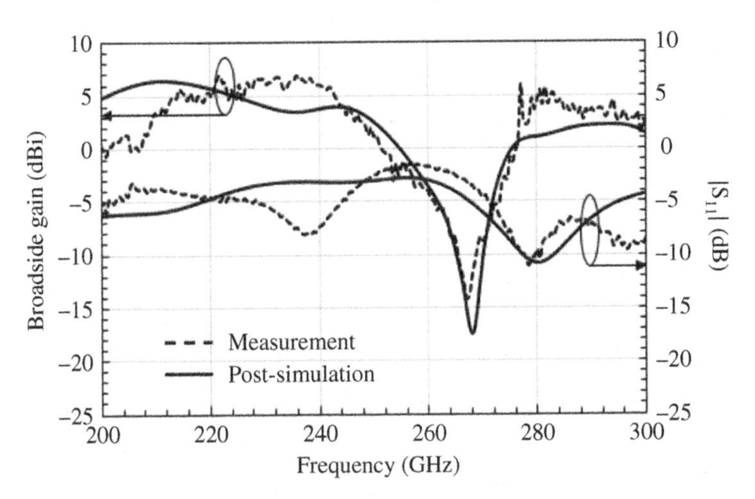


**Figure 7.40** Post-simulated and measured copolarization broadside realized gain and reflection coefficient of the AiP versus frequency (post-simulation with +40-μm shift on the vias positioning in the E-plane) [19].

Pushing the analysis of those results further, we observed that while the gain radiation pattern in the H-plane is in reasonable agreement with the simulated one, a strong discrepancy is observed between the simulated and measured radiation patterns of the E-plane [19]. Since the surface waves are travelling in the $y$–$z$ plane of the AiP (E-plane), the root cause of the observed discrepancies seems to be the tolerance accuracy of the grounded cavity.

An impact analysis of the fabrication tolerance was performed to identify the root cause of the issue. In Figure 7.40 the simulation was set with a cavity-width increase of +40 μm on each side of the cavity (in the E-plane direction). A fair simulation/measurement agreement was obtained, especially around the gain notch at 270 GHz. An X-ray of the fabricated antenna confirmed this dimensional issue for the manufactured cavity, validating the root cause of the observed problem. Moving to higher frequency, we become more sensitive to manufacturing tolerances and we are currently demonstrating that we are getting closer to organic packaging AiP limits.

## 7.5 Conclusion and Perspectives

AiP integration has been a hot topic for the past 10 years in an attempt to ease the development of silicon-based mmWave systems and make the technology available on the consumer market. To do this, various packaging substrate technologies have been evaluated. While early work used satisfactory ceramic technology, for cost reasons standard organic packaging technology was evaluated at mmWave frequency. This chapter has summarized the work performed by STMicroelectronics R&D, based in Crolles, and the University of Nice (Professor Cyril Luxey and Diane Titz's group) concerning the development of mmWave AiP modules. Below 100 GHz, both excellent antenna performance and agreement with simulations have been achieved. Equivalent results have been achieved by other research groups [20–26], confirming the suitability of organic packaging technology to support the integration of mmWave AiPs. Beyond the antenna performance, the developed AiP solutions have been assembled with silicon ICs in order to achieve a full system-demonstration up to 94 GHz. Below 100 GHz, AiP integrated in organic packaging technology is no longer a topic of research and we have already witnessed the introduction of the first mass-market product [27].

Our investigations were extended to the 120–140-GHz band to assess AiP achievable performance using industrial organic packaging technology. Once again, excellent performance and agreement with simulations were achieved. Moreover, the developed AiP module has been combined with a 55-nm BiCMOS IC and a 3D-printed plastic lens in order to achieve a functional 10 Gb/s link up to 5 m. Even at those high frequencies, we did not face any limitations when using existing industrial organic packaging technology.

Consequently, we moved to the upper band to achieve even higher data rates, the new sub-terahertz standard (IEEE802.15.3d) being discussed by the community. A 200–280-GHz AiP module has been designed and manufactured using standard organic packaging technology. While the achieved performance is acceptable and fair, for the first time we faced some limitations. Moving to higher frequencies, the dimensions of the antenna become smaller (patch width ~200 μm) and consequently more sensitive to the process accuracy (bearing in mind that most SAP processes offer minimum width and trace separation of ~20 μm). Beyond 200 GHz, we start to be limited by the capability of the technology to be able to go beyond an R&D demonstration and support real product development. However, it is likely that this limitation will not last long. Since digital ICs are continuously requiring more density, coreless organic packaging substrate technology is now focused on intense research to achieve minimum width and trace separation of ~2 μm [28]. Consequently, it is can be assumed that the accuracy issue faced by our first organic substrate-based AiP beyond 200 GHz should be solved by more aggressive design rules that will be available in more advanced organic packaging technologies.

However, beyond those incremental improvements, there is still some opportunity to perform research activity on more innovative and disruptive integration features. Moving to frequencies beyond 200 GHz, it becomes highly desirable to be able to embed waveguide structure since this can achieve better performance versus planar prototypes. Moreover, the prototype size is becoming compatible with targeted integration level. For instance, promising work has been performed using photo-imageable polymer-like material (especially SU8 resin [29]). This works has been further extended by JPL [30] using a more expensive process called DRIE on silicon wafer. From the research point of view, it would be very interesting to combine this approach with organic substrate packaging technology (e.g. using selective laser ablation). Such an approach would enable excellent planar antenna performance to be combined with a low-loss and high-performance waveguide feeding network in a cost-effective manner (along with the possibility of integrating silicon ICs). This could pave the way for a new wave of highly integrated mmWave and sub-terahertz systems operating beyond 200 GHz.

## References


1 F. Liu, C. Nair, V. Sundaram et al., Advances in embedded traces for 1.5 μm RDL on 2.5D glass interposers. *IEEE 65th Electronic Components and Technology Conference*, pp. 1736–1741, 2015.

2 D.M. Pozar, Microstrip antenna aperture-coupled to a microstripline. *Electronics Letters*, vol. 21, no. 2, pp. 49–50, January 1985.

3 R. Li, G. DeJean, M.M. Tentzeris et al., Radiation-pattern improvement of patch antennas on a large-size substrate using a compact soft-surface structure and its realization on LTCC multilayer technology. *IEEE Transactions on Antennas and Propagation*, vol. 53, no. 1, pp. 200–208, January 2005.

4 R. Pilard, D. Titz, F. Gianesello et al., HDI organic technology integrating built-in antennas dedicated to 60 GHz SiP solution. *IEEE International Symposium on Antennas and Propagation*, Chicago, IL, USA, 8–14 July 2012.

5 D. Titz, F. Ferrero, and C. Luxey, Development of a millimeter-wave measurement setup and dedicated techniques to characterize the matching and radiation performance of probe-fed antennas. *IEEE Antennas and Propagation Magazine*, vol. 54, pp. 188–203, 2012.
6 A. Bisognin, D. Titz, F. Ferrero et al., 3D printed plastic lens: enabling innovative 60 GHz antenna solution and system. *IEEE MTT-S International Microwave Symposium*, Tampa Bay, FL, USA, 1–6 June 2014.
7 A. Bisognin, D. Titz, F. Gianesello et al., BGA organic module for 60 GHz LOS communications. *International Symposium on Antennas and Propagation*, Okinawa, Japan, pp. 1038–1039, 2016.
8 A. Bisognin, D. Titz, F. Ferrero et al., Noncollimating MMW polyethylene lens mitigating dual-source offset from a Tx/Rx WiGig module. *IEEE Transactions on Antennas and Propagation*, vol. 63, issue 12, pp. 5908–5913, 2015.
9 J. Chen, L. Ye, D. Titz et al., A digitally modulated mm-Wave Cartesian beamforming transmitter with quadrature spatial combining. *IEEE International Solid-State Circuits Conference Digest of Technical Papers*, San Francisco, USA, 17–21 February 2013.
10 D. Titz, A. Bisognin, J. Chen et al., Antenna-array topologies for mm-Wave beamforming transmitter with quadrature spatial combining. *International Conference on Electromagnetics in Advanced Applications*, Aruba, 3–9 August 2014.
11 A. Townley, P. Swirhun, D. Titz et al., A 94 GHz 4TX-4RX phased-array for FMCW radar with integrated LO and flip-chip antenna package. *IEEE Radio Frequency Integrated Circuits Symposium*, pp. 294–297, 2016.
12 A. Townley, P. Swirhun, D. Titz et al., A 94-GHz 4TX–4RX phased-array FMCW radar transceiver with antenna-in-package. *IEEE Journal of Solid-State Circuits*, vol. 52, issue 5, pp. 1245–1259, 2017.
13 D. Liu, Md. R. Islam, C. Baks et al., A dual polarized stacked patch antenna for 94 GHz RFIC package applications. *IEEE Antennas and Propagation Society International Symposium*, pp. 1829–1830, 2014.
14 A. Bisognin, N. Nachabe, C. Luxey et al., Ball grid array module with integrated shaped lens for 5G backhaul/fronthaul communications in F-band. *IEEE Transactions on Antennas and Propagation*, vol. 65, issue 12, pp. 6380–6394, 2017.
15 A. Bisognin, D. Titz, F. Ferrero et al., Probe-fed measurement system for F-band antennas. *8th European Conference on Antennas and Propagation*, pp. 722–726, 2014.
16 Y. Álvarez, F. Las-Heras, and M. R. Pino, The sources reconstruction method for amplitude-only field measurements. *IEEE Transactions on Antennas Propagation*, vol. 58, no. 8, pp. 2776–2781, August 2010.

**17** N. Dolatsha, B. Grave, M. Sawaby et al., 17.8 A compact 130 GHz fully packaged point-to-point wireless system with 3D-printed 26 dBi lens antenna achieving 12.5 Gb/s at 1.55 pJ/b/m. *IEEE International Solid-State Circuits Conference*, pp. 306–307, 2017.

**18** E. Lacombe, F. Gianesello, A. Bisognin et al., Low-cost 3D-printed 240 GHz plastic lens fed by integrated antenna in organic substrate targeting sub-THz high data rate wireless links. *IEEE International Symposium on Antennas and Propagation & USNC/URSI National Radio Science Meeting*, pp. 5–6, 2017.

**19** E. Lacombe, F. Gianesello, A. Bisognin et al., 240 GHz antenna integrated on low-cost organic substrate packaging technology targeting high-data rate sub-THz telecommunication. *47th European Microwave Conference*, pp. 164–167, 2017.

**20** X. Gu, A. Valdes-Garcia, A. Natarajan et al., W-band scalable phased arrays for imaging and communications. *IEEE Communications Magazine*, vol. 53, no. 4, pp. 196–204, April 2015.

**21** S. Brebels, K. Khalaf, G. Mangraviti et al., 60-GHz CMOS TX/RX chipset on organic packages with integrated phased-array antennas. *Proceedings of the 10th European Conference on Antennas Propagation*, April 2016, pp. 1–5.

**22** G. Mangraviti, K. Khalaf, Q. Shi et al., A 4-antenna-path beamforming transceiver for 60 GHz multi-Gb/s communication in 28 nm CMOS. *IEEE International Solid-State Circuits Conference, Digital Technology Papers*, San Francisco, CA, USA, pp. 246–247, January/February 2016.

**23** A. Tomkins, A. Poon, E. Juntunen et al., A 60 GHz, 802.11ad/WiGig-compliant transceiver for infrastructure and mobile applications in 130 nm SiGe BiCMOS. *IEEE Journal of Solid-State Circuits*, vol. 50, no. 10, pp. 2239–2255, October 2015.

**24** C. Beck, H. Jalli Ng, R. Agethen et al., Industrial mmWave radar sensor in embedded wafer level BGA packaging technology. *IEEE Sensors Journal*, vol. 16, no. 17, pp. 6566–6578, September 2016.

**25** Y. Tsutsumi, T. Ito, K. Hashimoto et al., Bonding wire loop antenna in standard ball grid array package for 60-GHz short-range wireless communication. *IEEE Transactions on Antennas Propagation*, vol. 61, no. 4, pp. 1557–1563, April 2013.

**26** S. Li, T. Chi, J.S. Park et al., A fully packaged D-band MIMO transmitter using high-density flipchip interconnects on LCP substrate. *IEEE MTT-S International Microwave Symposium*, pp. 1–4, May 2016.

**27** Infineon, Project Soli information document. https://www.infineon.com/dgdl/Infineon-Google+Soli+FAQ+Document-FAQ-v03_00-EN.pdf?fileId=5546d4625d5945ed015d9845149e04c4.

**28** K. Oi, S. Otake, N. Shimizu et al., Development of new 2.5D package with novel integrated organic interposer substrate with ultra-fine wiring and high density bumps. *IEEE 64th Electronic Components and Technology Conference*, pp. 348–353, 2014.

**29** Y. Tian, X. Shang, Y. Wang et al., Investigation of SU8 as a structural material for fabricating passive millimeter-wave and terahertz components. *Journal of Micro/Nanolithography, MEMS, and MOEMS*, vol. 14, issue 4, December 2015.

**30** G. Chattopadhyay, T. Reck, C. Lee et al., Micromachined packaging for terahertz systems. *Proceedings of the IEEE*, vol. 105, issue 6, pp. 1139–1150, 2017.

# 8

# Antenna Integration in eWLB Package

*Maciej Wojnowski*[1*] *and Klaus Pressel*[2]

[1] *Infineon Technologies AG, Campeon 1-12, 85579 Neubiberg, Germany*
[2] *Infineon Technologies AG, Wernerwerkstraße 2, 93049 Regensburg, Germany*

## 8.1 Introduction

System-in-package (SiP) is a major trend in packaging today. It reflects a move toward integration of more functionality into a smaller volume. SiP enables heterogeneous integration of integrated circuits (ICs) along with sensors, micro-electromechanical (MEMS) components, and passive devices such as inductors, capacitor, filters and antennas. Another important trend for packaging is the continuing move toward higher frequencies. 5G high-speed wireless communication, millimeter-wave (mmWave) radar for autonomous driving, and high-resolution mmWave environment sensing and imaging are just a few examples of applications for future markets [1].

At mmWave frequencies, the size of a package becomes comparable to a wavelength. The parasitic wave effects such as impedance mismatch, signal reflections, crosstalk, and radiation can no longer be neglected even for very small chip-scale packages. This leads to large discontinuities at the chip–package and package–board interfaces that must be optimized. Moreover, a small wavelength at mmWave frequencies demands high-precision fabrication. Even small variations due to process tolerances can have a significant impact on the end-product performance. The use of traditional packages is therefore limited at mmWave frequencies.

Fan-out wafer-level packaging (FO-WLP) is a major trend to tackle both SiP integration and packaging of radio frequency (RF) and mmWave devices. In the following, embedded wafer-level ball grid array (eWLB) technology is presented, which was the first fan-out WLP technology introduced by Infineon in 2006 [2, 3].

*Corresponding author: Maciej Wojnowski; maciej.wojnowski@infineon.com

*Antenna-in-Package Technology and Applications*, First Edition.
Edited by Duixian Liu and Yueping Zhang.

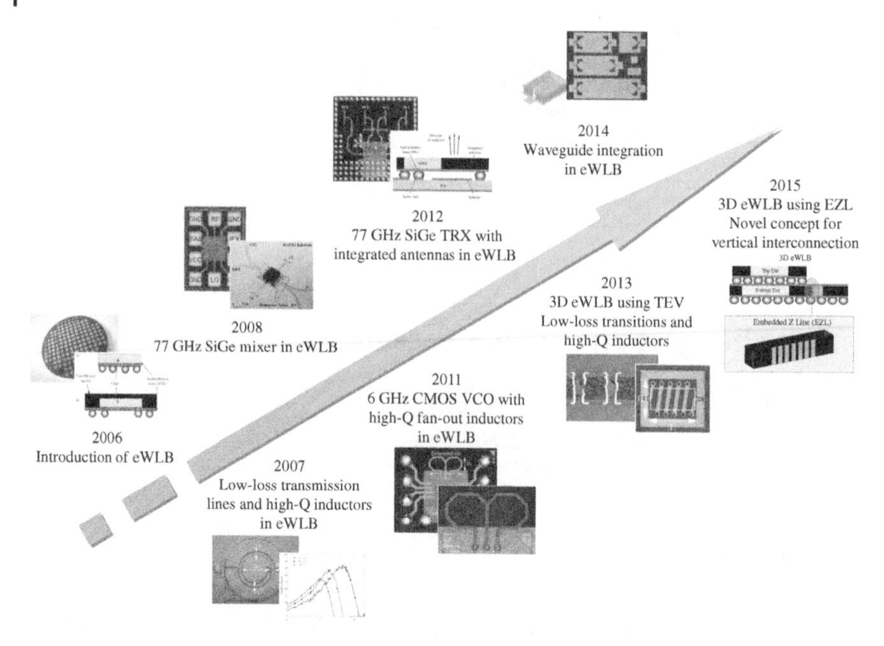


**Figure 8.1** Development of the eWLB since its introduction in 2006.

The eWLB is a modern assembly and packaging technology that offers attractive possibilities for mmWave systems. Figure 8.1 shows the development of the eWLB since its introduction in 2006. Originally introduced to overcome pin-count limitations of standard wafer level packages, the eWLB has been further developed since its introduction. Today it represents a universal 3D system integration platform for RF and mmWave applications.

## 8.2 The Embedded Wafer Level BGA Package

Wafer-level packaging (WLP) is defined as packaging of a component on the wafer level. Figure 8.2 (left) shows the classical WLP package with a thin-film redistribution layer (RDL) limited by the chip size. In the WLP a thin-film process can be applied on the entire front-end wafer to realize additional metallization layers above passivation. This RDL of the WLP is used to rearrange small chip pads located typically at chip edges into an array of larger solder ball pads over the chip surface. In the last step, solder balls are attached to the active side of the wafer. Thus, all chips are packaged together on the wafer level. After the wafer is diced, each chip is in a packaged format, ready for the subsequent tests and on-board assembly.

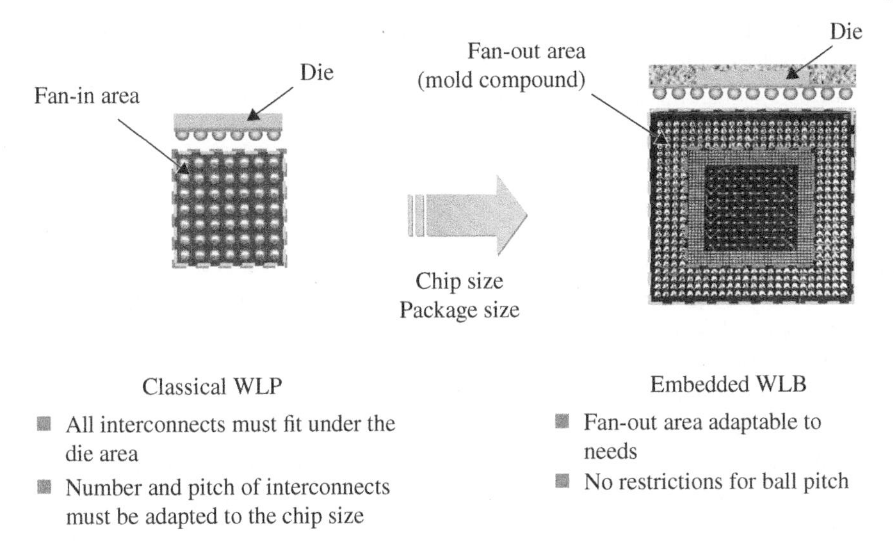


**Figure 8.2** Outline of a classical WLP with a fan-in area (left) and an eWLB package with a fan-out area (right).

WLP technology provides the ultimate size reduction because the package size is the chip size. In addition to redistribution purposes, the RDL can also be used for the realization of embedded passives, e.g. inductors or capacitors. Passives realized in the RDL take advantage of thick metals and low-permittivity (low-$k$) dielectrics. As a result, they usually show better performance compared to equivalent passives manufactured on a chip.

However, classical WLP techniques are always fan-in solutions. This means that all solder balls, which provide the interface to the board, are positioned under the chip area. For standard applications the pitch is typically larger than 0.4 mm, which restricts the possible number of package interconnects. Increasing the number of package pins requires additional space around the active area of the chip. As a result, classical WLP technologies cannot be used for devices with a high pin count, such as those used for digital data processing in communication and mobile applications. Another drawback of classical WLP packages is the close proximity of the structures realized in the RDL to the on-chip structures and lossy Si substrate. This close distance can change the electrical characteristics (detune) of the structures in the RDL and on the chip. It also can introduce crosstalk between the RDL and the chip. The latter is particularly pronounced for coils and transmission lines.

A solution that overcomes these limitations is the fan-out WLP package technology [4]. The eWLB was the first fan-out WLP technology when published by Infineon [2, 3]. The eWLB represents an innovative packaging concept that focuses on high input/output (I/O) count and high-performance RF as well as mmWave applications. The eWLB is based on a technology where the chips are embedded in a mold compound and a fan-out area for redistribution is generated. Figure 8.2 (right) illustrates the eWLB package with an additional fan-out area around the chip.

### 8.2.1 Process Flow for the eWLB

The basis of the eWLB technology is a molded reconfigured wafer concept. Unlike other wafer-level packaging technologies, thinned, diced, and tested chips are the starting point of the packaging process. Figure 8.3 shows the three main eWLB process blocks: (a) reconstitution, which leads to a so-called molded wafer, (b) redistribution, and (c) ball apply followed by singulation.

In the first step, the chips are rearranged face down on a metal carrier covered by a thermo-releasable adhesive tape. To increase the package area, the chips are relocated in a larger spacing than they had on the front-end wafer. The relocated chips are then encapsulated by compression molding. The encapsulation process (here based on molding) is the core process of the eWLB technology. It fixes the chips by filling the gaps with mold material. This reconfiguration creates an additional fan-out area around the chip. Afterward, the tape is peeled from the molded wafer. Next, a thin-film technology is applied on the molded reconfigured wafer to form the RDL. In the final process step, solder balls are attached and the molded

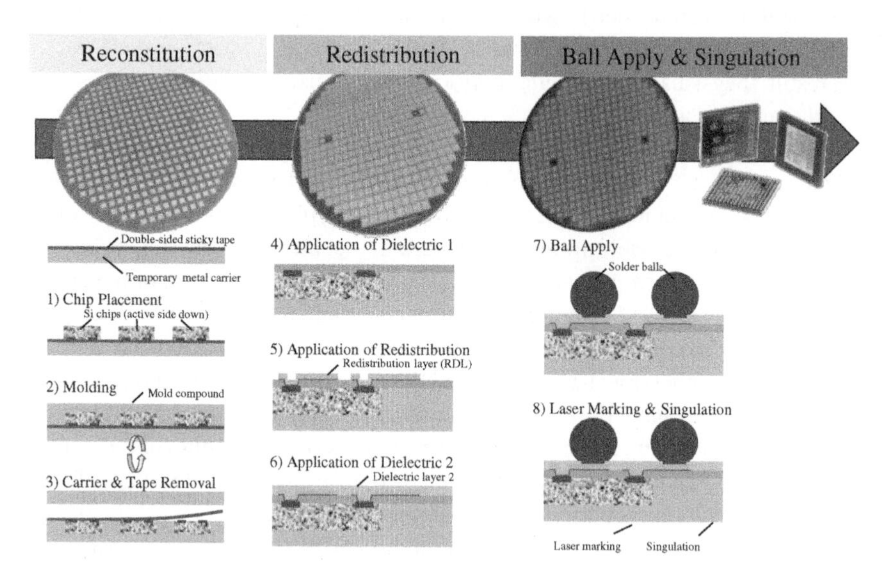


**Figure 8.3** Process flow for the fabrication of the eWLB package. The process flow is divided into the three parts: (a) reconstitution, (b) redistribution, and (c) ball apply together with singulation.

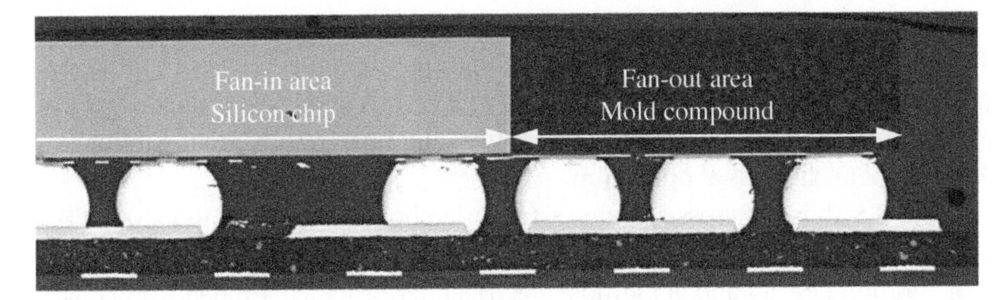


**Figure 8.4** Cross-section of a typical eWLB package with fan-in and fan-out areas.

wafer is diced. Figure 8.4 shows an example of a cross-section for an eWLB package with fan-in and fan-out areas together with the thin-film RDL. The fan-in area is limited by the chip size, whereas the fan-out area extends from the chip edge all the way out to the package edge. A detailed description of the process steps for the eWLB package can be found in [2, 3].

The eWLB technology combines in a unique way the state-of-the-art WLP processes with the possibility of fabricating a fan-out area. Parallel processing typical for front-end technologies is the main advantage of this technology. The additional fan-out area of the eWLB technology overcomes the restrictions of standard wafer level packages, such as the limited number of package pins possible at a given chip size and ball pitch. The thin-film technology offers the advantages of high metal-pattern resolution (small line/space), low temperature, and competitive costs. In addition, it enables very flexible and highly customizable package designs. The resulting eWLB package has a low profile, since the height of the package can be made comparable to the chip thickness (without contact balls to the board). Standard front-end grinding techniques can be used to reduce the thickness of the molded reconfigured wafer.

### 8.2.2 Vertical Interconnections in the eWLB

Three-dimensional (3D) integration provides additional flexibility for SiP integration. The key component for 3D integrations is a vertical interconnection. For the eWLB there have been various approaches investigated to realize vertical interconnections.

One example of vertical interconnection is the embedded via bar. This is typically a piece of laminate (bar) with pre-formed vias. The embedded via bars are placed on a metal carrier and encapsulated with mold compound together with Si chips. The achievable via aspect ratios and electrical characteristics depends on the laminate technology. Cost-effective FR-4 laminates as well as low-loss high-frequency laminates with high-resolution metallization for mmWave applications can be used. Instead of laminate material also via bars made of silicon with through silicon vias (TSVs) can be applied.

A second example of vertical interconnection is a through encapsulant via (TEV). TEVs are fabricated by laser drilling and subsequent copper filling by sputtering and electroplating. The resulting TEVs are cylindrical in shape and are non-filled inside. The diameter of TEVs is typically 100–150 μm and the thickness of the copper coating is about 10 μm.

Figure 8.5 shows a cross-section of (a) an embedded vias bar and (b) a TEV manufactured in the eWLB [5]. Both technologies are comparable with respect to integration density and achievable electrical performance. Embedded via bars

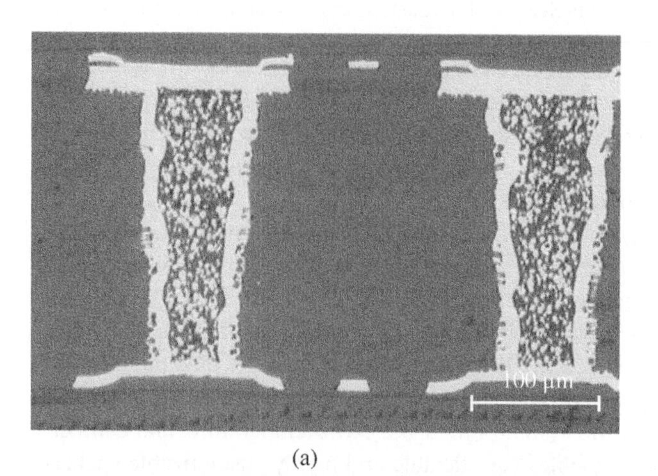


(a)

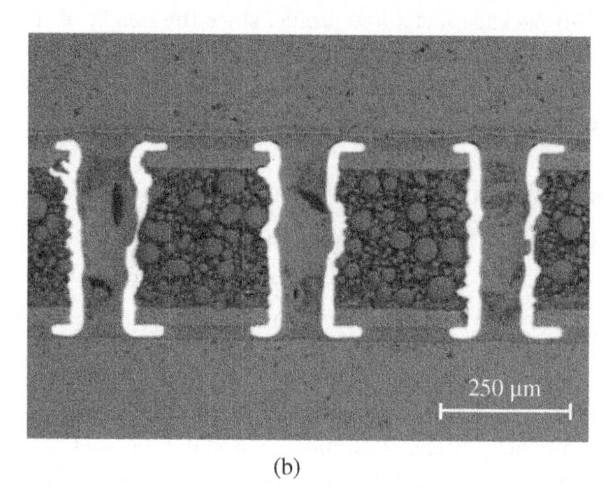


(b)

**Figure 8.5** Cross-section of vertical interconnections realized in the eWLB using (a) an embedded via bar and (b) a TEV (from [5], ©2019 IEEE, reprinted with permission).

offer less design freedom due to their fixed design but can have cost advantages. Depending on laminate technology for embedded via bars or manufacturing processes for TEVs, vertical interconnections of various diameters can be realized. On the one hand, thin vias of high aspect ratio allow compact design and the highest degree of miniaturization. On the other hand, thick vias offer lower resistance and reduced inductance but at the cost of increased size. Thick vias are an ideal fit for power/ground interconnections while thin vias are preferred solution for RF/high-speed signals. By selecting the diameter and the distance between the vias, the high-frequency performance of vertical transition can be optimized. However, the geometry of the transitions realized either using embedded via bars or using TEV remains limited to vertical cylinders of given diameter. Thus, they cannot compete with the RDL in terms of line/space resolution and design flexibility. One solution that overcomes these limitations is embedded Z-line (EZL) technology [6].

### 8.2.3 Embedded Z-Line Technology

The EZL vertical interconnect technology enables the realization of vertical interconnections of high metal-pattern resolution and a wide range of widths and pitches ranging from several micrometers up to hundreds of micrometers.

The EZLs are pre-fabricated bars of mold compound with interconnections realized using thin-film RDLs of an eWLB. The process flow for EZL technology starts with realizing horizontal traces onto a dummy molded wafer using a thin-film RDL of an eWLB. In the next step, the molded wafer is diced to obtain bars of mold compound with short interconnects on top. These bars are turned 90° and placed side by side to the Si chips and other components on the artificial wafer. Figure 8.6 shows a photograph of a system carrier with Si chips and EZL interconnects [6]. Figure 8.6a is prior to overmolding. Figure 8.6b is an X-ray computer tomography (CT) image of an EZL test structure.

The use of thin-film RDLs enables the realization of vertical interconnections of high resolution and a wide range of widths and pitches ranging from several micrometers up to hundreds of micrometers. Unlike embedded via bars and TEV, EZL technology is not limited to the direct vertical interconnection of a uniform cross-section. The EZL structure can vary in width along its length. This unique feature of EZLs allows the realization of both vertical and horizontal (skewed or stepwise) interconnections or complex signal/power distribution networks. Moreover, the high resolution of the RDL enables the realization of transmission lines and matching networks of precisely controlled electrical characteristics, as shown in Figure 8.7 [6].

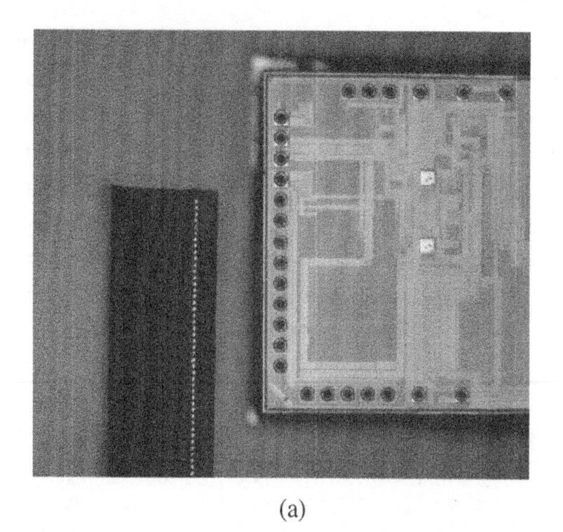

(a)

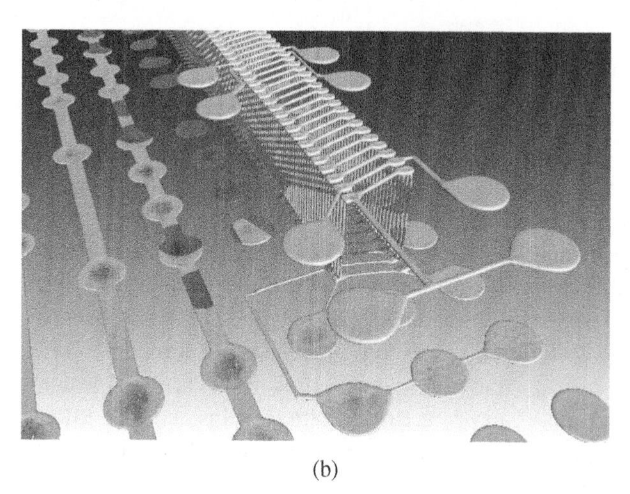

(b)

**Figure 8.6** (a) Photograph of a system carrier with Si dies and EZL interconnects prior to overmolding. (b) X-ray CT image of an EZL test structure with daisy chains (from [6], ©2019 IEEE, reprinted with permission).

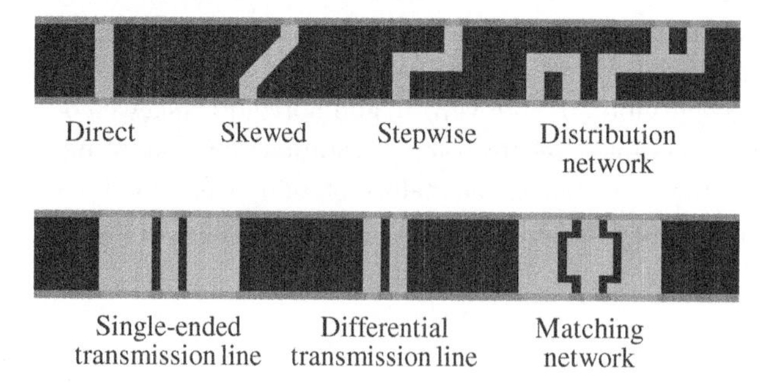


**Figure 8.7** A flexible EZL enables the realization of single vertical interconnections, distribution networks, and high-frequency transmission lines and matching networks (from [6], ©2019 IEEE, reprinted with permission).

## 8.3 Toolbox Elements for AiP in eWLB

The fan-out area of the eWLB is available for solder balls and interconnections. However, it also can be used for transmission lines and other planar components that can be realized using the thin-film RDL.

The key material of the eWLB technology is the mold compound used for encapsulation. It creates a carrier for all structures realized in the fan-out area and thus determines their electrical performance. The mold compound used in the eWLB shows very low loss ($\tan \delta = 0.004$). This property, in combination with the high metal-pattern resolution of the thin-film RDL, allows the realization of all basic toolbox elements required for antenna-in-package (AiP) integration, such as:

- transmission lines
- embedded passives and distributed RF circuits
- RF transitions to printed circuit boards (PCBs).

### 8.3.1 Transmission Lines

Transmission lines are fundamental building blocks in RF and mmWave systems. In packages, transmission lines are used for signal distribution between chip and board, and between components inside a package, e.g. between two chips or between a chip and a periphery. For AiP integration, transmission lines provide interconnection between chip and antenna elements, and can be used for impedance matching (e.g. $\lambda/4$ impedance transformer). They can also be used to realize distributed mmWave passive components in a package (e.g. couplers and baluns) to separate transmitted and received signals for monostatic radar systems. In all these applications, the key parameters of transmission lines are compact lateral dimensions, low attenuation, and a wide range of realizable impedances.

High resolution (small line/space) of the RDL metallization and low-loss mold compound enables the realization of low-loss transmission lines of precisely controlled characteristic impedance. For a single-layer RDL, standard coplanar waveguides (CPWs) for single-ended signals as well as coplanar strips (CPSs) for differential signals can be realized. For a double-layer RDL, a standard thin-film microstrip line (TFMSL) and various specialized lines are available, e.g. finite-width quasi-CPW with elevated signal conductor [7, 8]. Figure 8.8 shows schemes for the CPW, CPS, TFMSL and quasi-CPW transmission lines in the eWLB.

Table 8.1 lists the electrical parameters of CPW, CPS, TFMSL and quasi-CPW transmission lines simulated at 77 GHz for standard RDL metal-pattern resolution (min. line/space = 15 μm) and for assumed max. line/space = 100 μm [7, 8]. The second column in Table 8.1 gives the lateral dimensions of transmission lines for standard impedance ($Z_0 = 50\ \Omega$ for single-ended lines and $Z_0 = 100\ \Omega$ for differential lines). The last column in Table 8.1 provides the range of impedances $Z_0$

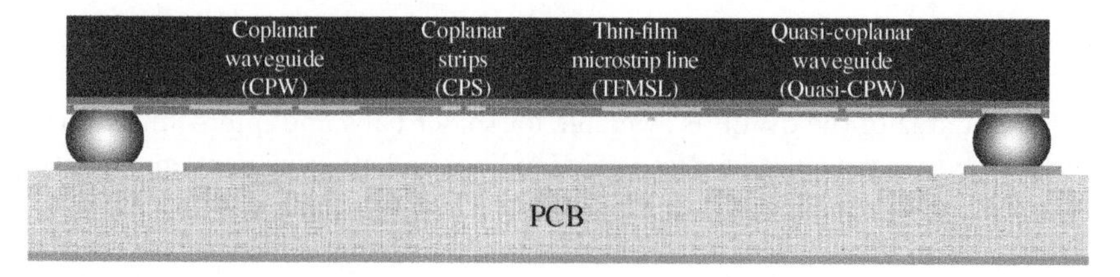


**Figure 8.8** Transmission lines available in the eWLB.

**Table 8.1** Parameters of transmission lines realizable in the eWLB for line/space = 15–100 μm simulated at 77 GHz [7, 8].

| Line type | Standard line dimensions w/s (μm) | Line attenuation $\alpha$ (dB/mm) | Relative effective permittivity $\epsilon_{r\,eff}$ | Characteristic impedance $Z_0$ (Ω) |
|---|---|---|---|---|
| CPW | 90/20 ($Z_0$ = 50 Ω) | 0.12–0.46 | 2.3–2.8 | 44–135 (0.9–2.7) |
| CPS | 45/20 ($Z_0$ = 100 Ω) | 0.13–0.43 | 2.3–2.7 | 76–219 (0.8–2.2) |
| TFMSL | NA ($Z_0$ = 50 Ω) | 0.66–0.71 | 2.9–3.0 | 11–41 (0.2–0.8) |
| Quasi-CPW | 30/20 ($Z_0$ = 50 Ω) | 0.12–0.55 | 2.7–3.1 | 14–135 (0.3–2.7) |

NA, not applicable.

(in Ω and in brackets relative to standard impedance) achievable for the assumed line/space range. The attenuation ranges from 0.12 to 0.46 dB/mm for CPW and CPS. This is significantly less than for transmission lines available on-chip (typically 1 dB/mm for a 50 Ω microstrip line). The range of realizable impedances includes 50 Ω for CPW and 100 Ω for CPS, and gives certain degree of freedom in designing distributed circuits. The TFMSL has the highest attenuation and the narrowest range of realizable impedances (not including 50 Ω) and is therefore of limited interest. The quasi-CPW line offers low attenuation and the widest range of impedances. Another advantage of the quasi-CPW line is a simple realization of connections between ground planes without vias. Such ground bridges are typically used to ensure the equipotentiality of the ground conductors on both sides of the signal line. This helps to suppress higher-order modes (e.g. coupled slot mode) when longer CPW lines or CPW junctions are used.

Figures 8.9 and 8.10 show the measured and simulated attenuation constant and relative effective permittivity of a 50 Ω CPW and a 100 Ω CPS realized in the eWLB, respectively [7]. The thin-film stack-up of the eWLB consists of two 10-μm thick dielectric layers and a single 8-μm thick copper metal layer in between. The lines have the dimensions as shown in Table 8.1. The CPW shows attenuation of 0.25 dB/mm at 60 GHz and 0.30 dB/mm at 80 GHz. The relative effective permittivity is 2.6 at 60 and 80 GHz. Ref. [9] presents a detailed analysis of the CPW

(a)

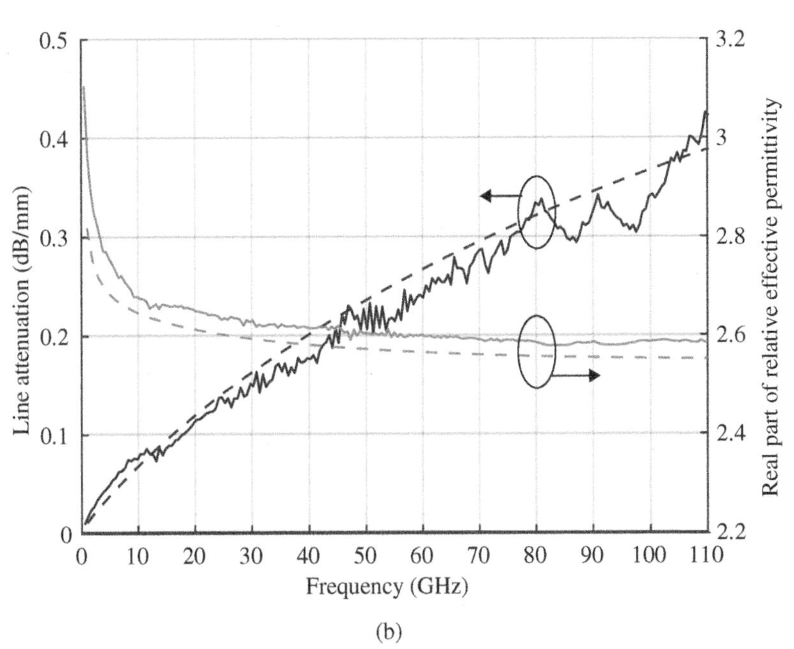


(b)

**Figure 8.9** (a) Photograph and (b) measured (solid line) and simulated (dotted line) attenuation and relative effective permittivity of a 50 Ω CPW realized in the eWLB (from [7], ©2019 IEEE, reprinted with permission).

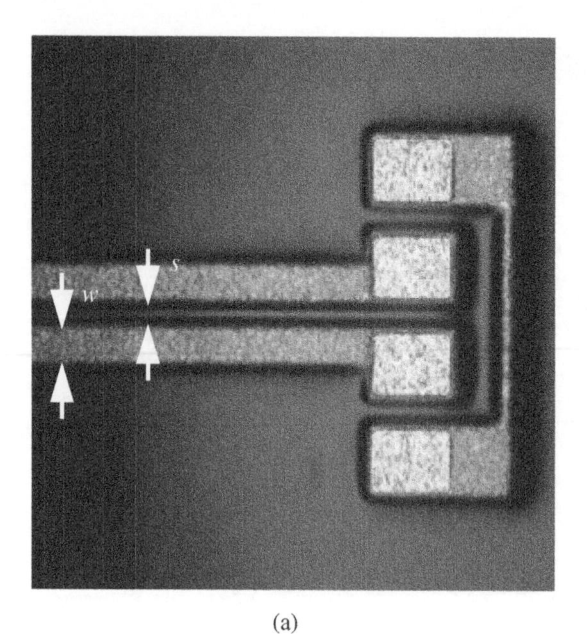


(a)

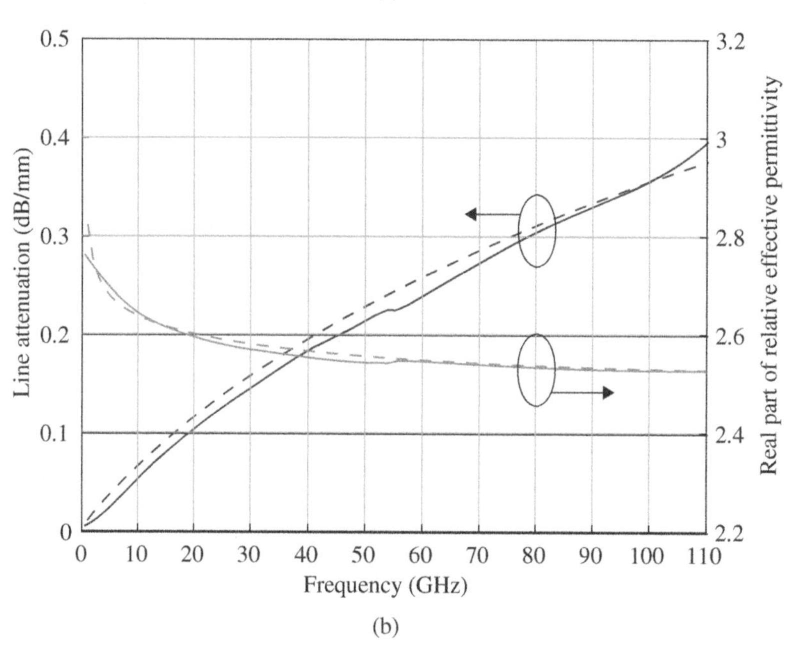


(b)

**Figure 8.10** (a) Photograph and (b) measured (solid line) and simulated (dotted line) attenuation and relative effective permittivity of a 100 Ω CPS realized in the eWLB (from [7], ©2019 IEEE, reprinted with permission).

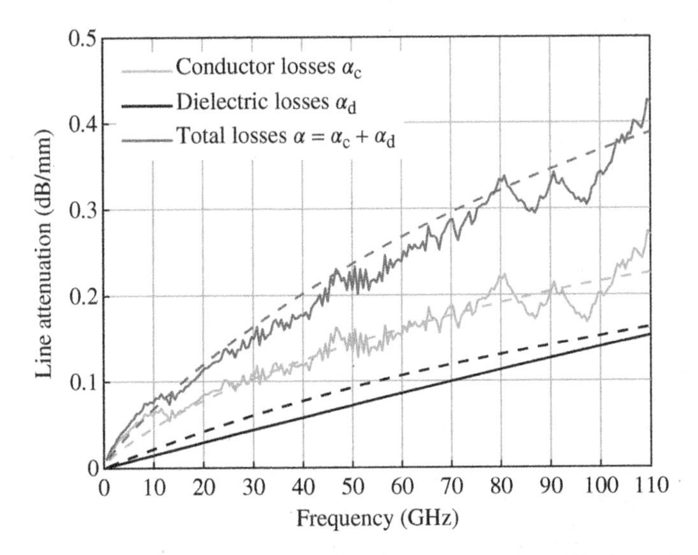


**Figure 8.11** Measured (solid line) and simulated (dotted line) contributions of the conductor and dielectric losses of a 50 Ω CPW in the eWLB (from [7], ©2019 IEEE, reprinted with permission).

realized in the eWLB in terms of resistance, inductance, capacitance, and conductance (RLCG) per unit length (p.u.l.) parameters and compares it to transmission lines fabricated in other common technologies.

Based on the measured distributed RLCG parameters, it is possible to separate the electrical effects of the conductors and dielectrics in the transmission line. Since the transmission line in the eWLB is a low-loss line, i.e. $R \ll \omega L$ and $G \ll \omega C$, the attenuation constant $\alpha$ can be expressed as a sum of the conductor and dielectric losses, $\alpha = \alpha_c + \alpha_d$. Figure 8.11 illustrates the measured and simulated frequency-dependent contributions of the conductor and dielectric losses to the attenuation of a 50 Ω CPW in the eWLB [7]. The attenuation at lower frequencies is mainly due to losses in conductors (90% at 1 GHz). However, as the frequency increases, the contribution of losses in dielectrics increases (already 30% at 10 GHz and 40% at 110 GHz). The electrical parameters of the CPS are very close to those of the CPW. This demonstrates that the performance of the transmission lines can be further enhanced by applying thin-film dielectrics of lower losses.

### 8.3.2 Passive Components and Distributed RF Circuits

The eWLB technology enables the realization of high-$Q$ passives and filters, as well as lumped and distributed mmWave circuits in the fan-out area. For AiP integration, these components can be used for impedance matching and realizing networks to divide/combine, filter, and separate transmitted and received signals.

The eWLB technology enables the realization of single- and double-layer planar inductors [9, 10]. Figure 8.12 shows a photograph of a single-layer spiral inductor realized in the fan-out area of the eWLB and the measured quality factor of four single-layer spiral inductors with different number of windings N. The thin-film

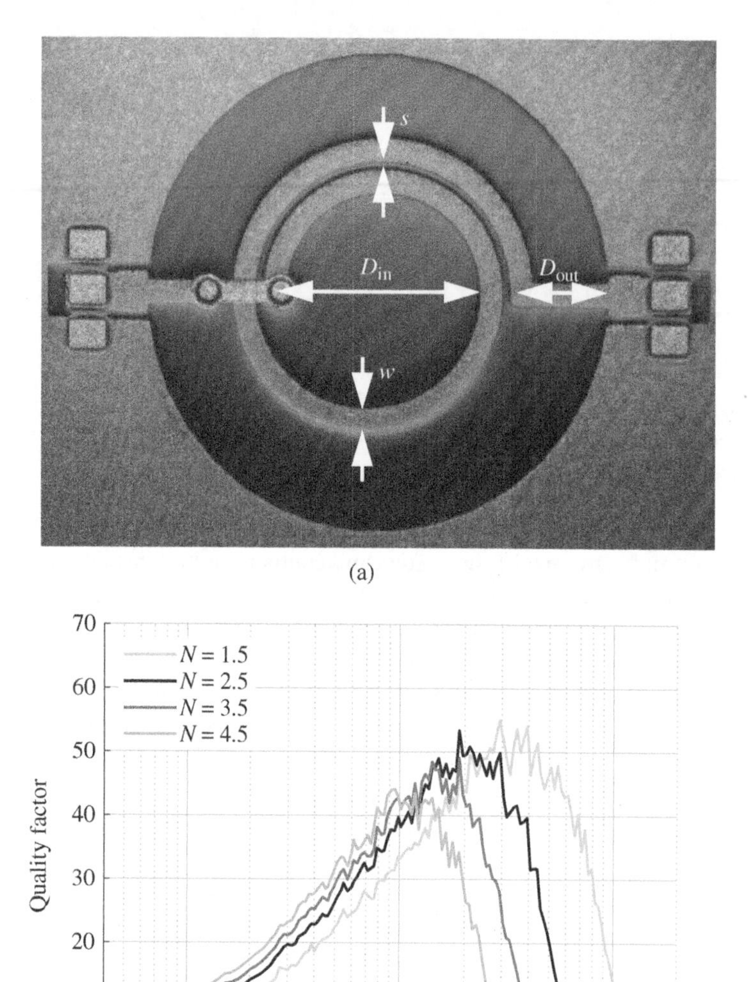


**Figure 8.12** (a) Photograph of a single-layer spiral inductor in the eWLB with N = 1.5 windings. (b) Measured quality factor of four single-layer spiral inductors in the eWLB with number of windings N ranging from 1.5 to 4.5.

stack-up consists of three 10-μm thick dielectric layers and two 5-μm and 6-μm thick metal layers in between. The metal stack-up consists of copper for the bottom RDL and copper/nickel/copper (3.5 μm/2 μm/0.5 μm) for the top RDL. The inductors are realized in the bottom RDL while the top RDL is used for underpasses. The manufactured inductors have a trace width $W$ of 50 μm and the spacing between turns $S$ is 20 μm, as shown in Figure 8.12(a). The dimensions of the inductors are listed in Table 8.2. The presented results demonstrate that spiral inductors integrated in the thin-film RDL of the eWLB offer significantly better performance compared to inductors integrated in standard on-chip technologies. Ref. [11] reports a 6-GHz voltage-controlled oscillator (VCO) chip realized in 65 nm complementary metal-oxide semiconductor (CMOS) technology with a fan-out eWLB inductor used for the LC resonant tank. A phase noise improvement of 9 dBc/Hz is achieved compared to the standard on-chip inductor.

The use of TEVs enables the realization of high-$Q$ 3D passives integrated inside the volume of the mold compound. Figure 8.13 shows photographs and the measured inductance of three different 3D solenoid inductors manufactured in the eWLB using TEVs [12]. The windings of the 3D inductors are composed of TEVs and RDL routing lines on both sides of the mold carrier. The dotted line in Figure 8.13b represents the simulated characteristics of the first inductor.

Like inductors, capacitors are fundamental components in RF applications. Larger value capacitors are commonly used for RF bypassing and DC blocking, whereas smaller value capacitors find usage in resonant circuits and filters. In the eWLB technology, it is possible to realize different types of capacitors. For small capacitance values (a few pF), standard metal–insulator–metal (MIM) capacitors are available. For very small capacitances (below 0.5 pF), interdigital capacitors can be used, especially at microwave frequencies [9]. By combining inductors and capacitors, resonant circuits can be realized. Figure 8.14 shows a photograph of a parallel LC resonant circuit and a 2.45-GHz bandpass filter manufactured in the eWLB. Such LC circuits can be used for antenna matching and signal distribution/filtering at lower frequencies, e.g. for radio frequency identification (RFID).

**Table 8.2** Layout parameters and measured performance of the single-layer spiral inductors realized in the eWLB. N is number of windings and SRF means self resonance frequency.

| $N$ | $D_{in}$ (μm) | $D_{out}$ (μm) | $L(Q_{max})$ (nH) | $Q_{max}$ | $f(Q_{max})$ (GHz) | SRF (GHz) | $R_{DC}$ (Ω) |
|---|---|---|---|---|---|---|---|
| 1.5 | 460 | 200 | 2.75 | 55 | 2.95 | 11.9 | 0.20 |
| 2.5 | 460 | 200 | 6.22 | 53 | 1.89 | 6.59 | 0.36 |
| 3.5 | 460 | 200 | 11.6 | 49 | 1.41 | 4.39 | 0.56 |
| 4.5 | 460 | 200 | 19.6 | 44 | 1.18 | 3.16 | 0.80 |

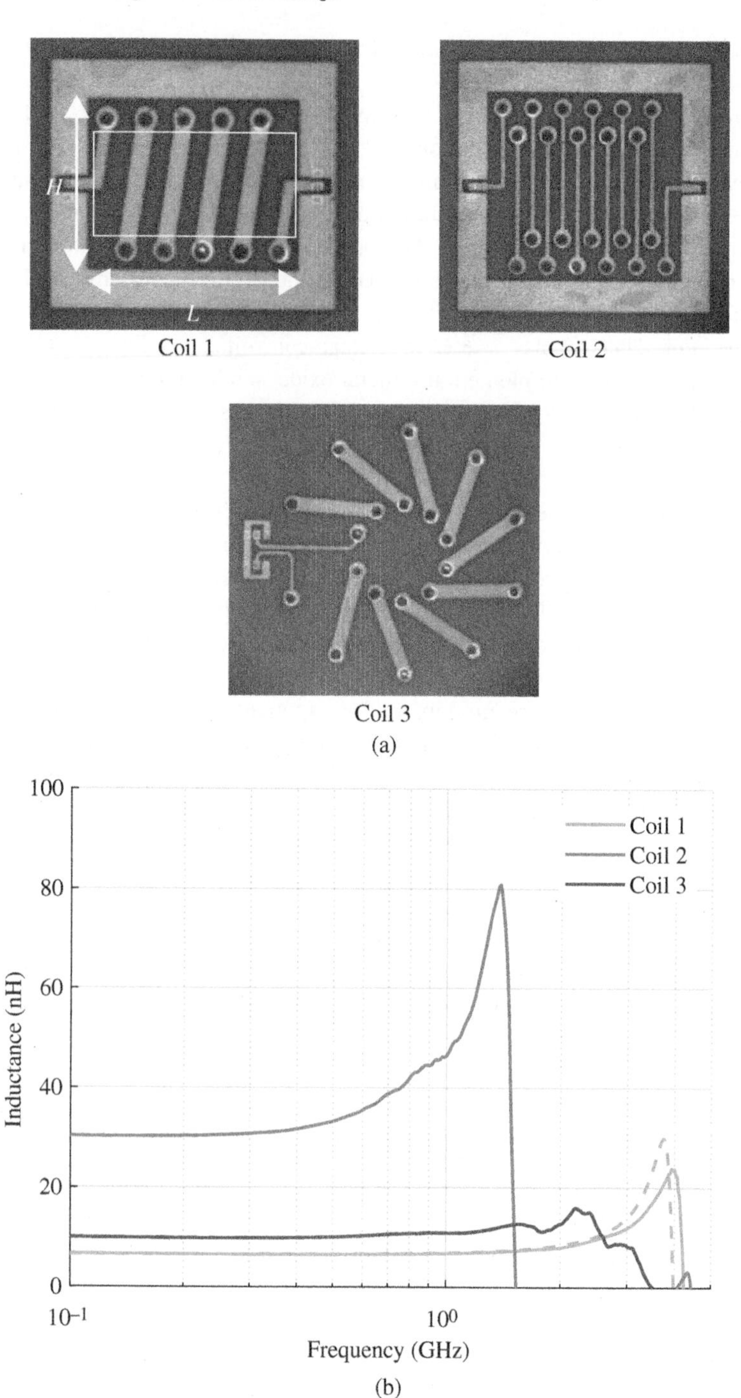


**Figure 8.13** (a) Photograph and (b) measured (solid line) and simulated (dotted line) inductance of 3D solenoid inductors realized in the eWLB using TEVs (from [12], ©2019 IEEE, reprinted with permission).

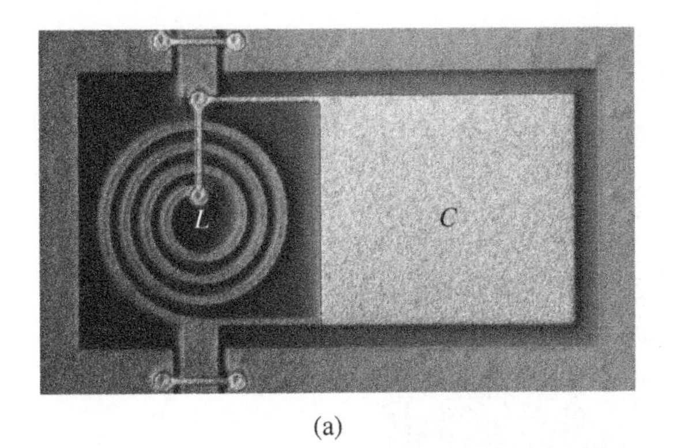


(a)

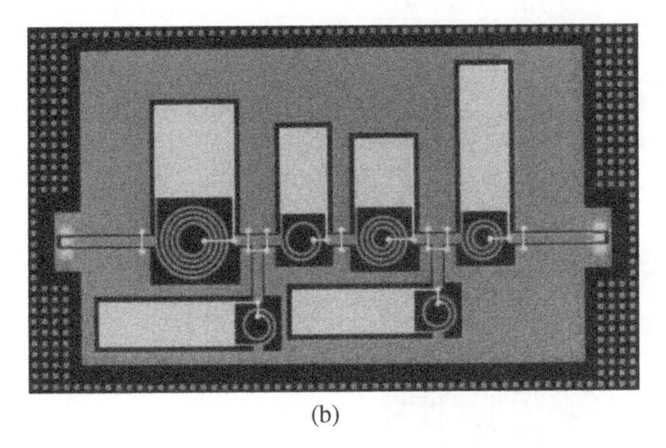

(b)

**Figure 8.14** Photographs of (a) a parallel LC resonant circuit and (b) a 2.45-GHz bandpass filter in the eWLB.

At mmWave frequencies, distributed circuits using transmission lines can be realized. Figure 8.15 shows photographs of 60-GHz branch-line (90° hybrid) and rat-race (180° hybrid) couplers realized in the eWLB. Both circuits are basic circuits for power dividing/combining. They are widely used to separate transmitted and received signals for monostatic radar systems. The presented branch-line coupler is realized using 50 and 35 Ω CPW transmission lines while the rat-race coupler uses 50 Ω CPW lines for coupler arms and 70 Ω SL lines for the coupler ring. In both structures, ground bridges in the top RDL are used.

Figure 8.16 shows design and simulated scattering parameters ($S$-parameters) of a 60-GHz rat-race coupler realized in the quasi-CPW technique in the eWLB [8].

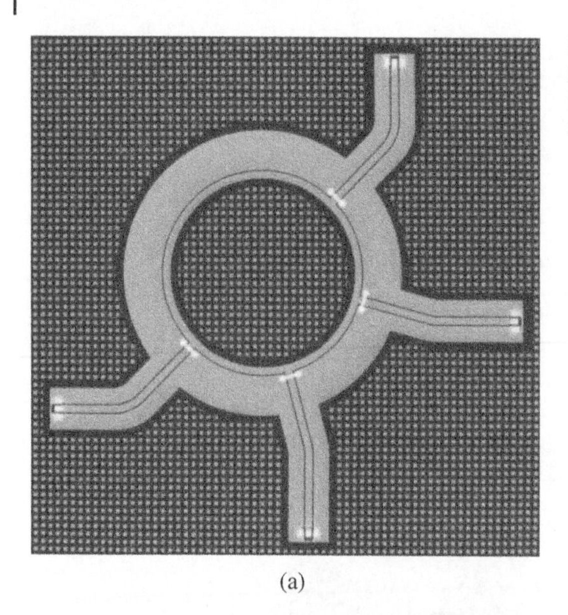

(a)

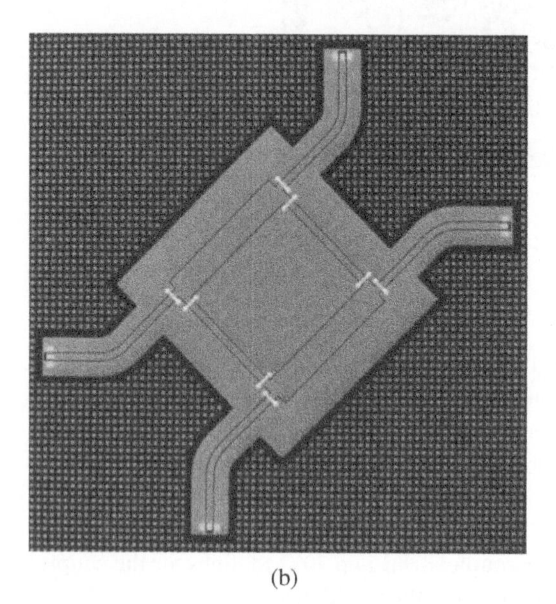

(b)

**Figure 8.15** Photographs of (a) a 60-GHz branch-line coupler and (b) a 60-GHz rat-race coupler in the eWLB.

In [13], different balun structures (CPW–CPS bridge, Marchand balun, double-Y balun) and branch-line couplers in CPW, CPS, and quasi-CPW techniques in the eWLB are investigated for monostatic radar transceivers for the 57–64-GHz industrial, scientific, and medical (ISM) band.

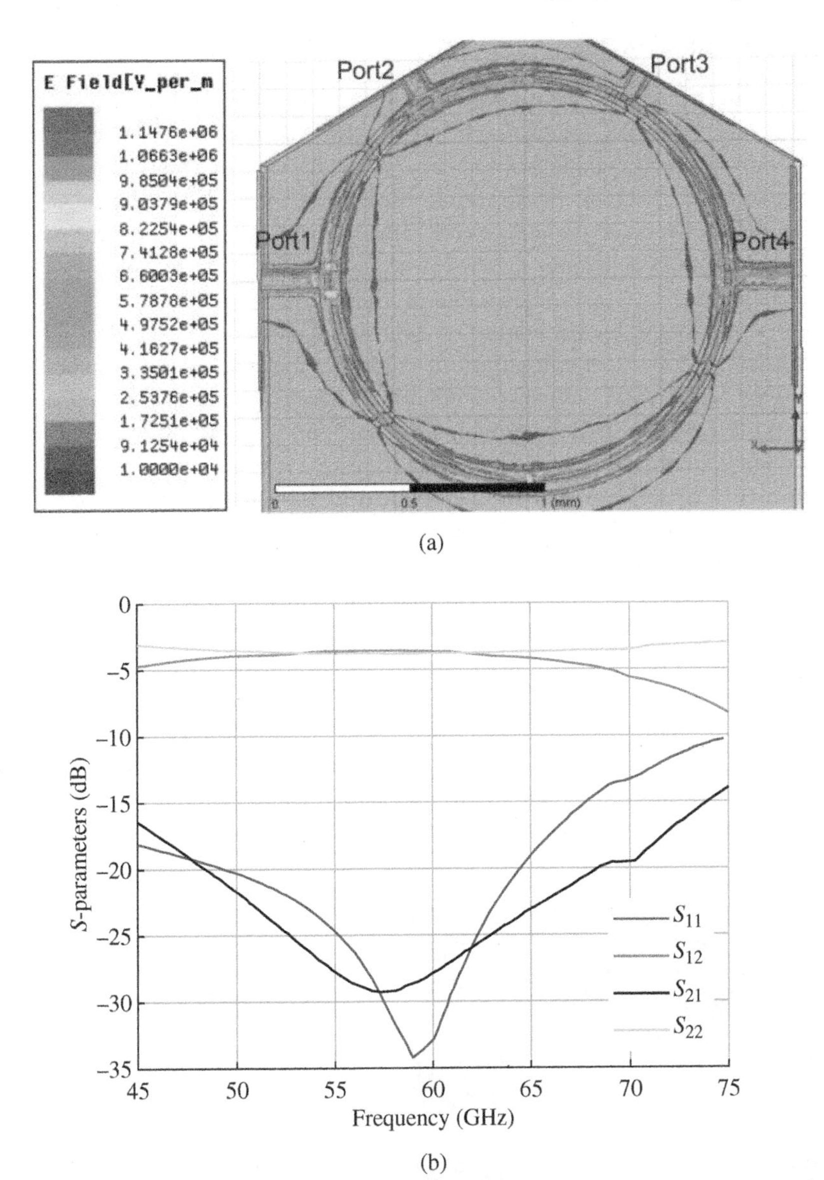


**Figure 8.16** (a) Design and (b) simulated $S$-parameters of a 60-GHz rat-race coupler realized using quasi-CPW transmission lines in the eWLB (from [8], ©2019 IEEE, reprinted with permission).

### 8.3.3 RF Transition to PCB

One of the most critical aspects in mmWave systems is signal reflections and losses at the chip–package–board interfaces. The mismatch and losses at these interfaces can reduce the gain and radiated power of antennas integrated on PCBs. AiP integration makes the RF transitions unnecessary, which is one of the main advantages behind AiP. However, in the particular case of antennas integrated in packages and using top or bottom PCB metallization for RF ground, such an RF transition may still be needed.

The eWLB allows the realization of very short interconnections between chip and PCB. This results in reduced parasitics and excellent electrical performance up to mmWave frequencies. Figure 8.17 shows a photograph of a 77-GHz radar receiver in the eWLB with very short single-ended transitions to the PCB [14]. Typically, for frequencies below 30–40 GHz no external (on PCB) or internal (in package) matching networks are required. At higher frequencies, the compensation and matching structures in the RDL can be realized. This allows the realization of RF transitions without external matching up to 60–100 GHz.

Figure 8.18 illustrates simulated *S*-parameters of a differential chip–package–board transition in an eWLB optimized for 77 GHz [7]. An insertion loss as low as −0.65 dB and a return loss below −16 dB at 77 GHz are achieved. A similarly good performance can be achieved for single-ended transitions. Ref. [15] reports a 77-GHz silicon germanium (SiGe) mixer assembled in the eWLB and showing performance comparable to bare die. Ref. [16] compares 24-GHz transitions in very thin quad flat no-lead (VQFN) and eWLB packages, demonstrating the superior performance and potential of the latter.

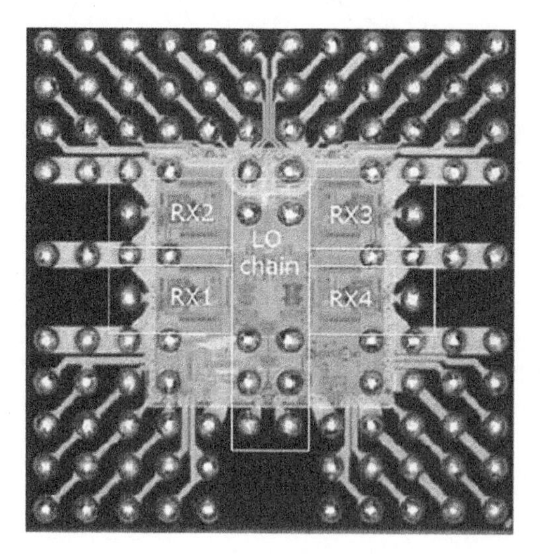


**Figure 8.17** Photograph of a four-channel 77-GHz radar receiver in the eWLB (from [14], ©2019 IEEE, reprinted with permission).

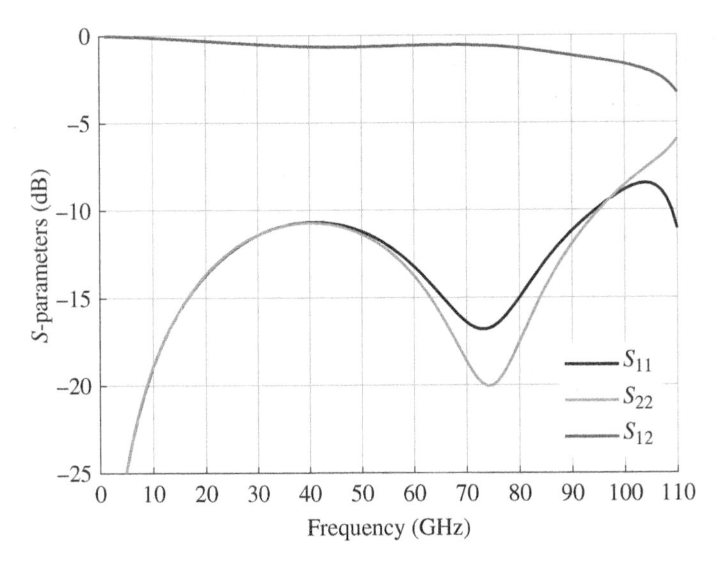


**Figure 8.18** Simulated $S$-parameters of an optimized differential chip–package–board transition in the eWLB (from [7], ©2019 IEEE, reprinted with permission).

### 8.3.4 Vertical RF Transitions

Vertical RF transitions are vertical interconnections of controlled characteristic impedance capable of transmitting RF and mmWave signals vertically through the package. This type of transition can be required for package-on-package modules. For AiP, vertical RF transitions can be used to feed antenna elements on top of a package. Similar to transmission lines, the key parameters of vertical RF transitions are compact lateral dimensions, low attenuation, and a wide range of realizable impedances.

For an eWLB package, vertical RF transitions can be realized using TEVs or EZLs. Figure 8.19a shows a single-ended vertical transition realized in the eWLB using TEVs [12]. It consists of three TEVs in a ground-signal-ground (GSG) configuration. The TEV pitch is 300 μm and the diameter of the TEV is about 100–150 μm. The distance between the TEVs determines the transmission-line impedance of the transition. By optimizing this distance, the RF transition can be matched to the system impedance (typically 50 Ω). Figure 8.19b shows the measured and simulated return and insertion loss of a cascaded connection of two single-ended RF transitions [12]. An insertion loss better than −0.5 dB and a return loss below −20 dB up to 40 GHz are achieved. A good matching of the vertical transition confirms the correctness of the diameter-to-pitch ratio selection.

Higher bandwidth and better performance can be achieved using EZL technology. Figure 8.20 compares the simulated $S$-parameters of vertical differential

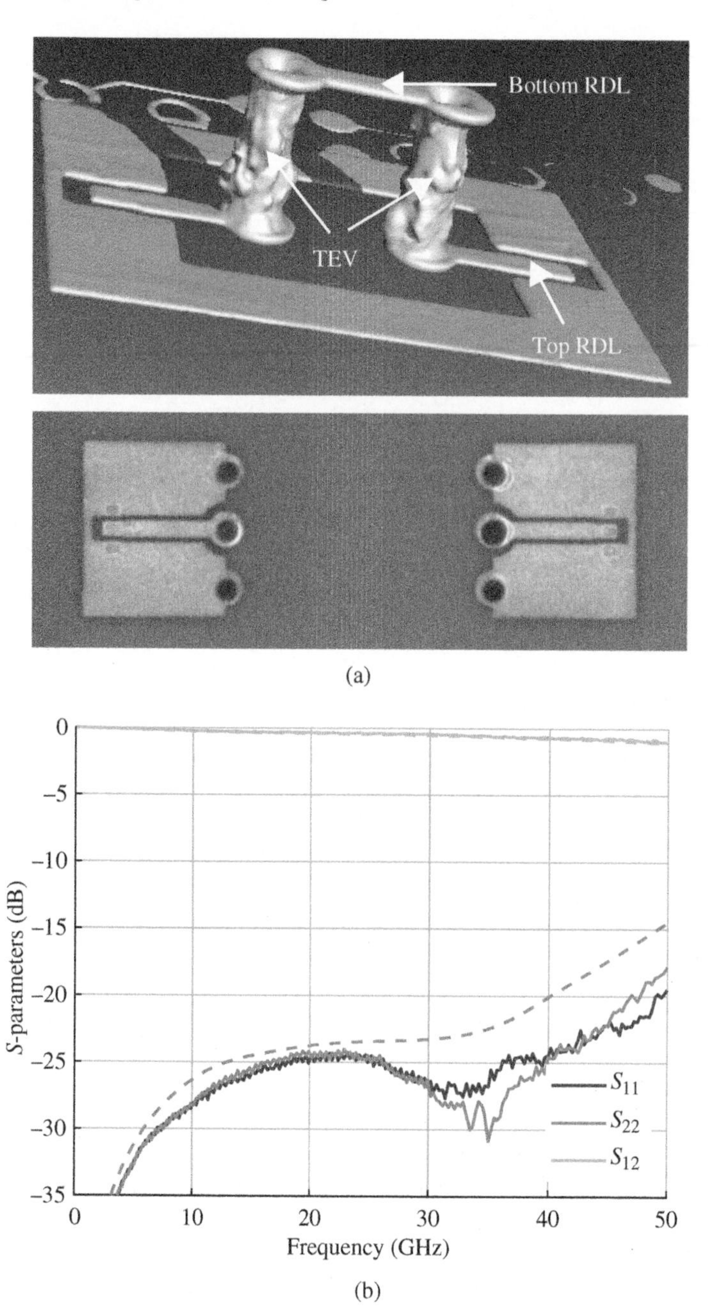


**Figure 8.19** (a) CT image and photograph and (b) measured $S$-parameters of a cascaded single-ended vertical transition realized in the eWLB using TEVs (from [12], ©2019 IEEE, reprinted with permission).

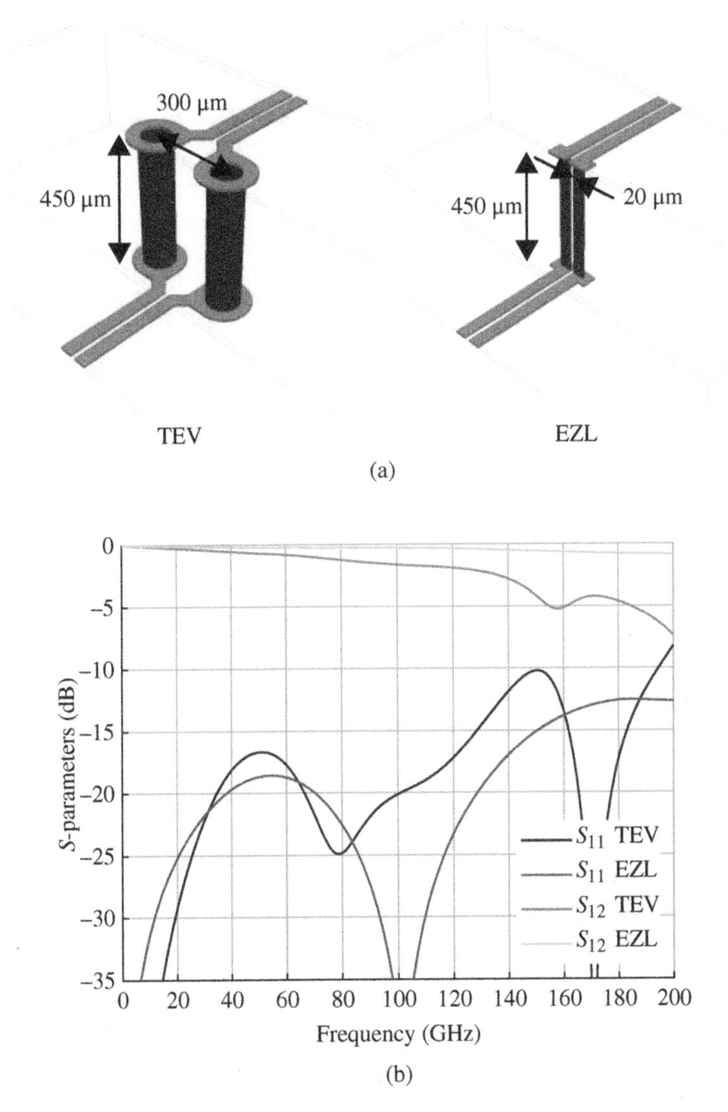


**Figure 8.20** (a) 3D model and (b) simulated *S*-parameters of a differential vertical transition realized in the eWLB using TEVs and ELZ (from [6], ©2019 IEEE, reprinted with permission).

transition realized using TEVs and EZLs [6]. The simulation is performed up to 200 GHz to capture the possible performance limits of the transitions. Both vertical transitions show very good performance up to approximately 100 GHz. However, above this frequency a significant increase of return and insertion loss is observed for the TEV transition. The reasons for this performance drop are the impedance

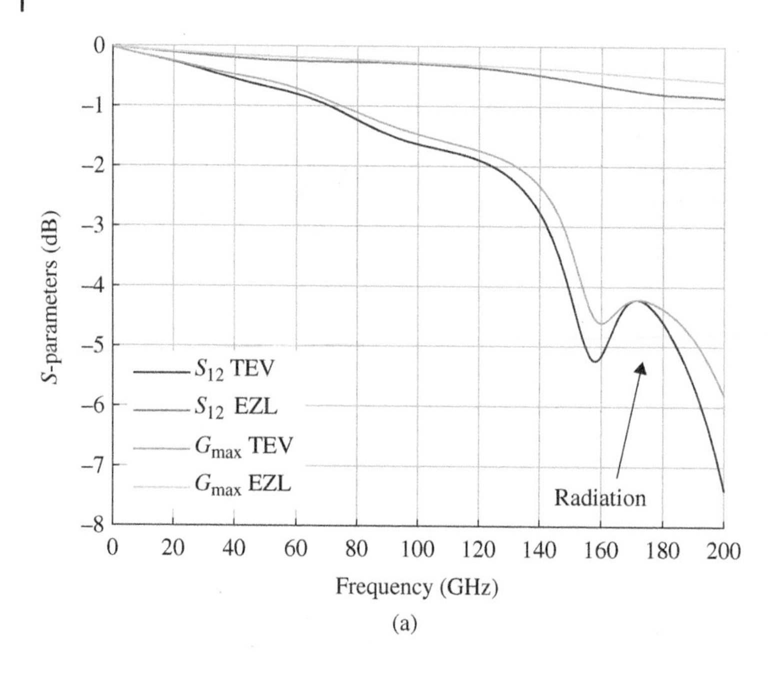


(a)

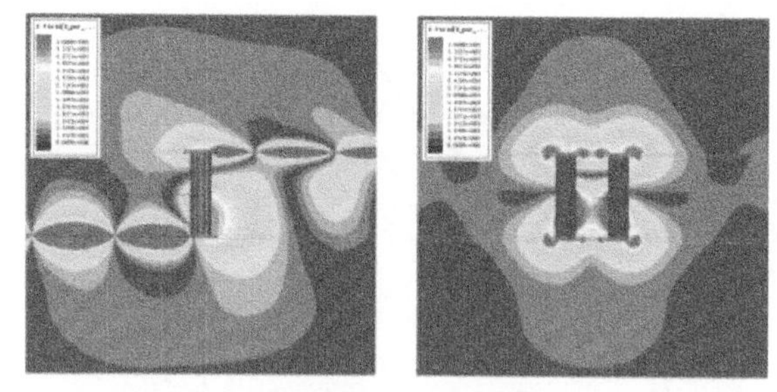

Transition using TEV

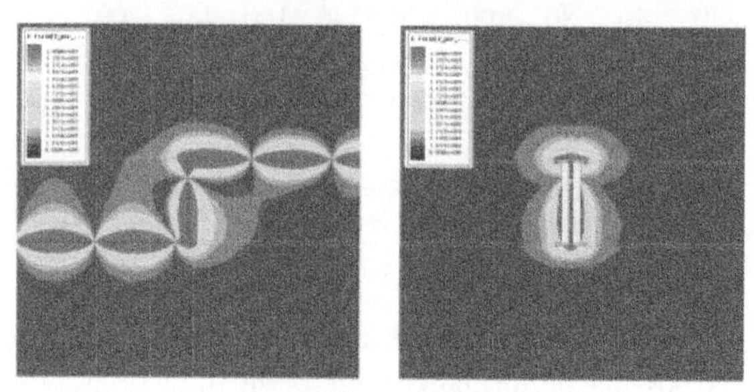

Transition using EZL

(b)

**Figure 8.21** (a) Simulated insertion loss $S_{21}$ and maximal available gain $G_{max}$ and (b) electric field distribution at 200 GHz of a high-frequency differential vertical interconnection realized using TEVs and EZLs (from [6], ©2019 IEEE, reprinted with permission).

discontinuity at the RDL/TEV interface and the radiation losses caused by large spacing between the TEVs. The length of the vertical interconnection approaches $\lambda/2$ at approximately 185 GHz and behaves at this frequency like a $\lambda/2$ antenna connected in series with the transmission line. On contrary, the transition realized using an EZL shows performance typical for a long uniform slightly mismatched transmission line.

Figure 8.21a shows the simulated insertion loss and maximal available gain $G_{max}$ of both analyzed transitions. It confirms that the transition realized using TEVs reaches its limit at frequencies above 100–150 GHz due to substantial intrinsic losses that cannot be compensated for using external matching. Figure 8.21b illustrates the electric field patterns simulated at 200 GHz for both analyzed transitions. Large discontinuity at the RDL/TEV interface and a wide pitch between the TEVs gives rise to radiation, which manifests itself in detached clouds of electric field above and below the transition. On the contrary, thanks to small spacing between the EZLs, the electric field remains trapped between the signal traces and no radiation occurs.

## 8.4 Antenna Integration in eWLB

In mmWave radar systems, antenna structures are usually realized on high-frequency laminates to which a transceiver chip is bound using flip-chip bumps or bondwires. The RF transition to the PCB can cause energy losses that in turn can reduce the gain and radiated power of antennas integrated on the PCB. The use of high-frequency laminate also increases the overall system cost.

Integration of AiP can increase system efficiency and lower cost. Moreover, the integration of AiP makes the RF transitions to the PCB unnecessary and hence significantly simplifies the design and assembly of mmWave radar systems.

Because of the small antenna dimensions at mmWave frequencies, it is possible to realize antenna inside a package. Figure 8.22 illustrates one concept of an antenna integration in an eWLB package [17]. The antenna is realized in the fan-out area of the eWLB. The transmission lines available in the RDL provide low-loss interconnections between a chip and an antenna. The solder balls are used for external power/ground and low-frequency signal contacts and provide mechanical support for the antenna.

The plain ground underneath the antenna structure is used as a reflector. The spacing between the antenna and the reflector should not exceed $\lambda/4$ to ensure maximum radiation in the direction perpendicular to the PCB and to avoid any grating lobes. This requirement is guaranteed with a typical height of solder balls of 200 μm, which corresponds to $\lambda/38$ at 40 GHz and $\lambda/15$ at 100 GHz. The position of the reflector on the top metallization of the PCB is a great advantage as it makes

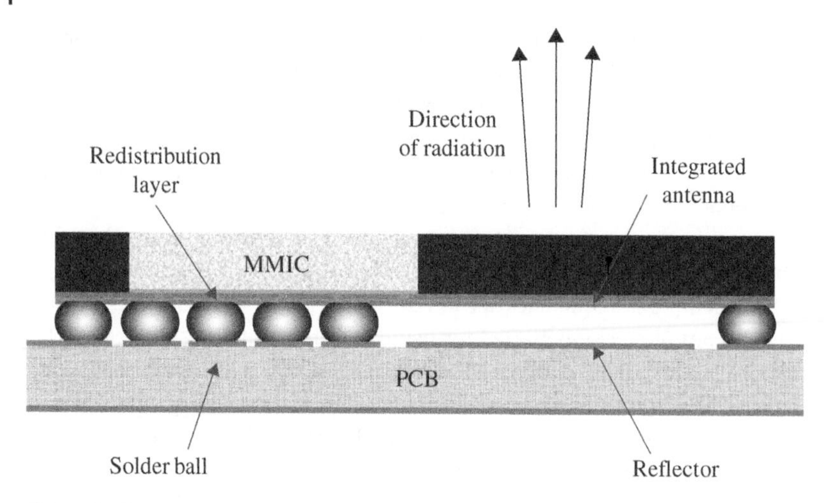


**Figure 8.22** Integration of the antenna in the eWLB (from [17], ©2019 IEEE, reprinted with permission).

the integrated antenna unconstrained by the actual PCB material. If a higher distance to the reflector is required, e.g. for a larger bandwidth, the bottom metallization of the PCB can be used.

### 8.4.1 Single Antenna

High metal-pattern resolution of the RDL of the eWLB and low-loss mold compound enables the realization of various types of mmWave-antennas. Figure 8.23 shows photographs of typical dipole, coplanar patch, and Vivaldi antennas manufactured in the RDL of an eWLB package [17]. Unlike dipole and coplanar patch antennas, the Vivaldi antenna radiates parallel to the PCB. Using a single antenna, directivities of about 8–9 dBi can be obtained. To increase the directivity, additional structures in the RDL (reflector and director elements) can be used.

The impedance matching and radiation characteristics of an antenna integrated in a package depend in general on package size, the position of antenna in the package, and the environment surrounding the antenna (Si chip, RDL, solder balls). Typically, the presence of a large Si chip close to the antenna will tilt the radiation beam. Another problem is related to excitation and propagation of surface waves. The surface waves are triggered at the antenna or at discontinuities in the feeding path. When excited, they propagate along the package surface, reflect at the package edges, and interfere with the main lobe, which reduces antenna radiation efficiency and deteriorates its radiation pattern. The first higher mode $TE_1$ will start to propagate when the mold compound thickness reaches approximately $\lambda/2$. Thus, the excitation of surface waves can be reduced by keeping the package

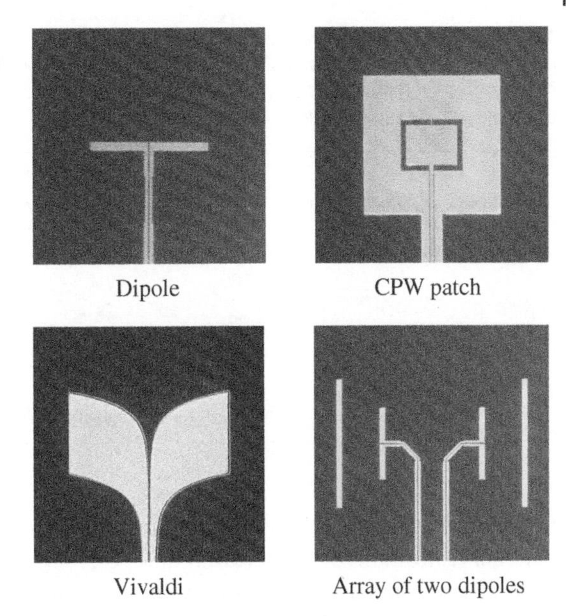


**Figure 8.23** Photographs of typical antennas manufactured in the RDL of the eWLB (from [17], ©2019 IEEE, reprinted with permission).

thickness smaller than $\lambda/2$. This can prevent triggering higher order surface modes but the fundamental $TM_0$ mode with zero cut-off frequency will always exist. Another way to reduce surface waves is to isolate the antenna element from the surrounding environment, e.g. by using a via fence made of TEVs. In general, the design and optimization of the AiP must take into account the position of the AiP and the interactions with package edges and surrounding structures.

### 8.4.2 Antenna Array

The size of a single radiating element is about $\lambda/2$ at the operating frequency. To build an antenna array with a broad-side radiation pattern (maximum radiation perpendicular to the PCB), the radiating elements are usually placed uniformly at a distance $d$ between $\lambda/2$ and $\lambda$. Table 8.3 summarizes the wavelengths at selected mmWave frequencies and the corresponding maximal number of radiating elements $N \times N$ that can be integrated in an $8 \times 8$ mm eWLB package for $d = \lambda$ [17]. The last two columns in Table 8.3 give the directivities of the resulting linear and area arrays estimated as:

$$D_i = \frac{8\pi}{\lambda^2} A_i, \tag{8.1}$$

where $A_{N\times1} = 8\ \text{mm} \times \lambda/2$ for a linear array ($N \times 1$) and $A_{N\times N} = 8 \times 8$ mm for an area array ($N \times N$) are the maximal effective areas of the antenna arrays. The directivity calculated by this equation corresponds to the rectangular aperture of area $A_i$ located on height $h \ll \lambda/4$ over an infinite ground plane. Thus, it provides only a

**Table 8.3** Wavelengths at selected frequencies for $\epsilon_{r\,\mathrm{eff}} = 2.5$ and the corresponding maximal number of antennas that can be integrated in an 8 × 8 mm eWLB package for spacing between antennas $d = \lambda$. The last two columns give the estimated directivities of the resulting linear and area arrays [17].

| Frequency (GHz) | Wavelength ($\lambda$) (mm) | Half of wavelength ($\lambda/2$) (mm) | Max. number of antennas $N \times N$ | Estimated directivity $D_{N\times1}$ (dBi) | Estimated directivity $D_{N\times N}$ (dBi) |
|---|---|---|---|---|---|
| 40 | 4.74 | 2.37 | 2 × 2 | 11.3 | 14.6 |
| 60 | 3.16 | 1.58 | 3 × 3 | 13.0 | 18.1 |
| 80 | 2.37 | 1.19 | 4 × 4 | 14.3 | 20.6 |
| 100 | 1.90 | 0.95 | 5 × 5 | 15.3 | 22.5 |
| 120 | 1.58 | 0.79 | 6 × 6 | 16.0 | 24.1 |

rough estimate of the achievable antenna performance. Directivities of particular antenna realizations can differ from those given in Table 8.3.

Because of the large number of radiating elements and high directivities, the eWLB enables the realization of compact systems with 2D/3D beamforming capabilities. The antennas in an array can be positioned at various angles to optimize the radiation pattern and to minimize the mutual coupling. In practice, the number of antennas that can be integrated in an eWLB package will be lower as a part of the package is used for chip and solder balls.

### 8.4.3 3D Antenna and Antenna Arrays

The EZL technology also allows integration of vertical antennas for mmWave applications. Figure 8.24 illustrates the concept of realizing vertical dipole antenna using the EZL in an eWLB package [6]. In contrast to the traditional concept using a horizontal RDL for the radiating element and a PCB metallization for the reflector, the solution using the EZL allows the reflector to be realized using the horizontal RDL. Figure 8.25 illustrates an alternative concept with the antenna realized on the top surface of an eWLB package. In this case, the EZL is used for vertical feeding and antenna matching. In both these concepts the integration of the reflector inside the package is a great advantage over the traditional configuration. It makes the radar system more compact, removes the need to allocate the PCB area below the antenna for the reflector, and makes the antenna characteristics less sensitive to package-board assembly tolerances. Moreover, the distance between the antenna and the reflector is set by the thickness of the mold compound. Thus, by adjusting its thickness, the antenna characteristics can be further optimized. This is particularly important for antenna integration above 100 GHz, which requires lower substrate thicknesses.

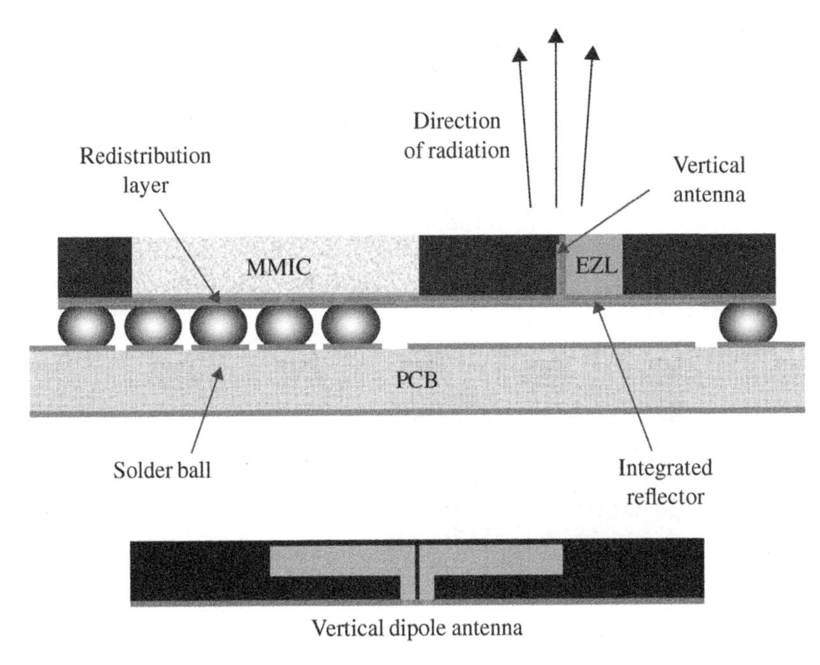


**Figure 8.24** Vertical dipole antenna using an EZL (from [6], ©2019 IEEE, reprinted with permission).

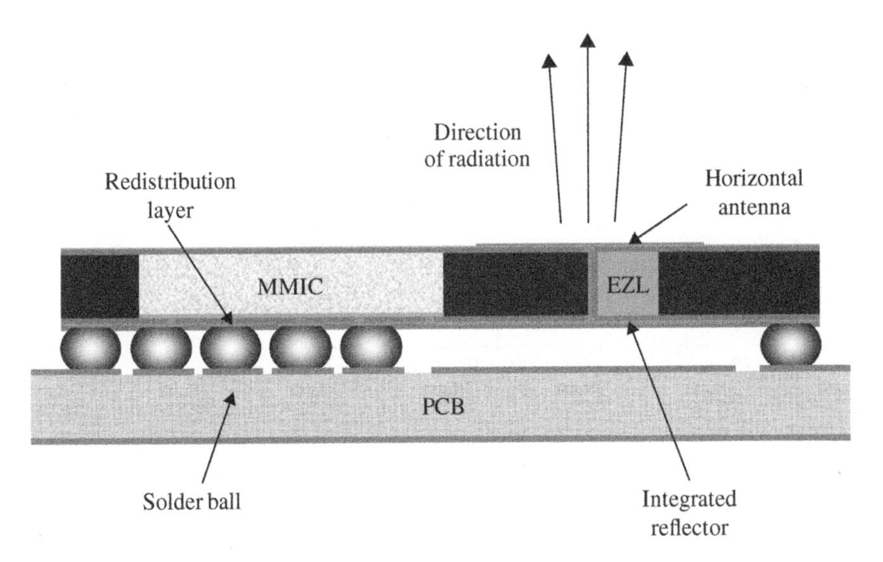


**Figure 8.25** Antenna realized on the package top surface using an EZL for vertical interconnection and antenna matching.

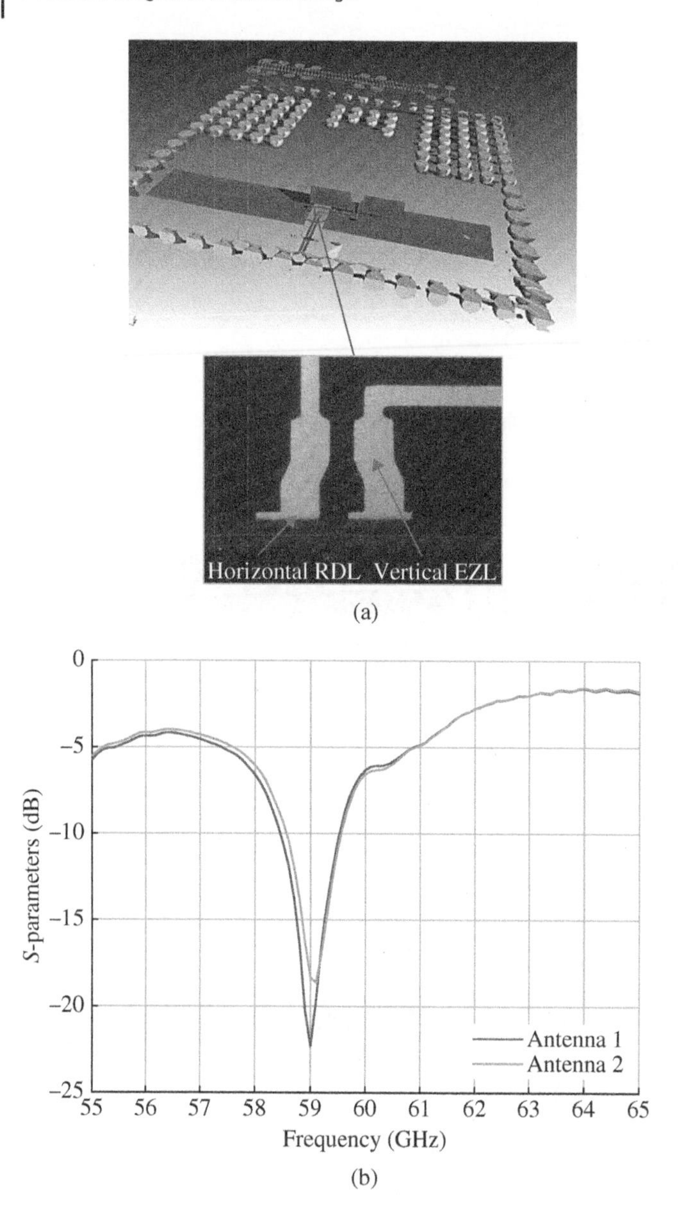


**Figure 8.26** (a) CT image and photograph of an EZL-based vertical dipole antenna integrated in the eWLB package. (b) Measured reflection loss of two samples of eWLB packages with vertical EZL dipole antennas (from [6], ©2019 IEEE, reprinted with permission).

Figure 8.26a shows a CT image and photograph of an EZL-based vertical dipole antenna integrated in the eWLB package [6]. A wide horizontal RDL plane below the dipole antenna used as the reflector is clearly visible. The bottom picture in Figure 8.26a shows a detailed view of the contact between the horizontal RDL and the vertical antenna.

Figure 8.26b shows the measured reflection loss of an eWLB package containing a vertical EZL dipole antenna similar to the one shown in Figure 8.26a. The two curves represent the measurement of two different samples of the same antenna design. Only a minor shift of the two measurements is observed, which demonstrates the very good reproducibility of this device.

## 8.5 Application Examples

In the following, three examples of eWLB AiP modules for industrial and automotive applications are presented. The first example shows an 8 × 8 mm two-channel AiP transceiver module for industrial sensor applications in the 60-GHz ISM band. Two variants for bistatic and monostatic configuration are realized with integrated dipole antennas. The second example presents an 8 × 8 mm four-channel 77-GHz transceiver for automotive applications. The module has a linear array of four dipole antennas. The third example presents a 14 × 14 mm six-channel 60-GHz transceiver for smart sensing and short-range communication. The module has two dipole transmit antennas and a 2 × 2 array of patch receive antennas.

### 8.5.1 Two-channel 60-GHz Transceiver Module

Figure 8.27 shows photographs of realized bistatic and monostatic transceiver AiP modules [18, 19]. The modules have a size of 8 × 8 mm and a footprint with a standard ball pitch of 0.5 mm. The transceiver chip is located in the bottom half of the package. In the upper half of the package the dipole antennas are realized using the RDL. The bistatic transceiver module uses two separate antennas for the transmit and receive channels. The monostatic transceiver module uses a single antenna and a differential rat-race coupler realized on chip is used to separate the transmitted and received signals.

In both variants, the interconnections between the transceiver chip and antennas are realized using 100 Ω differential CPS lines in the RDL. They provide low return loss and insertion loss of about −0.5 dB at 60 GHz. The reflector of the antenna is located on the top layer of the PCB. The impedance matching of antennas is obtained using $\lambda/4$ impedance transformers attached directly to the dipole antennas. The solder balls are used for DC supply and low-frequency signals. The solder balls located over the chip surface are used to improve heat transfer to

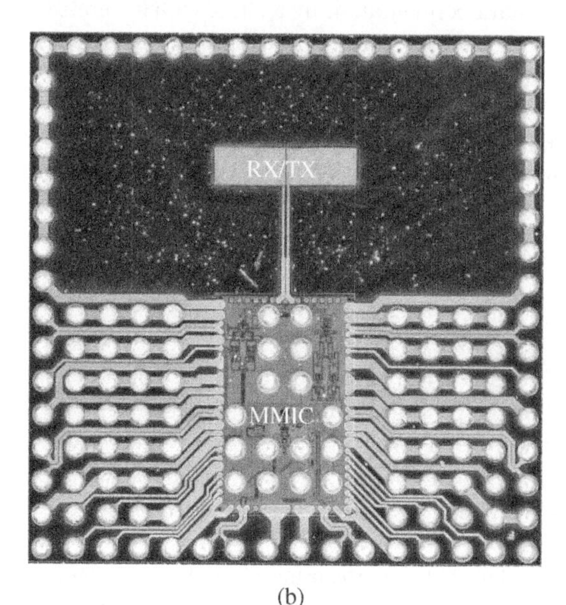


**Figure 8.27** Photographs of (a) the bistatic and (b) the monostatic transceiver eWLB modules with dipole antennas integrated in the RDL (from [19], ©2019 IEEE, reprinted with permission).

the PCB (so-called thermal balls). An RDL ground ring composed of solder balls and the RDL is realized around the antennas for a defined antenna pattern.

To make use of the available aperture area and to increase antenna gain, a $2 \times 1$ antenna array was developed for the monostatic variant, as shown in Figure 8.28. In this variant, the CPS lines in the RDL are used for signal distribution and the

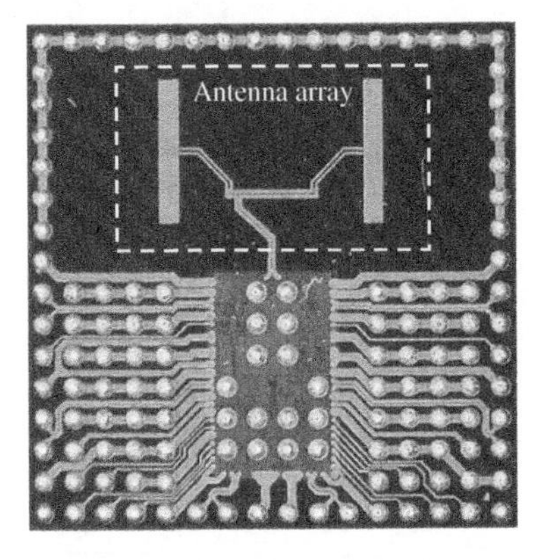


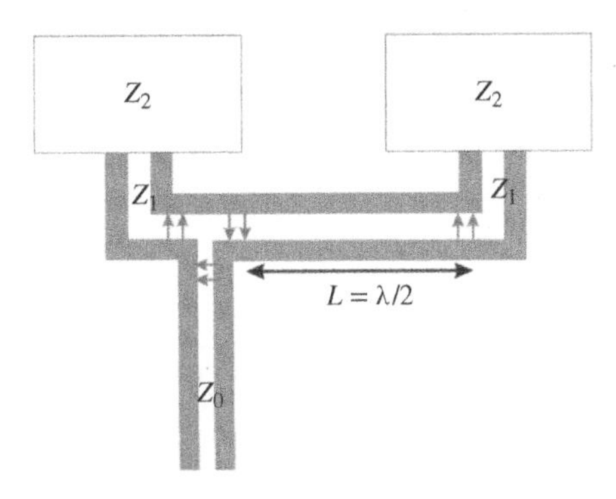


**Figure 8.28** Differential parallel feed system for two antenna elements for a 60-GHz monostatic transceiver (from [20], ©2019 IEEE, reprinted with permission).

matching network. Using a CPS junction with a 180° phase shift followed by a $\lambda/2$ CPS allows both antennas to be fed in phase.

The impedance matching and radiation characteristics of the realized dipole antennas depend in general on package size, the position of the antennas in the package, and the environment surrounding the antenna (Si chip, RDL, solder balls). Figure 8.29 illustrates a test package developed to measure the radiation pattern of the dipole antenna for the monostatic variant [21]. The position of the antenna in the test package and surrounding environment corresponds to a real device. The only difference is the antenna feeding realized through solder balls at the opposite side. Figure 8.30 shows the simulated and measured return loss and radiation pattern of the antenna in the test package [21]. The antenna shows reflection loss below −10 dB and gain of 8 dBi over a bandwidth of 5 GHz. The

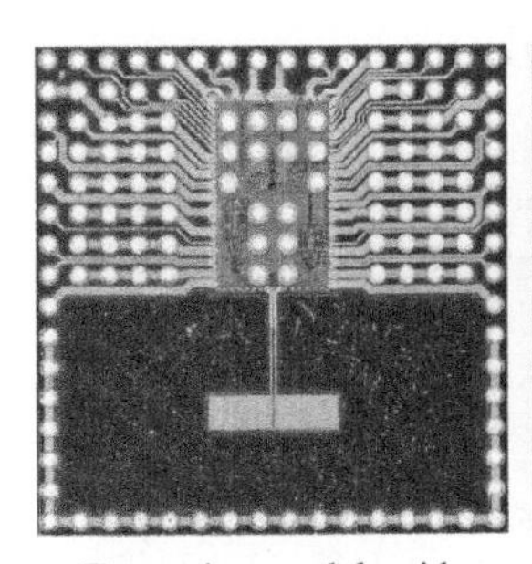


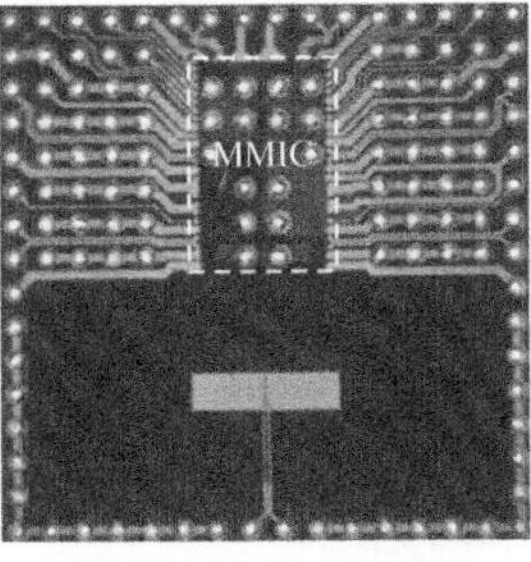

**Figure 8.29** Photographs of the eWLB package for a 60-GHz monostatic transceiver with integrated dipole antenna and test package developed to characterize in situ antenna characteristics (from [21], ©2019 IEEE, reprinted with permission).

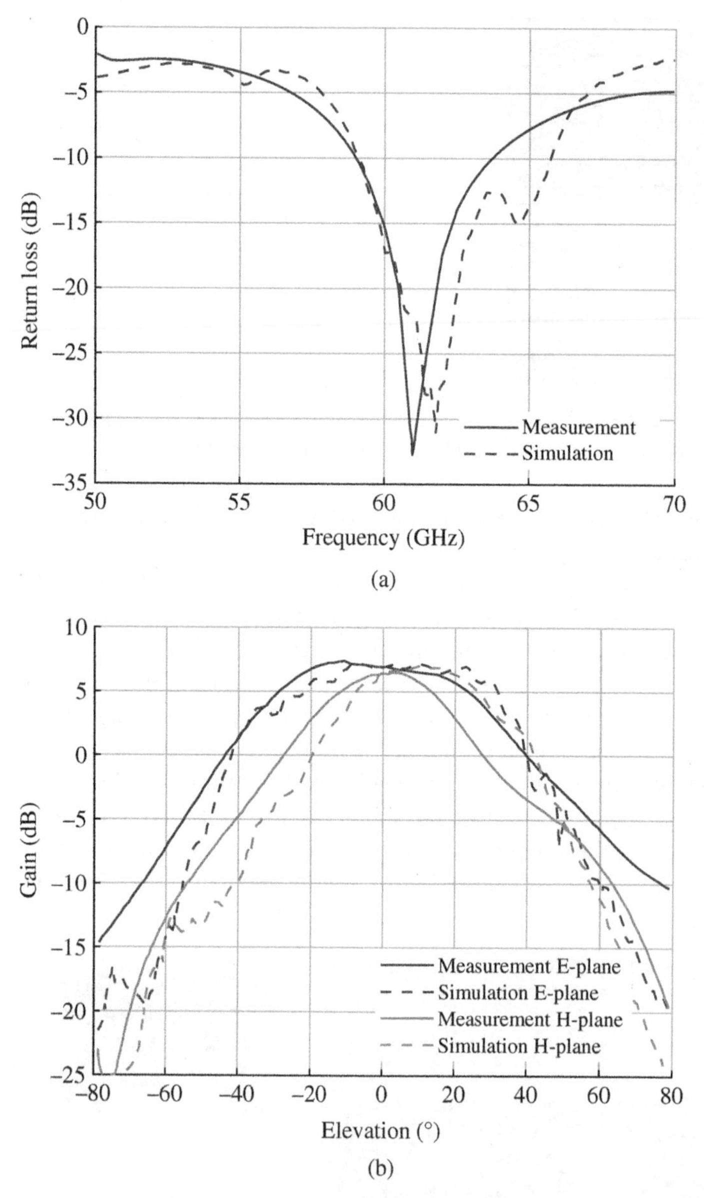


**Figure 8.30** Simulated (dotted line) and measured (solid line) (a) return loss and (b) radiation pattern of the single antenna shown in Figure 8.29 (from [21], ©2019 IEEE, reprinted with permission).

presented simulated and measured characteristics take into account all relevant environmental conditions and represent the antenna performance expected in the device with a connected transceiver chip. Because of the asymmetric placement of the antenna and the interaction with the Si chip, the main lobe is tilted by about 20° in the H-plane. A shift of the resonance peak in comparison to simulations can be observed, which is due to the tolerances of the antenna–reflector distance in the soldering process. In [22], the impact of the eWLB package dimensions and tolerances on antenna performance is analyzed.

Similar test packages were developed for the monostatic variant with a $2 \times 1$ antenna array (Figure 8.28) and for the bistatic configuration with two antennas (Figure 8.27b). Figure 8.31 shows the test package developed for the bistatic transceiver [20]. The developed test package also allows the isolation between the transmit and receive antennas to be measured. Figure 8.32 shows the corresponding simulated and measured return loss $S_{11}$, isolation $S_{12}$ and radiation pattern [20]. A single antenna achieves a matching of better than −10 dB over a 5-GHz bandwidth. The measured isolation at 61 GHz is 25 dB. The maximum measured antenna gain is 8 dBi. Similar to the monostatic variant, the main lobe is tilted by about 20° in the H-plane.

### 8.5.2 Four-channel 77-GHz Transceiver Module

Figure 8.33 shows a photograph of a single-chip four-channel 77-GHz transceiver in SiGe integrated in the eWLB package with four dipole antennas [17]. The module has size 8 × 8 mm and a footprint with a standard ball pitch of 0.5 mm. The transceiver chip occupies the bottom half of the package. In the upper half of the package the dipole antennas are realized using the RDL. The interconnections between the transceiver chip and antennas are realized using 100 Ω differential

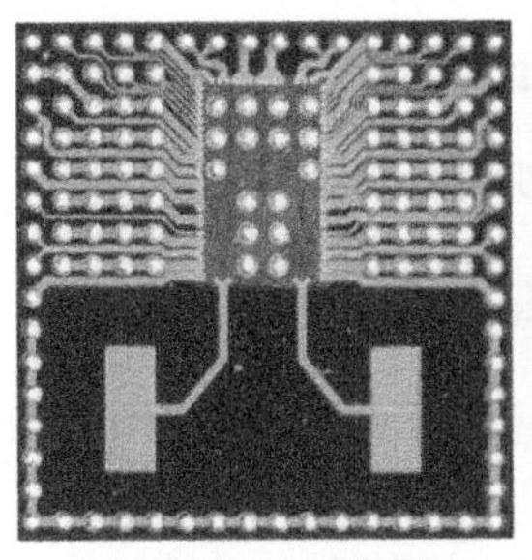

Transceiver module with integrated antenna array

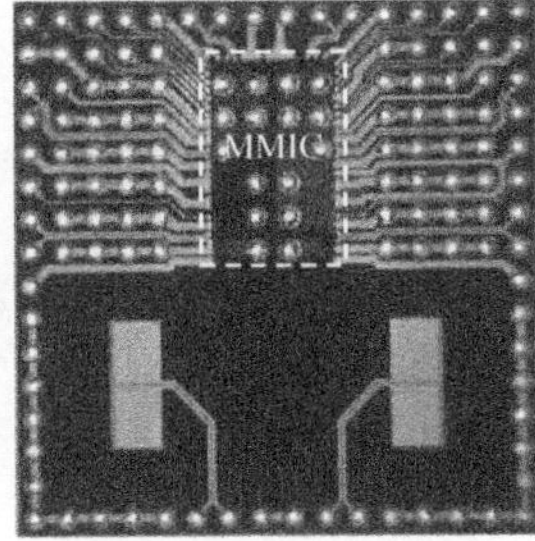


Test package for in-situ antenna array characterization

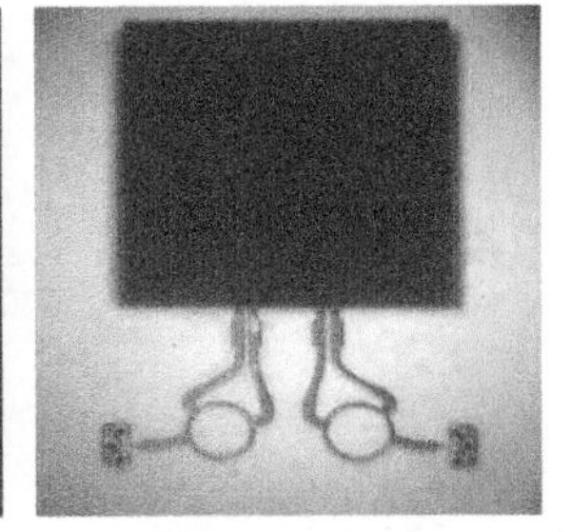

Test package mounted on RF board

**Figure 8.31** Photographs of the eWLB package for a 60-GHz bistatic transceiver with integrated two dipole antennas and a test package developed to characterize in situ antenna characteristics (from [20], ©2019 IEEE, reprinted with permission).

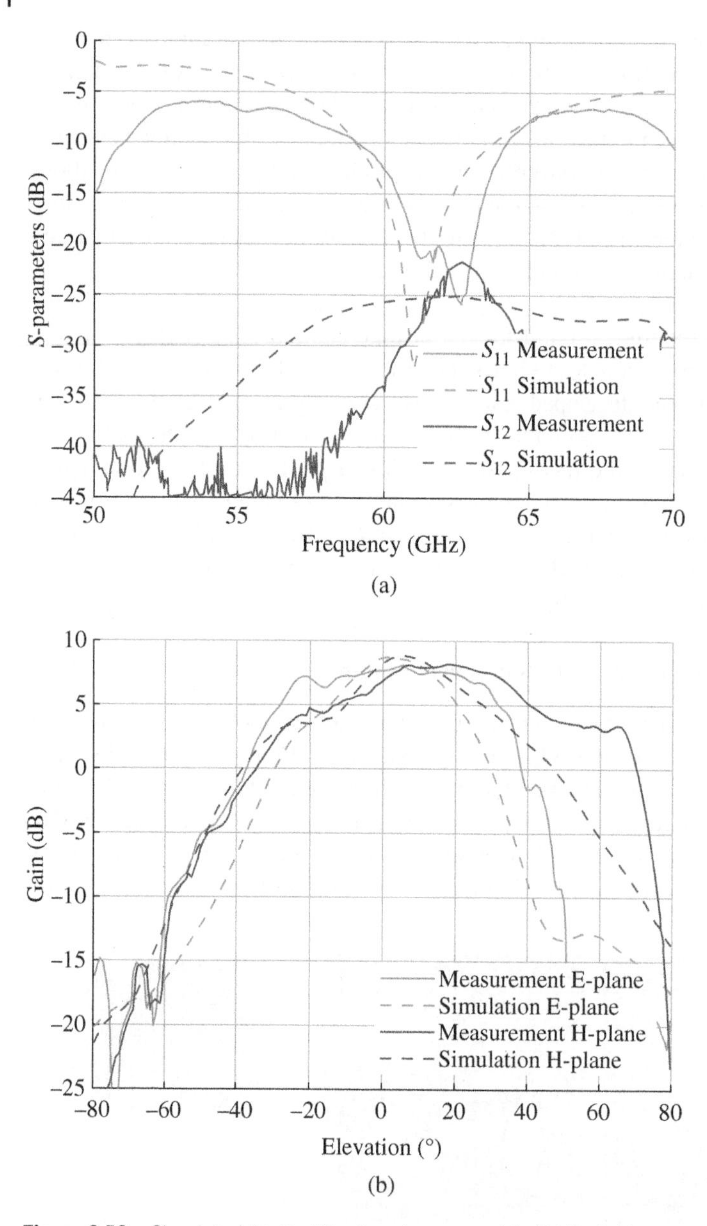


**Figure 8.32** Simulated (dotted line) and measured (solid line) (a) return loss and isolation and (b) radiation pattern of the single antenna shown in Figure 8.31 (from [20], ©2019 IEEE, reprinted with permission).

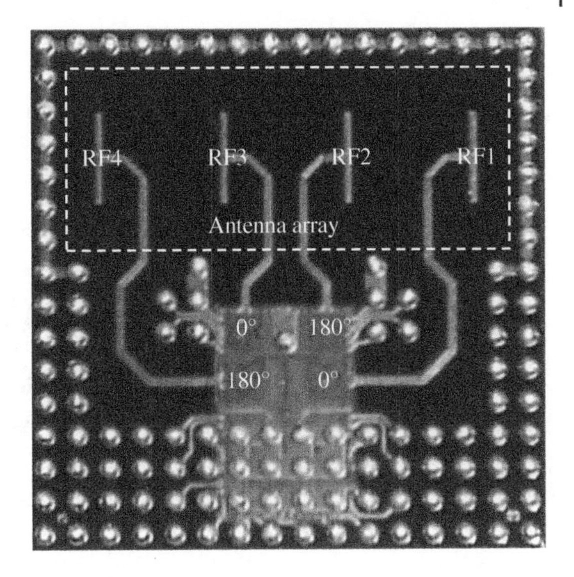


**Figure 8.33** Photograph of a 77-GHz eWLB module with four dipole antennas integrated in the RDL (from [17], ©2019 IEEE, reprinted with permission).

CPS lines. The impedance matching of antennas is obtained through $\lambda/4$ impedance transformers attached directly to the dipole antennas. A differential rat-race coupler is implemented on the chip to separate transmitted and received signals. This allows each dipole antenna to be used for both transmitting and receiving. The four $\lambda/2$ dipole antennas are placed at a distance close to $\lambda$ and are fed by 100 Ω differential CPS lines. The length of the feeding lines is optimized to guarantee in-phase excitation of all antennas in the array at 77 GHz. The simulated directivity of the linear array is 13.6 dBi, which agrees well with the estimated value of 14.3 dBi from Table 8.3.

To evaluate the manufactured 77-GHz eWLB module, a radar front-end has been realized. Figure 8.34a shows a photograph of a realized front-end consisting of a four-channel transceiver and a 19-GHz downconverter used for phase-locked loop (PLL) implementation [17]. The two multipin connectors provide connection to the baseband hardware. For comparison, Figure 8.34b shows a similar front-end realized using a standard technique. The bare-die transceiver is placed in a cavity and is connected using bondwires. To reduce the losses at the chip-board interface, on-board matching networks ($\lambda/4$ transformers) are used. The antenna array consisting of four differential series-fed patch antennas is realized on board. The integration of mmWave antennas and transmission lines on board requires the use of high-frequency laminates with high-resolution metallization and microvias. In contrast, the assembly shown in Figure 8.34a can be realized on any type of PCB. The only requirement is that the PCB area underneath the antenna is covered with metal.

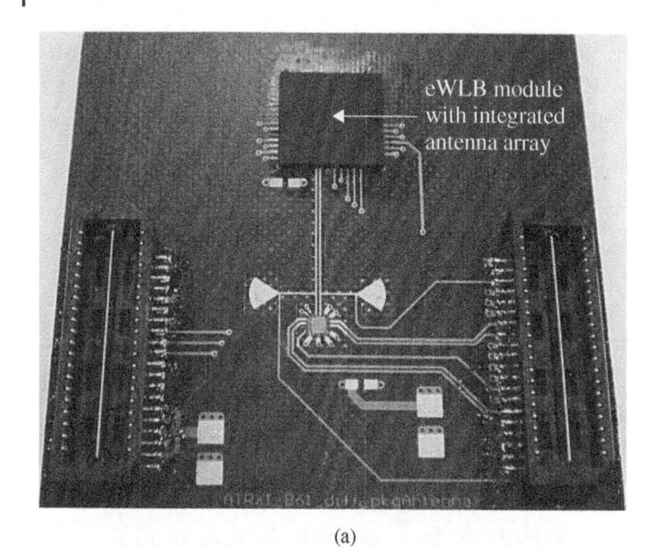


(a)

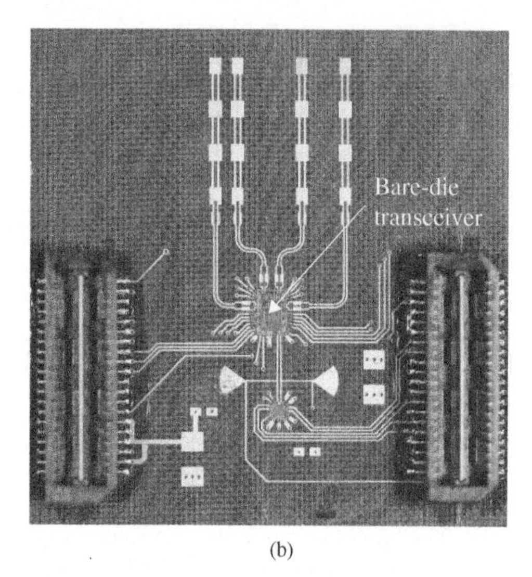


(b)

**Figure 8.34** Photograph of the front-end consisting of a four-channel transceiver chip and a 19-GHz downconverter. The overall size of the board is 4 × 5 cm: (a) eWLB module containing a transceiver chip with four dipole antennas integrated in the RDL and (b) a bare-die transceiver assembled using bondwires (from [17], ©2019 IEEE, reprinted with permission).

The radiation pattern of the assembled eWLB module is measured in an anechoic chamber over ±90° in the E- and H-planes, where the E-plane (azimuth) is parallel and the H-plane (elevation) is perpendicular to the dipoles. The free-space loss of 72 dB and the receiver antenna gain of 20 dBi have been accounted for in the measurements. Figure 8.35 shows the simulated and measured gain of the antenna array for different combinations of active TX channels [17]. The measured and simulated radiation patterns show reasonable agreement.

To verify the functionality of the front-end using the eWLB module, a multiple-input, multiple-output (MIMO) radar system has been built and tested in an anechoic chamber. By sequentially activating one of the four transmitters

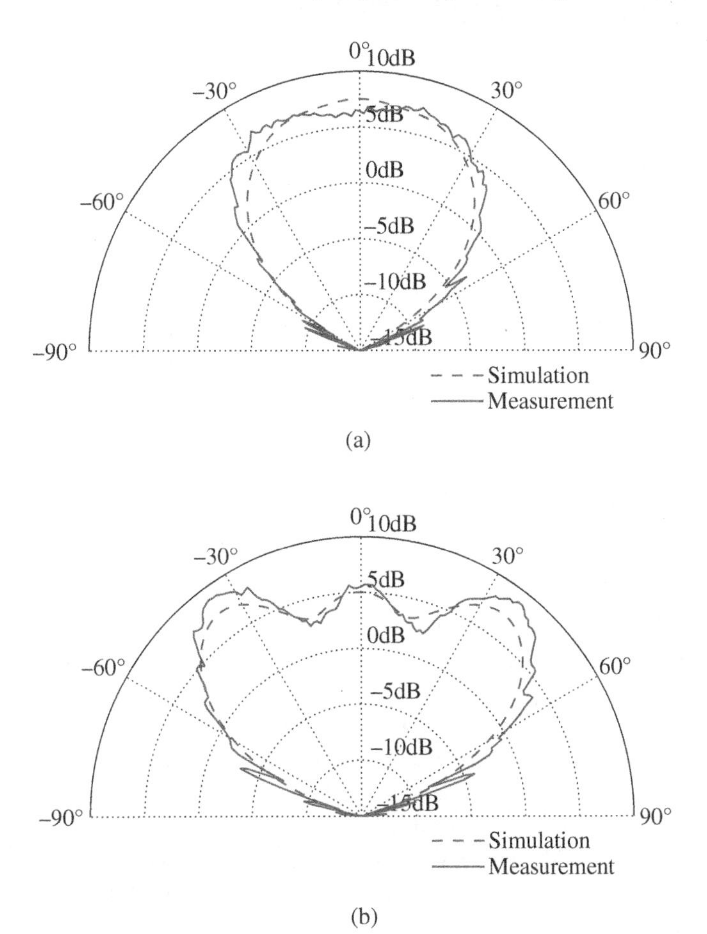


**Figure 8.35** Simulated and measured gain of the antenna array for (a) active TX channels 2 and 3 and (b) active TX channels 1, 2, 3, and 4 (from [17], ©2019 IEEE, reprinted with permission).

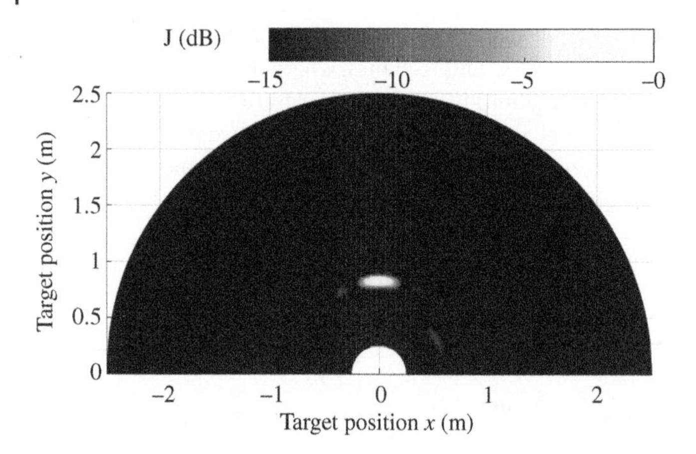


**Figure 8.36** Resulting power distribution using a target of approximately 10 m$^2$ RCS located at a distance of 0.82 m. The plot is normalized to the peak value (from [17], ©2019 IEEE, reprinted with permission).

while receiving with all four receivers, seven virtual antennas from only the four physically existing uniformly arranged array elements could be synthesized. To verify the detection capabilities, a target of approximately 10 m$^2$ radar cross-section (RCS) located in a distance of 0.82 m was used. Figure 8.36 illustrates the measured power distribution [17]. The plot is normalized to the peak value. As expected, the maximum of the power distribution occurs at the position of the target. The presented results verify the basic beamforming and target detection capabilities of the front-end using the eWLB module.

### 8.5.3 Six-channel 60-GHz Transceiver Module

Figure 8.37 shows a photograph of a single-chip six-channel 60-GHz transceiver in SiGe integrated in the eWLB package [23]. The module has size 14 $\times$ 14 mm and a footprint with a standard ball pitch of 0.5 mm. The transceiver chip occupies the bottom half of the package. On the left and right sides of the chip two transmit dipole antennas, TX1 and TX2, are realized. Similar to the first two presented examples, the dipole antennas are connected using 100 Ω differential CPS lines and matched using $\lambda/4$ impedance transformers. In the upper part of the package a 2 $\times$ 2 array of patch receive antennas, RX1–RX4, is realized. The orientations of the dipole receiver and patch transmitter antennas are aligned to provide the same polarizations and consequently higher signal levels when the radar front-end is in operation. The interconnections between the transceiver chip and receive antennas are realized using single-ended MSL lines having a signal conductor in the RDL and ground on the PCB. Unlike differential dipole antennas, which use PCB

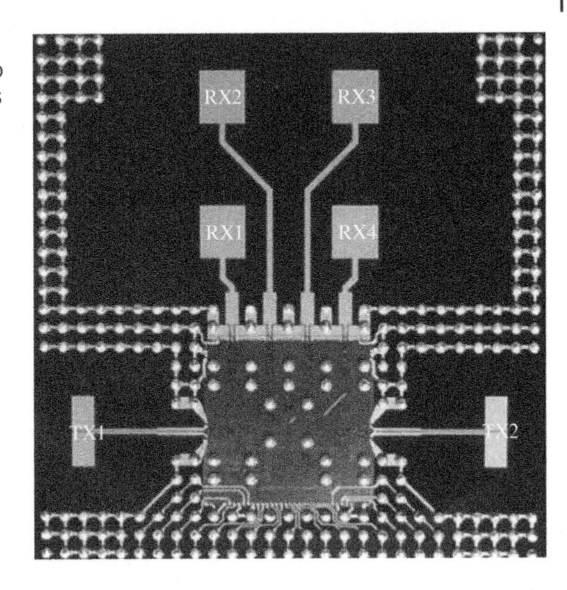


**Figure 8.37** Photograph of a 60-GHz eWLB module with two dipole and four patch antennas integrated in the RDL (from [23], ©2019 IEEE, reprinted with permission).

metallization for the reflector, a patch antenna uses PCB metallization for signal ground. Therefore, a low-loss and matched RF transition to PCB is required to provide a smooth interconnection between the chip and the MSL feeding lines.

Figure 8.38 illustrates the electromagnetic (EM) simulation environment used to model and optimize the antennas [23]. All four channels are matched and receive in phase. Thus, the two channels on the top, RX2 and RX3, show one $\lambda$ difference compared to the channels on the bottom, RX1 and RX4. Figure 8.39

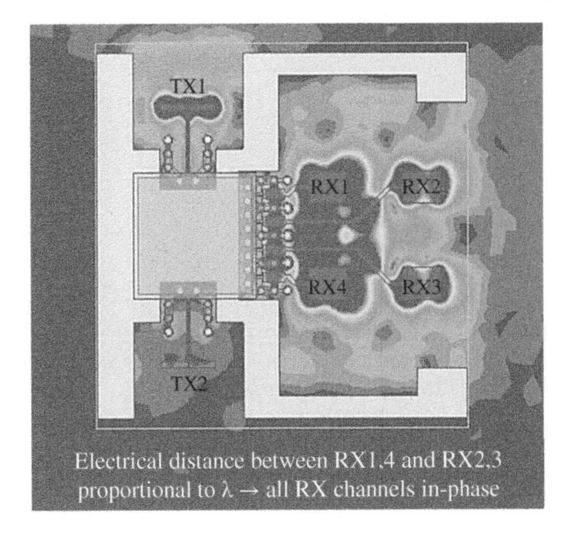


**Figure 8.38** EM simulation environment (from [23], ©2019 IEEE, reprinted with permission).

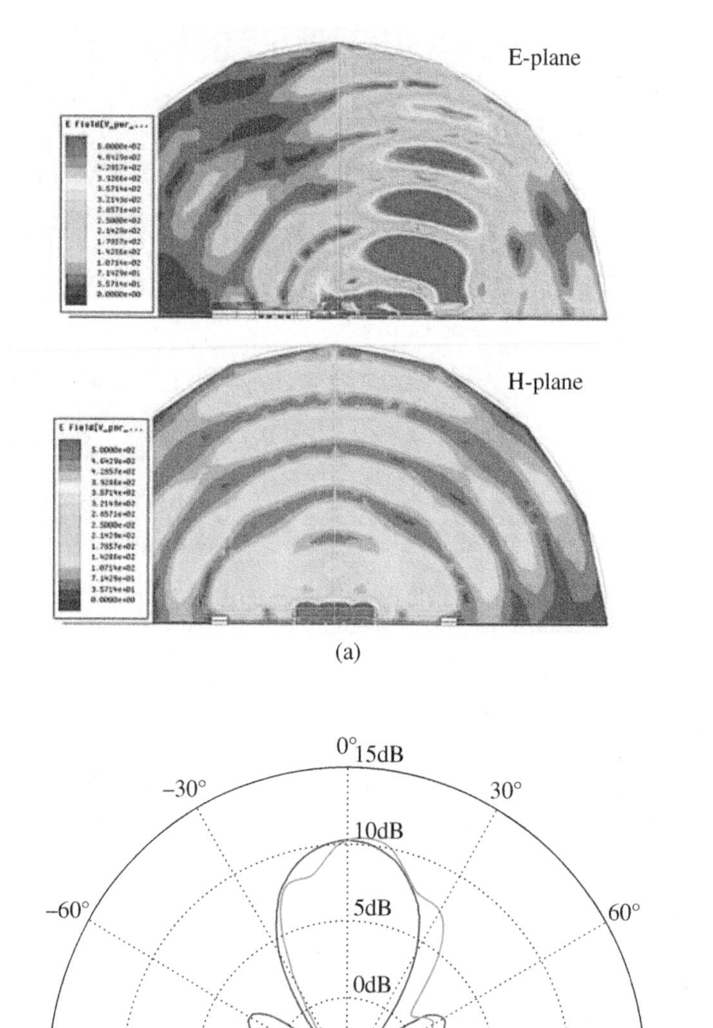


**Figure 8.39** (a) Electric field distribution in the E- and H-planes at 60 GHz and (b) the corresponding radiation pattern for all four receive antennas in phase (from [23], ©2019 IEEE, reprinted with permission).

shows the simulated electric field distribution in the E- and H-planes at 60 GHz and the corresponding radiation pattern for all four receive antennas in phase [23]. The gain of 2 × 2 configurations is about 10 dBi. The asymmetric behavior in the E-plane radiation pattern is due to the additional one $\lambda$-long feeding line at RX2 and RX3. A slight tilt of the radiation beam caused by the Si chip is visible. Figure 8.40 presents the results of the same simulations for the single transmit antenna [23]. The radiation pattern shows a gain of about 7 dBi.

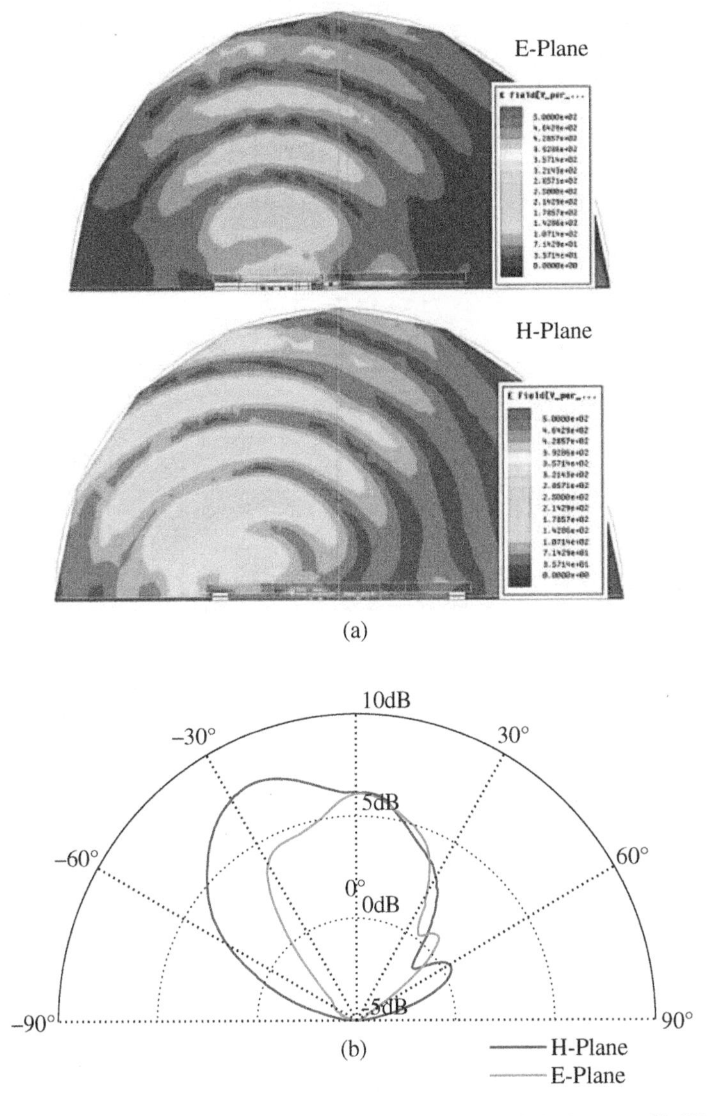


**Figure 8.40** (a) Electric field distribution in the E- and H-planes at 60 GHz and (b) the corresponding radiation pattern for the single transmit antenna (from [23], ©2019 IEEE, reprinted with permission).

The required bandwidth of the antennas to operate is 57–64 GHz. Figure 8.41 shows the simulated receiver and transmitter antenna return losses [23]. They represent the reflection seen from the chip and take into account the frequency dependent response of the transceiver. Figure 8.42 shows the simulated isolation between transmitter and receiver channels [23]. The large distance between

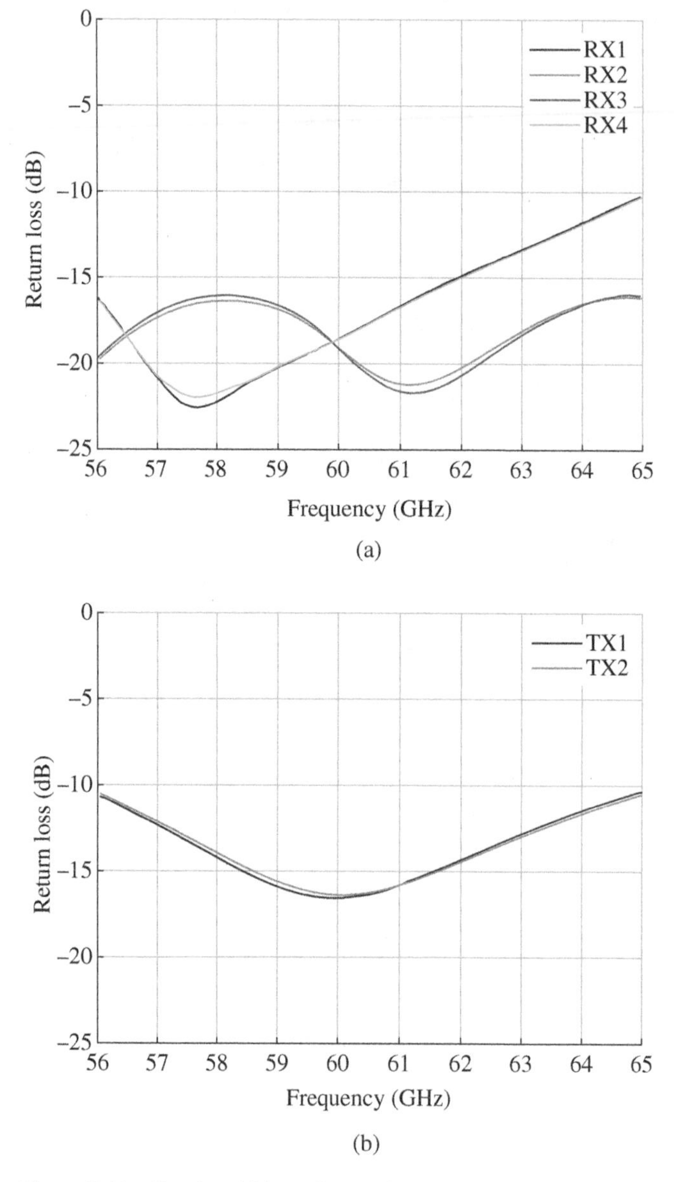


**Figure 8.41** Simulated (a) receiver and (b) transmitter antenna return loss (from [23], ©2019 IEEE, reprinted with permission).

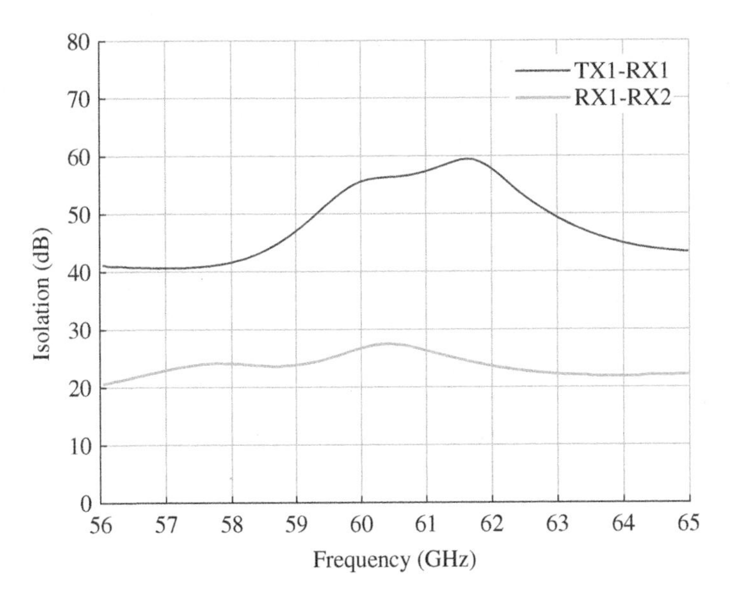


**Figure 8.42** Simulated isolation between transmitter and receiver channels (from [23], ©2019 IEEE, reprinted with permission).

transmitter and receiver channels and the ground wall realized using solder balls helps to increase the isolation. The achieved isolation is better than 40 dB between the RX1 and TX1 channels and higher than 20 dB between RX antennas.

## 8.6 Conclusion

Accurate manufacturing process tolerances of the eWLB technology which do not exist in other package technologies open up innovative capabilities for the next generation of mmWave applications. The research presented here on integrated antennas demonstrates the design and performance potential of eWLB packaging technology.

## Acknowledgement

This research work was funded by the German Bundesministerium für Bildung und Forschung (BMBF) in the projects CoSiP, RoCC, JEMSiP, V3DIM/3DIM3, and SiPoB-3D. The CoSiP project ran in the EUREKA MEDEA+ program, the projects V3DIM/3DIM3 and SiPoB-3D ran in the EUREKA CATRENE program. JEMSiP was also funded by the ENIAC JU.

## References


1 Heterogeneous Integration Roadmap. Available on-line: https://eps.ieee.org/technology/heterogeneous-integration-roadmap.html, 2019.

2 M. Brunnbauer, E. Furgut, G. Beer et al., An embedded device technology based on a molded reconfigured wafer. In *56th Electronic Components and Technology Conference*, pages 1–5, May 2006. doi: 10.1109/ECTC.2006.1645702.

3 M. Brunnbauer, E. Fürgut, G. Beer et al., Embedded wafer level ball grid array eWLB. In *8th Electronics Packaging Technology Conference*, pages 1–5, December 2006. doi: 10.1109/EPTC.2006.342681.

4 J. H. Lau, *Fan-Out Wafer-Level Packaging*. Springer, Singapore, 2018. ISBN 978-981-10-8883-4.

5 M. Wojnowski, K. Pressel, G. Beer et al., Vertical interconnections using through encapsulant via (TEV) and through silicon via (TSV) for high-frequency system-in-package integration. In *IEEE 16th Electronics Packaging Technology Conference*, pages 122–127, December 2014. doi: 10.1109/EPTC.2014.7028413.

6 M. Wojnowski, K. Pressel, and G. Beer, Novel embedded Z line (EZL) vertical interconnect technology for eWLB. In *IEEE 65th Electronic Components and Technology Conference*, pages 1071–1076, May 2015. doi: 10.1109/ECTC.2015.7159727.

7 M. Wojnowski, R. Lachner, J. Böck et al., Embedded wafer level ball grid array (eWLB) technology for millimeter-wave applications. In *IEEE 13th Electronics Packaging Technology Conference*, pages 423–429, December 2011. doi: 10.1109/EPTC.2011.6184458.

8 M. PourMousavi, M. Wojnowski, K. Pressel et al., Passive Components using Quasi-CPW in eWLB Package. In *19th International Conference on Microwaves, Radar Wireless Communications*, volume 2, pages 709–712, May 2012. doi: 10.1109/MIKON.2012.6233595.

9 M. Wojnowski, M. Engl, M. Brunnbauer et al., High frequency characterization of thin-film redistribution layers for embedded wafer level BGA. In *9th Electronics Packaging Technology Conference*, pages 308–314, December 2007. doi: 10.1109/EPTC.2007.4469826.

10 M. Wojnowski, V. Issakov, G. Knoblinger et al., High-Q embedded inductors in fan-out eWLB for 6 GHz CMOS VCO. In *IEEE 61st Electronic Components and Technology Conference*, pages 1363–1370, May 2011. doi: 10.1109/ECTC.2011.5898689.

11 M. Wojnowski, V. Issakov G. Knoblinger et al., High-Q inductors embedded in the fan-out area of an eWLB. *IEEE Transactions on Components, Packaging and Manufacturing Technology*, vol. 2, no. 8, pp. 1280–1292, August 2012. ISSN 2156-3950. doi: 10.1109/TCPMT.2012.2186963.

12 M. Wojnowski, G. Sommer, K. Pressel et al., 3D eWLB horizontal and vertical interconnects for integration of passive components. In *IEEE 63rd*

*Electronic Components and Technology Conference*, pages 2121–2125, May 2013. doi: 10.1109/ECTC.2013.6575873.

**13** P. Schmidbauer, M. Wojnowski, R. Weigel et al., Concepts for a monostatic radar transceiver front-end in eWLB package with off-chip quasicirculator for 60 GHz. In *IEEE 20th Electronics Packaging Technology Conference*, pages 8–12, December 2018. doi: 10.1109/EPTC.2018.8654332.

**14** C. Wagner, J. Bock, M. Wojnowski et al., A 77 GHz automotive radar receiver in a wafer level package. In *IEEE Radio Frequency Integrated Circuits Symposium*, pages 511–514, June 2012. doi: 10.1109/RFIC.2012.6242334.

**15** M. Wojnowski, M. Engl, B. Dehlink et al., A 77 GHz SiGe mixer in an embedded wafer level BGA package. In *58th Electronic Components and Technology Conference*, pages 290–296, May 2008. doi: 10.1109/ECTC.2008.4549984.

**16** E. Seler, M. Wojnowski, G. Sommer et al., Comparative analysis of high-frequency transitions in embedded wafer level BGA (eWLB) and quad flat no leads (VQFN) packages. In *IEEE 14th Electronics Packaging Technology Conference*, pages 99–102, December 2012. doi: 10.1109/EPTC.2012.6507059.

**17** M. Wojnowski, C. Wagner, R. Lachner et al., A 77-GHz SiGe single-chip four-channel transceiver module with integrated antennas in embedded wafer-level BGA package. In *IEEE 62nd Electronic Components and Technology Conference*, pages 1027–1032, May 2012. doi: 10.1109/ECTC.2012.6248962.

**18** R. Agethen, M. PourMousavi, H.P. Forstner et al., 60 GHz industrial radar systems in silicon-germanium technology. In *IEEE MTT-S International Microwave Symposium Digest*, pages 1–3, June 2013. doi: 10.1109/MWSYM.2013.6697731.

**19** C. Beck, H.J. Ng, R. Agethen et al., Industrial mmWave radar sensor in embedded wafer-level BGA packaging technology. *IEEE Sensors Journal*, vol. 16, no. 17, pp. 6566–6578, September 2016. ISSN 1530-437X. doi: 10.1109/JSEN.2016.2587731.

**20** M. PourMousavi, M. Wojnowski, R. Agethen et al., Antenna array in eWLB for 61 GHz FMCW radar. In *Asia-Pacific Microwave Conference Proceedings*, pages 310–312, November 2013. doi: 10.1109/APMC.2013.6695129.

**21** M. PourMousavi, M. Wojnowski, R. Agethen et al., Antenna design and characterization for a 61 GHz transceiver in eWLB package. In *European Microwave Conference*, pages 1415–1418, October 2013. doi: 10.23919/EuMC.2013.6686932.

**22** M. PourMousavi, M. Wojnowski, R. Agethen et al., The impact of embedded wafer level BGA package on the antenna performance. In *IEEE-APS Topical Conference on Antennas and Propagation in Wireless Communications*, pages 828–831, September 2013. doi: 10.1109/APWC.2013.6624913.

**23** I. Nasr, R. Jungmaier, A. Baheti et al., A highly integrated 60 GHz 6-channel transceiver with antenna in package for smart sensing and short-range communications. *IEEE Journal of Solid-State Circuits*, vol. 51, no. 9, 2066–2076, September 2016. ISSN 0018-9200. doi: 10.1109/JSSC.2016.2585621.

# 9

# Additive Manufacturing AiP Designs and Applications

*Tong-Hong Lin, Ryan A. Bahr, and Manos M. Tentzeris*

*The School of Electrical and Computer Engineering, Georgia Institute of Technology, Atlanta GA 30332, USA*

## 9.1 Introduction

Additive manufacturing (AM) technologies [1] are emerging fabrication methods that are expanding rapidly in both industrial applications and academic research. Compared with traditional subtractive manufacturing technologies, which remove materials in unwanted regions, AM technologies only deposit materials at desired regions. Due to the reduction of materials used, AM can reduce the costs by two main vectors: it significantly reduces material usage during structural formation and it reduces waste byproduct materials such as etchants, solvents, and coolants. The most common subtractive fabrication method for electronics is photolithography, as shown in Figure 9.1. The process often begins with copper cladding a substrate, and then adhering a thin layer of photoresist onto said copper. A mask needs to be fabricated, which is then used in the light exposure process to either weaken or strengthen bonds via exposed light, depending on if it is a positive or negative resist, respectively. Thereafter, chemical etching is utilized to remove unwanted materials (in this case, the copper pattern), and finally one more chemical wash is used to remove the remaining photoresist to achieve a patterned circuit. While the process has been utilized successfully for countless years, the advancement of digital fabrication, the so-called "fourth industrial revolution" has enabled the exploration of alternative technologies that would have previously been determined as too complex.

The most popular AM technologies are inkjet printing and three-dimensional (3D) printing. Inkjet printing utilizes liquid materials or "inks" with different properties for a wide range of applications. The nozzles or the substrate are moving while inks are jetting out so that the inks can be deposit on the desired regions. Extensive research on how to use different inks to print components

*Antenna-in-Package Technology and Applications*, First Edition.
Edited by Duixian Liu and Yueping Zhang.

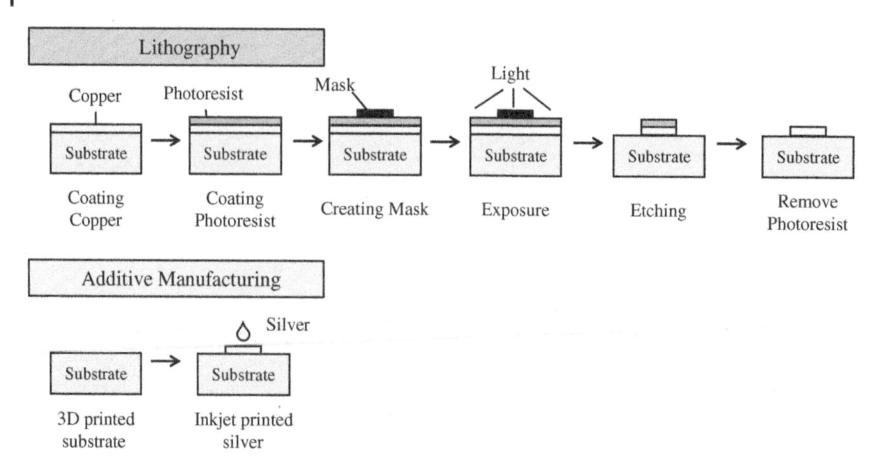


**Figure 9.1** Comparisons between subtractive and additive manufacturing processes.

such as sensors, passives, origami, microfluidics, flexible electronics, transistors, and solar cells has been proposed [2–4]. For 3D printing, the most popular two methods for radio frequency (RF) and millimeter-wave (mmWave) components are stereolithography (SLA) 3D printing and digital light projection (DLP) 3D printing since they can achieve a smoother surface and thus reduce the surface roughness [5]. Furthermore, different AM techniques can be combined to realize a fast and large-scale prototyping method with higher material and structural flexibility. As demonstrated in Figure 9.1, ideally, the circuit fabrication process can be reduced to only two steps by using both inkjet and 3D printing. The first step is to 3D print the substrate and the second step is to inkjet print the conductive patterns for electronics.

Recently, the emerging high-frequency fifth-generation (5G) and mmWave wireless communications have resulted in size reduction of the RF components and given rise to increased interest in system-on-package (SoP) design, which incorporates integrated circuits (ICs) with RF peripheral components such as filters, couplers, and antennas in one package [6, 7]. Thus, research involving exploiting AM techniques to achieve 3D structures for performance improvement and fully AM SoP modules has seen a drastic increase in interest [8]. In [9], inkjet-printed ramp structures with transmission lines on the top were used as the die-to-package interconnects. It has been proved that this structure can support lower parasitics and better physical reliability [10]. In [5], 3D-printed ramp structures were used as inter-layer interconnects with lower parasitics compared with vias. A 3D-printed AiP design using origami folding is presented in [11]. Fully AM SoP modules, including AiP, ICs, and interconnects from the AiP to the module, wee first investigated in [12, 13].

In section 9.2, we provide a detailed introduction to different AM technologies, including inkjet printing, and discuss a variety of 3D printing technologies and their relevance to RF applications. The material characterization methods are given in section 9.3. We then introduce a hybrid fabrication process by integrating 3D printed packaging substrates and inkjet printing conductive traces in section 9.5. A broadband 5G AiP design realized with AM and the respective measurement results is introduced in section 9.6. Finally, the summary and future visions of realizing complex 3D AiP and SoP design using AM are included in section 9.7.

## 9.2 Additive Manufacturing Technologies

### 9.2.1 Inkjet Printing

Inkjet printing may not be the first thought to come to mind when 3D printing is discussed, but by utilizing this historically significant process to repeatedly print layers over and over on the same area, 3D structures can be formed. An advantage of inkjet printing is that it has demonstrated its scalability through centuries of use, often utilizing multiple cartridges, each of which can contain over 1000 drop-on-demand nozzles dispensing material of picoliter volume, at a rate that can exceed 20 000 times per second. Typical cartridges are in the 1–10 pL range, corresponding to a drop diameter size of roughly 30 μm [14], with a layer thickness between 0.5 and 2 μm. Materials need to be within a certain viscosity (often between 1 and 30 cPs) and surface energy, with various solutions consisting of metallic nanoparticles, surfactants, and polymers enable a wide choice of formulated materials [15]. The resolutions of this technology are relevant to structures beyond 5G, and with the ability to utilize multiple cartridges of different materials, similar to how a consumer inkjet printer utilizes multiple color cartridges, it can be readily utilized to print multiple materials simultaneously, ideal for fabricating anything from circuit boards to interconnected multichip modules.

### 9.2.2 FDM 3D Printing

Improvements in 3D printing within the last decade have caused increased interest in utilizing automated fabrication methodology across a variety of fields, enabling new, complex designs that may previously have been cost or time prohibitive. The wide set of technologies that make up 3D printing have an even wider variety of unique characteristics, but all methodologies tend to consolidate on automation of intricate designs, consisting of hundreds if not millions of individually patterned layers.

When most people think of 3D printers, they associate the technology quite often with the low-cost models available at schools across the world. These fused-deposition modeling (FDM) 3D printers, while extremely useful, only begin to scratch the surface of the variety of AM tools available and are associated with being slow and producing designs with visible layers. While these primitive printers are overall still a useful tool, in order to be beneficial to industry there needs to be significant improvements in areas enabling scalability, reduced cost, and increased accuracy.

FDM printers rely on the extrusion of thermoplastics through a heated nozzle in a layer by a laser deposition process. The polymers are extremely low cost as they utilize common bulk plastics such as polylactic acid (PLA), acrylonitrile butadiene styrene (ABS), and polyethylene terephthalate (PET), and can even utilize these plastics in a recycled form. The quality of the print is often quantified by the resolution of the printer, with the minimum feature size determined and equal to the nozzle diameter (often between 100 and 400 μm), as well as the layer height (typically between 20 and 100 μm). While smaller nozzles and layer heights are possible, due to the reliance of often a single nozzle to deposit material, the print time increases relatively linearly as each variable is adjusted, significantly increasing fabrication times compared to other fabrication methodologies. The technique does excel at being able to integrate multiple nozzles and technologies, including direct write tools, where a variety of liquids and pastes can be deposited simultaneously alongside the FDM printing, enabling the deposition of silver conductive pastes, ultrasonically depositing copper wiring, or spraying aerosol particles of various materials that can enable electronically functional prints [16]. The versatile nature, wide variety of materials, including bulk/pure polymers, and resolution enable this technology to provide a low-cost prototyping tool for electrically large devices, such as dielectric lenses, or circuits that ideally are below mmWave frequencies due to the surface roughnesses associated with the process. While FDM is known for lower resolution and frequency structures, variations of the technology enable feature sizes below 13 μm, which is relevant to sub-terahertz (sub-THz) structures [17].

### 9.2.3 SLA 3D Printing

When a person thinks of "high resolution" and extremes of manufacturing technologies, lasers often come to mind. Similarly, there exist many AM technologies which utilize optics and lasers, enabling resolutions that can mimic component semiconductor manufacturing. Laser-based processes such as selective laser sintering (SLS) and selective laser melting (SLM) generally consist of depositing a bed of powder, acting as both the building material and support material, and sintering or melting particles at the point of exposure. The parts generated have a roughness that is often directly related to the size of the particles utilized, often in the 10–60 μm range, and the majority of the particles that are not sintered can be recycled into future prints.

Instead of sintering polymers together, utilizing the thermal properties of lasers, the selective exposure of light can be used to crosslink photopolymer resins. SLA 3D printing often utilizes an ultraviolet (UV)-based laser to crosslink a layer of liquid resin, solidifying the material selectively, layer by layer. The resolution of SLA is often limited by the diameter of the laser in use, determining the feature size, as well as the properties of the photopolymer resin. While a laser diameter of 100 μm is typical, an extreme variation of this technology, called two-photon SLA, can enable printing of structures as small as 100 nm, which has been utilized in bio-micro-electromechanical systems (MEMS) devices and photonic metamaterials [18]. As with most technologies, the smaller the feature size, the longer print times take. A typical 3D printer using the SLA technique from FormLabs is shown in Figure 9.2.

Similar to the maskless lithography technique within the semiconductor realm, a variation of SLA 3D printing utilizes DLP or liquid crystal display (LCD) screens for digital masks and a photonic curing light source. SLA-DLP/LCD utilizes a light source (between UV and visible light depending on the photopolymer photoinitiator being utilized) that is exposed onto a build plate that is submerged in a vat of

**Figure 9.2** The FormLabs SLA printer has seen widespread adoption due to a low price point and ease of use. It originated from a crowd-funding campaign.

resin. The build plate changes height per layer, and a new pattern is exposed until the 3D model is complete. The resolution of SLA-DLP/LCD is near the size of the pixel of the mask (often between 20 and 100 μm but dependent on the photonics of the system), though sub-pixel resolution can be achieved with characterized techniques. These mask-based techniques offer one significant advantage, and a critical one at that: scalability. As long as all the parts fit within the build area, printing one part or a thousand will take an identical amount of time, the time dependency is on the height of the design. Additionally, SLA offers extremely low surface roughness across the planes of the print, with measured roughness below 100 nm root mean square (RMS) [5].

There is no standardized nomenclature to differentiate SLA techniques, and often when SLA is discussed without specifying, it is implied that it is a laser-based system that does not utilize two-photon polymerization (TPP). The original SLA patent covers exposure of resins with masks or motorized light sources for 3D fabrication, covering practically all variations. Those variations include the use of DLP, LCD, and TPP. From a process standpoint, DLP and LCD both are similar to maskless lithography, exposing an entire layer simultaneously, while TPP is a laser direct-write method that enables high resolution, as low as 100 nm. When these are discussed, they will be referred to as SLA-DLP, SLA-LCD, and SLA-TPP for DLP, LCD, and TPP techniques, respectively. Recent advances have enabled a combination of some of these technologies, as well as adaptive optics, which enable the resolution of SLA-TPP when necessary, but offer bulk exposure and printing utilizing SLA-DLP. Ignoring the poor high-frequency electrical properties of current SLA materials, which will be discussed further later, the resolution and scalability make this a promising technology for anything from traditional devices to next-generation devices, all the way up to photonic structures.

AM is a significant multidisciplinary process, and almost every fabrication technique requires an understanding of mechanical properties, thermal properties, algorithms for path or pattern generation, and, often most significantly, a very wide breath of material design and properties. While a variety of technologies have been compared, this does not serve as an all-inclusive list and new methodologies are rapidly being developed, with some of the latest innovations utilizing techniques such as continuous liquid interface printing (CLIP), holography, 4D printing, reverse tomography, and more [19, 20].

## 9.3 Material Characterization

There is often a significant disconnect between the majority of RF and material science engineers, and often the materials designed for AM and 3D printing have a focus on specifications related to printability, strength, flexibility, temperature, and a wide variety of other features. An advantage to 3D printing techniques that utilizes relatively bulk materials, such as FDM, SLS, and SLM, is that the RF material properties are relatively well known. There may be some concerns with the crystalline structure or any internal voids affecting the density and thereby effective properties, though techniques such as scanning electron microscopy (SEM) and scanning acoustic microscopy (SAM) can be beneficial to verify the properties of such prints. Other printing techniques that require specially formulated materials, such as inkjet printing, aerosol jet printing, and SLA, have ever-evolving materials with novel properties, and often require constant characterization with every variation of a material.

### 9.3.1 Resonator-based Material Characterization

While there is a wide variety of methods to characterize dielectric materials, two primary techniques are utilized for characterization of materials for electromagnetic purposes, consisting of either resonant structures or transmissive structures. Resonant structures are often utilized for single frequency measurements with the increased accuracy of loss measurements compared to transmissive techniques. One methodology is the utilization of a split post dielectric resonator (SPDR), as shown in Figure 9.3, which enables measurement of thin sheets (i.e. up to 600 μm thick for 15-GHz structures), where the resonant frequency variation and quality factor determine the dielectric permittivity and loss, respectively [21]. If the 3D printed substrate can be metalized, ring resonators can be patterned directly onto

**Figure 9.3** Photograph of a commercially available split post dielectric resonator.

the substrate, offering an integrable approach where characterization structures can be included during fabrication [22].

### 9.3.2 Transmissive-based Material Characterization

Alternatively, transmissive material characterizations enable broadband measurement of materials, though often at a reduced accuracy compared to resonant structures. For planar materials such as those used in printed circuit boards (PCBs), the National Institute of Standards and Technology (NIST) has several recommended methodologies, one of which is using two different lengths of microstrip transmission lines for loss and dielectric constant measurements. In practice, manufacturing can introduce minor air gaps which can lower the effective dielectric constant if the substrate thickness is thinner than the substrate thickness utilized during characterization. The NIST guidelines for dielectric characterization offer significant in-depth comparison between characterization techniques relevant to the wide variety of 3D printing methodologies.

Another commonly used broadband transmissive-based characterization technique is that of waveguides, where samples are fitted into a waveguide shim, and the scattering parameters are measured for the reflection and transmission values utilized with the Nicholson–Ross–Weir (NRW) derivations to calculate the dielectric properties, as shown in Figure 9.4. As samples are ideally nearly a quarter wavelength in thickness and need to fill a waveguide, this can lead to large samples at low frequencies. Overmoding, avoiding half-wavelengths, and reducing

**Figure 9.4** Material characterization using waveguides.

or preventing any air gaps in the shim must also be accounted for, which may be difficult depending on the tolerances of the 3D printer at higher frequencies [23]. Regardless of whichever method is utilized, any characterization technique should be verified to satisfy the Kramers–Kronig relation.

While bulk materials have often been characterized, dielectric material properties contain a wide variety of influences from the printing technology where that characterization of the printed products is always beneficial. For example, with SLA it has been shown that the post-print curing methodology can have a significant effect on the dielectric properties of the 3D print, with a variation of loss between 0.0196 and 0.0368 depending on the utilization of a thermal cure during UV curing [24]. With other materials, such as particle suspensions in polymers or other dielectrics, charges can develop between interfaces, known as Maxwell–Wagner–Sillars polarization, that can contribute to a significant increase in dielectric losses [25]. Ideally, homogenous, non-conductive, non-polar molecular materials, if possible, should be utilized to reduce the dielectric losses. While the electromagnetic properties of 3D printing are often secondary to the manufacturing quality, novel materials focused on electromagnetic materials have been developed, ranging from new low-loss materials specially formulated for printing to modifying existing low-loss materials in a manner that makes them compatible with available 3D printers [26]. For example, bulk thermoplastics for FDM printing such as polypropylene (PP) and ABS enable low loss and very low cost materials that can enable complex 3D designs ranging from gradient materials to dielectric lenses relative to mmWave frequencies [24]. On the other hand, many photopolymer materials, from inks to SLA, are currently quite lossy. While the loss may not be optimal for production-grade devices, relatively lossy materials still enable demonstration of novel concepts and rapid prototyping. The utilization of ceramic-loaded photopolymers can enable sintering and firing of the prints, burning off the lossy components attributed to the photopolymer, thereby offering 3D printing of ceramic designs that includes benefits such as reduced loss and higher permittivities. Additionally, almost all 3D printing techniques can be metallized with electroless or electroplating techniques, which means complex, low-cost, and low-weight metallic coated structures can be realized, ideal for RF devices of almost all frequencies. With the recent explosion of 3D printing, including the utilization of 3D printers dedicated to replacing traditional circuit board milling for rapid automation of multilayer circuit boards, there has been a significant renewed interest in creating low-loss materials for large-scale AM.

## 9.4 Recent Advances in AM for Packaging

While the relevance of a variety of AM techniques towards RF design has been discussed, it has only been in relatively recent years that high-resolution conductive traces with AM technologies have enabled fabrication of reliable mmWave conductive structures. Precise tools, including the nScrypt direct write and Optomec aerosol jet printers, enable resolutions down to 10 μm. Additionally, with the rapid evolution of nanomaterials, the ability to create solutions of conductive nanoparticle inks enables the printing of functional circuits, including materials specially formulated for transistors, capacitors, inductors, piezoelectric materials, functionalized materials for chemical sensing, and more. At the beginning, relatively low frequencies only in the low megahertz regime were explored, but this quickly grew towards frequencies related to radio-frequency identification (RFID) and Wifi. The explosion of AM has enabled structures with frequencies as high as the photonic region, enabling the fabrication of structurally complex photonic metamaterials.

### 9.4.1 Interconnects

An important component in any high-frequency circuit are the interconnects between different components. From the packaging perspective, there are two critical interconnects to realize the SoP design: die-to-packaging interconnects and inter-layer interconnects. While easy to overlook, they can cause performance degradation of a system that may turn a ground-breaking design into architecture that underperforms. Design engineers are often limited to the manufacturing process for the design of interconnects that are utilized between layers. This is extremely important at mmWave and and sub-terahertz frequencies, where the parasitics of the interconnects can cause significant losses or contribute to unwanted resonance structures.

A significant benefit to AM is an extremely wide breath of fabrication design choices for engineers, and therefore a variety of controlled interconnect structures can be fabricated, enabling reduced losses. 3D printed devices can enable suspended transmission lines where the predominant dielectric is air, such as coaxial or microstrip lines [28]. The most prominent benefit of utilizing air as a dielectric is the reduction of dielectric loss, relevant to mmWave and higher frequencies. Stretchable coaxial lines have been realized with AM in conjunction with liquid metals [29]. In [9], a mmWave die-to-package interconnect using inkjet printing is demonstrated and characterized to 40 GHz. Furthermore, bow-tie AiP design is used to demonstrate a direct connection from die to AiP design, as shown in Figure 9.5. A process flow for fully AM smart encapsulation including through-mold vias (TMVs) is shown in Figure 9.6 [27]. The TMVs are fabricated with inkjet and 3D printing, and can be used to provide inter-layer interconnects between silicon die and AiP at different layers, as shown in Figure 9.7. This topology demonstrates the lowest loss in this category for mmWave applications up to 90 GHz [5]. These technologies, by offering non-planar geometries, enable reduced loss structures compared to traditional technologies, which becomes extremely beneficial in next-generation wireless communication topologies.

**Figure 9.5** Inkjet printed on-package mmWave bow-tie antenna samples [9].

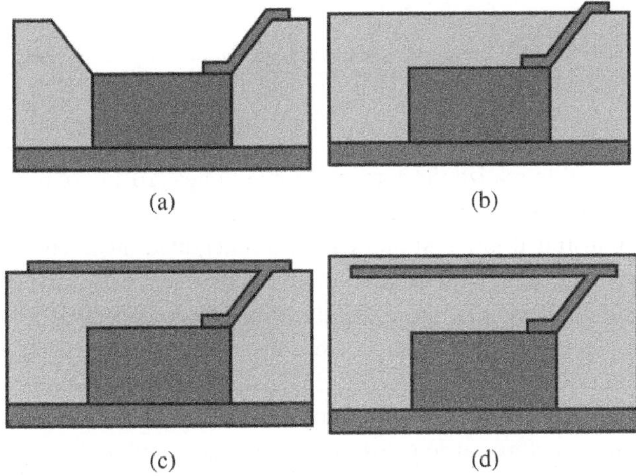


**Figure 9.6** "Smart" wireless encapsulation process flow. (a) 3D print partial encapsulation with die and inkjet print sloped TMV. (b) Cap partial encapsulation with photopolymer resin leaving exposed TMV interconnecting to the embedded die. (c) Inkjet-printed antenna (or other passive component) for SoP devices. (d) 3D print final encapsulation (from [27], © 2018 IEEE, reprinted with permission).

### 9.4.2 AiP

The interconnects can be utilized to connect to a variety of components. One component that is extremely beneficial to directly connect to are the antennas because it is ideal to integrate antennas as close to the package as possible to reduce the loss. AiP enables the integration of antennas that may be too large to include on the bare die, while minimizing connection losses that result from moving the antennas onto the board or other support circuitry. The integration of the antennas has been demonstrated directly in the epoxy molding of traditional ICs for AiP mmWave applications at 30 GHz [30]. Similar demonstrations have been shown to print antennas directly connected to bare die, enabling a reduction of size and losses by realizing a full SoP solution [31]. Similarly, these same

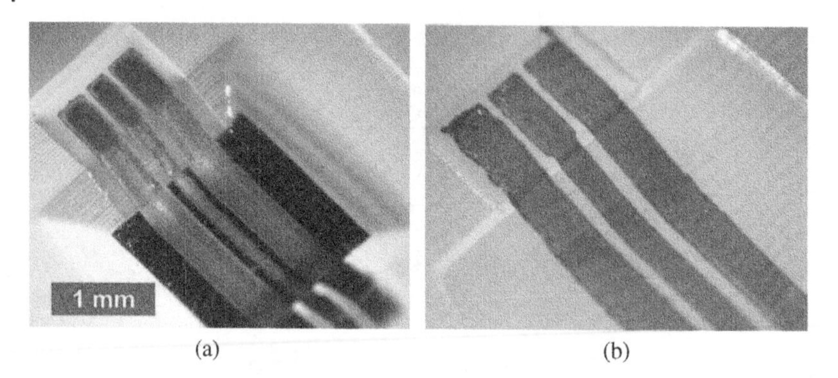


**Figure 9.7** (a) Fully-printed 3D TMV-integrated partial encapsulation with silicon die. (b) TMV-integrated encapsulation (from [27], © 2018 IEEE, reprinted with permission).

fabrication techniques used to realize the previously mentioned AiPs have been used to demonstrate functional systems, including a 90-GHz beamforming array [32]. AM has also demonstrated low cost fabrication of integrated metal coated plastic horn antennas for flip chip applications, enabling increased performance for sub-THz systems [33]. Naturally, with the printing of complex dielectric structures, electrically large structures such as dielectric lenses, ranging from a simple elliptical lens all the way to gradient material lenses, can be incorporated directly into the package [34]. It becomes that apparent that AM technologies that enable complex 3D structures at low cost are here to stay and will enable novel designs for next-generation mmWave and sub-terahertz structures.

## 9.5 Fabrication Process

### 9.5.1 3D Printing Process

After comparing the pros and cons of the different 3D printed technologies, DLP 3D printing has been adopted since it can achieve both high resolution and large-scale manufacturability. The substrate used is FormLabs FLGR02 Flexible resin. The material can be cured around 405 nm and thus an off-the-shelf projector is utilized in an open-source DLP printer, exposing broadband patterned white light to project the individual greyscale images that make up the layers of the designs and cure the material, layer by layer. The layer resolution is 50 μm. The light exposure time of each layer is a significant parameter in determining the quality of the print and should be tuned with respect to the final irradiance incident to the

**Figure 9.8** (a) The effects of exposure time on the 3D printed substrate. (b) Demonstration of large-scale fabrication (from [12], © 2018 IEEE, reprinted with permission).

photopolymer. As shown in Figure 9.8a, not enough light exposure will result in serious defects on the samples, including incomplete holes or layers that are too thin. On the other hand, too much light exposure will close the holes, overcure through layers, and otherwise deviate from the design parameters. For this specific scenario, the light exposure time for each layer was set to 13 s after testing with different setups.

One of the key advantages of DLP is large-scale manufacture, as introduced in section 9.2.3. Since the images of multiple samples can be projected and cured simultaneously, the fabrication time is similar for one sample and multiple samples. As demonstrated in Figure 9.8b, two samples are fabricated simultaneously with the same fabrication time as a single sample. This can be extended to multiple samples as long as they can be fitted onto the build plate. The combination of high resolution and scalability makes this technology extremely attractive, as a future for large-scale fabrication may be more realizable compared to many AM alternatives. Furthermore, with this open-source setup, the whole build plate can be taken off and moved to the inkjet printer while maintaining the relative positions of all samples, enabling rapid inkjet deposition of a metallic pattern that covers all 3D printed samples simultaneously, without needing alignment between the various samples.

The material used is flexible, as the flexible packaging substrate enables resistance of damage from shocks and vibrations. Thus, it can be used in a wide range of applications, such as wearables and the Internet of Things (IoT). To demonstrate the flexibility, a sample with thickness equal to 1.7 mm is folded onto a 14.3 mm diameter torque wrench, as shown in Figure 9.9. The packaging substrate material is characterized using the waveguide method introduced in section 9.3.2. The dielectric constant is 2.8 and the loss tangent is 0.03.

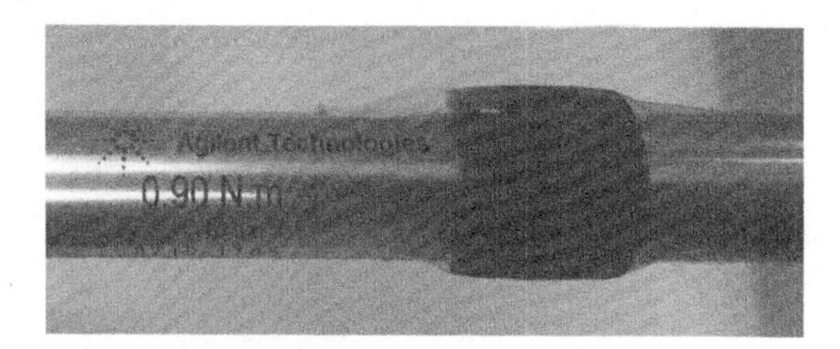


**Figure 9.9** Flexibility of the 3D printed substrate (from [12], © 2018 IEEE, reprinted with permission).

### 9.5.2 Inkjet Printing Process

After 3D printing the packaging substrate, it is then transferred to the inkjet printer to print conductive traces. There are two main problems that need to be solved while combining 3D printing and inkjet printing technologies: ink adhesion on the 3D printed substrate and coefficient of thermal expansion (CTE) differences between the 3D printed and inkjet printed materials. Often, the ink adhesion on 3D printed material is terrible without any surface treatment, as shown in Figure 9.10a. This causes severe conductivity drop, delamination, and performance degradation. Thus, surface treatment is often necessary before inkjet printing materials onto 3D printed substrates. A couple of surface treatment methods have been tested, and in this scenario a 30 s exposure of UV ozone enables the wettability desired. The comparison of with and without surface treatment is shown in Figure 9.10. The ink spreads much more uniformly with surface treatment and thus the overall conductivity and performance can be improved significantly.

Since the silver conductive ink has to be sintered for the silver particles to combine and enhance the conductivity, the CTE mismatch between the 3D printed materials and liquid inks becomes an important issue. As demonstrated in Figure 9.11a, due to CTE mismatch between materials, micro-cracks will appear and spread all over the patterns. This is fatal when micro-cracks break any traces. In order to solve the problem, a thin layer of SU-8 is printed on the substrate as a buffer layer before printing the silver trace. SU-8 provides a thick

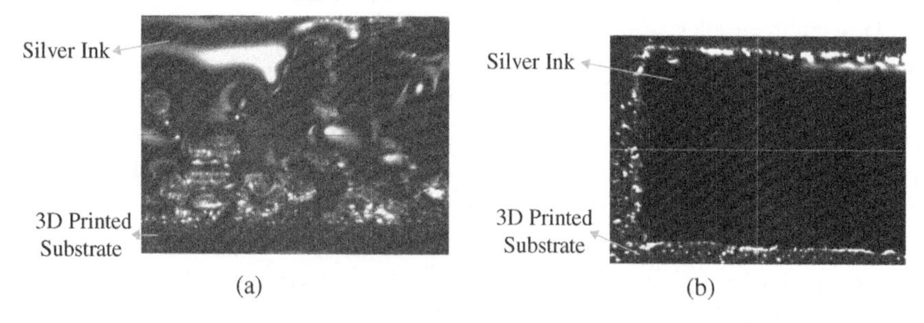


**Figure 9.10** Ink adhesion on the 3D printed substrate (a) without surface treatment and (b) with surface treatment (from [12], © 2018 IEEE, reprinted with permission).

**Figure 9.11** (a) Inkjet printing of a silver trace without an SU-8 layer and (b) inkjet printing of a silver trace with an SU-8 buffer layer (from [12], © 2018 IEEE, reprinted with permission).

film dielectric, and the SU-8 2000 polymer series by MicroChem (Newton, MA, USA) is modified to adjust to the viscosity necessary for inkjet printing. As shown in Figure 9.11b, significant improvement can be observed with a buffer SU-8 layer. Furthermore, the thin SU-8 layers, measured at 7 μm, can act as passivation layers, reducing the surface roughness of the 3D print, which has a staircase effect across graded 3D structures that corresponds to the 50 μm layer height.

### 9.5.3 AiP Fabrication Process

The procedure of fabricating AiP utilizing 3D and inkjet printing is summarized in Figure 9.12. The flexible package substrate is first printed with the 3D printer.

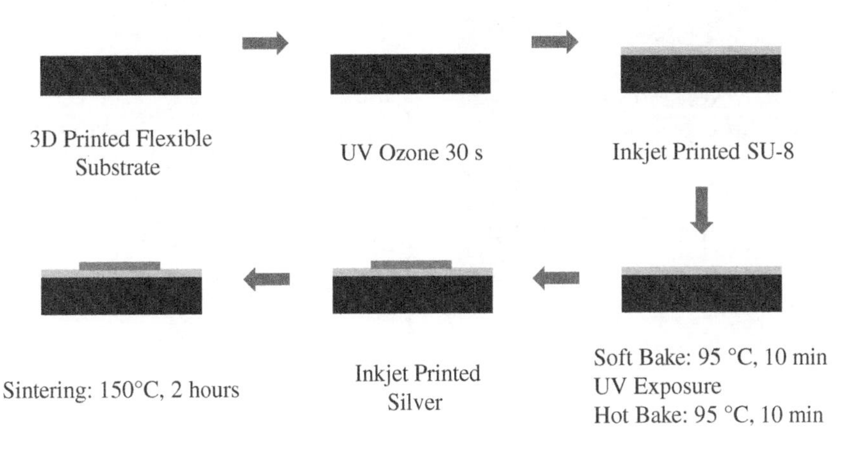


**Figure 9.12** 3D and inkjet printing fabrication process (from [12], © 2018 IEEE, reprinted with permission).

Thereafter the surface treatment is done by exposure to UV ozone for 30 s for ink adhesion improvement. A layer of SU-8 is then inkjet printed to smooth the surface roughness as well as act as a CTE buffer layer. The SU-8 is soft baked at 95 °C for 10 min, exposed under UV light for 2 min and hot baked at 95 °C for 10 min. Thereafter, the silver conductive traces are printed on top of the SU-8 layer. Finally, the sample is put into an oven to be sintered at 150 °C for 2 hours.

## 9.6 AiP and SoP using AM Technologies

### 9.6.1 AiP Design

After finishing characterization of the inkjet printing procedure, the 3D printing procedure, and integration of both techniques, they can be used to fabricate AiP design. The AiP design proposed here is a broadband Yagi antenna covering the 5G bands, n257, n258, and n261 which are from 24.25 to 29.5 GHz. 5G wireless communication supports a much broader operation band to provide faster data transmission. Hence, broadband performance for the antenna is necessary to reduce the number of antenna and minimize the 5G wireless communication module size. The proposed broadband AiP design is shown in Figure 9.13a and the physical dimensions are summarized in Table 9.1. The antenna dimensions are 8.57 × 11.8 mm. The connectors used to interface are Southwest Model 1092-03A-5 end launch connectors for operation up to 40 GHz. The dimensions of the connectors are 12.7 × 19.92 mm. Since the size of connector is in the same

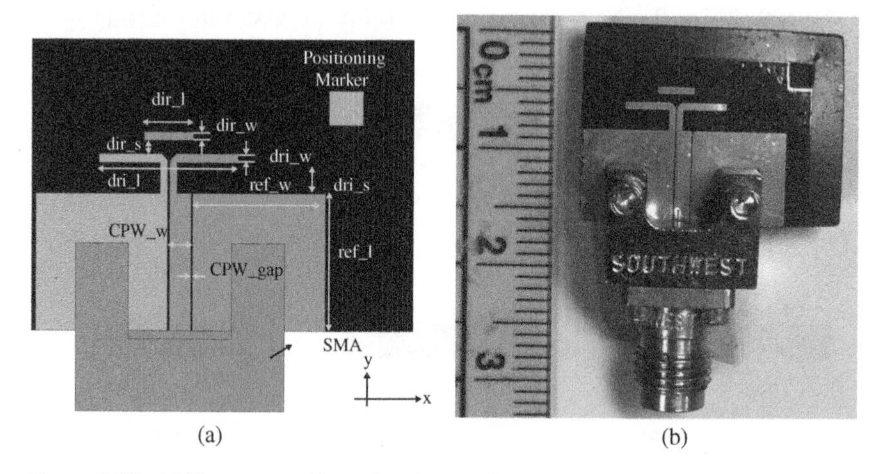


Figure 9.13 (a) The proposed broadband on-package antenna. (b) The inkjet and 3D printed broadband on-package antenna (from [12], © 2018 IEEE, reprinted with permission).

**Table 9.1** Physical dimensions of the broadband AiP design.

| Parameter | Dimension (mm) | Parameter | Dimension (mm) |
|---|---|---|---|
| dir_l | 3 | dir_w | 0.6 |
| dir_s | 0.8 | dri_w | 0.6 |
| dri_l | 8.57 | dri_s | 1.8 |
| ref_w | 8 | ref_l | 8 |
| CPW_w | 1.3 | CPW_gap | 0.17 |

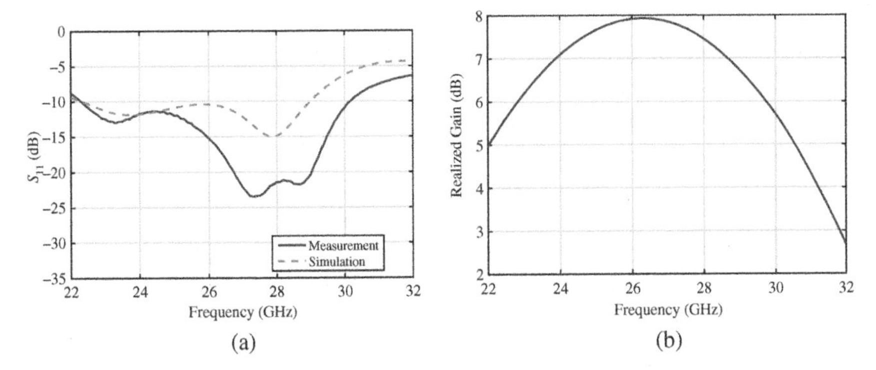


**Figure 9.14** (a) Measured and simulated $S_{11}$. (b) Measured gain of the proposed AiP design (from [12], © 2018 IEEE, reprinted with permission).

order as the antenna and they are near each other, the connectors may therefore induce non-negligible effects, and this has to be taken into consideration while performing the full-wave simulation and design of the antenna. The positioning marker is placed both on the 3D printed substrate and the inkjet printed pattern to help align both structures. The thickness of the substrate is 1.7 mm. The total size excluding the positioning marker is 11.8 × 17.8 mm. It can be further reduced to 5 × 9 mm which is $0.44\lambda_0 \times 0.79\lambda_0$ after excluding the structure for the connector.

The proof-of-concept sample using AM techniques is shown in Figure 9.13b. A transparent layer of SU-8 for CTE buffering and surface smoothing can be observed. The measured $S_{11}$ of this broadband AiP design is shown in Figure 9.14a. The full-wave simulation results using Ansoft HFSS are also included for comparison and good agreement can be observed. The differences between simulated and measured results are because of the loading effects of the end launch connector on the feeding network. Despite the differences, the general trend and the location of zeros are the same for the simulated and measured results. The measured $S_{11}$ is smaller than −10 dB from 22.4 to 30.1 GHz. The fractional bandwidth is 29.3%.

The bandwidth of the antenna is broad enough to provide good radiation from 24.25 to 29.5 GHz, applicable for the 5G band. In addition to matching, antenna gain is also an important perspective. Thus, the realized gain of this antenna is measured with a three-antenna setup. First, the $S_{21}$ of the standard, professionally characterized transmitted and received antenna is measured. Then the standard received is replaced with this test antenna while keeping all other setups the same to measure the $S_{21}$. Since the gain of the characterized transmitted and received antenna is already known across their bands, the realized gain of the test antenna can be calculated by comparing the two $S_{21}$ results. The broadband measured results are shown in Figure 9.14b. The measured realized gains are larger than 5 dBi from 22 to 30.5 GHz, which can cover the entire 5G band around 28 GHz. Moreover, the differences in realized gain within this operational band are smaller than 3 dB. The normalized radiation patterns of this AiP design are also measured and shown in Figure 9.15. Five different frequencies, 22, 24, 26, 28, and 30 GHz are measured to demonstrate the broadband behavior of this antenna. The antenna rotation plane is the same $x$–$y$ plane as shown in Figure 9.13a. The 0° is the end-fire, positive $y$, direction. Since the antenna is a Yagi antenna, the main-beam direction is expected to be at end-fire, 0°. As demonstrated in Figure 9.15, the main-beam directions of these five frequencies are all close to 0° as expected. This work can be used as successful proof of utilizing AM techniques to realize AiP design at mmWave range with good performance.

### 9.6.2 SoP Design

In order to realize a SoP design, we need to have inter-layer interconnects to connect the AiP design on the top layer to other ICs and structures on other layers. In conventional subtractive manufacturing, the inter-layer connections are vias. This can induce severe parasitics at high-frequency applications such as 5G and mmWave. With AM, 3D inter-layer interconnects utilizing ramp structures for inter-layer broadband transmission lines can be realized with lower parasitics [5]. The SoP design is shown in Figure 9.16a. There are three sets of ramp. One is for coplanar waveguide (CPW) lines for the IC input, the second is for CPW lines for the IC output, and the final one is for the DC bias. The ramp height is 1.2 mm, with a slope angle equal to 30°. Since the interconnects are now transmission lines (CPW lines) rather than inductive parasitics like vias, the overall parasitics can be much smaller. The end launch connector and the positioning marker are also included in the figure. The fabricated proof-of-concept sample is shown in Figure 9.16b. Since the layer height resolution for this structure is 10 μm, the ramp is a staggered structure with height equal to 10 μm. This causes a conductivity

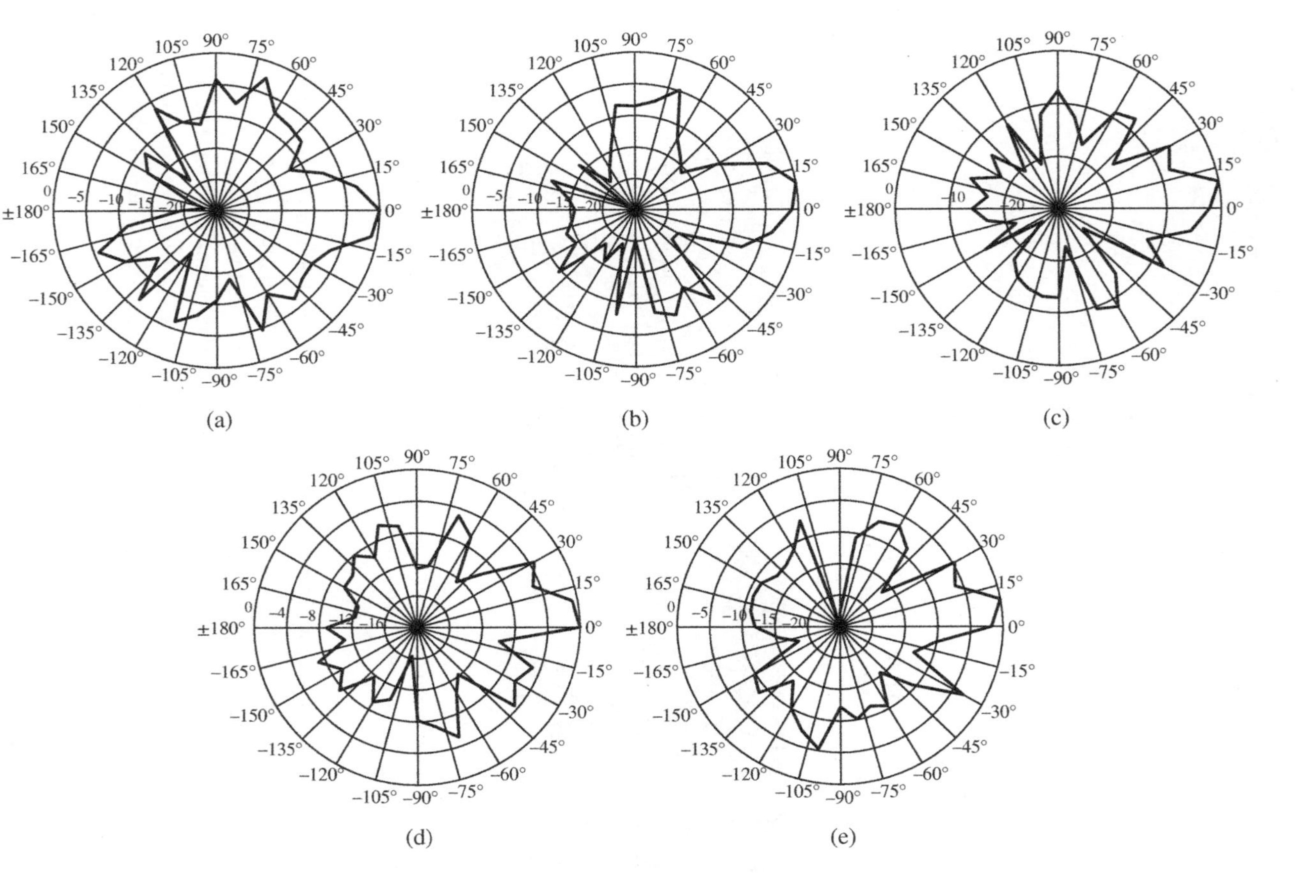


**Figure 9.15** Measured radiation pattern of the proposed broadband on-package antenna at (a) 22 GHz, (b) 24 GHz, (c) 26 GHz, (d) 28 GHz, and (e) 30 GHz (from [12], © 2018 IEEE, reprinted with permission).

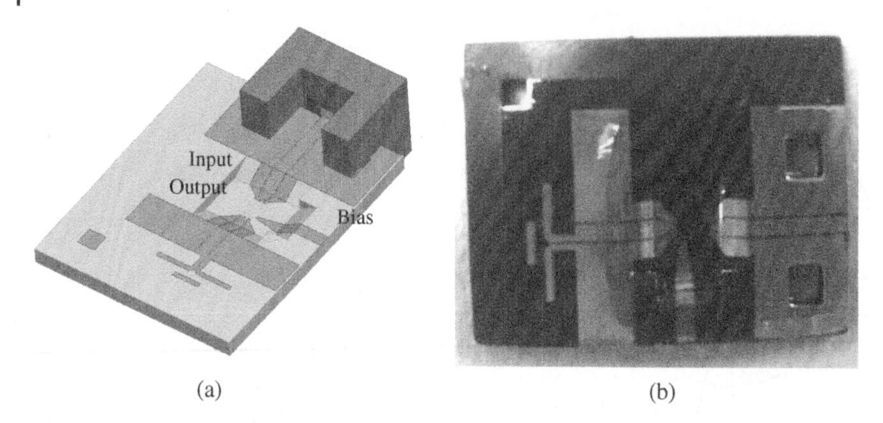


**Figure 9.16** (a) SoP design using 3D inter-layer connections. (b) Proof-of-concept samples (from [12], © 2018 IEEE, reprinted with permission).

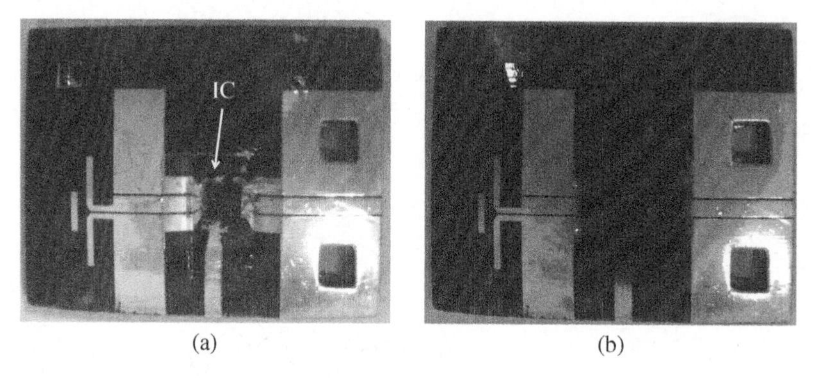


**Figure 9.17** SoP design (a) with IC attached and (b) sealed with flexible material (from [12], © 2018 IEEE, reprinted with permission).

issue for the silver trace, similar to the previously discussed 50 μm layers. Three layers of SU-8 are therefore printed before the silver to smooth the staggered structure. The IC is attached to the silver trace using silver conductive epoxy, as shown in Figure 9.17. The structure after the final layers of flexible material have been deposited, sealing the system, is shown in Figure 9.17b. The structure size can be reduced significantly by removing the connector and placed the antenna on the top of the IC, as shown in Figure 9.18.

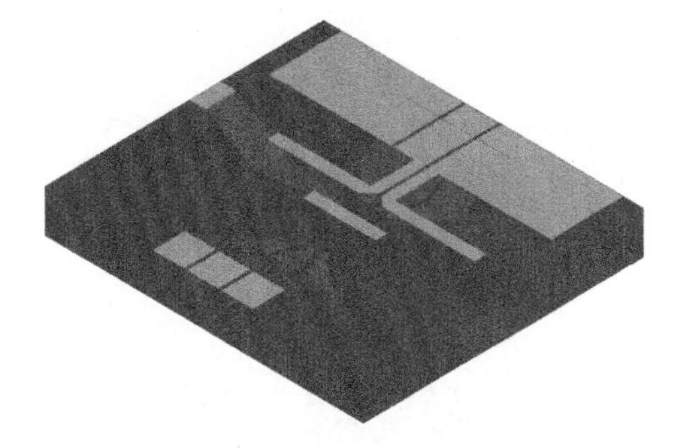

**Figure 9.18** Miniaturized 3D and inkjet printed SoP design (from [12], © 2018 IEEE, reprinted with permission).

## 9.7 Summary and Prospect

In this chapter, we introduced a variety of AM technologies, including inkjet printing, FDM, and variations of SLA 3D printing, and their relations to next-generation millimeter wave devices focusing on the possibilities that are being developed for next-generation systems. In order to utilize these technologies, these ever-evolving technologies need additional characterization data, specifically electrical properties for relative permittivity and loss tangents for the wide bandwidths that relate to this new generation of wireless communication. With this background in mind, we discussed a hybrid printing process that exploits 3D printing in conjunction with inkjet printing, enabling a wide range of features. Dielectric buffer layers are demonstrated in applying SU-8 to improve surface roughness and CTE mismatch, as well as fine conductive features with inkjet-printed silver conductive lines. Utilizing this proposed hybrid technology, a broadband 5G AiP design was demonstrated, with an on-package, broadband antenna that extends across the 5G band from 22.4 to 30.1 GHz, with 5 dBi of gain. Similar to how many semiconductor manufacturers are integrating more 3D structures in semiconductor design for increased performance, the SoP design discussed demonstrates 3D structural inter-layer connections which can offer significantly increased performance with reduction in parasitics, a key limiting factor in many systems. This is just the first of many future demonstrations of what AM can achieve, as well as a demonstration of increased functionality density as packages become smarter by integrating previously separated single function structures into a single multifunctional structure.

Additive manufacturing is still, relatively, in its infancy. The technology has gained a foothold in the manufacturing industry for its low cost, rapid prototyping,

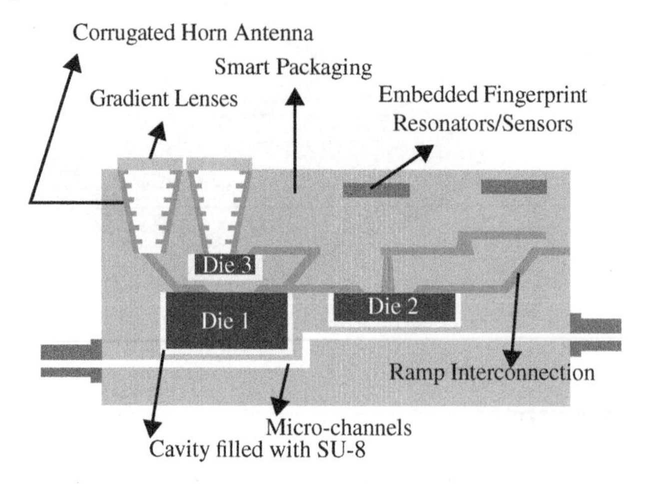


**Figure 9.19** AM SoP module design.

and complex design capabilities. Utilizing the same mindset, it is expected the future manufacturing technologies for SoP may integrate AM to enable complex SoP designs at reduced costs, shorter prototyping times, integrated structures, and higher performance connections, such as the smart AM SoP module design shown in Figure 9.19. A variety of features can be achieved in these smart SoP modules, including microvias, through-mold, through-, blind, and buried vias. Variable depth cavities can be printed to embed ICs of different thicknesses, while transmission lines can be printed on non-orthogonal ramp structures for inter-layer interconnects between different height ICs. More complex smart functionality, such as integrated microfluidic channels, can enable embedded heat dissipation or sensing, and enable more compact, higher power designs relevant to next-generation semiconductor substrates that need additional thermal considerations, such as GaN high-electron-mobility transistors. All the while, a vertical AiP with various dielectric lens structures can be directly integrated, including designs that exploit the gradient materials that 3D printing can offer, such as Luneburg lenses and impedance matching layers, offering increased gain for antenna arrays. As AM continues to evolve, it is only a matter of time before it becomes a more refined process that can be adopted in sectors of industrial electronics manufacturing. At the time of writing, PCB manufacturers are utilizing inkjet printing to create mask layers on traditional circuit boards. Many engineers are likely unaware that the solder masks on the circuits they have ordered today may have been produced utilizing AM technologies for printed polymer dielectric layers. In a similar manner, we hope that one day engineers may, without realizing it, have AiP structures that were only made possible through AM technologies that offer complex, multifunctional structures that were previously cost prohibitive with current technologies.

## References

1 H. Bikas, P. Stavropoulos, and G. Chryssolouris, Additive manufacturing methods and modelling approaches: A critical review. *International Journal of Advanced Manufacturing Technology*, vol. 83, no. 1, pp. 389–405, 2016.
2 J.G. Hester, S. Kim, J. Bito et al., Additively manufactured nanotechnology and origami-enabled flexible microwave electronics. *Proceedings of the IEEE*, vol. 103, no. 4, pp. 583–606, 2015.
3 E. Tekin, P.J. Smith, and U.S. Schubert, Inkjet printing as a deposition and patterning tool for polymers and inorganic particles. *Soft Matter*, vol. 4, pp. 703–713, 2008.
4 S.H. Ko, H. Pan, C.P. Grigoropoulos et al., All-inkjet-printed flexible electronics fabrication on a polymer substrate by low-temperature high-resolution selective laser sintering of metal nanoparticles. *Nanotechnology*, vol. 18, no. 34, p. 345 202, 2007.
5 B.K. Tehrani, R.A. Bahr, W. Su et al., E-band characterization of 3D-printed dielectrics for fully-printed millimeter-wave wireless system packaging. In *IEEE MTT-S International Microwave Symposium*, 2017, pp. 1756–1759.
6 R.R. Tummala, SOP: What is it and why? a new microsystem-integration technology paradigm - Moore's law for system integration of miniaturized convergent systems of the next decade. *IEEE Transactions on Advanced Packaging*, vol. 27, no. 2, pp. 241–249, 2004.
7 T. Lin, P.M. Raj, A. Watanabe et al., Nanostructured miniaturized artificial magnetic conductors (AMC) for high-performance antennas in 5G, IoT, and smart skin applications. In *IEEE 17th International Conference on Nanotechnology*, 2017, pp. 911–915.
8 T. Lin, J.G. Hester, A. Eid et al., 5G wireless modules enabled by additive manufacturing technologies. *IEEE Future Networks Technical Focus*, vol. 3, no. 1, 2019.
9 B.K. Tehrani, B.S. Cook, and M.M. Tentzeris, Inkjet-printed 3D interconnects for millimeter-wave system-on-package solutions. In *IEEE MTT-S International Microwave Symposium*, 2016, pp. 1–4.
10 T. Lin, A. Eid, J. Hester et al., Novel additively manufactured packaging approaches for 5G/mm-wave wireless modules. In *IEEE 69th Electronic Components and Technology Conference*, Las Vegas, NV, USA, 2019, pp. 896–902.
11 S. Zhen, R.M. Bilal, and A. Shamim, 3D printed system-on-package (SoP) for environmental sensing and localization applications. In *International Symposium on Antennas and Propagation*, 2017, pp. 1–2.
12 T. Lin, R. Bahr, M. Tentzeris et al., Novel 3D-inkjet-printed flexible on-package antennas, packaging structures, and modules for broadband 5G applications. In *IEEE 68th Electronic Components and Technology Conference*, 2018, pp. 214–220.

**13** T. Lin, S. Daskalakis, A. Georgiadis et al., Achieving fully autonomous system-on-package designs: An embedded-on-package 5G energy harvester within 3D printed multilayer flexible packaging structures. In *IEEE/MTT-S International Microwave Symposium*, 2019, pp. 1375–1378.

**14** W. Su, B.S. Cook, Y. Fang et al., Fully inkjet-printed microfluidics: A solution to low-cost rapid three-dimensional microfluidics fabrication with numerous electrical and sensing applications. *Scientific Reports*, vol. 6, pp. 241–249, 2016.

**15** P. Calvert, Inkjet printing for materials and devices. *Chemistry of Materials*, vol. 13, no. 10, pp. 3299–3305, 2001.

**16** Y. Yan, Design methodology and materials for additive manufacturing of magnetic components. PhD dissertation, Virginia Polytechnic Institute and State University, 2017.

**17** D. Desai, Passive planar terahertz retroreflectors. PhD dissertation, New Jersey Institute of Technology, 2018.

**18** A. Sadeqi, H.R. Nejad, and S. Sonkusale, 3D printed metamaterials for high-frequency applications. in *Terahertz, RF, Millimeter, and Submillimeter-Wave Technology and Applications XII*, vol. 10917, 2019, pp. 33–37.

**19** M. Shusteff, A.E.M. Browar, B.E. Kelly et al., One-step volumetric additive manufacturing of complex polymer structures. *Science Advances*, vol. 3, no. 12, 2017.

**20** J. Kimionis, M. Isakov, B.S. Koh et al., 3D-printed origami packaging with inkjet-printed antennas for RF harvesting sensors. *IEEE Transactions on Microwave Theory and Techniques*, vol. 63, no. 12, pp. 4521–4532, 2015.

**21** *Split Post Dielectric Resonators (SPDR)*, http://www.qwed.com.pl/resonators_spdr.html.

**22** R. Bahr, T. Le, M.M. Tentzeris et al., RF characterization of 3D printed flexible materials – NinjaFlex Filaments. In *European Microwave Conference*, 2015, pp. 742–745.

**23** I. Zivkovic and A. Murk, Free-space transmission method for the characterization of dielectric and magnetic materials at microwave frequencies. In *Microwave Materials Characterization*, S. Costanzo, (ed.), InTech Open, 2012, ch. 5, pp. 73–90.

**24** P.I. Deffenbaugh, R.C. Rumpf, and K.H. Church, Broadband microwave frequency characterization of 3D printed materials. *IEEE Transactions on Components, Packaging and Manufacturing Technology*, vol. 3, no. 12, pp. 2147–2155, 2013.

**25** M. Samet, V. Levchenkol, G. Boiteuxl et al., Electrode polarization vs. Maxwell–Wagner–Sillars interfacial polarization in dielectric spectra of materials: Characteristic frequencies and scaling laws. *Journal of Chemical Physics*, vol. 142, no. 19, p. 194–703, 2015.

26 M. Lis, M. Plaut, A. Zai et al., Polymer dielectrics for 3D-printed RF devices in the Ka band. *Advanced Materials Technologies*, vol. 1, no. 2, p. 1600 027, 2016.

27 B. Tehrani, R. Bahr, D. Revier et al., The principles of "smart" encapsulation: Using additive printing technology for the realization of intelligent application-specific packages for IoT, 5G, and automotive radar applications. In *IEEE 68th Electronic Components and Technology Conference*, 2018, pp. 111–117.

28 M. Craton, J.A. Byford, V. Gjokaj et al., "3D printed high frequency coaxial transmission line based circuits. In *IEEE 67th Electronic Components and Technology Conference*, 2017, pp. 1080–1087.

29 J. Shen, D.P. Parekh, M.D. Dickey, and D.S. Ricketts, 3D printed coaxial transmission line using low loss dielectric and liquid metal conductor. In *IEEE/MTT-S International Microwave Symposium*, 2018, pp. 59–62.

30 B.K. Tehrani, B.S. Cook, and M.M. Tentzeris, Post-process fabrication of multilayer mm-wave on-package antennas with inkjet printing. In *IEEE International Symposium on Antennas and Propagation USNC/URSI National Radio Science Meeting*, 2015, pp. 607–608.

31 B.K. Tehrani and M.M. Tentzeris, Substrate-independent system-on-package antenna integration with inkjet printing. In *IEEE International Symposium on Antennas and Propagation*, 2016, pp. 827–828.

32 J. Kimionis, S. Shahramian, Y. Baeyens et al., Pushing inkjet printing to W-band: An all-printed 90-GHz beamforming array. In *IEEE/MTT-S International Microwave Symposium*, 2018, pp. 63–66.

33 A. Standaert and P. Reynaert, A 400 GHz transmitter integrated with flip-chipped 3D printed horn antenna with an EIRP of 1.26dBm. In *IEEE/MTT-S International Microwave Symposium*, 2018, pp. 141–144.

34 R. Bahr, X. He, B. Tehrani et al., A fully 3D printed multi-chip module with an on-package enhanced dielectric lens for mm-Wave applications using multimaterial stereo-lithography. In *IEEE/MTT-S International Microwave Symposium*, 2018, pp. 1561–1564.

# 10

# SLC-based AiP for Phased Array Applications

*Duixian Liu and Xiaoxiong Gu*

*Thomas J. Watson Research Center, IBM, 1101 Kitchawan Rd. Route 134, Yorktown Heights, NY 10598, USA*

## 10.1 Introduction

The increasing capabilities of high-speed silicon technologies have made millimeter-wave (mmWave) frequencies attractive for low-cost applications. An overview of the capabilities can be found in [1]. Promising applications could be high data rate wireless personal area networks (WPANs) at 60 GHz [2], automotive radars at 76–77 or 78–81 GHz [3], imaging at 94 GHz (W-band) [4, 5], and recently mmWave fifth-generation (5G) technology [6]. In the area of imaging, the availability of a wide bandwidth at the W-band, as well as the presence of a low-absorption atmospheric window, makes W-band frequencies particularly attractive. A number of imaging and point-to-point wireless link applications require highly directional transceivers, the ability to rapidly scan in two dimensions, and support for dual-polarized operation to meet performance needs. One of the current major research and development areas is around 5G wireless communication technologies. The objectives for developing 5G cellular networks include higher capacity, higher data rate, lower end-to-end latency, massive device connectivity, reduced cost, and consistent quality of experience provisioning [7]. For 5G radio access infrastructure such as base stations, one of the key enabling technologies is mmWave beamforming [8, 9], which can utilize a flexible allocation of the spectrum and implement an adaptive phased-array antenna to multiplex messages for several devices on a time division basis, focusing the radiated energy toward the intended directions while minimizing intra- and inter-cell interference. Table 10.1 lists some of the state-of-the-art phased-array designs in these fields with at least 64 antenna elements [10–14].

These mmWave applications subsequently require small and low-profile packaged high-performance systems available at moderate to low cost. Such mmWave

*Antenna-in-Package Technology and Applications*, First Edition.
Edited by Duixian Liu and Yueping Zhang.

**Table 10.1** State-of-the-art phased-array survey.

| | IBM [10] | IBM [11] | Qualcom [12] | Broadcom [13] | Nokia [14] | UCSD [15] |
|---|---|---|---|---|---|---|
| Process | 0.13 μm SiGe BiCMOS | 0.13 μm SiGe BiCMOS | 28 nm LP-RF CMOS 1P7M | 28 nm/ 40 nm CMOS | 0.18 μm SiGe BiCMOS | SBC18H3 SiGe BiCMOS |
| Frequency band | 28 GHz | 94 GHz | 28 GHz | 60 GHz | 94 GHz | 28 GHz |
| Polarizations/module | 2 (64H, 64V) | 2 (64H, 64V) | 2 (8H, 8V) | 1 (48V) | 1 (24V) | 1 (4V) |
| Antennas/module | 8 × 8 TRX | 8 × 8 TRX | 4 × 4 TRX | 8 × 6 TRX | 16 TX + 8 RX | 2 × 2 TRX |
| RFICs/module | 4 | 4 | 2 | 3 | 1 | 1 |
| Modules/phased array | 1 | 1 | 2x8 | 2x3 | 16 | 16 |
| Beam steering | ±50° horizontal ±50° vertical | ±15° horizontal ±15° vertical | ±60° horizontal ±60° vertical | ±60° azimuth ±10° elevation | Not available | ±50° azimuth ±25° elevation |
| AiP materials | Organic | Organic | Organic | Ceramic | Organic | Organic |
| AiP technology | Buildup | Buildup | Buildup | LTCC | PCB | PCB |

SiGe, silicon-germanium; LP-RF, low-power radio frequency; BiCMOS, bipolar complementary metal oxide semiconductor; CMOS, complementary metal oxide semiconductor; H, horizontal; V, vertical; TRX, transceiver; TX, transmitter; RX, receiver; RFIC, radio frequency integrated circuit; AiP, antenna-in-package; LTCC, low-temperature co-fired ceramic; PCB, printed circuit board.

systems require not only radio frequency integrated circuits (RFIC), but also a wide range of high-quality passive components, such as antennas, switches, and other devices, which must be densely packaged with high-performance interconnects. It is well known that electronic device packages play a key role in the function of any semiconductor product. Electronic packages conduct signals through metal in the form of wires, contacts, foils, plating, and solders. The package provides means for insulating circuit nets from each other, environmental protection, and physical support for attached circuits. Packaging of mmWave components is particularly challenging because of the associated complexity in both the electrical and physical design and fabrication. The small wavelength at mmWave frequencies often demands high-precision machining, accurate alignment, or high-resolution photolithography. Combining a package integrated antenna approach with a mmWave radio architecture with a baseband modulated interface, however, offers significant benefits in terms of reducing the routing of high-frequency signals; packages of highly integrated radios with in-phase/quadrature baseband output and package-integrated antennas do not have to conduct frequencies higher than the baseband [16]. Of course, integrating an antenna into a package is not an easy task. There are also many papers discussing integrating antennas directly into RFICs [17–25]. A significant drawback of this approach is that directly integrated antennas typically show very poor performance, with antenna gain ranging from −15 to −7 dBi [17, 18]. Micromachined antennas or antennas using high-resistivity silicon have gains ranging from −4 to 5 dBi [19–25], but the impedance bandwidth these designs support is generally still too narrow for practical applications. These antennas are also not cost effective due to the substantial increase in chip area they demand. Therefore, for mmWave frequency bands below 100 GHz, integrating the antennas on an RFIC package is a practical choice [26].

A variety of innovative substrate technologies have been explored for mmWave antenna design and system integration to date. Multiple demonstrations of antenna arrays have been reported for the 60-GHz band. This work began in earnest with the exploration of mmWave antenna integration associated with 60-GHz standardization efforts that started around 2006. Multiple technologies were explored for this application, including using liquid crystal polymer-based board processes [27, 28], organic high-density interconnect (HDI) [29], surface laminar circuit (SLC) substrates [10, 11, 30–33], glass substrates [34], high- and low-temperature co-fired ceramic substrates (HTCC, LTCC) [35–37], silicon substrates [38, 39], and molding-compound-based wafer-level substrates [40]. Similarly, numerous substrate options have been explored for use in other mmWave frequency bands. In any practical case of implementing a mmWave antenna array module or antenna-in-package (AiP), antenna performance needs to be optimized to achieve the application-driven desired gain, bandwidth, and radiation pattern. At the same time, antenna design must take into account not only substrate material properties, but also system and circuit requirements driving array size (i.e. the number of elements and pitch) and interconnect flexibility (e.g. for wiring power supplies and control signals), all in the context of constraints driven by the demands of thermo-mechanical compatibility, integrated circuit (IC) assembly, and board integration. This chapter will focus on large-array AiP designs based on SLC technology at 28 and 94 GHz.

## 10.2 SLC Technology

IBM pioneered direct chip attach flip-chip processes originally for computer-driven applications, including the development of controlled collapse chip connection (C4) technology. With cost as the driver, the company developed SLC technology during the 1980s and early 1990s in IBM Japan; the goal of this technology was to address high input/output (I/O) count and C4 flip chip assembly issues, and to create an alternative to low-temperature co-fired ceramic (LTCC) packaging technology in the form of a lower cost organic package [41, 42]. Coupling C4 and SLC technologies allowed efficient assembly of ICs on laminate substrates, where previously only ceramic substrates were used to package ICs. SLC, in effect, applies thin-film-type processes that effectively provide for fine lines, spaces, and microvias to a core board (e.g. FR-4) to produce the required wiring density for C4 flip chips. In addition, since the key concept behind SLC is the separation of the substrate and electrical wiring layers, SLC allows packages to be designed to be specific to a particular user. For example, the material of the substrate can be changed to have better thermal conductivity (by adding metal) or lighter material can be used for weight reduction, or varying dielectric thicknesses used for specific electrical requirements. This customizability started the age where electronics packaging adapts much more to user-specific requirements, just as application-specific integrated circuits (ASICs) themselves provide specific, customized functions to a specific user. SLC technology became the basis of today's very popular low-cost organic package substrates with build-up layers vertically connected through microvias to support flip chips [43].

There are two main components of SLC technology [44]: the core substrate and the SLC for the signal wiring. The core substrate is made using an ordinary glass epoxy panel. By contrast, the SLC layers are sequentially built up on the top and bottom of the core, with the dielectric layers made of photosensitive epoxy and the conductor plane of copper plating (semi-additive technique). In general, a package substrate with 12 layers [e.g. two core layers and ten build-up layers (5-2-5)] and 10 μm line width and spacing is more than adequate to support the requirements of most chips. Figure 10.1 shows a typical 3-2-3 SLC stack-up. The stack-up consists of three portions: core, top build-up, and bottom build-up. As shown in Figure 10.1,

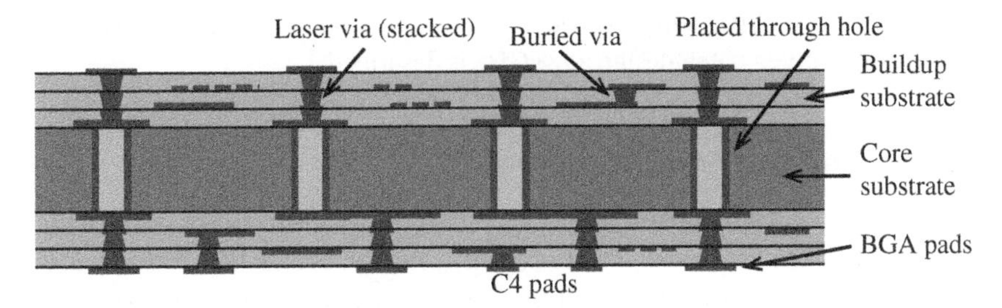


**Figure 10.1** A 3-2-3 SLC cross-section.

one very important feature or requirement is that the package has to be top-bottom symmetric from a layer count perspective; the number of build-up layers above and below the core layer must be the same.

## 10.3 AiP for 5G Base Station Applications

Several mmWave bands from 10 GHz up to 100 GHz are currently being considered for use by 5G mobile communication systems, with enabling very wide transmission bandwidths as a key opportunity driving consideration of these frequencies [45, 46]. For example, recent communication channel measurements, radio prototypes, and test bed developments are focused on the 28, 30.5, 39, 60, and 73 GHz frequency bands [45, 47–49]. To account for different propagation attributes in these mmWave bands, approaches to the design and implementation of antenna arrays, as well as to integration with active mmWave circuits, are critically important to achieving beam forming and steering with sufficient power gain and spatial resolution for high-throughput communication applications.

Typical requirements for antennas in a mmWave-based mobile link include matching the target frequency with sufficient bandwidth and balancing the antenna gain and radiation patterns for optimal beam forming and beam scanning. For example, a 28-GHz antenna array with 520-MHz bandwidth, 18-dBi array gain for an 8×4 array, and ±30° of scanning range is reported in [45]. In this section, we investigate antenna designs considering both manufacturability and these general performance requirements, and a 30.5-GHz antenna design and associated characterization results are presented. The antenna design is based on SLC technology and supports both horizontal and vertical polarizations to enable realization of a large phased-array module with multiple simultaneous beams for 5G applications. A probe-fed dual-polarized (horizontal and vertical polarizations) stacked patch antenna has been studied in [30]. Here we will describe an aperture-coupled cavity-backed dual-polarized patch antenna at 30.5 GHz that provides much improved performance with respect to the probe-fed patch antenna. Finally a demonstrated dual-polarized 8×8 phased array based on the aperture-coupled patch design at 28 GHz is described.

### 10.3.1 Package and Antenna Structure

A sketch of a typical AiP-based phased-array stack-up is shown in Figure 10.2. The AiP contains all the antenna elements as well as interfaces to both the RFICs and the application board. A single or multiple (for a scaled phased array) RFICs can be attached to the bottom of the AiP, typically through C4 balls. The AiP is attached to the application board, usually through ball grid array (BGA) balls. Unlike an ordinary RFIC package where heat is removed from the package by attaching a heat sink to the top of the package, the heat sink here is attached to the bottom of the application board. The heat from the RFICs is transferred to the heat sink through thermal vias.

A detailed example of an AiP stack-up for an aperture-coupled stacked patch antenna is illustrated in Figure 10.3, with major components identified. The stack-up consists of four portions: the bottom build-up, core, top build-up, and a superstrate/cover. The core layer in this work contains four metal layers separated by three substrate layers of the same dielectric material; this material has a dielectric constant of 4.6 and a loss tangent of 0.01. The top and bottom build-ups each consist of five metal layers, separated by a substrate material with a dielectric constant of 3.4 and a loss tangent of 0.01. Each build-up contains five substrate layers. The patch is implemented on the bottom side of the superstrate. The patch on the top side of the superstrate is also intended to address manufacturability requirements such as reducing cover warpage [33]. The antenna apertures are on top metal 6 (TM6) and the horizontal and vertical polarization feed lines are on TM5 and TM7, respectively (similar in topology to the antenna design in [50]). A via ring intended to prevent parallel-plate mode propagation in the package connects the TM3 and TM6 ground planes.

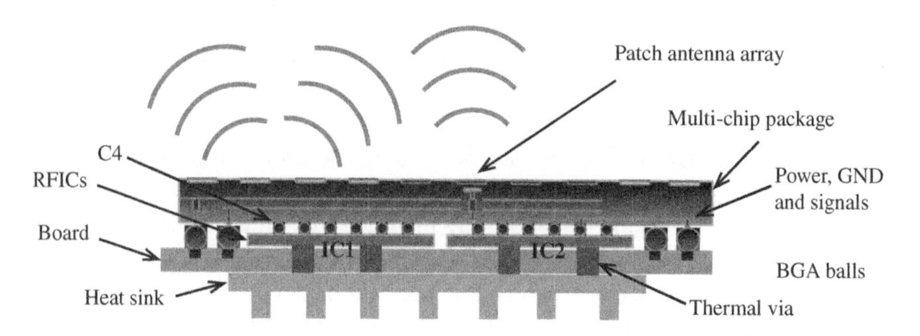


**Figure 10.2** A typical AiP-based phased array stack-up.

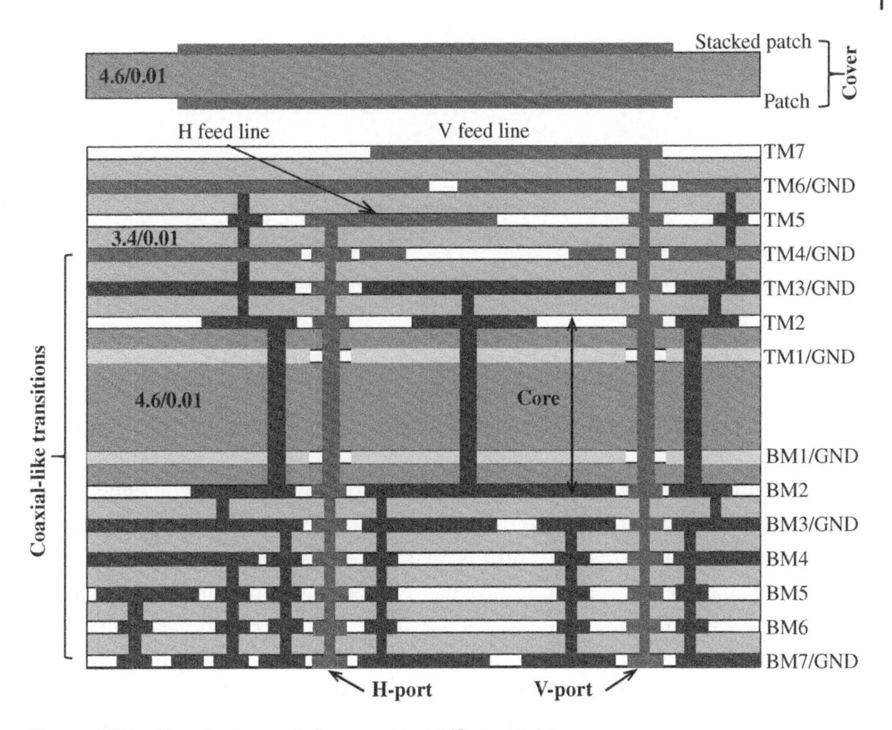


**Figure 10.3** The stack-up of the proposed AiP structure.

For the dual-polarized aperture-coupled stacked patch antenna design, the horizontal polarization feed line runs from TM5 to bottom metal 7 (BM7) with a coaxial-like structure, while the vertical polarization feed line runs from TM7 to BM7, also with a coaxial-like structure. Since the stack-up functions as both an antenna package and an RFIC package, beside the antenna and its feed lines, additional metal layers primarily on the bottom build-up are used to implement connectivity for low-frequency functions as needed by the RFIC chip.

## 10.3.2 AiP Design Considerations

### 10.3.2.1 Surface Wave Effects

RFIC package materials usually have a dielectric constant higher than 3.0, therefore strong surface waves are easily excited in the package environment. Reducing surface-wave excitation from microstrip antennas (typically patch antennas) can be beneficial for several reasons. First, the reduction of surface-wave excitation will increase the radiation efficiency of the antenna. Low antenna efficiency not only reduces the antenna gain but also increases the heat in the RFIC package, especially for large phased arrays. Second, the reduction of surface-wave

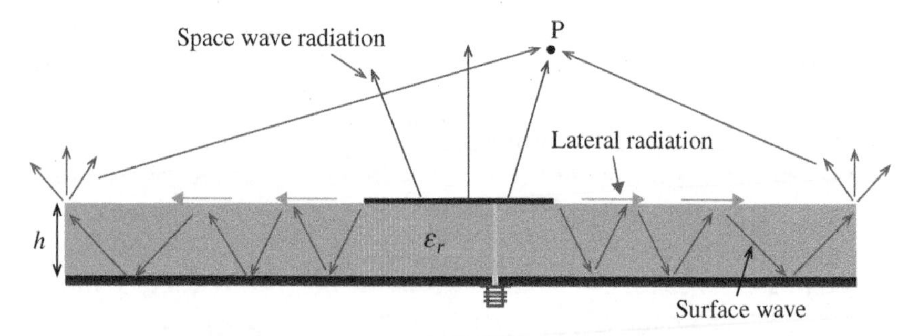


**Figure 10.4** Waves related to a patch antenna.

excitation will result in less diffraction from the edges of the substrate or ground plane supporting the antenna, resulting in less back radiation, less interference with the main beam pattern in the forward region [51] as illustrated in Figure 10.4, and less element gain variation for array application [52]. Furthermore, reduced surface-wave excitation usually results in reduced coupling between adjacent antenna elements, critical for improving phased array performance through stabilizing active antenna impedance and reducing blind spots in scanning.

One way to design a high-performance patch antenna is to use an electrically thick air cavity or foam materials with a relative dielectric constant close to 1. This technique is widely used in wideband and high efficiency patch antenna designs. Another approach for minimizing surface-wave excitation is to use a synthesized substrate that lowers the effective dielectric constant of the substrate either under or around the patch [20, 53]. Electromagnetic band gap (EBG) substrates recently have also been investigated for antenna applications [54]. EBG substrates have periodic structures with frequency bands within which no surface waves can propagate. However, for phased-array applications in the package environment, EBG or foam material methods are difficult to apply. Therefore, air cavity-backed superstrate patch antenna designs are used in this study. Superstrate antennas with proper design can have weak surface waves [55]. The air cavity and superstrate combination in this design is optimized for weak surface-wave excitation under the constraints of package fabrication requirements.

#### 10.3.2.2 Vertical Transitions

Unlike regular patch antenna arrays where antennas are built with one, two or three substrate layers, resulting in metal structures being present on all metal layers, the AiP here contains many substrate layers, e.g. 16 in the proposed build-up. The metal layers below the core layer may be densely populated with low-frequency structures for RFIC connectivity, but some metal layers above the core layer will need many fewer metal structures from the antenna design point of view, such as metal layers TM5 and TM7. However, insufficient metal density on a layer may violate the build-up fabrication rules. To address this requirement, grounded metal rings on these layers can be added to increase metal density; such structures may also improve antenna performance if designed properly.

Since antennas are placed above the core layer while RFIC chips are placed below the AiP, RF feed lines have to go through vertical transitions, especially through core layers. This requirement may affect the antenna impedance matching, increase coupling between antennas as well as between antenna and other structures, and potentially may also excite parallel-plate modes in the AiP. We can use coaxial-like structures to provide reliable and high-performance vertical transitions. This approach is especially important for the vertical transition through the core substrate layer, where via diameter and spacing are large, and the substrate is much thicker and with a higher dielectric constant than the build-up layers. Figure 10.5 shows the coaxial-like structure from TM4 to BM7 for this transition study. The number of ground vias along a circle of 375 μm radius for the core section, three shown in the figure, can go from two to six.

From [30], the electric field around the center conductor of the coaxial section is very tightly contained inside the ground vias (considered as the outside shield of a coaxial cable) if six ground vias are used. The difference between six and eight ground vias is very small, so six vias should provide a high-performance transition. However, the circumferential spacing between grounded vias could be too small for fabrication depending on fabrication rules and other layout requirements. For practical applications, three or four (preferred) vias are a good choice. Beside the fabrication requirements, integration density and AiP layout are also important factors in determining the number of ground vias. The two rings on BM2 and TM2 provide a transition from the core ground vias to the build-up ground vias. Six ground vias are used from BM2 to BM3 and TM2 to TM3, respectively. Three build-up ground vias are used in the other parts of the vertical transition.

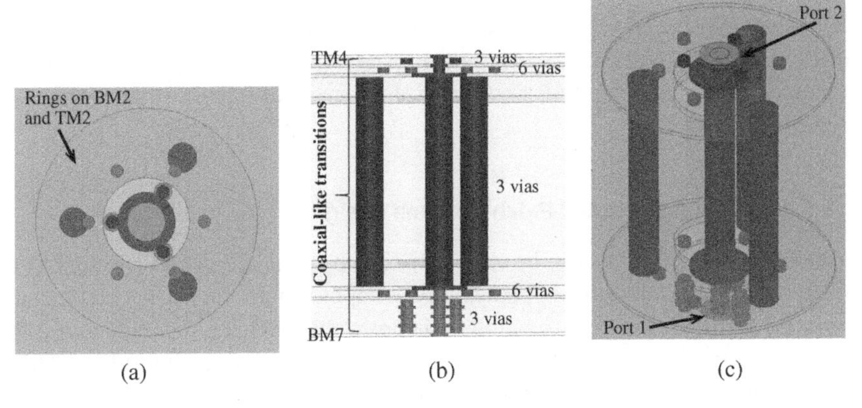


**Figure 10.5** The vertical transition: (a) 2D view, (b) cross-section, and (c) 3D view.

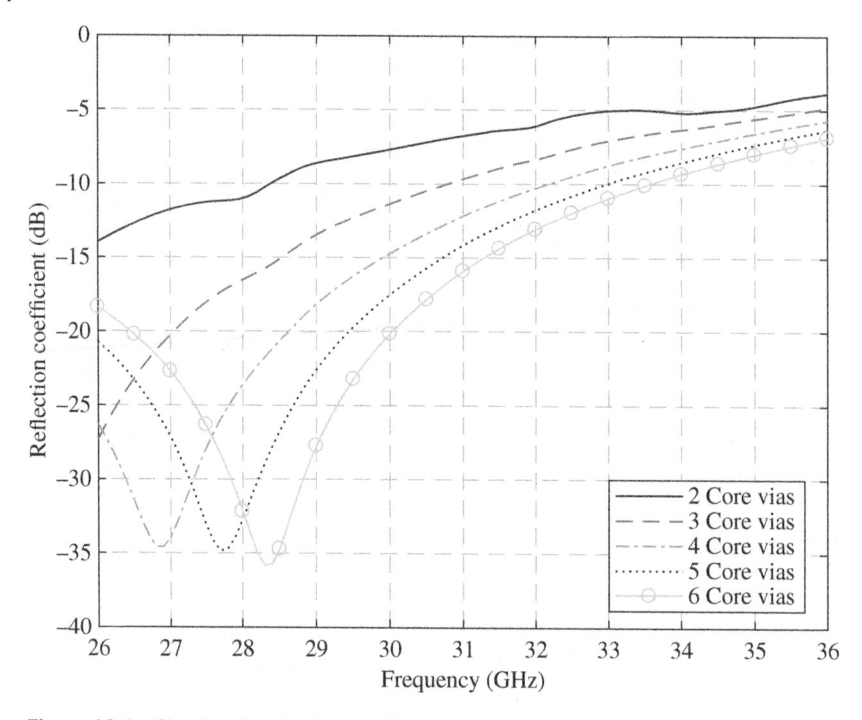


**Figure 10.6** Simulated reflection coefficient (S11) of the transitions.

Figures 10.6 and 10.7 show the simulated ports S11 (reflection coefficient) and S21 in dB (negative of insertion loss) of the transitions normalized to 50 Ω, respectively. Figure 10.6 indicates that the transition does not have a 50 Ω impedance at the high end of the frequency range. As the number of ground core vias increases, the transition section almost behaves as a 50 Ω coax. According to Figure 10.7, the transition produces about 0.9 and 0.5 dB losses at 31 GHz for three and four grounded vias, respectively. The transition loss is due to material loss and electromagnetic (EM) wave leakage. At least three ground core vias were used in this work.

### 10.3.3 Aperture-coupled Patch Antenna Design

Many antenna structures can be used for package applications. However, for the phased-array antennas used in many mmWave applications, the aperture-coupled or probe-fed patch antenna is a preferred choice since the antenna radiators and feed lines can be separated from each other by the antenna ground plane. This arrangement is especially true for aperture-coupled patch antennas. In this context, the antenna radiators and feed lines can be optimized almost independently.

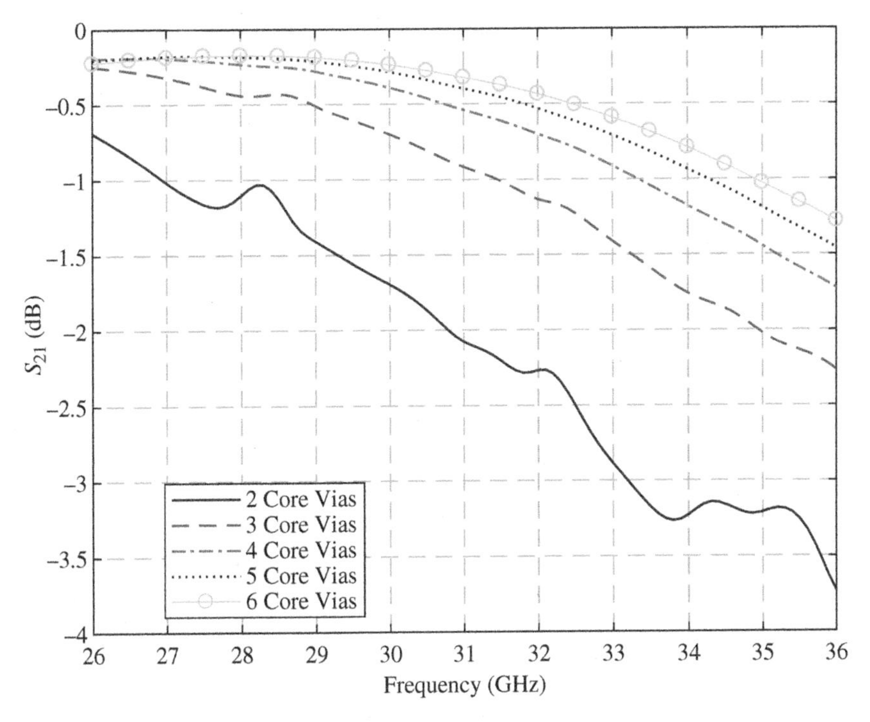


**Figure 10.7** Simulated S12 of the transitions.

This feature is especially important for RFIC package applications, since not only the antenna feed line layout but also routing for many low-frequency passive components must be taken into account in the designs. From the studies in [30], the probe-fed patch antenna has a narrow bandwidth in the package environment, therefore this design approach is not suitable for wideband communication applications. In the following sections, we will concentrate on wideband aperture-coupled patch antenna designs [57–59] in the AiP environment for phased-array applications.

Compared to probe-fed patch antennas, aperture-coupled patch antennas have other advantages: (1) they exhibit very low cross-polarization levels (theoretically zero cross-polarization in principle planes) [60], making them well suited to circularly or dually polarized antenna designs, (2) they can be implemented with an air cavity, and (3) they can utilize a near-resonant or resonant aperture. The last two capabilities can widen the patch antenna bandwidth. The aperture-coupled patch antenna can generally be impedance matched with an aperture that is well below resonant size, thus limiting the level of back radiation to about 20 dB below the main lobe. However, a patch antenna coupled in this fashion is capable of only about 5% bandwidth, owing to the fact that the small coupling aperture limits the antenna substrate thickness that may be used. By using a thick antenna substrate with a low dielectric constant, a bandwidth of 20–25% can be achieved [61]. However, because of the thick antenna substrate, a larger slot size is needed to obtain the necessary coupling to impedance match the antenna, resulting in a higher level of back radiation. The back radiation issue can be easily resolved in AiP designs due to the fact that additional metal layers are available and complex via structures can be implemented. One can also say that the wide bandwidth in this case is due to two radiating elements: patch and aperture, beside an air cavity.

Figure 10.8a shows a portion of the proposed antenna design concentrating on the antenna and feed structures [62]. The cross aperture and feed line structures are similar in topology to the one in [50] for dual polarizations. Both patches have a 2.8×2.8 $mm^2$ size. Figure 10.8b shows the antenna optimization parameters of the antenna aperture and feed lines. Since there is another ground plane on TM4 behind the antenna ground plane on TM6 that has the cross aperture, the via ring connecting the two ground planes is used to reduce the leakage from back radiation. As a result, antenna efficiency will not be reduced and parallel-plate modes can be suppressed. The substrate and antenna ground plane size is 20×20 $mm^2$. For this prototype design, there are three grounded core vias along a circle of radius 375 μm. There are six grounded build-up vias from BM2 to BM3 and TM2 to TM3 also along a circle of radius 375 μm. In addition, there are three grounded build-up vias from BM3 to BM7 along a circle of radius 175 μm. These ground vias emulate the outside shield of a coaxial cable. As indicated in the previous section,

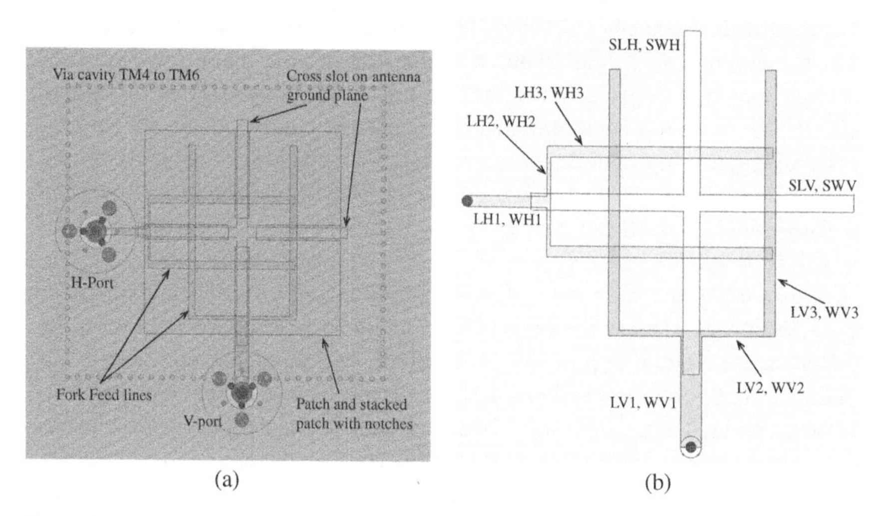


**Figure 10.8** The aperture-coupled stacked patch antenna structure: (a) antenna structure and (b) aperture and feed line dimension parameters.

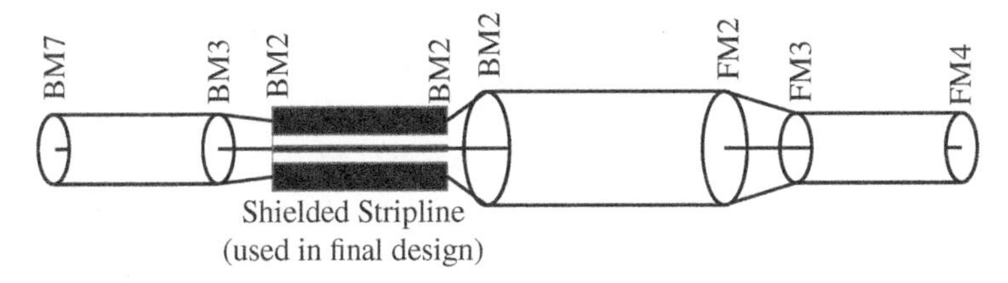


**Figure 10.9** A sketch of the complete vertical feed line sections.

at least three grounded vias are required for reliable performance. Figure 10.9 shows a sketch of the complete vertical feed line sections. In general, the simulated coaxial sections do not have 50 Ω characteristic impedance due to manufacturing constraints, so the final antenna matching is achieved by adjusting the fork feed dimensions, the patch sizes, and the aperture dimensions. The vias for the ring are formed along a $6 \times 6\,mm^2$ box. The ring size, defining the patch separation for array applications, is 0.6 wavelengths at 30 GHz.

Figure 10.10 shows the front side picture of the fabricated prototype design without the cover. The AiP area is $6.85 \times 6.85\,cm^2$. This is the same size and

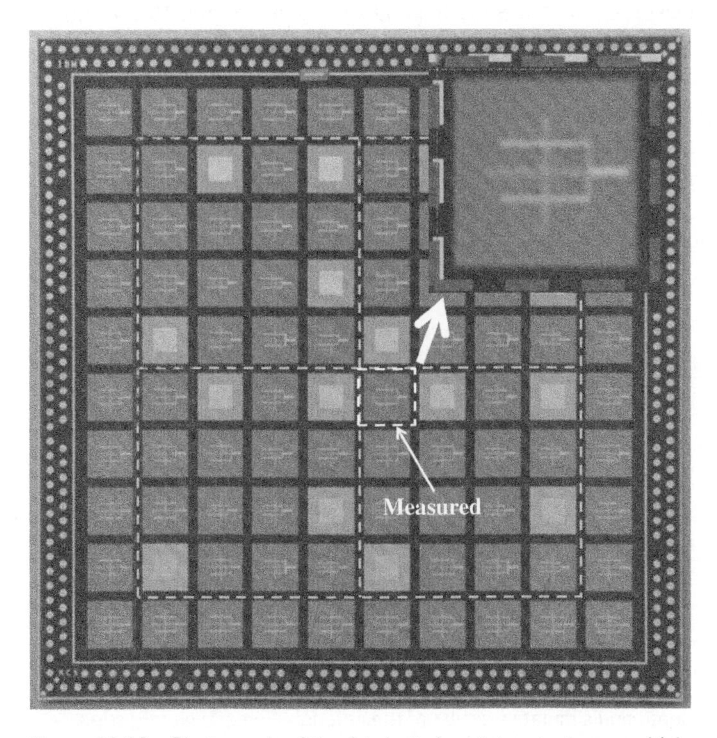


**Figure 10.10** Photograph of the fabricated prototype antenna vehicle.

topology as the 28-GHz phased array presented later. The AiP is divided into four quadrants, each of which contains 16 antenna elements, as shown in the large dashed-line box in the lower-right quadrant of the front side picture. Thirty-six dummy elements are placed around the edges of the package. Multiple antenna design variations were included in this prototyping vehicle for both probe-fed and aperture-coupled (excluding the cover) dual-polarized patch antennas. Specifically, we designed four variations of probe-fed patch antennas (described in [30]) and 12 variations of aperture-coupled patch antennas. Here we will concentrate on one aperture-coupled patch antenna design, located in the small dashed-line box shown in Figure 10.10. The 16 element designs were then copied to the other three quadrants.

The fabricated antenna elements were measured in a probe-based antenna chamber designed for 60-GHz measurements [63], but refitted with WR28 components for this work. Figure 10.11a shows the simulated (dashed lines) and measured (solid lines) reflection coefficients of the antenna for the horizontal and vertical ports. Minor modifications were made to the original design using feedback from EM simulation tool results to meet package layout requirements. The simulated results were generated using the structure used in the layout. The simulation and measurement results are in good agreement. Figure 10.11b shows the simulated and measured antenna gains. The simulation results were generated using an isolated antenna, while the measurements were generated using an antenna surrounded by other antenna structures. The nearby structures cause the antenna vertical polarization gain drop in the 26–28 GHz frequency range due to coupling between the structures and the antenna. The antenna peak gains are 6.6 dBi for the horizontal polarization (HP) and 7.2 dBi for the vertical polarization (VP), respectively. Table 10.2 shows the comparison between the probe-fed and

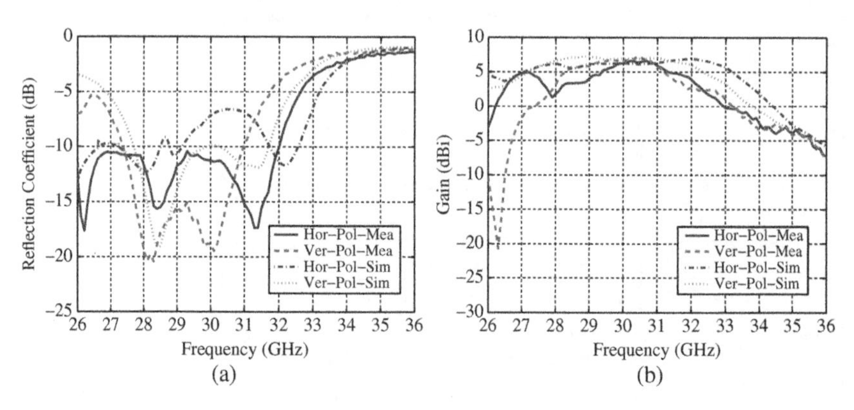


**Figure 10.11** Measured and simulated results of the antenna element: (a) reflection coefficient and (b) realized gain.

**Table 10.2** Measured antenna performance comparison.

| Antenna type | | Range (GHz) | Bandwidth (GHz) | Gain (dBi) |
|---|---|---|---|---|
| Probe-fed patch[30] | HP | 30.09–30.81 | 0.72 | 4.4 |
| | VP | 30.10–30.78 | 0.68 | 4.3 |
| Aperture-coupled patch | HP | 25.87–32.01 | 6.14 | 6.6 |
| | VP | 27.38–30.98 | 3.61 | 7.2 |

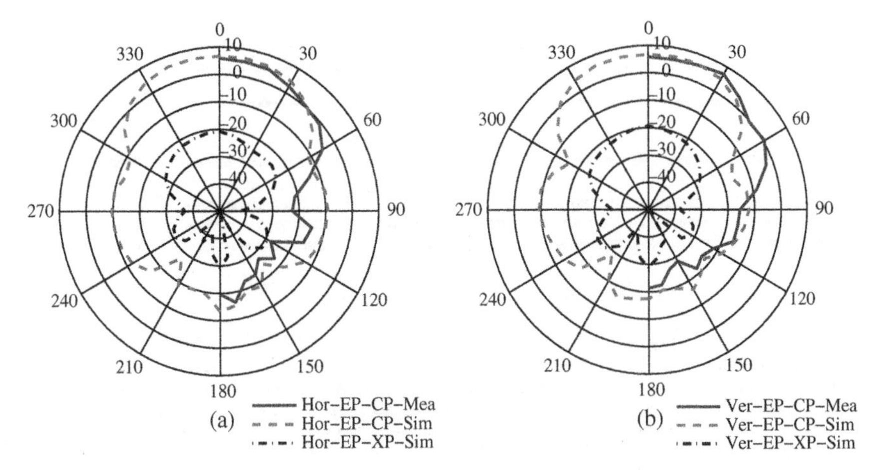


**Figure 10.12** Measured and simulated E-plane radiation patterns: (a) horizontal port and (b) vertical port.

the aperture-coupled patch antenna performance. The aperture-coupled patch antenna has both wider 10-dB return loss bandwidth and higher gain than the probe-fed patch antenna.

Figure 10.12 shows the measured and simulated E plane (EP) co-polarization (CP) and cross-polarization (XP) radiation patterns at 31 GHz for the horizontal and vertical ports, respectively. Due to the probe-based measurement setup, radiation patterns can only be measured with a 180° range. In both cases, the simulated cross-polarization of the antenna is more than 25 dB below the simulated co-polarization. The cross-polarizations of the antenna were not measured since performing that evaluation demands measurement setup changes.

### 10.3.4 28-GHz Aperture-coupled Cavity-backed Patch Array Design

Figure 10.13 illustrates the AiP phased-array module layer stack-up, assembly breakout, and the mounting concept with four ICs and printed circuit board

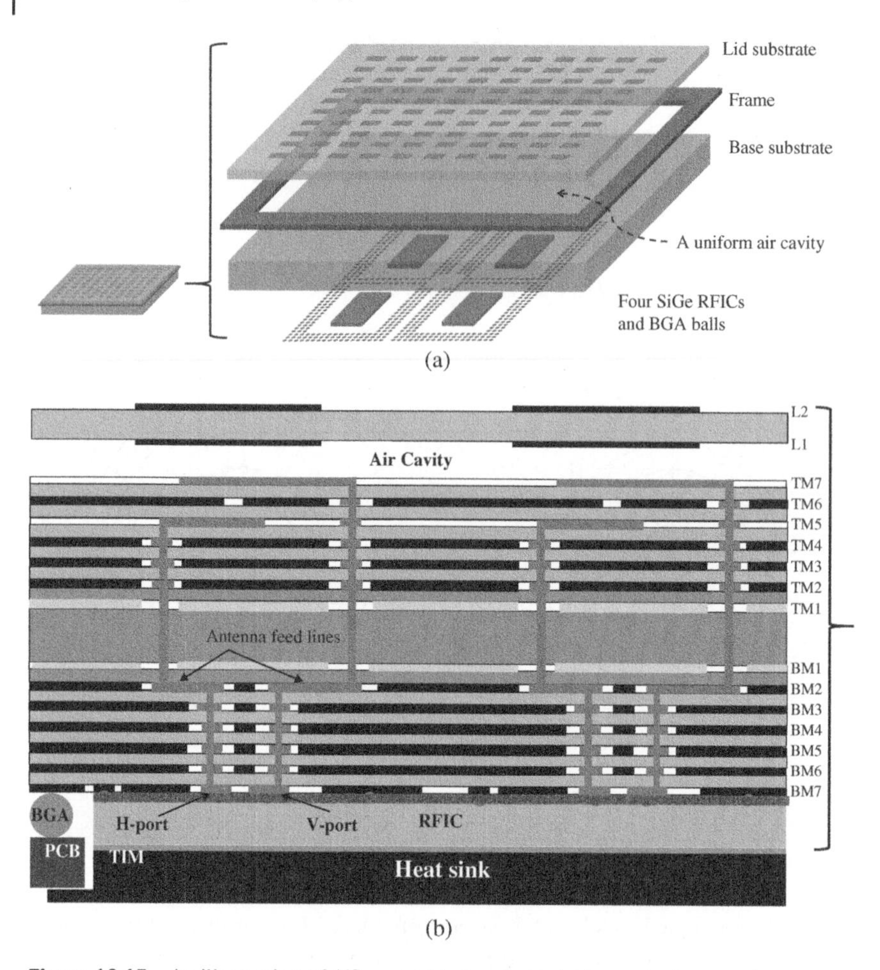


**Figure 10.13** An illustration of AiP assembly breakout: (a) layer stack-up and (b) board-level mounting concept (from [33], © 2017 IEEE, reprinted with permission).

(PCB) [33]. The assembled package consists of a two-layer lid/cover substrate, a two-layer frame used to form the air cavity, and a 14-layer base substrate as discussed previously. A dual-polarized aperture-coupled patch structure with air cavity is implemented for the antenna array. Both H and V polarization signals are input on the bottom C4 pads from ICs, fan out on the BM2 layer as shown in Figure 10.14, and reach feed structures located on the TM5 and TM7 layers, respectively. The stacked patches are implemented using L1 and L2 layers in the lid substrate. A uniform air cavity is formed between the lid and base substrates in the package to increase antenna bandwidth and gain, and, at the same time, reduce the impact of surface waves on coupling between antenna elements.

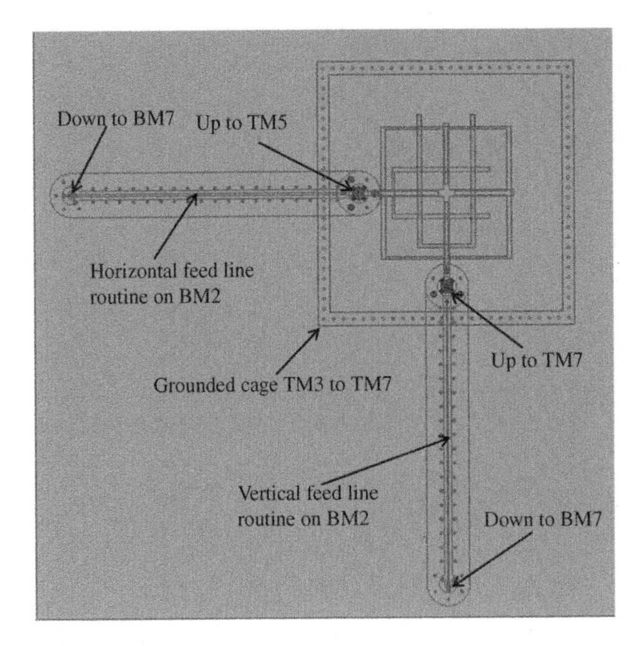


**Figure 10.14** An illustration of antenna feed line routing.

The RFICs are flip-chip attached to the bottom side of the package before being under-filled for C4 protection. The package is further mounted to a second-level PCB via BGA balls at the bottom. To enable cooling of the RFICs, there are cut-outs in the PCB in the region underneath each die that allow fitting of a heat sink with pedestals. Thermal interface material (TIM) is applied at the junctions between the ICs and the heat sink.

Figure 10.15 illustrates the top and bottom views of the antenna array package [33]. The package dimensions are $70 \times 70 \times 2.7\,\text{mm}^3$. The RFIC size is $10.6 \times 15.6\,\text{mm}^2$. There are in total 100 ($10 \times 10$ with a 5.9 mm pitch) antenna elements in the package: 64 active elements fed by four ICs and 36 dummy elements. The dummy elements are match-load terminated and placed around the periphery of the package to maintain uniformity of the radiation patterns of the active elements. On the bottom of the package base, 655 BGA balls at a 1.27-mm pitch provide the ICs with four power supplies, ground connections, digital controls, IF, and LO signals.

### 10.3.5 Passive Antenna Element Characterization

Direct probe measurements of individual antenna elements at the front-end signal C4 pads of a bare package were performed in the probe-based antenna chamber. Reflection coefficients for both H and V polarization are plotted in

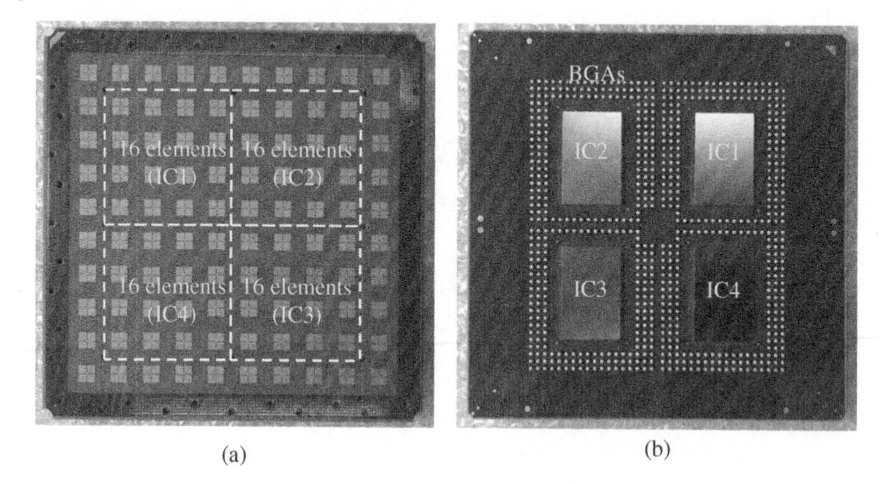


**Figure 10.15** Top view (a) and bottom view (b) of a fully assembled antenna array package with mounted BGA solder balls and four transceiver ICs (from [33], © 2017 IEEE, reprinted with permission).

Figure 10.16 [33]. The measured bandwidths of the antennas at the −10 dB reflection coefficient are 3.7 GHz for H polarization and 3.3 GHz for V polarization, respectively, which is typical for all the elements. Measured radiation patterns in the E plane for H polarization and in the H plane for V polarization are also shown in Figure 10.16. In the boresight direction, the measured antenna gains are 3 and 4 dBi for H and V polarizations, respectively. The measured element patterns maintain a broad shape suitable for wide-range beam-steering, showing gains of more than 0 dBi over a 100° range in the H plane and over an 80° degree range in the E plane off-boresight direction. Furthermore, the measured element patterns demonstrate model-to-hardware correlation with the simulation results obtained from a full-wave EM model of the entire package.

### 10.3.6 Active Module Characterization of 64-element Beams

Fully assembled antenna array modules with ICs and BGA balls were screened using a test board with a custom-made socket. Digital control functionality and dual-polarized radiation performance were verified for all elements prior to soldering the module to the PCB. External IF and LO signals were connected to the four ICs via on-board splitters and baluns, as shown in Figure 10.17.

A functional module was characterized in an antenna chamber with a 1.84 m distance between the module and a receiving horn antenna (Figure 10.17) [10, 33]. A field-programmable gate array (FPGA) board and a fan were connected to the test board for digital programming and air cooling. Two motors were mounted to

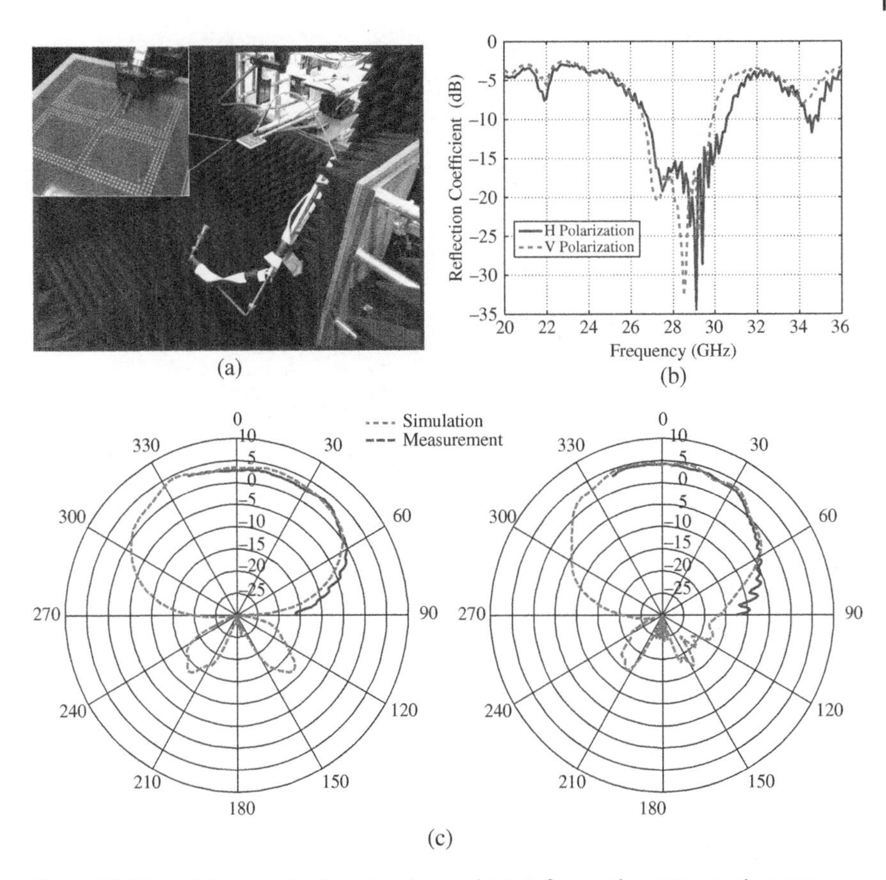


**Figure 10.16** (a) Antenna testing chamber and setup for passive antenna element probing measurement. (b) Measured frequency matching of a single element for H and V polarization. (c) Model-to-hardware correlation of radiation patterns: H plane for V polarization (bottom left) and E plane for H polarization (bottom right).

the fixture to provide rotation capability in both azimuth and elevation angles to support radiation pattern measurement.

Figure 10.18 shows the normalized boresight H-polarization patterns in the E and H planes, respectively, for the 64-element array in the TX mode. For the measurement here, an identical power amplifier bias setting is applied to all the front-end elements. Without element amplitude tapering or phase calibration, 12° half-power beam width, lower than −12 dB side levels, and over 20 dB deep notches between the main and side lobes are observed, which is expected for a uniform array. With realistic amplitude tapering, the side lobe level has been reduced to around −20 dB, which is close to practically achievable limits. Figure 10.19 plots two measured 64-element H-polarization patterns steered to ±40° off-boresight

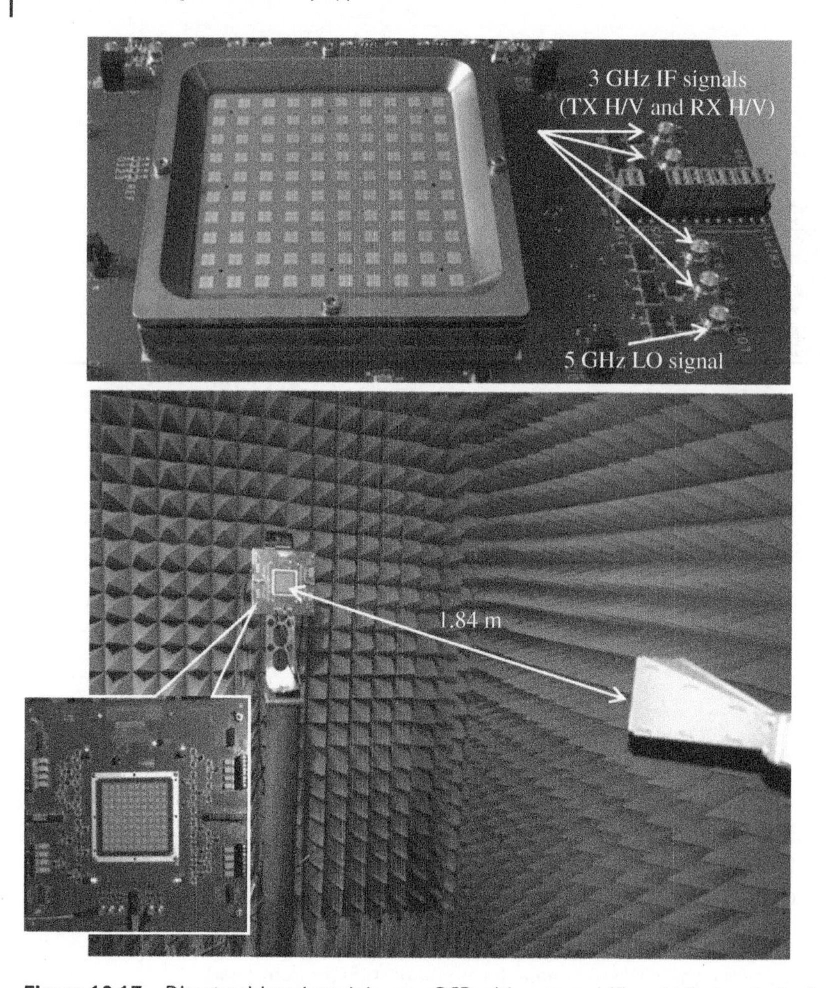


**Figure 10.17** Direct soldered module on a PCB with external IF and LO signals (top). Over-the-air active module measurement and characterization in an antenna chamber (bottom) (from [10], © 2017 IEEE, reprinted with permission).

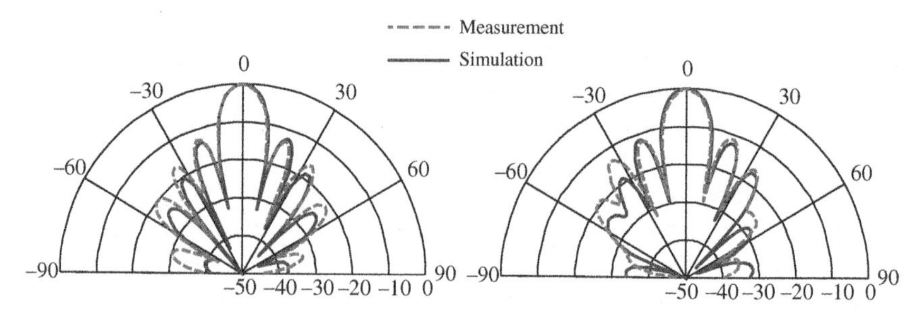


**Figure 10.18** Measured and simulated 64-element H-polarization beams at boresight direction with model-to-hardware correlation: H-plane pattern (left) and E-plane pattern (right).

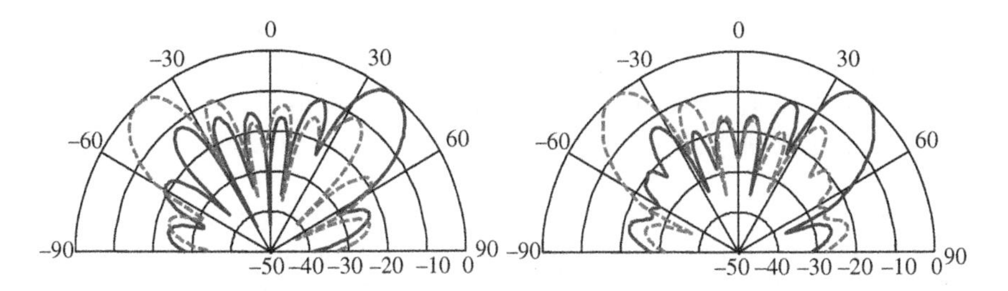


**Figure 10.19** Measured 64-element H-polarization beam patterns steering to ±40° after normalization: H-plane (left) and E-plane (right).

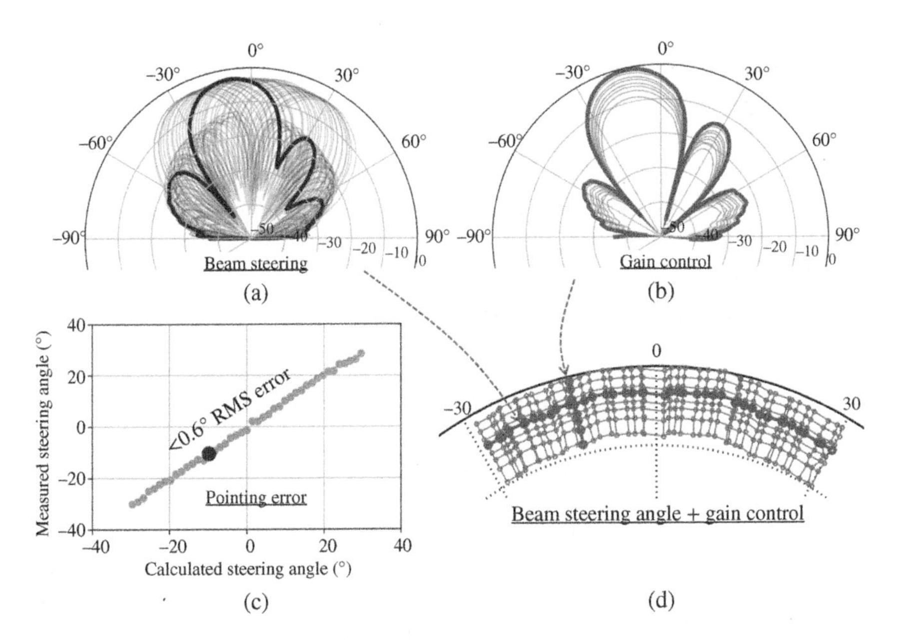


**Figure 10.20** (a) Measured beam steering example at a fixed variable gain amplifier (VGA) setting. (b) A gain control example at a fixed phase setting. (c) Uncalibrated steering angle versus calculation error. (d) Beam pointing directions for uncalibrated 16-element beam steering precision between ±30° with 8-dB VGA control (from [10], © 2017 IEEE, reprinted with permission).

in the E and H planes, respectively, with lower than −10 dB side lobe levels. The measured patterns agree with the full package EM simulation, validating the overall antenna-IC-package co-design and implementation.

Figure 10.20 shows the performance of the AiP phased array [10]. To demonstrate the accuracy of the phase and gain control, as well as the matching among elements, the beamforming tests presented here were made without gain or phase calibration, i.e. uniform excitations (amplitude and phase) were applied to all elements for the broadside beams. For beam steering, mathematically computed phase shifts for each element were linearly translated to phase settings using a 4.9° average phase shifter step. All beam measurements have a 1° angular resolution.

The beam steering results shown in the top left plot of Figure 10.20 demonstrate a 1.4° beam steering resolution across a range of ±30° using uniform phase gradients in the phased array. Forty-three different 16-element beam directions are shown here. As shown in the bottom right plot, each data point along a given arc on the polar plot slice represents a beam pointing direction at a given gain control level. Thus the top left plot corresponds to the middle arc in the bottom right plot. This indicates that changing the beam direction will not affect the element gain control. The antenna element excitation amplitude/gain and phase can be controlled independently. This independence is also demonstrated in the top right plot. Here the beam direction is fixed, but the excitation amplitude of the elements is changed over a 9 dB range, represented by the data points on the radial axis in the bottom right plot. Without any gain or phase calibration, the RMS pointing error across all directions shown in the bottom left plot in Figure 10.20 is less than 0.6°.

### 10.3.7 28-GHz AiP Phased-array Conclusion

Silicon-based mmWave transceivers and phased-array AiPs have matured, and telecom providers have begun to utilize mmWave frequencies for mass deployment in 5G cellular systems. We have presented a 28-GHz phased-array AiP module with 64 dual-polarized antenna elements and fully integrated dual polarized silicon-based transceivers. Furthermore, we have demonstrated simultaneous dual-polarized beams in both TX and RX. Due to the orthogonal phase and amplitude control in the transceivers and antenna element gain and phase uniformity in the AiP design, we have also demonstrated an advancement in beam steering and tapering precision without the need for element-to-element calibration. Measurement results of an assembled phased array module that contains an AiP and four transceiver ICs show over 54 dBm equivalent isotropic radiated power (EIRP) in TX mode and up to ±60° scanning range. Extensive measurements across process and temperature prove the suitability of this design for 5G cellular communications. We expect that the demonstrated phased-array advancement will not only impact mmWave mobile communication systems, but also open up new opportunities for imaging applications, and for the development of complex and adaptive multi-function systems.

## 10.4 94-GHz Scalable AiP Phased-array Applications

Millimeter-wave scaled phased arrays support many antenna elements (typically from tens to hundreds) by utilizing multiple ICs. Module-level integration techniques are leveraged to couple radiating elements and their associated substrates to the supporting ICs [11, 31, 64, 65]. Such scalable phased arrays are attractive for several emerging applications such as 5G, backhaul communications, and imaging. A higher number of antenna elements in an array leads to narrower beamwidth and higher EIRP, both of which are desirable features in communications and imaging systems [65]. Phased-array scaling is advantageous for three main reasons [64]: (i) it provides a path for supporting antenna arrays with a larger number of elements than a single IC can support, (ii) it provides flexibility in terms of the beamforming implementation, and (iii) it enables forming arrays of different dimensions (e.g. square arrays or rectangular arrays) based on the same unit ICs. The increased level of complexity of these designs, however, brings several challenges. Some of the factors that increase the level of difficulty with respect to phased arrays with one IC include (i) an increased number of transitions between ICs and substrate/packaging materials, (ii) an increased number of assembly steps, (iii) stringent area constraints on the ICs to fit within an area compatible with that of the antennas they support, and (iv) increased digital control complexity.

Millimeter-wave links are now an integral part of the wireless backhaul infrastructure, particularly at E-band frequencies (71–76, 81–86, and 92–94 GHz) [66]. Note that these frequencies are a subset of the W-band frequency range. E-band links are currently implemented by a combination of single-element transceivers and antennas with high gain (i.e. >30 dBi) and consequently require large form factors. Mechanical alignment is required for these antennas, as such, links are established and maintained in a single fixed direction, as shown in the top example in Figure 10.21a [11].

Monolithic phased arrays intended for indoor applications at 60 GHz have been demonstrated to support links with data rates in excess of 5 Gb/s at distances of ~10 m employing only 16 antennas [67] and 200 m line-of-sight wireless gigabit alliance (WiGig) systems with 288 antennas (144-element phased array) [13]. These and other similar results have motivated research on highly integrated phased arrays at higher frequencies and with a larger number of elements. Silicon-based scalable phased arrays at the W-band offer the possibility of attaining a similar coverage range as current fixed-beam solutions with the additional advantage of dynamic steerability, at the cost of complexity and power

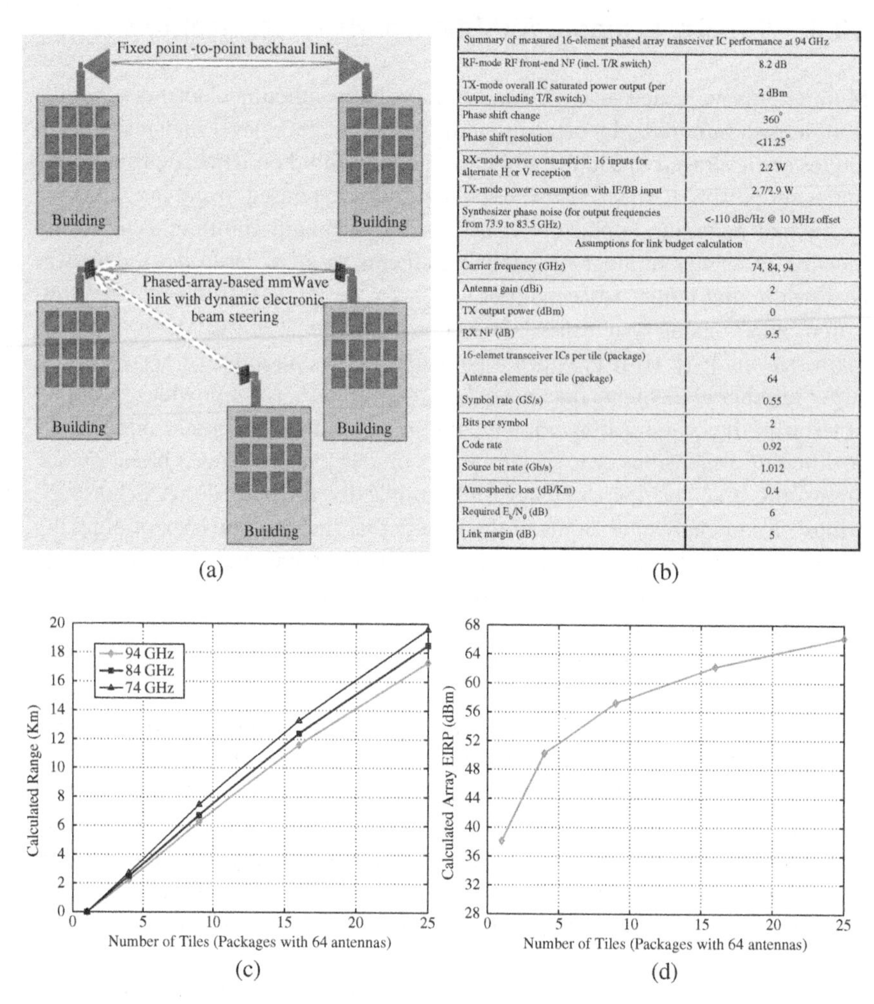


| Summary of measured 16-element phased array transceiver IC performance at 94 GHz | |
|---|---|
| RF-mode RF front-end NF (incl. T/R switch) | 8.2 dB |
| TX-mode overall IC saturated power output (per output, including T/R switch) | 2 dBm |
| Phase shift change | 360° |
| Phase shift resolution | <11.25° |
| RX-mode power consumption: 16 inputs for alternate H or V reception | 2.2 W |
| TX-mode power consumption with IF/BB input | 2.7/2.9 W |
| Synthesizer phase noise (for output frequencies from 73.9 to 83.5 GHz) | <-110 dBc/Hz @ 10 MHz offset |
| **Assumptions for link budget calculation** | |
| Carrier frequency (GHz) | 74, 84, 94 |
| Antenna gain (dBi) | 2 |
| TX output power (dBm) | 0 |
| RX NF (dB) | 9.5 |
| 16-elemet transceiver ICs per tile (package) | 4 |
| Antenna elements per tile (package) | 64 |
| Symbol rate (GS/s) | 0.55 |
| Bits per symbol | 2 |
| Code rate | 0.92 |
| Source bit rate (Gb/s) | 1.012 |
| Atmospheric loss (dB/Km) | 0.4 |
| Required $E_b/N_0$ (dB) | 6 |
| Link margin (dB) | 5 |

**Figure 10.21** (a) Illustration of two different types of mmWave backhaul link, with fixed high-gain antennas (top) and with a phased array (bottom). (b) Summary table of measured transceiver IC performance [68] and assumptions for link budget calculation. (c) Estimated range for a 1 Gb/s terrestrial data link at 74, 84, and 94 GHz, constructed using scalable phased arrays of various sizes. The estimates assume quadrature phase shift keying (QPSK) (2b/symbol) in 800 MHz RF bandwidth, low-density parity check (LDPC) code (1369, 1260) with code rate $R = 0.92$, and bit error rate (BER) $= 10^{-7}$ with 5 dB implementation loss and 0.4 dB/km atmospheric loss. (d) Calculated EIRP as a function of the number of tiles (from [11], © 2015 IEEE, reprinted with permission).

consumption. Electronic steerability would not only eliminate the need for mechanical alignment, but would also open the possibility of dynamic backhaul networking, as shown in the bottom example in Figure 10.21a. The scalability of a unit cell array with a moderate number of elements is key, since different links and usage scenarios may require a different number of elements.

To illustrate the potential of a scalable phased array at W-band, the table in Figure 10.21b introduces the link budget considerations for a 1 Gb/s link formed with tiles of 64 antenna elements supported by four 16-element phased array ICs. This link budget calculation considers the measured phased-array IC performance reported in [68], which is also shown in Figure 10.21b. It should be noted that the noise figure (NF) and output power performance used in the link budget calculation is slightly worse (1.5–2 dB) than that reported for an IC at room temperature to account for performance degradation at higher temperatures. Figure 10.21c shows the potential link range as a function of the number of tiles for three different E-band frequencies. It can be observed that a 10km+ range is potentially achievable with a $4 \times 4$ array of 64-element tiles (1024 elements total), and such a unit tile array would occupy an area smaller than $16.2 \times 16.2\,\text{mm}^2$ at 94 GHz. The calculated array EIRP as a function of the number of tiles is plotted in Figure 10.21d.

There are three key aspects to consider in scalable phased-array designs [64]: (i) the integration and assembly of the phased-array antenna module, consisting of ICs and one or more substrates stacked either vertically or laterally, (ii) the circuit-level architecture in terms of the frequency conversion, power combining/splitting, and beamforming functions across the array to enable module-level scaling using either RF or hybrid (digital + RF) beamforming, and (iii) the digital control architecture and digital-to-RF co-integration for coordinated control of multiple ICs at the module level, which is critical to beam-steering agility. This section will concentrate on the AiP portion of the first reported SiGe-based scalable phased-array module to illustrate the challenges and initial solutions.

### 10.4.1 Scalable Phased-array Concept

Our approach to realizing a dual-polarized scalable phased array is illustrated in Figure 10.22. This approach leverages the AiP approach described in Figure 10.23. Each transceiver IC contains 16 dual-polarized RF phase-shifting transceiver front-ends. RF-path phase shifting was selected to achieve minimum hardware and power consumption at the IC level. All of the mmWave functions as well as frequency conversion and LO generation are integrated monolithically [68]. Each package houses four ICs and includes 64 dual-polarized antennas. The antennas are placed at a $\sim\lambda/2$ pitch at 94 GHz in both the $x$ and $y$ dimensions, and the antennas on the perimeter are placed $\sim\lambda/4$ away from the package edge. By tiling the packages adjacent to one another on a PCB, phased arrays of large aperture

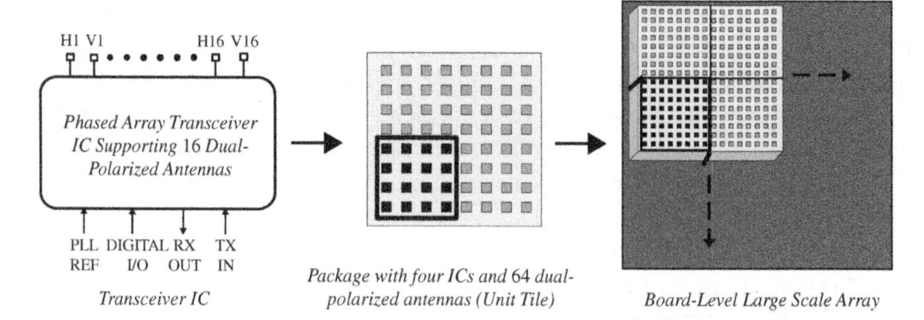


**Figure 10.22** Scalable phased-array concept (from [11], © 2015 IEEE, reprinted with permission).

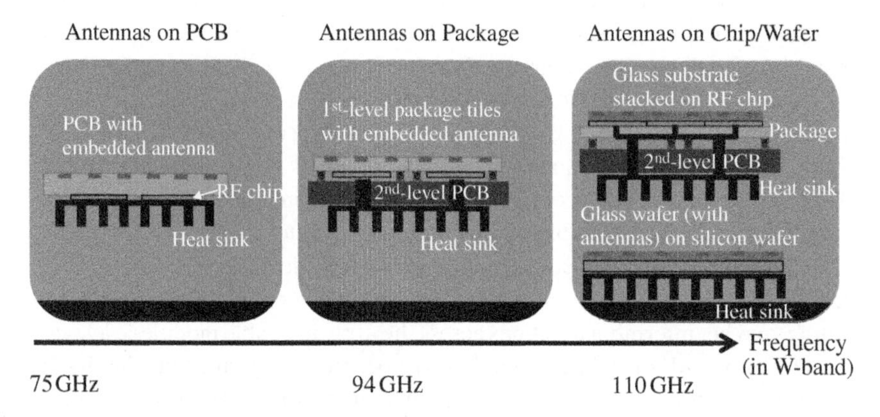


**Figure 10.23** Antenna and package options for W-band scalable phased array (from [11], © 2015 IEEE, reprinted with permission).

can be created to support long-distance communication and high-resolution imaging. The first two steps of this approach to implementing scalable phased arrays (transceiver IC and package integrating ICs and antennas) have been demonstrated in hardware [11, 31] and are described later in more detail.

The concept of tiling small arrays (subarrays) to form large arrays at low frequency is relatively simple to execute [69, 70]. However, forming large uniform arrays at mmWave frequencies using small AiP units (typically $8 \times 8 = 64$ dual-polarized antenna unit) is not straightforward and at the W-band it is even more challenging. Due to the very small wavelength at the W-band and AiP fabrication technology requirements, an edge antenna element in an AiP should be greater than $\lambda/4$ (0.798 mm at 94 GHz) from the edge of the AiP. The distance (or gap) between the adjacent edges of two neighboring AiPs in a tiled array should be at least $\lambda/2$ (1.596 mm) due to AiP fabrication and AiP placement tolerances.

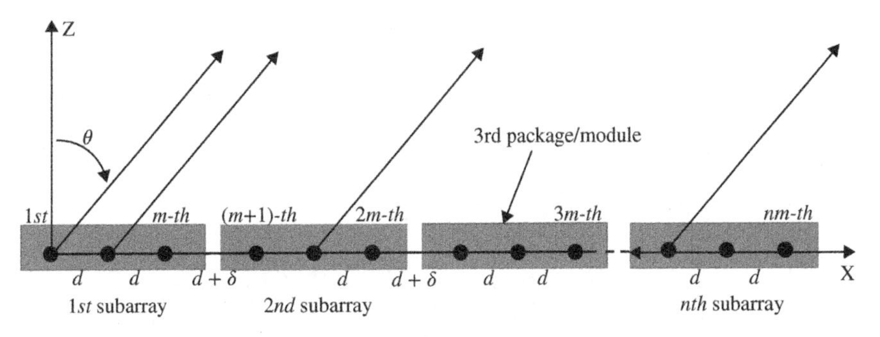


**Figure 10.24** A nonuniform linear array with *mn* isotropic radiators.

As a result, center-to-center antenna element spacing between the two adjacent elements on the two AiPs would be greater than $1\lambda$. Figure 10.24 illustrate the case for a linear array with $n$ AiPs (subarrays or packages/modules) and each AiP contains $m$ antenna elements, where $d$ is the center-to-center element spacing on an AiP while $d + \delta$ ($\delta = 0$ for an uniform array) is center-to-center element spacing between the two adjacent AiPs. The array factor for the whole array and the first beam null location are given by:

$$AF = 20\log_{10}\left(\frac{\sin\left(\frac{m\beta d}{2}\sin\theta\right)}{m\sin\left(\frac{\beta d}{2}\sin\theta\right)}\cdot\frac{\sin\left(\frac{n\beta(md+\delta)}{2}\sin\theta\right)}{n\sin\left(\frac{\beta(md+\delta)}{2}\sin\theta\right)}\right) \tag{10.1}$$

$$\theta_{FN} = \sin^{-1}\left(\frac{\pm 2\pi}{n\beta(md+\delta)}\right) \tag{10.2}$$

Where the overall array factor includes a subarray factor and a tile/unit array factor. For a linear uniform array ($\delta = 0$) with $nm$ elements, the array factor for the whole array and the first beam null location are:

$$AF = 20\log_{10}\left(\frac{\sin\left(\frac{nm\beta d}{2}\sin\theta\right)}{nm\sin\left(\frac{\beta d}{2}\sin\theta\right)}\right) \tag{10.3}$$

$$\theta_{FN} = \sin^{-1}\left(\frac{\pm 2\pi}{nm\beta d}\right) \tag{10.4}$$

Figure 10.25 shows the array factors for the cases $n = 6$, $m = 8$, and $d = 0.5\lambda$, with $\delta = 0$, $0.5\lambda$, $1.0\lambda$, and $1.5\lambda$, respectively. As expected, increasing the $\delta$ value will lower the first null location (the array's physical size has also been increased), cause non-monotonic side lobe levels, and increase the side lobe level values. The reason for this side lobe level behavior is the misalignment between the grating

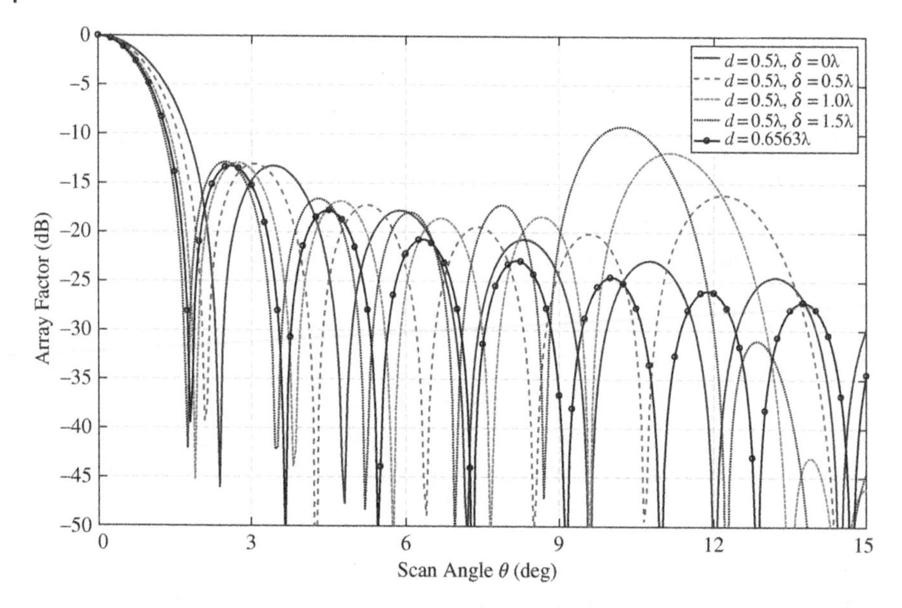


**Figure 10.25** Array simulation for a nonuniform linear array with 48 isotropic radiators.

lobes of the tile array factor and the zeros of the element array factor [70]. The array physical size for the $\delta = 1.5\lambda$ case is $nmd + (n-1)\delta = 31.5\lambda$. For comparison, the last curve is for the case where a uniform array also has $nmd_{ave} = 31.5\lambda$ size with an average element spacing $d_{ave} = 31.5\lambda/(nm) = 0.6563\lambda$. Note that the side lobe level is not increased in the last case. It is therefore beneficial to use more than one RFIC for a package/module; we use four RFICs per package in our 94-GHz designs.

### 10.4.2 94-GHz Antenna Prototype Designs

Figure 10.26a illustrates a cross-sectional view of the assembled multi-chip module with embedded antennas on a system board [31, 32]. The patch antenna array is on the top of the package. Four SiGe ICs are flip-chip attached to the bottom of the package. The module is mounted to a system board via a pogo-pin-based interposer, which allows air cooling and supports easy removal for module screening. A detailed zoom-in view of the AiP cross-section is shown in Figure 10.26b with major components identified. The stack-up consists of three portions: the core, the top build-up, and the bottom build-up. The top and bottom substrate build-ups have the same dielectric properties, and each portion contains five metal layers. The core portion features two metal layers. The overall

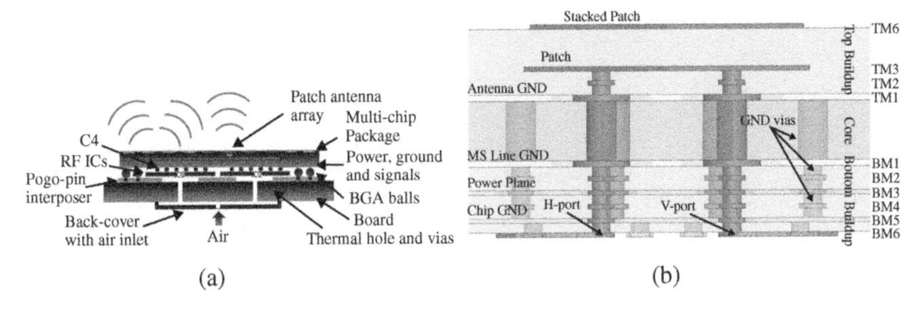


**Figure 10.26** (a) Illustration of a W-band phased-array MCM module and (b) a package substrate stack-up (from [31], © 2014 IEEE, reprinted with permission).

package therefore has a 5-2-5 layer structure. The total layer count is 12 metal layers (BM1–BM6 and TM1–TM6) with six layers in the top and bottom portions, respectively. The patch is formed on the TM3 metal layer, while the stacked patch is formed on the TM6 metal layer. The probe-fed stacked patch antenna has two feeds since dual polarizations are required in this design. The antenna feed lines go from the TM3 metal layer, passing through the antenna ground plane on TM1, a ground plane on BM1, the power plane on BM3, the chip ground plane on BM5, and finally ends on the BM6 metal layer. Since the antenna feed lines have to go vertically, coax-like sections are used for the vertical portions from the TM1 to BM5 metal layers. For the final package design, an antenna feed line goes from the TM3 metal layer to the BM2 metal layer first, routes on the BM2 layer to a position just above the corresponding RFIC antenna port, then connects from that position vertically to the antenna port, similar to the one shown in Figure 10.13b. Other than the antenna feed lines, the structure has additional layers, as shown in Figure 10.26b, to emulate the ground and power planes of the RFIC. The antenna and the RFIC chip ground planes are connected together using core and build-up vias.

Figure 10.27a shows the top view of an antenna prototype structure. Note that in order to make the antenna feed lines easier to implement, the patches are rotated 45°. For the purpose of antenna evaluation, the antenna feed lines in this prototype are much longer than those required in actual applications. Since probe measurements are required, the feed lines end in a ground-signal-ground layout. Figure 10.27b shows the actual antenna layout for fabrication. Due to metal fill requirements, an additional 12 patches are placed around the antenna patch with a $1/2\lambda$ separation (similar to an array layout). The patch placements are also arranged in a triangular fashion to increase the element spacing in order to mitigate layout challenges. Outside of the patch regions, a cheesed metal plate is used.

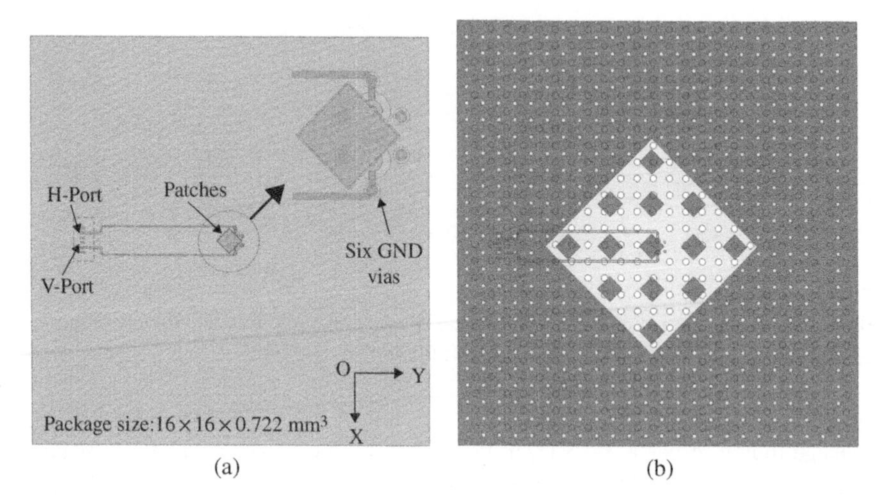


**Figure 10.27** (a) Prototype antenna design and (b) the actual layout of the antenna test vehicle.

### 10.4.3 94-GHz Antenna Prototype Evaluation

Figure 10.28 shows measured and simulated antenna reflection coefficients or $S_{11}$ in dB. The two results have similar characteristics, although the simulated result shows better impedance match and wider bandwidth as compared to the measured result. We attribute the disagreement primarily to two factors: (i) the difference between the designed and fabricated prototypes, referring to Figure 10.27b, and (ii) effects driven by inherent manufacturing tolerances. The first of these factors arises because manufacturing requirements demand that metal layers be filled with copper in numerous areas which are empty in the simulation model. This discrepancy affects the antenna hardware-to-simulation correlation to some degree.

Figures 10.29a and 10.29b show the measured and simulated radiation patterns of the phi and theta components in the $y$–$z$ plane (see Figure 10.27a) for the H port and V port, respectively. The measured patterns agree well with the simulated results. Since the antenna chamber is configured for probe-based antenna measurement, where the transmitting antenna under testing cannot rotate and the receiving standard gain horn antenna can only rotate on the $x$–$z$ and $y$–$z$ planes, E and H plane polarizations were not measured. Measurements of individual antenna test structures show ~3 dBi peak gain in both polarizations and ~8 GHz bandwidth [32].

### 10.4.4 94-GHz AiP Array Design

The multi-function dual-polarization phased-array transceiver IC supports both radar and communication applications at the W-band [68]. Thirty-two receive

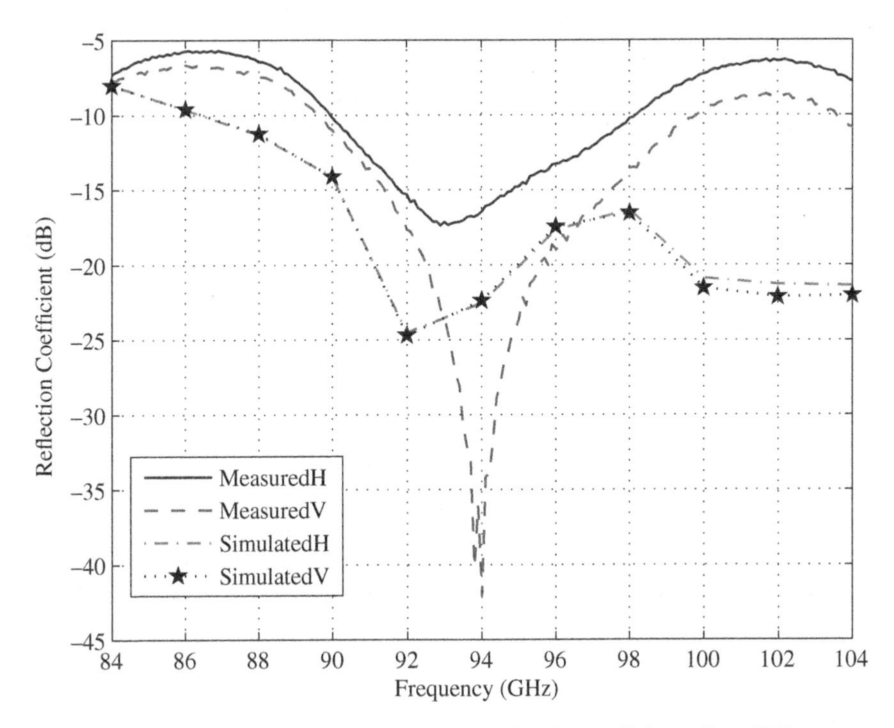


**Figure 10.28** Measured and simulated antenna reflection coefficients (from [30], © 2007 IEEE, reprinted with permission).

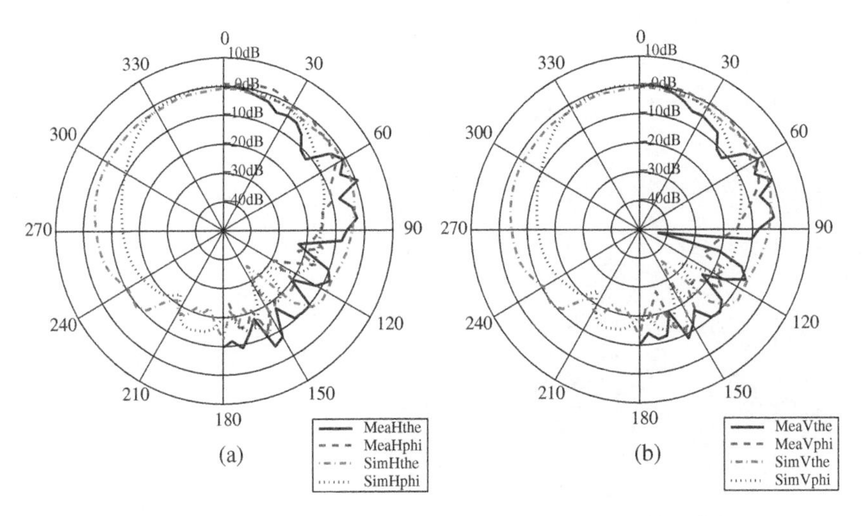


**Figure 10.29** Measured and simulated radiation patterns of the phi and theta components for (a) the horizontal port and (b) the vertical port.

elements and 16 transmit elements with dual outputs are integrated to support 16 dual-polarized antennas in a package. The IC includes two independent 16:1 combining networks, two receiver down-conversion chains, an up-conversion chain, a 40 GHz PLL, an 80 GHz frequency doubler, extensive digital control circuitry, and on-chip IF/LO combining/distribution circuitry to enable scalability to arrays at the board level. The fully integrated transceiver was fabricated in IBM's SiGe BiCMOS 8HP 0.13 μm process, occupies an area of 6.6 ×6.7 $mm^2$, and operates from 2.7 V (analog/RF) and 1.5 V (digital) supplies. Multiple operating modes are supported, including the simultaneous reception of two polarizations with a 10 GHz IF output, transmission in either polarization from an IF input, or single-polarization transmission/reception from/to in-phase and quadrature baseband signals.

Iterations of circuit-package-antenna co-design were performed in the context of severe physical dimension constraints to support array scalability at the package and board levels. Figure 10.30 shows a close-up view of antenna patches at the top of the package. One hundred (10 × 10) patch structures at 1.6 mm spacing ($\lambda/2$ at 94 GHz) cover the surface of the package. To support all required functionality, the IC area is very close to that required for 16 (4 × 4) antennas with $\lambda/2$ spacing. As a result, a multi-chip 16.2 × 16.2 × 0.75 $mm^3$ package containing four SiGe-based RFICs and a 292-pin 0.4 mm-pitch BGA was designed to achieve

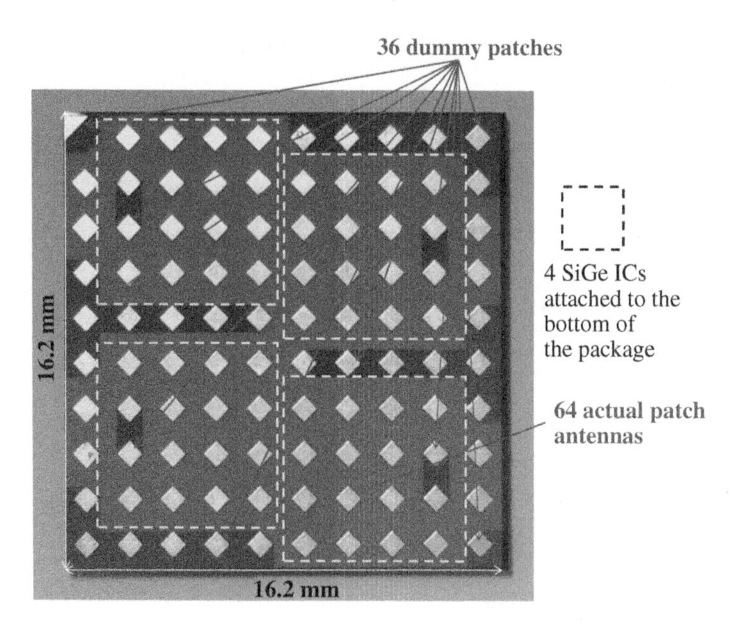


Figure 10.30 A close-up view of the four-chip package with actual patch antennas and dummies (from [11], © 2015 IEEE, reprinted with permission).

as high an array fill factor as possible [31]. The multi-chip package approach also mitigates the board-level integration risks compared to the single-chip package approach. Out of the 100 patches for each package, 64 are actual dual-polarized patch antennas and 36 are dummy structures (which do not have actual antenna features other than the surface patches). Therefore, the effective array fill factor is 64%. The dummy structures are placed at pseudo-randomized locations to minimize the impact of the reduced fill factor on side lobes. The copper balance in terms of metal percentage per layer also increases with the inclusion of dummies, which improves the manufacturability of the package. Figure 10.31 illustrates an array simulation in MATLAB with 1024 isotropic radiators based on the antenna layout pattern, which is equivalent to tiling 16 (4 × 4) 94 GHz packages. Notice that by choosing the patch locations carefully, empty rows or columns of active radiation elements can be avoided. The simulated radiation patterns are plotted in Figures 10.32a and 10.32b for 0° and 30° scan angles, respectively.

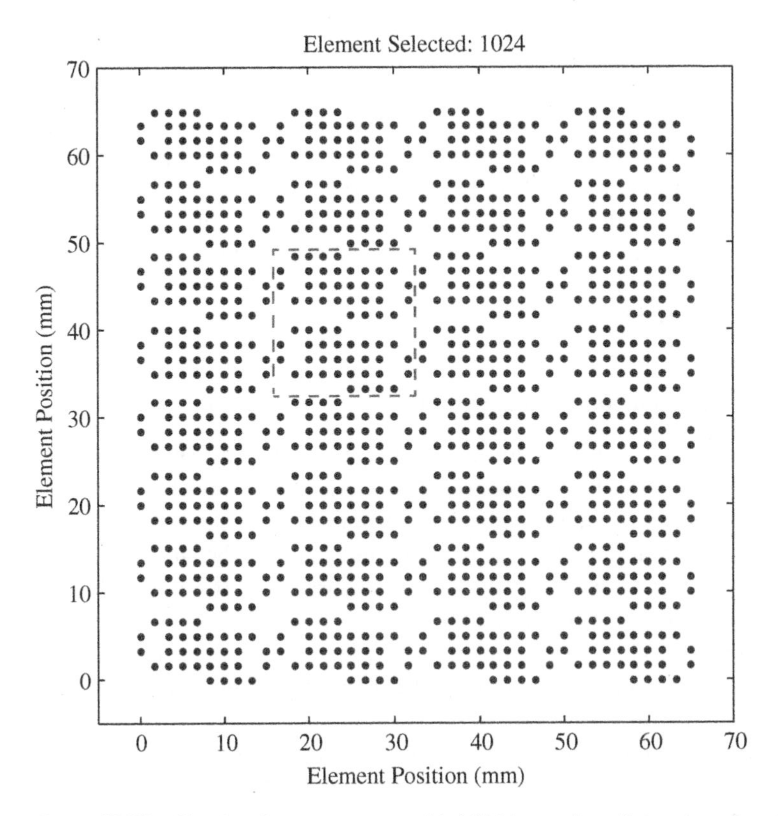


**Figure 10.31** Simulated antenna array with 1024 isotropic radiators based on the 94-GHz package antenna placement pattern (as highlighted) (from [11], © 2015 IEEE, reprinted with permission).

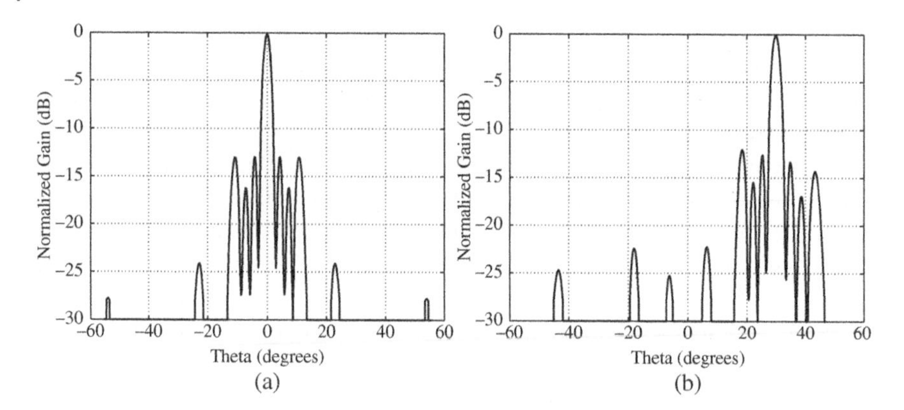


**Figure 10.32** Simulated 1024-element radiation patterns with beams at (a) 0° and (b) 30°.

High-speed differential signals are routed from the inner BGA row as short microstrip pairs on the BM6 layer to avoid via transitions, whereas low-speed single-ended signals are routed from the outer BGA row as striplines. Furthermore, two groups of voltage supply pins as well as ground pins are placed evenly on the periphery to ensure good power distribution to the chip. The front-end C4s for the W-band antenna feed are laid out using a 225 μm pitch GSGSG configuration. In order to minimize the RF antenna feed line length, the locations of these C4s were optimized together with the circuit layout for the front-end, core, and digital macros.

### 10.4.5 Package Modeling and Simulation

Full-wave 3D simulation was performed to optimize and verify the package layout. Figure 10.33 shows a finite element method-based HFSS model for the entire package with 128 ports defined for all 64 of the embedded antennas.

Due to the small wavelength at the target frequencies and the large number of antennas in the package, it is computationally expensive to perform the full-wave analysis of the package. In particular, the adaptive volume mesh generates ~5.3 million tetrahedra. The simulation was performed using 10 Xeon processors. It took approximately 9 hours to analyze one frequency with 146 GB peak memory usage.

Figure 10.34 plots the simulated $S_{11}$ in dB of the 16 antennas associated with one of the four chips. Both horizontal and vertical polarizations are shown. Notice that most antennas here show better than 10 dB return loss at 94 GHz. The only two antennas exhibiting relatively higher reflection levels are the ones which are intentionally placed with a $\lambda/2$ offset to randomize the antenna array, as shown

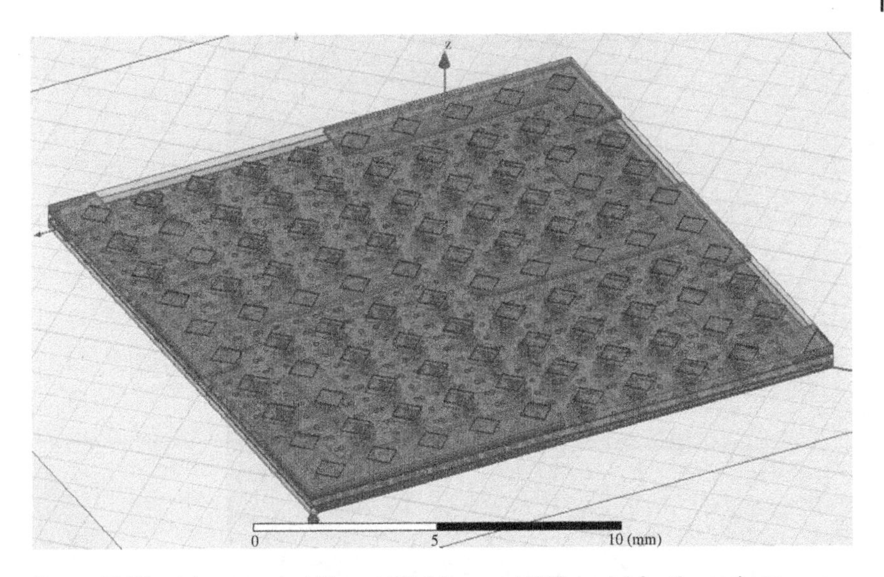


**Figure 10.33** A large-scale 128-port 3D full-wave HFSS model for the package simulation (from [31], © 2014 IEEE, reprinted with permission).

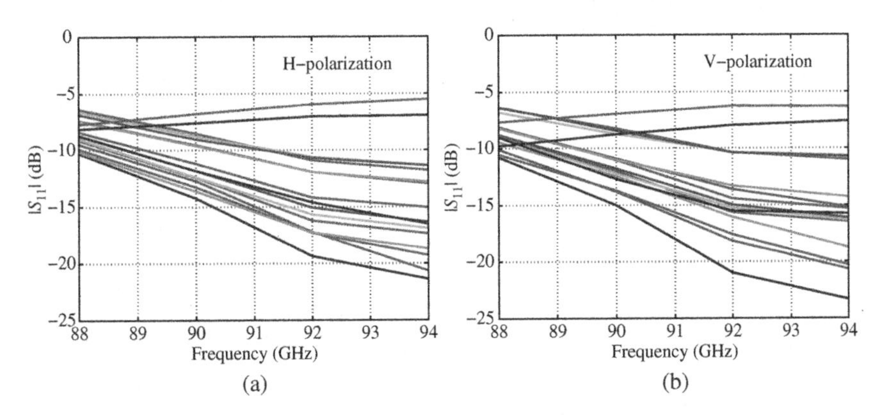


**Figure 10.34** Simulated reflection coefficients (both polarizations) of the 16 elements associated with one SiGe IC (from [31], © 2014 IEEE, reprinted with permission).

in Figure 10.35. As a trade-off, these two antennas have more complex and longer feed lines to enable proper impedance matching.

All the antennas associated with the other three SiGe ICs exhibit similar characteristics. Antenna gain was simulated as well and the results are plotted below along with the relevant measurement to show model-to-hardware correlation.

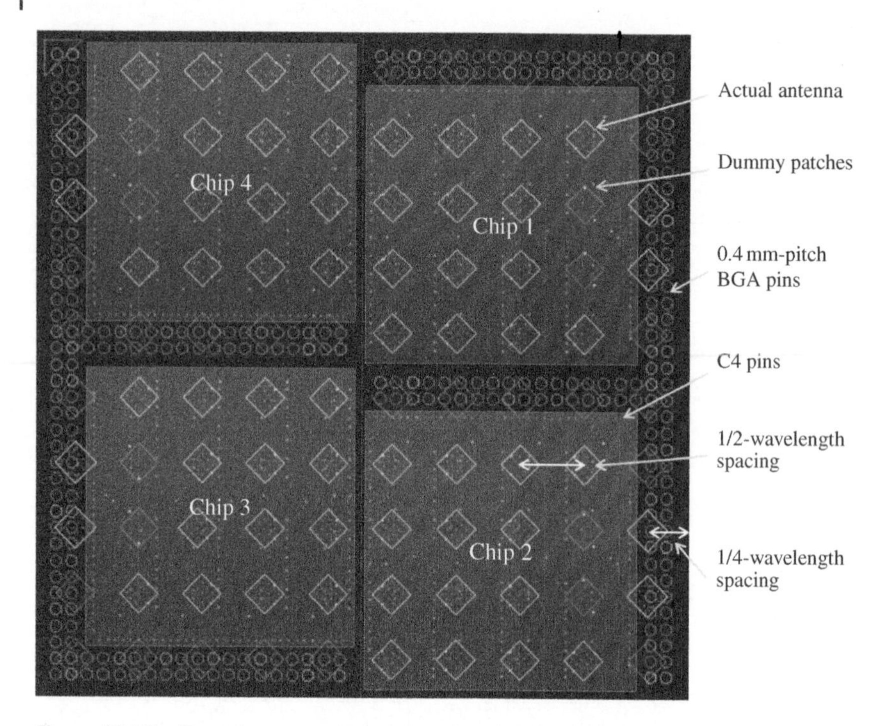


**Figure 10.35** Phased-array package layout housing four SiGe ICs with 64 dual-polarization antennas (from [31], © 2014 IEEE, reprinted with permission).

### 10.4.6 Package Assembly and Test

The IC package assembly was performed using standard flip-chip attach processes with lead-free solder reflow and underfill. Figure 10.36 shows a picture of the fully assembled packages with both front and back sides. The antenna patches, four SiGe ICs, and BGA solder balls are clearly visible.

Assembled packages were subsequently tested and screened in a socketed evaluation board, as shown in Figure 10.37. A high-speed pogo pin test socket with air cooling allows functional verification of assembled packages by monitoring synthesizer locking and voltage/current consumption of each power supply.

Figure 10.38 shows a fully assembled package mounted on a test board using a polymer cover which has a center window to allow measurement of the output power and radiation pattern of each antenna element. Sub-miniature push-on connectors are populated on the board to provide PLL reference and IF signals to the four ICs. In addition, a daisy-chain configuration is implemented so only one PLL reference input and one IF input are required for antenna pattern and radiated power measurement.

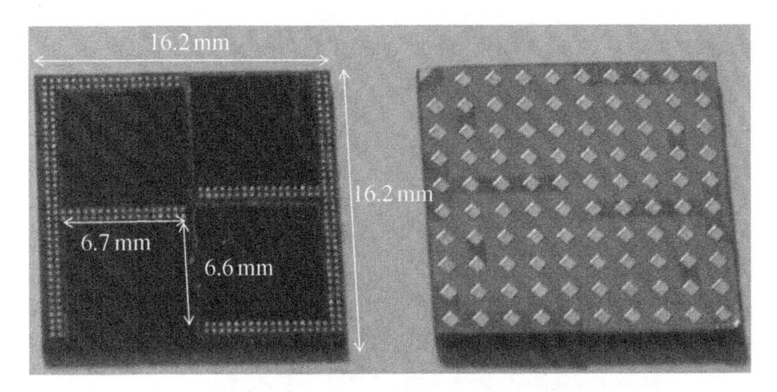


**Figure 10.36** Flip-chip assembled packages with 64 integrated antennas (package and IC dimensions included) (from [31], © 2014 IEEE, reprinted with permission).

**Figure 10.37** A socket with high-speed pogo pins for package screening and test (from [31], © 2014 IEEE, reprinted with permission).

Figure 10.38 An assembled package mounted to the test board for antenna radiation pattern and power measurement (from [31], © 2014 IEEE, reprinted with permission).

### 10.4.7 Antenna Pattern and Radiated Power Measurement

The radiation pattern of each individual antenna in the package has been measured in both polarizations using the setup shown in Figure 10.39. Two motors are used to control the rotation of the package and the receiving horn (i.e. azimuth and elevation angles), respectively.

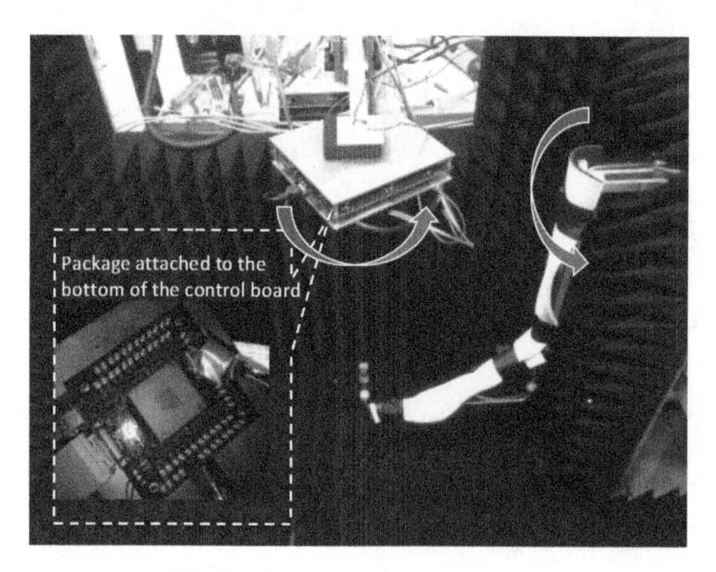


Figure 10.39 W-band antenna chamber measurement setup (from [31], © 2014 IEEE, reprinted with permission).

Figure 10.40 plots an example of the radiation patterns of one antenna element. In this experiment, the array is operated in TX mode, activating one element and one polarization at a time. The measured radiation patterns are in good agreement with simulations and the measured cross-polarization isolation is ~15 dB. The measured antenna gain varies between −5 and +2 dBi across elements. This variability is comparable to that measured in state-of-the-art packages for V-band applications fabricated in organic substrates [56, 71].

EIRP for individual elements in TX mode was measured across frequency. The measured average EIRP across 64 elements is over −1 dBm from 90 to 94 GHz, with some antenna elements radiating as much as 4 dBm EIRP at 92 GHz, as shown in Figure 10.41. Spatial power combining experiments have been performed with the

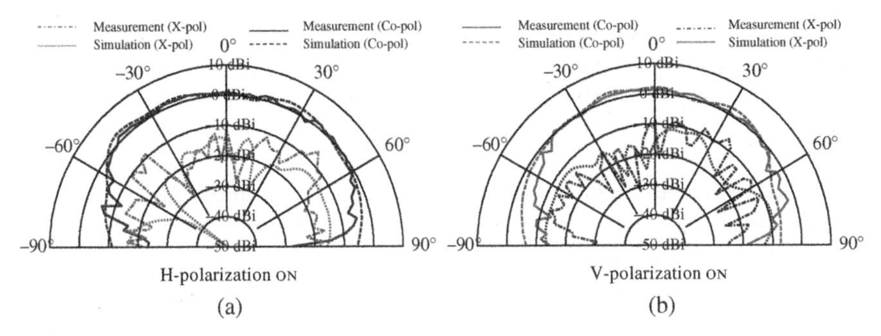


Figure 10.40 Model-to-hardware correlation of antenna radiation patterns for one element in package (from [31], © 2014 IEEE, reprinted with permission).

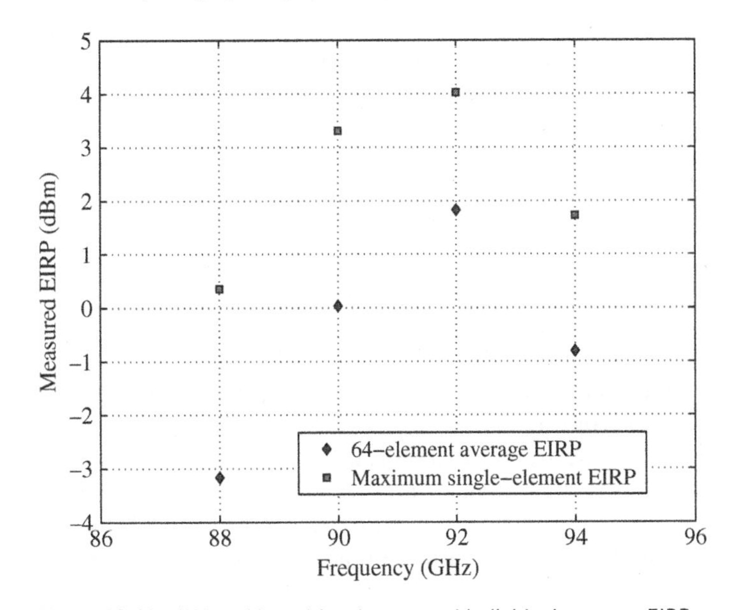


Figure 10.41 W-band board-level measured individual antenna EIRP.

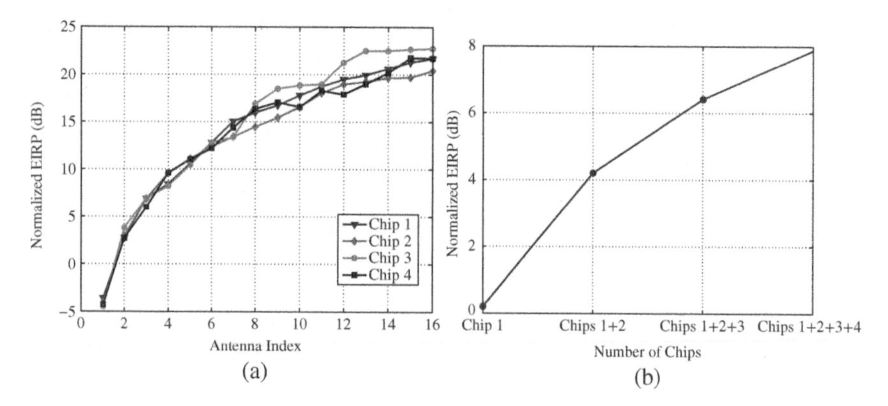


**Figure 10.42** (a) Measured spatial power combining of each IC (16 elements) normalized by the average power and (b) spatial power combining of four ICs at the package level normalized by the average EIRP of four ICs (from [31], © 2014 IEEE, reprinted with permission).

assembled module in TX mode at 90 GHz. For each of the four ICs, 16 TX elements are activated sequentially, adjusting their phase for optimum power combining. The resultant sequential increase in total EIRP is recorded. Figure 10.42a presents the results of this experiment. In the plot, the results are normalized with respect to the average EIRP of each IC (each subset of 16 elements). As can be observed, in the four cases a spatial power combining gain of 20–23 dB is measured, close to the expected ideal value of 24 dB with 16 identical elements. The maximum EIRP for each group of 16 elements is ~20 dBm. Next, we demonstrate the spatial power combining of all 64 elements in the package. Here, only one PLL reference input and one TX IF input are employed, and these signals are distributed through the integrated daisy-chain circuitry. To minimize temperature variation effects on the measurement, we keep all four ICs on and maintain constant supply voltage and current for this test. In this case, switching the control register to enable or disable the TX output does not change the overall power consumption (12.3 W) of the module. Furthermore, we adjust the IF VGA of every IC to equalize EIRP after power combining. The EIRP variation of the four ICs is well controlled to within 1 dB. Under these conditions, we sequentially adjust the phase difference of the 16 TX elements of each IC. Figure 10.42b shows that a total 8 dB increase of EIRP is achieved after power combining the four ICs. Here, the expected gain with ideal antennas and 100% fill factor is 12 dB.

Figure 10.43 shows the measured radiation patterns after spatial power combining of all 64 elements for both H and V polarizations. Good correlation with a simulated ideal radiation pattern is shown in [31]. In addition to the broadside radiation patterns, patterns with 15° beam steering are also demonstrated for both

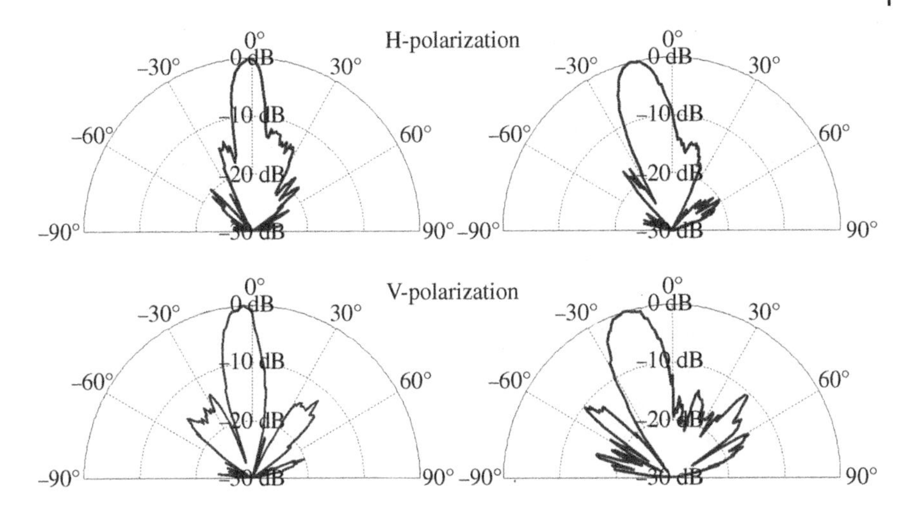


**Figure 10.43** Measured 64-element array radiation patterns and beam steering (from [31], © 2014 IEEE, reprinted with permission).

polarizations with side-lobe levels lower than 10 dB. A wider steering angle (e.g. 30°) is achievable at the expense of a higher side lobe level, which in turn can be overcome by using different tapering techniques.

## Acknowledgment

The 94-GHz phased-array work presented in this chapter was partially funded by the DARPA Strategic Technology Office (STO) under contract no. HR0011-11-C-0136 (Si-Based Phased-Array Tiles for Multifunction RF Sensors, DARPA Order no. 8320/00, Program Code 1P30). The views, opinions, and/or findings contained in this presentation are those of the authors or presenters and should not be interpreted as representing the official views or policies, either expressed or implied, of the Defense Advanced Research Projects Agency or the Department of Defense. The 28-GHz phased array presented in this chapter was co-developed by IBM Research and Ericsson. The authors thank Ericsson's S. Vecchiattini, R. Lindman, E. Pucci, M. Wahlen, A. Ladjemi, and A. Malmcrona for technical and management support. The authors would also like to thank IBM's M. Soyuer, D. Friedman, A. Valdes-Garcia, and S. K. Reynolds for mmWave project leadership and management support, and B. Sadhu and C. Baks for technical support. The authors are grateful for A. Valdes-Garcia's critical but very insightful comments and suggestions that enhanced the chapter's quality.

## References

1 S.K. Reynolds, B.A. Floyd, U.R. Pfeiffer et al., Progress toward a low-cost millimeter-wave silicon radio. In *Proceedings of the IEEE Custom Integrated Circuits Conference*, pp. 563–570, San Jose, CA, USA, September 2005.
2 P. Smulders, Exploiting the 60 GHz band for local wireless multimedia access: prospects and future directions. *IEEE Communications Magazine*, vol. 40, no. 1, pp. 140–147, January 2002.
3 W.J. Fleming, Overview of automotive sensors. *IEEE Sensors Journal*, vol. 1, no. 4, pp. 296–308, December 2001.
4 T. Hirose, M. Sato, T. Oki, et al., Development of a 94-GHz passive millimeter-wave imaging sensor. *IEIC Technical Report*, vol. 106, no. 403, pp. 35–40, 2006.
5 D. Notel, J. Huck, H. Essen et al., A demonstrator approach for the technology of a passive millimeter wave imaging system and the related image processing. *Infrared and Millimeter Waves and 13th International Conference on Terahertz Electronics*, vol. 1, pp. 319–320, September 19–23, 2005.
6 J. Thompson, X. Ge, H. Wu et al., 5G wireless communication systems: prospects and challenges. *IEEE Communications Magazine*, pp. 62–65, February 2014.
7 A. Gupta and R.K. Juha, A survey of 5G network: architecture and emerging technologies. *IEEE Access*, vol. 3, pp. 1206–1233, August 2015.
8 E. Dahlman, G. Mildh, S. Parkvall et al., 5G radio access. *Ericsson Review Journal*, pp. 1–6, June 18, 2014.
9 F. Boccardi, R.W. Heath Jr., A. Lozano et al., Five disruptive technology directions for 5G. *IEEE Communications Magazine*, pp. 74–80, February 2014.
10 B. Sadhu, Y. Tousi, J. Hallin et al., A 28-GHz 32-element TRX phased-array IC with concurrent dual-polarized operation and orthogonal phase and gain control for 5G communications. *IEEE Journal of Solid-State Circuits*, vol. 52, issue 12, pp. 3373–3391, December 2017.
11 X. Gu, A. Valdes-Garcia, A. Natarajan et al., W-band scalable phased arrays for imaging and communications. *IEEE Communications Magazine*, vol. 53, issue 4, pp. 196–204, April 2015.
12 J.D. Dunworth, A. Homayoun, B-H. Ku et al., A 28GHz Bulk-CMOS dual-polarization phased-array transceiver with 24 channels for 5G user and basestation equipment. *IEEE International Solid-State Circuits Conference*, pp. 70–72, San Francisco, CA, February 11–15, 2018.
13 T. Sowlati, S. Sarkar, B.G. Perumana et al., A 60-GHz 144-element phased-array transceiver for backhaul application. *IEEE Journal of Solid-State Circuits*, vol. 53, no. 12, pp. 3640–3659, December 2018.
14 S. Shahramian, M. Holyoak, A. Singh et al., A fully integrated scalable W-band phased-array module with integrated antennas, self-alignment and self-test.

*IEEE International Solid-State Circuits Conference*, pp. 74–76, San Francisco, CA, February 11–15, 2018.

**15** K. Kibaroglu, M. Sayginer, T. Phelps et al., A 64-element 28-GHz phased-array transceiver with 52-dBm EIRP and 812-Gb/s 5G link at 300 meters without any calibration. *IEEE Transactions on Microwave Theory and Techniques*, vol. 66, no. 12, pp. 5796–5811, December 2018.

**16** U.R. Pfeiffer, J. Grzyb, D. Liu et al., A chip-scale packaging technology for 60-GHz wireless chipsets. *IEEE Transactions on Microwave Theory and Techniques*, vol. 54, no. 8, pp. 3387–3397, August 2006.

**17** K. Kihong, H. Yoon, and K.K. O, On-chip wireless interconnection with integrated antennas. *IEEE International Electron Devices Meeting*, pp. 485–488, December 2000.

**18** Y.P. Zhang, M. Sun, and L.H. Guo, On-chip antennas for 60-GHz radios in silicon technology. *IEEE Transactions on Electron Devices*, pp. 1–5, July 2005.

**19** E. Öjefors, E. Sönmez, S. Chartier et al., Monolithic integration of an antenna with a 24 GHz image-rejection receiver in SiGe HBT technology. In *Proceedings of the 35th European Microwave Conference*, Paris, France, Oct. 2005.

**20** I. Papapolymerou, R.F. Drayton, and L.P. Katehi, Micromachined patch antennas. *IEEE Transactions on Antennas and Propagation*, vol. 46, no. 2, pp. 275–283, February 1998.

**21** M. Zheng, Q. Chen, P. Hall et al., Broadband microstrip patch antenna on micromachined silicon substrates. *Electronics Letters*, vol. 34, no. 1, pp. 3–4, January 1998.

**22** D. Neculoiu, A. Muller, P. Pons et al., The design of membrane-supported millimeter-wave antennas. In *Proceedings of the International Semiconductor Conference*, CAS, vol. 1, pp. 65–68, January 2003.

**23** P. Caudrillier, A. Takacs, O. Pascal et al., Compact circularly polarized radiating element for Ka-band satellite communications. In *Proceedings of the IEEE Antennas and Propagation Society International Symposium*, vol. 1, pp. 18–21, June 2002.

**24** D. Neculoiu, P. Pons, M. Saadaoui et al., Membrane supported Yagi-Uda antennae for millimetre-wave applications. *IEEE Proceedings on Microwaves, Antennas and Propagation*, vol. 151, no. 4, pp. 311–314, August 2004.

**25** Q. Chen, V. Fusco, M. Zheng et al., Micromachined silicon antennas. In *Proceedings of the International Conference on Microwave and Millimeter Wave Technology*, pp. 289–292, Beijing, China, August 1998.

**26** Y.P. Zhang and D. Liu, Antenna-on-chip and antenna-in-package solutions to highly integrated millimeter-wave devices for wireless communications. *IEEE Transactions on Antennas and Propagation*, vol. 57, no. 10, pp. 2830–2841, October 2009.

**27** D.G. Kam, D. Liu, A. Natarajan et al., Organic packages with embedded phased-array antennas for 60-GHz wireless chipsets. *IEEE Transactions on Components, Packaging and Manufacturing Technology*, vol. 1, no. 11, pp. 1806–1814, November 2011.

**28** A.L. Amadjikpé, D. Choudhury, C.E. Patterson et al., Integrated 60-GHz antenna on multilayer organic package with broadside and end-fire radiation. *IEEE Transactions on Microwave Theory and Techniques*, vol. 61. no. 1, pp. 303–315, January 2013.

**29** R. Pilard, D. Titz, F. Gianesello et al., HDI organic technology integrating built-in antennas dedicated to 60 GHz SiP solution. *Proceedings of the IEEE Antennas and Propagation Society International Symposium*, July 2012.

**30** D. Liu, X. Gu, C.W. Baks et al., Antenna-in-package design considerations for Ka-band 5G communication applications. *IEEE Transactions on Antennas and Propagation*, vol. 65, no. 12, pp. 6372–6379, December 2017.

**31** X. Gu, D. Liu, C. Baks et al., A compact 4-chip package with 64 embedded dual-polarization antennas for W-band phased-array transceivers. *Proceedings of the IEEE Electronic Components & Technology Conference*, Orlando, Florida, May 27–30, 2014.

**32** D. Liu, MdR. Islam, C. Baks et al., A dual polarized stacked patch antenna for 94 GHz RFIC package applications. *Proceedings of the IEEE Antennas and Propagation Society International Symposium*, pp. 1829–1830, July 2014.

**33** X. Gu, D. Liu, C. Baks et al., A multilayer organic package with 64 dual-polarized antennas for 28GHz 5G communication. *IEEE MTT-S International Microwave Symposium*, pp. 1899–1901, Honololu, Hawai, June 4–9, 2017.

**34** T. Kamgaing, A.A. Elsherbini, T.W. Frank et al., Investigation of a photodefinable glass substrate for millimeter-wave radios on package. *Proceedings of the IEEE International Electronic Components and Technology Conference*, pp. 1610–1615, May 2014.

**35** D.G. Kam, D. Liu, A. Natarajan et al., LTCC packages with embedded phased-array antennas for 60 GHz communications. *IEEE Microwave Wireless Components Letters*, vol. 21, no. 3, pp. 142–144, March 2011.

**36** F.F. Manzillo, M. Ettorre, M.S. Lahti et al., A multilayer LTCC solution for integrating 5G access point antenna module. *IEEE Transactions on Microwave Theory and Techniques*, vol. 64, no. 7, pp. 2272–2283, July 2016.

**37** J. Lantéri, L. Dussopt, R. Pilard et al., 60 GHz antennas in HTCC and glass technology. In *Proceedings of the 4th European Conference on Antennas and Propagation*, April 2010.

**38** O.E. Bouayadi, L. Dussopt, Y. Lamy et al., Silicon interposer: a versatile platform towards full-3D integration of wireless systems at millimeter-wave frequencies. *Proceedings of the IEEE Electronic Components and Technology Conference*, pp. 973–980, May 2015.

**39** N. Hoivik, D. Liu, C.V. Jahnes et al., High-efficiency 60 GHz antenna fabricated using low-cost silicon micromaching techniques. In *Proceedings of the IEEE Antennas and Propagation Society International Symposium*, pp. 5043–5046, Honolulu, Hawaii, June 10–15, 2007.

**40** M. Wojnowski and K. Pressel, Embedded wafer level ball grid array (eWLB) technology for high-frequency system-in-package applications. *Proceedings of the International Microwave Symposium*, June 2013.

**41** Y. Tsukada, S. Tsuchida, and Y. Mashimoto, Surface laminar circuit packaging. *Proceedings of the IEEE Electronic Components and Technology Conference*, pp. 22–27, May 18–20, 1992.

**42** W. Greig, *Integrated Circuit Packaging, Assembly and Interconnections*. Springer Science & Business Media, 2007, pp. 253–254.

**43** J.H. Lau and S.W.R. Lee, *Microvias for Low Cost, High Density Interconnects*. McGraw-Hill, New York, 2001.

**44** J.H. Lau, Recent advances and new trends in flip chip technology. *Journal of Electronic Packaging*, vol. 138, pp. 1–23, September 2016.

**45** W. Roh, J. Seol, J. Park et al., Millimeter-wave beamforming as an enabling technology for 5G cellular communications: theoretical feasibility and prototype results. *IEEE Communications Magazine*, pp. 106–113, February 2014.

**46** Ericsson. (2016, Apr.). 5G Radio Access - Capabilities and Technologies. [Online]. Available: https://www.ericsson.com/assets/local/publications/white-papers/wp-5g.pdf.

**47** M.K. Samimi and T.S. Rappaport, 3-D millimeter-wave statistical channel model for 5G wireless system design. *IEEE Transactions on Microwave Theory and Techniques*, vol. 64, no. 7, pp. 2207–2225, July 2016.

**48** T.S. Rappaport, F. Gutierrez Jr., E. Ben-Dor et al., Broadband millimeter-wave propagation measurements and models using adaptive-beam antennas for outdoor urban cellular communications. *IEEE Transactions on Antennas and Propagation*, vol. 61, no. 4, pp. 1850–1859, April 2013.

**49** N. Kathuria and S. Vashisht, Dual-band printed slot antenna for the 5G wireless communication network. *International Conference on Wireless Communications, Signal Processing and Networks*, pp. 1815–1817, Chennai, India, 2016.

**50** P.A. Serra, P. Nepa, G. Manara et al., A wide-band dual-polarized stacked patch antenna. *IEEE Antennas and Wireless Propagation Letters*, vol. 6, pp. 141–143, 2007.

**51** V.R. Komanduri, D.R. Jackson, J.T. Williams et al., A general method for designing reduced surface wave microstrip antennas. *IEEE Transactions on Antennas and Propagation*, vol. 61, no. 6, pp. 2887–2894, June 2013.

**52** D. Liu and S. Reynolds, Package size effects on the gains of patch antennas and finite patch arrays. In *Proceedings of the IEEE Antennas and Propagation Society International Symposium*, pp. 103–104, Orland, Florida, July 7–13, 2013.

**53** G.P. Gauthier, A. Courtay, and G.H. Rebeiz, Microstrip antennas on synthesized low dielectric-constant substrate. *IEEE Transactions on Antennas and Propagation*, vol. 45, no. 8, pp. 1310–1314, August 1997.

**54** D.F. Sievenpiper, L. Zhang, R.F.J. Broas et al., High-impedance electromagnetic surfaces with a forbidden frequency band. *IEEE Transactions on Microwave Theory and Techniques*, vol. 47, no. 11, pp. 2059–2074, November 1999.

**55** N.G. Alexopoulos and D.R. Jackson, Fundamental superstrate (cover) effects on printed circuit antennas. *IEEE Transactions on Antennas and Propagation*, vol. 32, no. 8, pp. 807–816, August 1984.

**56** D.G. Kam, D. Liu, A. Natarajan et al., Organic packages with embedded phased-array antennas for 60-GHz wireless chipsets. *IEEE Transactions on Components, Packaging and Manufacturing Technology*, vol. 1, no. 11, pp. 1806–814, November 2011.

**57** S.D. Targonski and D.M. Pozar, Design of wide-band circularly polarized aperture coupled microstrip antennas. *IEEE Transactions on Antennas and Propagation*, vol. 41, pp. 214–220, February 1993.

**58** S. Targonski, R. Waterhouse, and D. Pozar, Design of wide-band aperture-stacked patch microstrip antennas. *IEEE Transactions on Antennas and Propagation*, vol. 46, pp. 1245–1251, September 1998.

**59** D. Liu, J.A.G. Akkermans, H. Chen et al., Packages with integrated 60-GHz aperture-coupled patch antennas. *IEEE Transactions on Antennas and Propagation*, vol. 59, no. 10, pp. 3607–3616, October 2011.

**60** D. Pozar. (1996, May). A Review of aperture coupled microstrip antennas: history, operation, development, and applications. [Online]. Available: http://www.ecs.umass.edu/ece/pozar/aperture.pdf.

**61** F. Croq, A. Papiemik, and P. Brachat, Wideband aperture coupled microstrip subarray. In *IEEE Antennas and Propagation Symposium Digest*, pp. 1128–1131, May 1990.

**62** D. Liu, X. Gu, C.W. Baks et al., An aperture-coupled dual-polarized stacked patch antenna for multi-layer organic package integration. *Proceedings of the IEEE Antennas and Propagation Society International Symposium*, July 2019.

**63** T. Zwick, C. Baks, U.R. Pfeiffer et al., Probe based MMW antenna measurement setup. *Proceedings of the IEEE Antennas and Propagation Society International Symposium*, vol. 3, pp. 747–750, June 2004.

**64** A. Valdes-Garcia, B. Sadhu, X. Gu et al., Scaling millimeter-wave phased arrays: challenges and solutions. *2018 IEEE BiCMOS and Compound Semiconductor Integrated Circuits and Technology Symposium*, San Diego, CA, USA, October 14–17, 2018.

**65** S. Jeon, Y. Wang, H. Wang et al., A scalable 6-to-18 GHz concurrent dual-band quad-beam phased-array receiver in CMOS. *IEEE Journal of Solid-State Circuits*, vol. 43, no. 12, pp. 2660–2773, December 2008.

**66** Z. Pi and F. Khan, An introduction to millimeter-wave mobile broadband systems. *IEEE Communications Magazine*, pp. 101–107, June 2011.

**67** A. Valdes-Garcia, S. Reynolds, A. Natarajan et al., Single-element and phased-array transceiver chipsets for 60-GHz Gb/s communications. *IEEE Communications Magazine*, pp. 120–131, April 2011.

**68** A. Valdes-Garcia, A. Natarajan, D. Liu et al., A fully-integrated dual-polarization 16-element W-band phased-array transceiver in SiGe BiCMOS. *Proceedings of the IEEE Radio Frequency Integrated Circuits Symposium*, pp. 375–378, June 2013.

**69** A.M. Puzella, J.M. Crowder, P.S. Dupuis et al., Tile sub-array and related circuits and techniques. US patent 7348932B1, March 25, 2008. Available: https://patentimages.storage.googleapis.com/33/53/25/7e3aee0e883d0a/US7348932.pdf.

**70** Z. Yang, K. Browning, and K.F. Warnick, High efficiency planar arrays and array feeds for satellite communications. In *Proceedings of the Utah NASA Space Grant Consortium Fellowship Symposium*, pp. 1–6, May 2015.

**71** X. Gu, D.G. Kam, D. Liu et al., Enhanced multilayer organic packages with embedded phased-array antennas for 60-GHz wireless communications. *63rd IEEE Electronic Components and Technology Conference (ECTC)*, pp. 1650–1655, May 2013.

# 11

# 3D AiP for Power Transfer, Sensor Nodes, and IoT Applications

*Amin Enayati*[1], *Karim Mohammadpour-Aghdam*[2], *and Farbod Molaee-Ghaleh*[2]

[1] *Emerson & Cuming Anechoic Chambers, Antwerp Area, Belgium*
[2] *School of Electrical and Computer Engineering, University of Tehran, Iran*

## 11.1 Introduction

Antennas possess some basic characteristics that make it difficult to integrate them together with other parts of a communication system. The most important characteristic of an antenna, especially when integrated solutions for mass production are of interest, is the physical limit imposed on the antenna dimensions. Because of the wavelength-dependent nature of electromagnetic radiation, the physical dimension of an antenna is in inverse relation to the frequency it functions at [1]. This that means at frequency bands lower than a few gigahertz the antenna will consume a big proportion of physical available working place in a wireless device. Moreover, the use of more than one sub-communication system in a single device is becoming more popular and necessary than before. Therefore, a miniaturized antenna with the highest performance and integration capacity is in high demand. Antenna-in-package (AiP) solutions are good candidates to achieve this goal and they find use in various wireless systems such as Internet of Things (IoT) devices, radio-frequency identification (RFID) tags, sensor nodes, radio telemetries, and wireless power transfer devices. In this way, the antenna is no longer a separate component within the wireless device but is integrated in the package. This new insight into the concept of electrically small antenna design is taking wireless system miniaturization and integration to a new level. Note that antenna miniaturization results in bandwidth and efficiency loss [1–6]. At first glance, there is a clear paradox between more miniaturized antenna and large bandwidth/high efficiency, which are both also required in wireless device antennas.

*Antenna-in-Package Technology and Applications,* First Edition.
Edited by Duixian Liu and Yueping Zhang.

By increasing the operating frequency, the corresponding wavelength becomes comparable to the total radio dimensions, making a completely integrated solution feasible.

## 11.2 Small Antenna Design and Miniaturization Techniques

Antenna miniaturization has been a hot topic for many years. Since the start of radio communications, having a small all-purpose antenna has always been one of the designer's challenges. Nowadays, the need for multifunctioning systems has become the motivation for miniaturizing wireless systems such as cell phones, sensor nodes, IoT devices, short- and long-range communication tools, and transmission and navigation systems. This wide range of applications urges antenna designers to rise to the challenge of realizing increasingly smaller and multifunctional antennas.

### 11.2.1 Physical Bounds on the Radiation *Q*-factor for Antenna Structures

It is known that when the physical size of an antenna is reduced, the antenna cannot radiate effectively [6]. This effectiveness is described by the radiation quality factor (*Q*-factor), which is proportional to the stored electromagnetic energy in the space around the antenna and inversely related to the radiated power. For an antenna designer, it is ideal to have a small antenna with a radiation *Q*-factor near zero that means the antenna would radiate most of the received energy and store the minimum energy in the space. However, there is a lower bound on the radiation *Q*-factor of a single-mode antenna which is proportional to the antenna dimensions and operating frequency. Much of the research addressing this issue has been conducted in the literature and important studies will be reviewed in the following sections.

#### 11.2.1.1 Lower Bounds on Antenna Enclosed in a Sphere: Chu, McLean, and Thal Limits

After Wheeler published his research on fundamental limits for small antennas in 1947 [6], Chu used the same concept but with more precision [1]. He calculated the radiation *Q*-factor and its minimum limit for a single-mode antenna enclosed in a sphere with radius $a$ using full-wave spherical vector functions and expanding the normalized impedance of spherical modes as [1]. In 1994, McLean calculated the limit using spherical fields and Poynting vectors [7]. McLean not only introduced

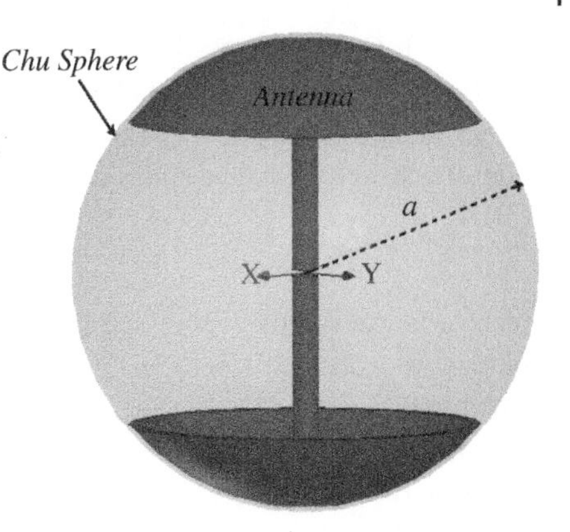


**Figure 11.1** Chu sphere of radius $a$. The Chu sphere is the minimum circumscribing sphere enclosing the antenna with a maximum dimension $2a$ (from [10], reprinted with permission).

a simple way to recalculate what Chu had already found, but also made Chu's limit more precise. He considered the minimum limit to be the limit of the dominant mode and then, by using a simple analysis, calculated the minimum limit for the *Chu antenna* to be:

$$Q_{lb}^{sphere} = \frac{1}{(ka)^3} + \frac{1}{ka} \tag{11.1}$$

where $k$ is the wave number. He presumed that this Chu antenna has no loss and no reactive energy stored inside the Chu sphere.

Thal [8] tried to take into account the ignored energy inside the Chu sphere and introduced a new limit which was almost 50% higher than the lower bound. He presumed that the electrical current is only present at the surface of the Chu sphere. By calculating and expanding the normalized impedance of the outwards and inwards traveling wave, Thal achieve new expressions for the minimum $Q$-factor. Hansen and Collin verified the accuracy of $Q$ predicted by Thal [9].

#### 11.2.1.2 Lower Bounds on Antenna Enclosed in an Arbitrary Structure: Gustafsson–Yaghjian Limit

In real life, for many antennas like dipoles and planar antennas, it is better to enclose the antenna in a cylinder or a cube, as enclosing it in a Chu sphere will cause a huge difference between the lower bound limit and the calculated $Q$. It is therefore useful to consider the enclosing structure of an antenna in the calculation of the lower bound limit. In particular, the planar shape is of great importance since this type of antenna is now commonly used in many portable devices, such as the IoT and sensors.

Gustafsson and his fellow researchers extracted the radiation $Q$ of small antennas using the scattering properties of small particles involving polarization dyads and calculated the minimum limit for an arbitrary antenna enclosed by a prism [11]. Considering an omni-directional radiation pattern with maximum directivity of 1.5 and the smallest magnetic scattering that are acceptable for many small antennas, the new lower bound limit that is applicable to an arbitrary enclosing structure can be simplified to:

$$Q_{lb}^{st} = \frac{1.5}{(ka)^3}(\gamma_1^n + \gamma_2^n + \gamma_3^n)^{-1} \tag{11.2}$$

where $a$ is the radius of the Chu sphere and $\gamma_1^n$, $\gamma_2^n$, and $\gamma_3^n$ are the normalized polarizability eigenvalues to $\pi a^3$ on the physical topology of the antenna in three principal directions [12]. Unfortunately, there are no closed-form expressions for these and they can be calculated from simulation for any 3D structure. This value is unit for a sphere and 0.42 for a disk with radius $a$. For planar rectangular (printed circuit board (PCB)-type) small antennas in which the substrate thickness is very small in comparison with the antenna size, the third eigenvalue is neglected and the others have been computed by numerical simulations of a perfect electric conductor (PEC) sheet for various length to width ratios $\xi = l/w$. Figure 11.2 shows this lower bound, normalized to McLean limit, in three different cases: single pure linear polarization along the length and width of the rectangle and dual polarization. According to this figure, it is interesting to note that the minimum $Q$ for single pure polarization, e.g. along the length, is not for a square with $\xi = 1$ but for a rectangle with a length to width ratio of $\xi = 1.84$, and the least value of $Q$ is for dual polarization case. An antenna designer can use these ratios to obtain a planar antenna with the minimum possible $Q$-factor.

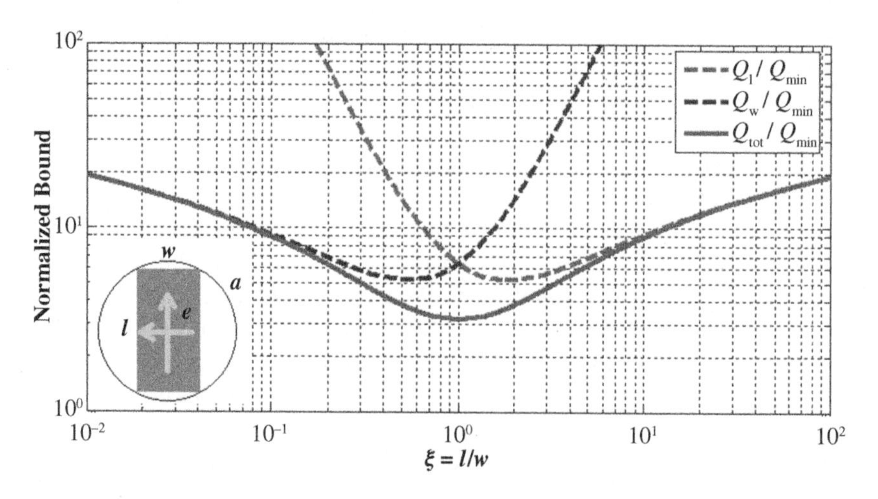


**Figure 11.2** Normalized (to McLean) boundary of a rectangular PCB-type antenna with dual-polarization capability (from [10], reprinted with permission).

In 2010, Yaghjian modified the Gustafson limit and calculated the lower bound limit of $Q$ for single polarized antenna with an arbitrary enclosing structure as [13]:

$$Q_{lb}^{st} = \frac{1.5}{(ka)^3} \frac{1}{\gamma_j^n} \left(1 - \frac{V/V_s}{\gamma_j^n}\right) \tag{11.3}$$

where $V$ and $V_s$ are the volumes of the antenna and the Chu sphere, respectively, and $\gamma_j^n$ is the polarizability eigenvalue in one of the principal directions, normalized to $\pi a^3$, as in Gustafsson's equations. Correction of Yaghjian returns to the effect of the secondary magnetic source, which is caused by primary electric current sources (flowing in small loops and thus creating magnetic dipoles). If the sources are only global electric currents, the second term vanishes. As a conclusion, *if antenna designers can employ both electric sources and secondary magnetic currents in an effective way, they can reduce the stored energy inside the sphere and therefore reduce the radiation Q.*

## 11.2.2 Figure of Merit for Antenna Miniaturization

### 11.2.2.1 Relation between Q-factor and Antenna Input Impedance

Wheeler [6] states that the $Q$ of an antenna is inversely related to its bandwidth, ($Q \approx 1/B$, where $B$ is fractional bandwidth). He puts the condition that $Q \gg 1$, without any precise proof. Rhodes has also proved same relation for a planar dipole by calculating $Q$ and $B$ separately [14]. He also predicted such a relation to be valid for other antennas as well. In 2000, Geyi used the complex Poynting vector in the frequency domain to describe the relation between $Q$ and $B$ more rigorously [15]. Geyi successfully found a relation between the antenna's $Q$-factor and its input impedance, employing Foster's reactance theorem. Yaghjian and Best proposed an accurate equation for $Q$ in terms of input impedance for arbitrary antennas [16]. They also expressed $Q$ in terms of the VSWR bandwidth [17]. They finally evaluated their theoretical results with measurements for a number of antennas [18]:

$$Q_z \approx \frac{\omega_0}{2R_{in}(\omega_0)} |Z_{in}'(\omega_0)| \tag{11.4}$$

The radiation $Q$ in terms of impedance bandwidth can then simply be written as:

$$Q_z = \frac{\frac{s-1}{\sqrt{s}}}{B} \tag{11.5}$$

which is valid for $Q \gg 1$ [16]. If $s = \text{VSWR} < 2:1$ is an acceptable bandwidth criterion, we have $Q_z \approx 0.7/B$.

#### 11.2.2.2 Antenna Efficiency Effect on the Radiation *Q*

The antenna radiation efficiency, $\eta_r$, is simply defined as the ratio of the radiated power to the power entering the antenna port. Also, the conjugate match factor (CMF) is defined as $\eta_m = CMF = 1 - |\Gamma|^2$, where $\Gamma$ is the input reflection coefficient of the antenna. Therefore, a total efficiency can be defined as $\eta_s = \eta_r\eta_m$. When considering the radiation $Q$, we assumed 100% efficiency, i.e. no loss in the radiating structure, a lossless matching network, and no reflection at the matching network port. So, in the case of losses and mismatch, we need to modify the lower bound expressions and include the antenna total efficiency to have realistic limits. So, the lower bound limit becomes:

$$Q_{lb} = \eta_s Q_{lb}^{sphere} \tag{11.6}$$

#### 11.2.2.3 Cross-polarization Effect on Antenna Radiation *Q*

In the lower bound limit as set by Chu, a pure TE or TM mode is present. However, in the case of an elliptical polarization, both TE and TM modes are present, with different powers. In practical antennas, a cross-polarization parameter can be defined which is equal to the ratio of the non-expected cross-polarization to the one expected and indicated as $\chi$. With this definition, the radiated power is obtained from the superposition of the dominant modes, like $TM_{01}$, and the one normal to it, like $TE_{01}$. As Thal pointed out [8], the minimum $Q$ is calculated as the summation of these two $Q$s as:

$$\frac{1}{Q_{\min}} \approx \frac{1}{Q_{TE}} + \frac{1}{Q_{TM}} = (1+\chi)\frac{1}{Q_{lb}} \tag{11.7}$$

#### 11.2.2.4 Figure of Merit Definition

The relative distance between the exact radiation quality for a real antenna and the enclosing structure's lower bound limit can be considered a structure penalty that is directly related to the ability of the antenna designer to minimize the $Q$ for a specific circumscribing structure, for example a planar structure. Therefore, we define the figure of merit (FoM) of an antenna as its design penalty [10]:

$$FoM = \frac{Q_{ant}}{Q_{lb}^{st}} \tag{11.8}$$

The closer this figure is to 1, the more the antenna is miniaturized. In fact, with this definition, we can de-embed the constraint coming from the circumscribing shape. Indeed, it is possible to compare an antenna in a planar circumscribing shape with another antenna in a 3D spherical circumscribing shape.

### 11.2.3 Antenna Miniaturization Techniques

This section deals with miniaturization, which is a crucial challenge that has been addressed by many researchers. Miniaturization is the main drive in the attempts to find the lower value for radiation $Q$. In practice, miniaturization techniques are used for antennas with pre-determined basic structure and they can be categorized as miniaturization through geometrical shaping of the antenna and miniaturization through material loading. Various methods related to these categories are discussed in the following sections. These methods are applicable for a majority of 2D and 3D antennas.

#### 11.2.3.1 Miniaturization through Geometrical Shaping of the Antenna

In the first category, the miniaturization is realized through changes in antenna shape and topology. For instance, by folding a wire antenna, its volume can be effectively reduced. Wheeler has studied a number of such antennas [19]. Many others have also used this technique to miniaturize antennas [20, 21]. In [22], a number of small resonant antennas optimized using a genetic algorithm are presented. In [23], fractal structures have been implemented for size reduction. The different techniques related to this category are slot loading, pin or plate loading, inductive loading, and capacitive loading.

##### *11.2.3.1.1 Slot Loading*

In this technique, a number of slots is implemented in the conducting surface of the antenna. By doing this, a controlled re-routing of the current is realized and a longer electrical length for the main current can be created, so the resonant frequency occurs at lower values. This method is applicable for both 2D and 3D antennas and depending on the length and shape of the slot(s), a size reduction of 10–40% is achievable. Figure 11.3 shows a rectangular patch with two latitude slots at the two ends of the patch which is loaded by five slots.

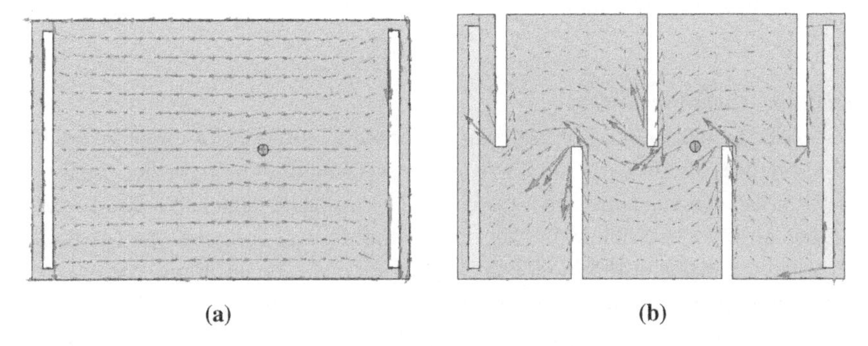


**Figure 11.3** Slotting in a rectangular patch antenna to shape the current flow and reduce the resonance frequency. (a) A rectangular patch antenna with two slots for dual-band operation. (b) A rectangular patch antenna with five slots for electrical length reduction of 37% for the first and 32% for the second operation band (from [10], reprinted with permission).

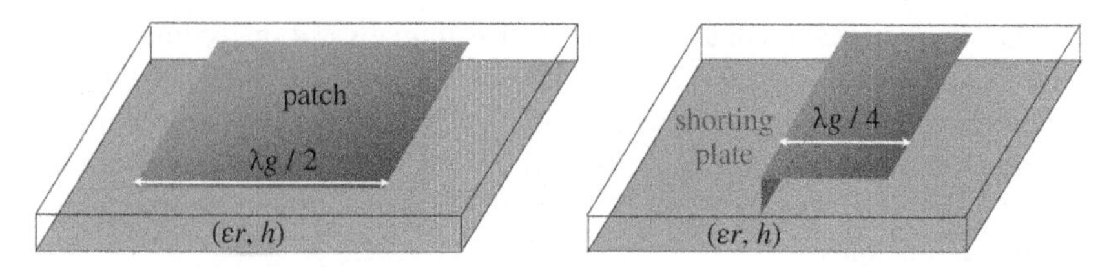


**Figure 11.4** Shorting plate used to half the patch size (from [10], reprinted with permission).

#### *11.2.3.1.2 Pin or Plate Loading*

Maybe the simplest miniaturization method is the use of a ground plate. A monopole antenna is the compacted equivalent of a dipole. The use of the ground plate reduces the resonant length of the antenna from $\lambda/2$ to $\lambda/4$. This "ground plate" imaging can also be obtained by using a smaller shorting plate or pin that connects a part of the radiating element to the ground. Inverted L antenna (ILA), loop feed array (LFA), and planar inverted F antenna (PIFA) topologies are based on this technique.

#### *11.2.3.1.3 Inductive Loading*

Inductive loading is the process of inserting inductors in the antenna structure. Obviously, this method is useful only if the antenna input impedance is capacitive or is supposed to be connected to a load that has a capacitive input impedance. Examples are the ordinary mode helix antenna, which is a "helical" monopole antenna, and the meander line antenna, which is realized by zigzagging the current flow of a planar dipole or monopole. Space-filling antennas use a specific algorithm to fill 3D or 2D spaces. Best has studied a number of such antennas, including the inductance loading effects [24]. His summary is shown in Figure 11.5.

#### *11.2.3.1.4 Capacitive Loading*

This method implements capacitors in the antenna structure in order to reach self-resonance for otherwise inductive antennas, or in order to make the antenna input impedance capacitive so that it can be conjugate matched with a load that has an inductive behavior. The most common way is to use one (or two) flat or curved plates at the monopole (or dipole) ends. For example, consider a simple monopole antenna with known dimension. If this monopole is loaded by a disk with radius $b$, in such a way that the enclosing Chu sphere has a constant radius, the resonance frequency will be reduced without increasing the

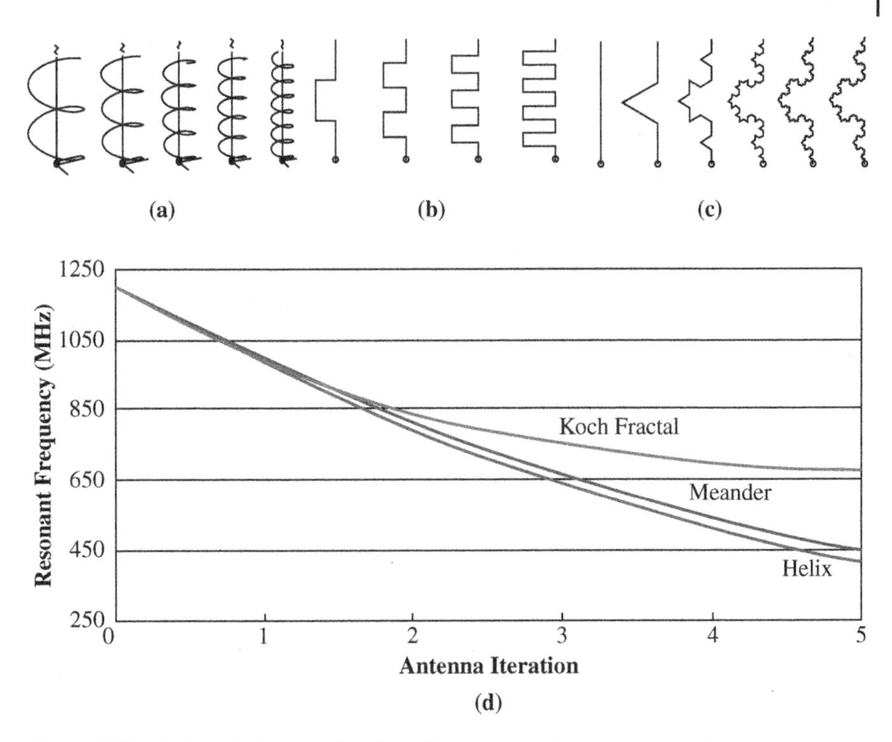


**Figure 11.5** Various inductance loadings for a monopole antenna to reduce resonance frequency: (a) ordinary mode helix antenna, (b) meanderline antenna, (c) Koch fractal antenna, and (d) comparison (from [24], © 2003 IEEE, reprinted with permission).

antenna dimensions. Figure 11.6 shows the resonance frequency and FoM of this antenna in terms of disk radius.

#### 11.2.3.2 Miniaturization through Material Loading

Antenna miniaturization techniques through material loading can be categorized as loading with dielectric or ferromagnetic materials, loading with lumped-element circuits, and loading by meanderlines.

##### *11.2.3.2.1 Loading through Dielectric or Ferromagnetic Materials*

An easy method of miniaturization is to use dielectrics with high permittivity or ferromagnetic materials with high permeability [25]. This reduces the effective wavelength and thus makes the antenna smaller by allowing resonance at smaller dimensions. This technique is well known and commonly used in patch

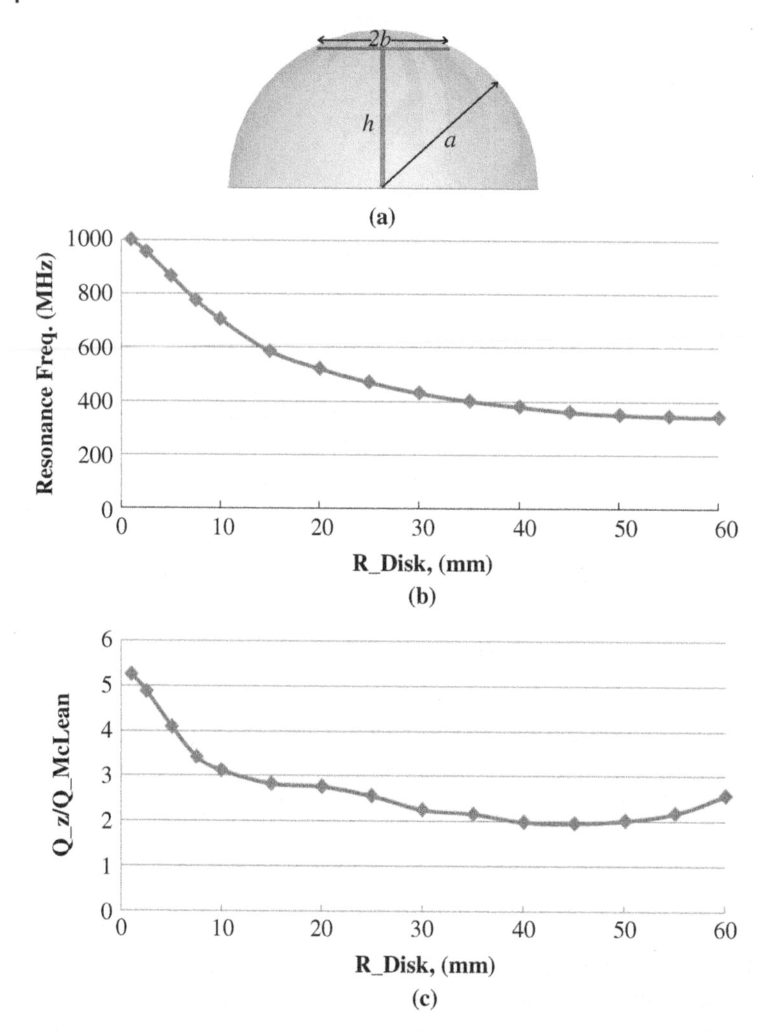


**Figure 11.6** (a) Loaded monopole antenna with a fixed circumscribing sphere. (b) Resonance frequency versus disk radius. (c) FoM = $Q_z/Q_{McLean}$ at resonance frequency versus disk radius (from [10], reprinted with permission).

antennas, since they are already built on substrates. In [26] and [27], a mixture of dielectric and magnetic materials is suggested for antenna miniaturization without efficiency degradation.

#### *11.2.3.2.2 Loading through Metamaterials*

A recent thriving method is the use of artificial materials to miniaturize the antennas. A metamaterial substrate is made of small electromagnetic circuits placed close to each other in such a way that a different equivalent permittivity and permeability are obtained compared to the primary host substrate material. In [28] a reactive impedance was applied for a specific bandwidth in order to design a miniaturized antenna. In [29] a metamaterial substrate which acts as a perfect magnetic conductor (PMC) was used in an integrated miniaturized antenna design. Stuart showed that if the area around the antenna is filled by a metamaterial with a negative permittivity, an excellent size reduction is achievable [30]. However, he reminds us that minimum limits cannot be exceeded and the FoM of the proposed antenna is 1.47.

#### 11.2.3.2.3 Loading through Lumped-Element RLC Circuits

In this technique, RLC circuits are embedded within the antenna in order to manipulate the current flow of the antenna in a desired way. Many such structures are introduced in [31–33].

#### 11.2.3.2.4 Miniaturization by Slow-wave Meanderline Loading

The implementation of delay lines for time delay and phase tuning purposes is a known method in tunable antennas and a meanderline is a block to realize such delays. As shown in Figure 11.7a, a meanderline is composed of a conductive plate and a number of parallel transmission lines (TLs) placed above it. In such structure, there are two types of TL. The first type is closer to the metal plate, which gives a small characteristic impedance, and the second type is parallel to the first one, but at a higher position with respect to the metal plate, so has a higher characteristic impedance. These two types of TL are connected to each other by short vertical conductors. The periodic change in low and high impedance creates a slow-wave TL and it takes much longer for a wave to propagate through such a structure in comparison to a normal microstrip line with the same length [34]. The electrical length of the antenna can therefore be increased by using meanderlines.

Antenna designers may modify their design in order to load it with meanderlines. Figure 11.7b shows a meanderline loaded monopole antenna which has a pair of horizontal radiators and a single vertical one [36, 37]. Each meanderline is connected to a vertical element from one end and a horizontal one from the other end.

In [35], a novel spherical monopole antenna which has been miniaturized through a couple of meanderline sections is presented. Thanks to the antenna geometry and the meanderline loading technique, the exact $Q$-factor of the antenna approaches the Thal lower bound. This antenna has tunable capability by using switches in the meanderline circuits and controlling the electrical length of meanderlines (see Figure 11.8b,c).

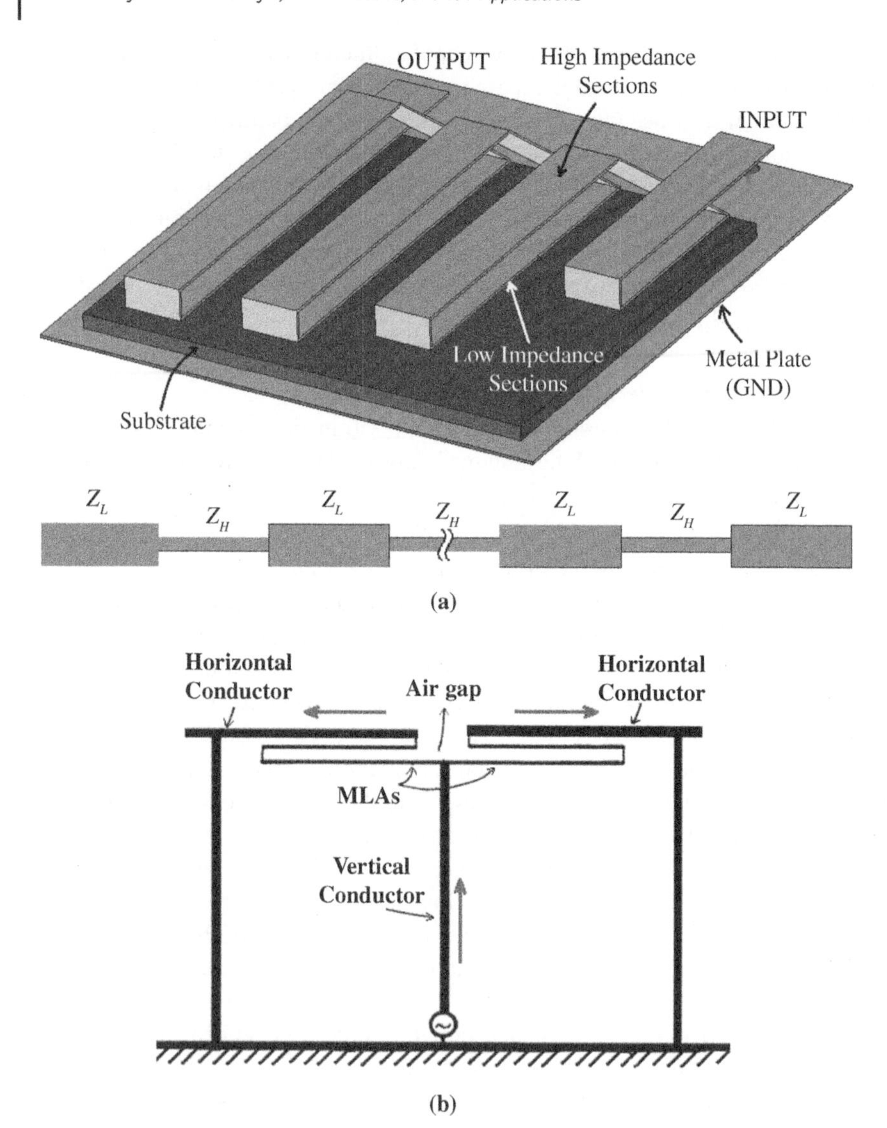


**Figure 11.7** (a) Perspective presentation of a slow-wave meanderline and its equivalent circuit (from [35], © 2011 IEEE, reprinted with permission). (b) A monopole schematic loaded with two meanderline sections (from [10], reprinted with permission).

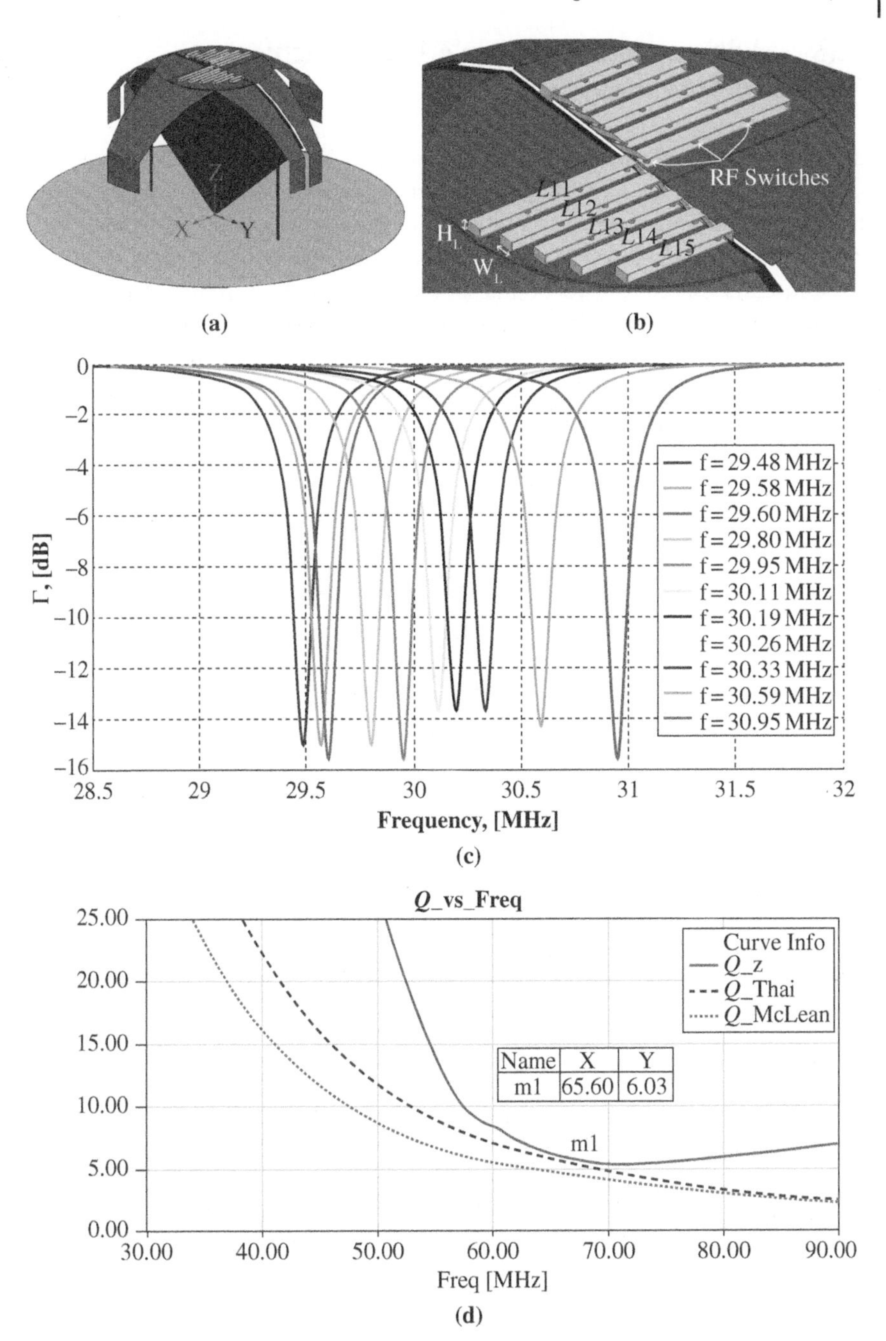


**Figure 11.8** Meanderline loaded tunable spherical antenna: (a) antenna geometry, (b) meanderline blocks with switches placement, (c) tunable behavior of the antenna for different states of switches, (d) antenna $Q$-factor in comparison with Thal's and McLean's limits versus frequency (from [35], © 2011 IEEE, reprinted with permission).

## 11.3 Multi-mode Capability: A Way to Achieve Wideband Antennas

Multi-mode antennas are interesting because they can cover wider impedance bandwidth than single-mode antennas and can show a frequency bandwidth that exceeds the corresponding bandwidth of the lower bounds. As shown, the bandwidth of a single-mode antenna is inversely related to the radiation $Q$-factor of the antenna. Since the lower bound is valid for single-mode resonant antennas, the inverse relationship between the radiation $Q$-factor and the bandwidth does not hold in the case of multi-mode antennas and their bandwidths are not restricted by the same bound. This is why multi-mode resonant antennas offer the promise of higher bandwidth. In this way the designer may control the resonant frequencies of two or more different modes with similar radiation

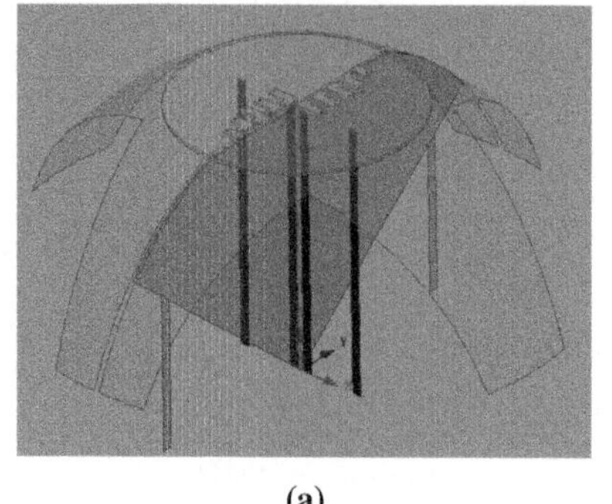

(a)

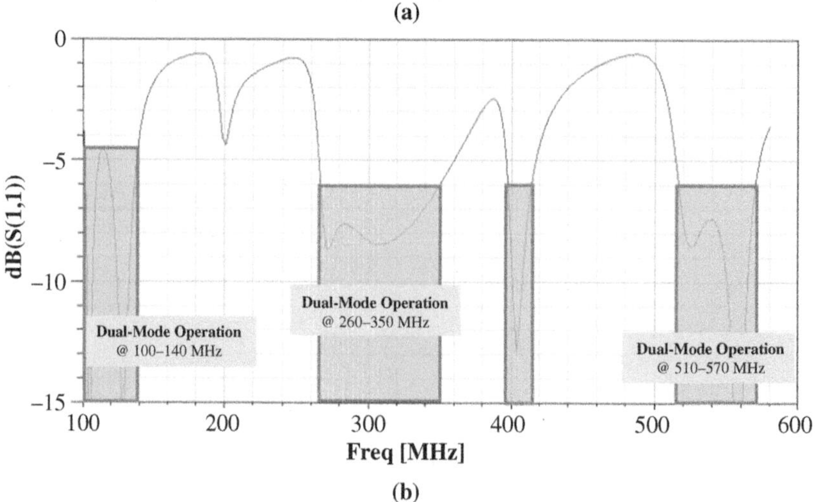


(b)

**Figure 11.9** A dual-mode miniaturized antenna: (a) antenna geometry and (b) dual-mode operation (from [40], © 2017 IEEE, reprinted with permission).

properties, such as radiation pattern and polarization, and put them close together to achieve wideband coverage. Also, if the resonant frequency of each mode can be electrically tuned, the bandwidth may increase to larger values.

Characteristic mode analysis (CMA) provides antenna designers with insights into the natural current modes of antenna structures and their radiation properties. It can therefore be a useful tool to evaluate and design wideband antennas. Three different 3D multi-mode miniaturized antennas have been designed in [38–40] using CMA.

## 11.4 Miniaturized Antenna Solutions for Power Transfer and Energy Harvesting Applications

Conventional wireless communication devices have limited energy storage capacity that their frequent charging or replacement can be costly and even impossible in some applications. A promising approach to extend the lifespan of these devices is to deliver energy for them via wireless power transfer (WPT) technologies. There are three categories for WPT technologies: inductive coupling, magnetic resonant coupling [41], and RF-based WPT. The first two categories are based on near-field electromagnetic waves, which operate in the range of several centimeters and require perfect alignment of the receiver and transmitter circuits. However, RF-based WPT simultaneously enables wireless charging and data transmission over hundreds of meters. Also, the required energy can be harvested from environmental energy sources such as wind, solar, and RF signals radiated by access points, base stations, etc. [42]. Implementation and integration of the antenna is an important issue in the cost and size reduction of WPT and energy harvesting systems.

In [43], a method to construct true 3D antennas for RF power harvesting in millimeter-scale sensors has been fully described. In this approach, the power-transfer efficiency can be increased by 22 dB over a traditional single-coil setup at the worst-case orientations for the sensor. A full system is presented in [44] to showcase the feasibility of utilizing origami-folding principles for energy harvesting applications. In this work, additive manufacturing technology was utilized to build a 3D package for RF electronics which has on-package antennas for RF signal reception.

An interesting approach to designing an energy harvesting system is to use more than one energy source. The design of a dual (solar + electromagnetic) energy harvesting system for wireless sensor applications has been discussed in [45]. The proposed harvester consists of a dual port slot antenna which operates at 2.4 GHz industrial, scientific and medical (ISM) band and a solar cell on a 3D printed package.

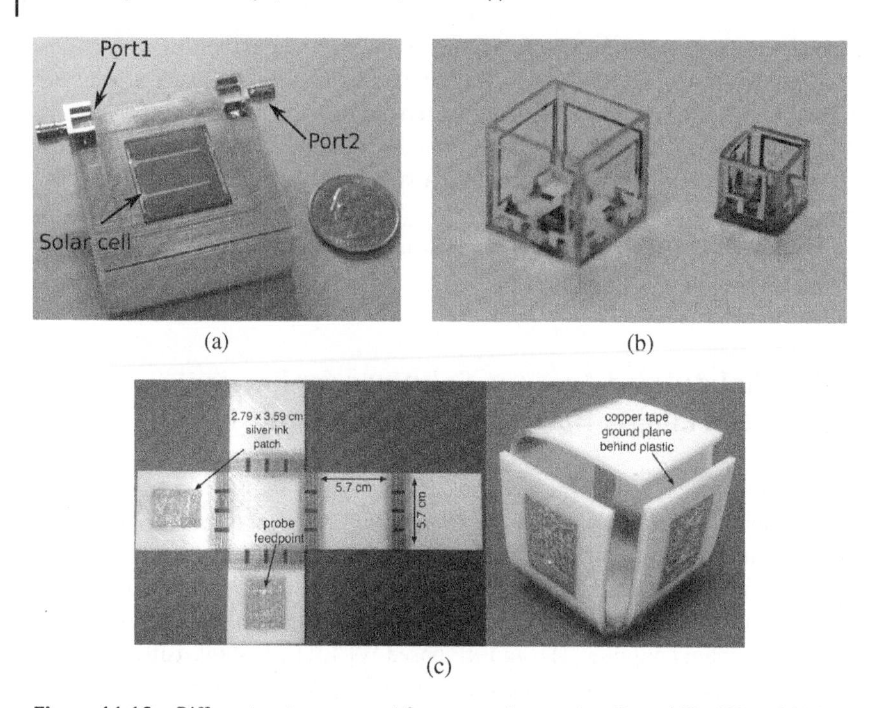


**Figure 11.10** Different antennas used for energy harvesting (from [43–45], © 2014, 2015, 2017 IEEE, reprinted with permission).

## 11.4.1 Integrated Antenna Design Challenges for WPT and Scavenging Systems

### 11.4.1.1 Conjugate Impedance Matching

The first issue in a WPT or scavenging system design is the impedance matching consideration. According to the free-space formula for a radio link [46], conjugate matching is important to deliver maximum scavenged power to the receiver circuit and to increase the maximum range:

$$S_{\min} = \frac{P_t G_t G_r \rho \tau}{(4\pi)^2 (d_{\max}/\lambda)^2} \tag{11.9}$$

where $\lambda$ is the wavelength, $G_t$ is the gain of the transmit antenna, $P_t$ is the transmitted power, $G_r$ is the gain of the receiving antenna, $S_{min}$ is the minimum threshold power necessary to provide sufficient power to the receiver, $\rho$ is the polarization efficiency, and $\tau$ is the CMF, which is given by [46]:

$$\tau = \frac{4R_c R_a}{|Z_c + Z_a|^2}, \quad 0 \le \tau \le 1 \tag{11.10}$$

where $Z_c = R_c + jX_c$ is the input impedance of the receiver and $Z_a = R_a + jX_a$ is the antenna impedance.

The designer must attend to the impedance of the circuit in order to obtain the best CMF. To obtain low-cost devices, it is preferable not to use external matching networks involving lumped components, therefore the matching mechanisms have to be embedded within the antenna layout.

#### 11.4.1.2 Antenna Structure Selection

Considering the requirement for conjugate matching, the impedance behavior of the scavenging circuit determines the overall structure of the antenna, which should be designed for any arbitrary input impedance. Also, the available area for the antenna may force some size reduction consideration, so the antenna structure should have miniaturization capability to fit in a small area. In addition, the designer must attend to other constraints such as radiation pattern requirements (e.g. omni-directional coverage) and realized gain or radiation efficiency.

### 11.4.2 Small Antenna Structure that can be Optimized for Arbitrary Input Impedance

#### 11.4.2.1 Basic Antenna Structure

Among several kind of antennas used for scavenging applications, there are two simple structures as a start point to design a miniaturized antenna that can be optimized for arbitrary input impedance. These structures have three main parts: the feeding loop, the extension parts which work as the modified dipole or loop, and the connection between these two parts. In the structure shown in Figure 11.11a, the feeding loop and the radiating parts are combined by a direct connection, while in the structure in Figure 11.11b they are connected by mutual coupling. Both of the proposed structures permit a nearly independent tuning of the resistance and reactance of the antenna by choosing the proper geometrical parameters.

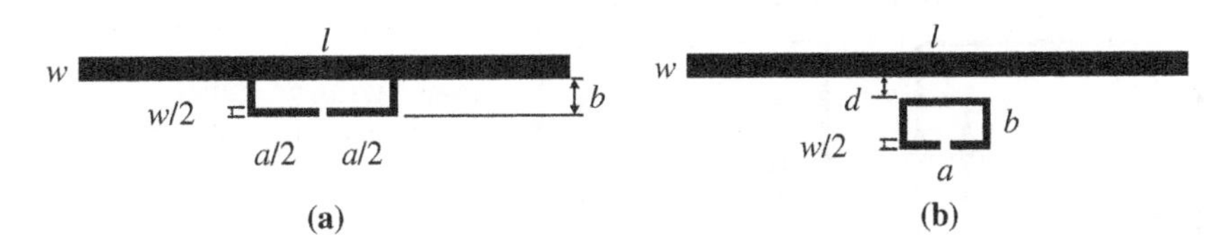


**Figure 11.11** Scavenging antenna structure. (a) The feeding loop is directly connected to the radiating parts. (b) The feeding loop is electromagnetically coupled to the radiating parts (from [47], © 2012 IEEE, reprinted with permission).

#### 11.4.2.2 Antenna Size Reduction by Folding

After the selection of the basic antenna shape, the available area forces us to use some miniaturization techniques. In general, it is well known that antenna miniaturization yields layouts with reduced bandwidth [4, 5].

A meander antenna is an extension of the basic folded antenna which includes a large number of folded elements in various linear patterns. Folding the elements in a meander form produces resonances at frequencies much lower than the resonances of a single-element antenna of equal length [48]. Figure 11.12 shows an inductive-shape meandered dipole. In this antenna in-phase coupled lines mainly control the radiation resistance and the real part of the input impedance while adjacent opposite-phase lines give storage of electric energy and loss [49]. Also, the overall conductor length affects the inductance of the input impedance.

The effect of the number of folds on antenna behavior has been studied on a planar dipole. By introducing folds along the dipole strips, the antenna is miniaturized while its length is tuned to resonate at 915 MHz. The simulation results are shown in Figure 11.13. As expected, the folded antenna has a narrower bandwidth and a lower efficiency than a straight dipole.

As a rule of thumb, *adding a fold causes a drop in gain and bandwidth, but introduces a size reduction, so the designer should minimize the number of folds in order to fit the antenna in the specified area.*

#### 11.4.2.3 Final Antenna Structure and Parameter Analysis

Figure 11.14 shows the general form of a scavenging antenna with a directly connected feeding loop and multi-turn meandered dipole arms. This topology has a large miniaturization potential and forms a better realization to obtain an arbitrary impedance than other topology which feeding loop is electromagnetically coupled to the radiating parts. The shape of the feeding loop has to be optimized in every design to match any particular impedance, while the number of turns or folds in the dipole arms is adjusted to fit the antenna in a small specified area. In fact, the required inductive reactance to overcome any occurring capacitance can be achieved by shaping the extension parts like an inductor in a rectangular area.

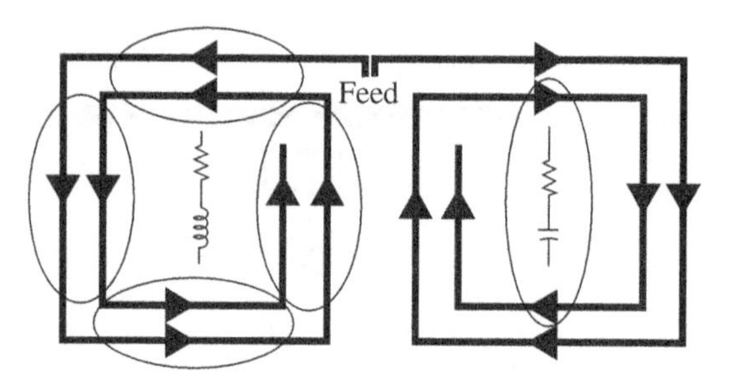


**Figure 11.12** Geometry of an inductive-shape meander antenna having multiple folds (from [47], © 2012 IEEE, reprinted with permission).

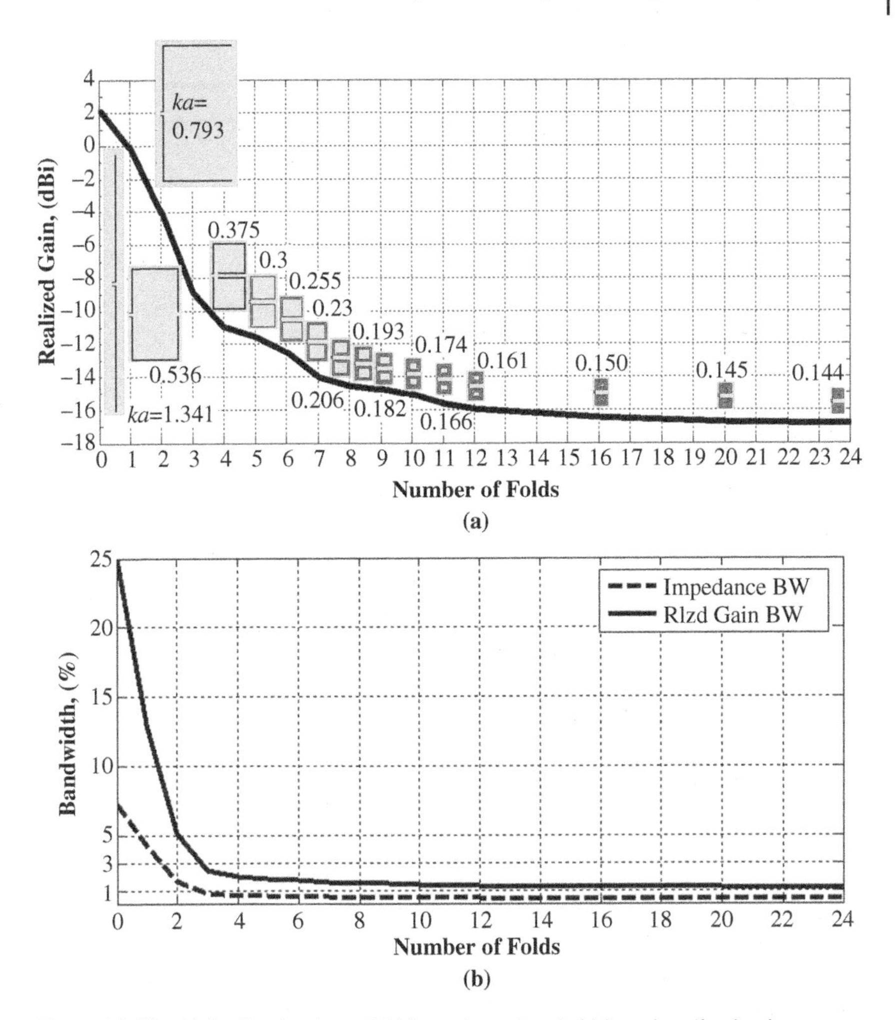


**Figure 11.13** (a) Realized gain and (b) impedance bandwidth and realized gain bandwidth versus the number of folds along the arms of the dipole (from [47], © 2012 IEEE, reprinted with permission).

In this structure, antenna input impedance is controlled by the loop length. This influence has been studied for a prototype antenna at 915 MHz with dimensions of $9 \times 18\,\text{mm}^2$. In this study, the radiating parts were fixed and the loop length was swept from 5 mm ($0.016\lambda$) to 120 mm ($0.424\lambda$). As shown in Figure 11.15a, both the real and imaginary parts of the input impedance increase with increasing loop length. This indicates that for a high value of the impedance, the antenna designer must consider a large loop length.

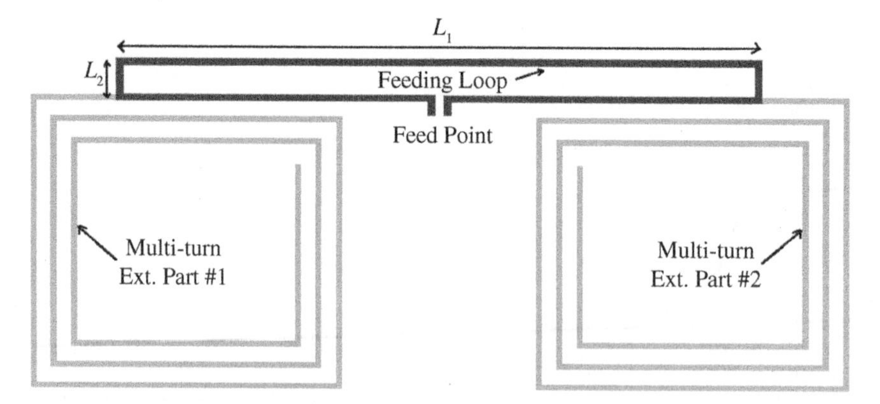


**Figure 11.14** General multi-turn loop-dipole structure (from [10], reprinted with permission).

Another parameter is the conjugate match gain (CMG) of the antenna (i.e. the realized gain of the antenna when it is conjugate matched to its input impedance). As shown in Figure 11.15b, the CMG is not too sensitive to the loop length and it increases only by 1.5 dB for the whole range of loop length increments from $0.016\lambda$ up to $0.424\lambda$. This implies that, as expected, the feeding loop mainly determines the input impedance and does not make a significant contribution to the radiation. This is an interesting result for antenna designers because they can focus on the feeding loop to reach the desired input impedance without worrying about the antenna gain.

### 11.4.3 Example of an AiP Solution for On-chip Scavenging/UWB Applications

An RFID system identifies the tag's unique ID or detailed information programmed in the tag through an RF link. RFID systems are widely used in applications such as asset monitoring, access control, supply chain monitoring, etc. [50]. The most limiting factor in cost and size reduction in current RFID systems is the implementation and integration of passive components like large antennas and decoupling capacitors, so miniaturization of the antennas is needed for this application.

The power of the active RFID tag used during sensing and transmission is provided by the electromagnetic energy of the incoming RF wave at different ISM frequencies such as 915 MHz, 2.45 GHz, or 5.8 GHz. In this example, the RFID tag uses UWB impulse radiation for the transmission to achieve a high data rate along with a low transmit power consumption.

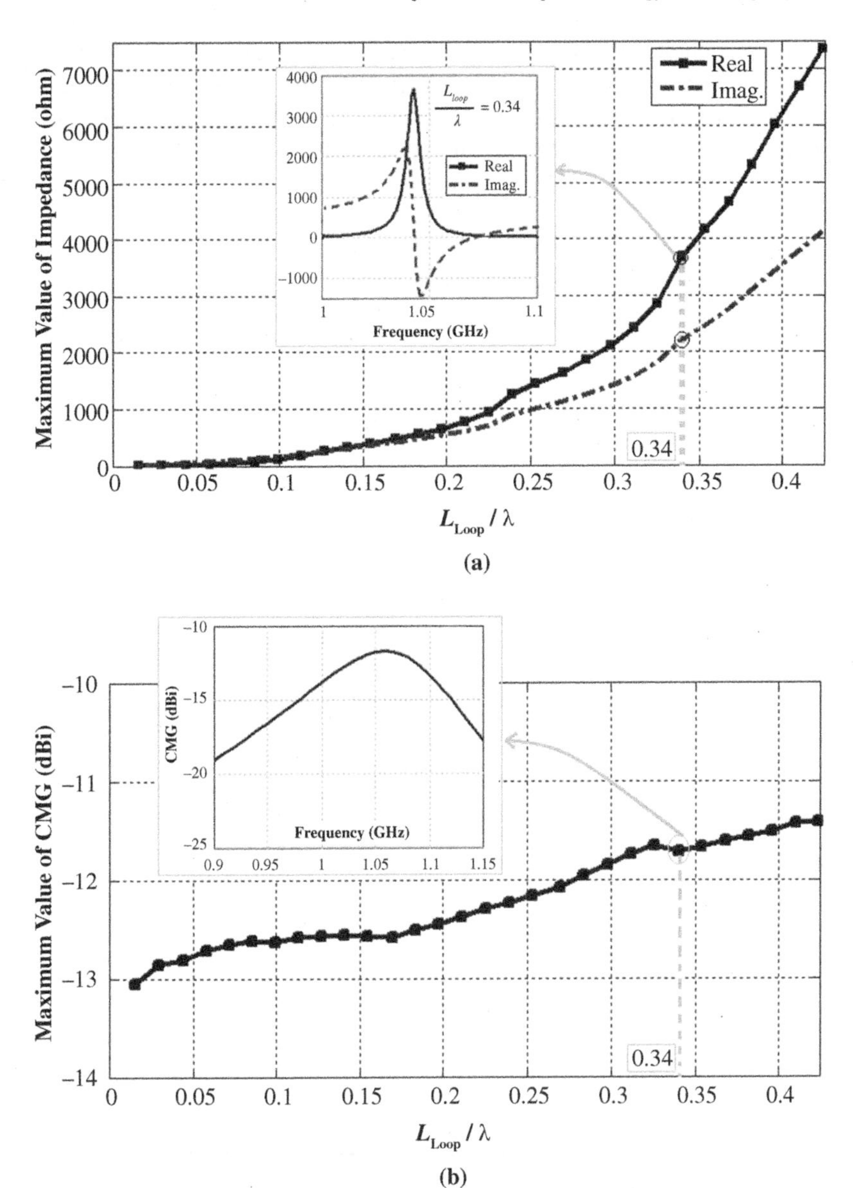


**Figure 11.15** (a) Maximum achievable value for the input impedance and (b) the CMGn versus the normalized loop length (from [47], © 2012 IEEE, reprinted with permission).

Implementation of antennas on a chip provides further advantages such as low cost, low complexity, and high reliability. Nonetheless, there are several challenges to integrate a system on a semi-conductor substrate, such as undesirable radiation into the substrate and high substrate conductivity, which causes energy loss. On-chip antenna with reasonable gain and efficiency are therefore not easy to implement [51, 52].

Figure 11.16 shows two antennas with the required characteristics given in Table 11.1 on complementary metal oxide semiconductor (CMOS) technology. With the specified available area, which is the area of two dies in CMOS

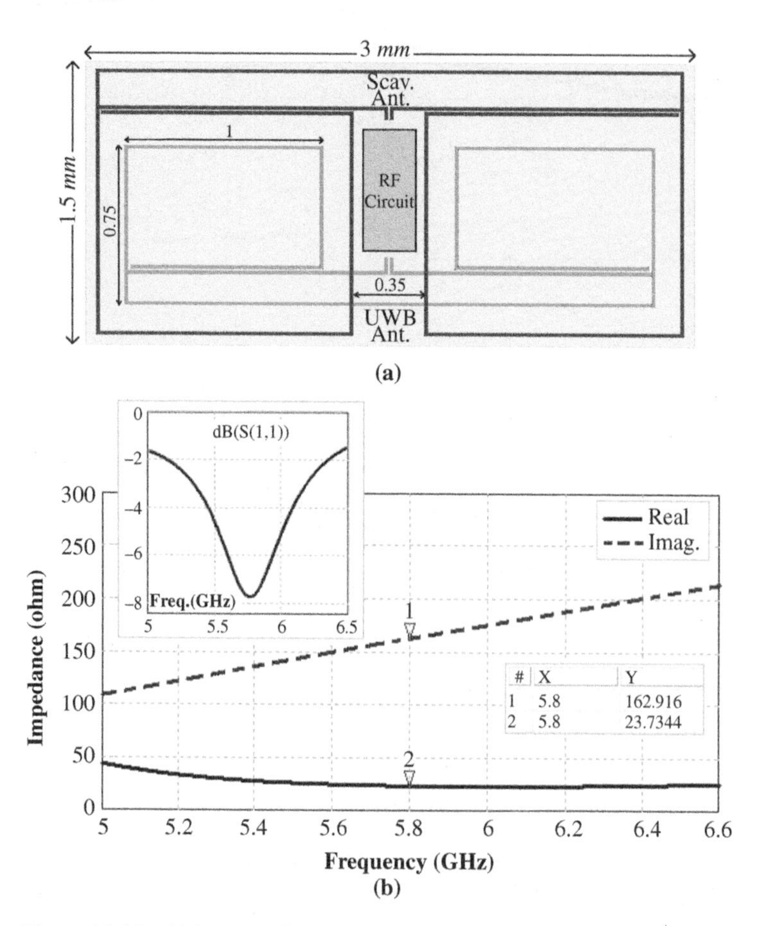


**Figure 11.16** (a) Integrated on-chip scavenging and UWB antennas. Real and imaginary parts of the input impedance versus the frequency (the insert is the reflection coefficient versus the frequency) for (b) scavenging and (c) transmitting antenna (from [10], reprinted with permission).

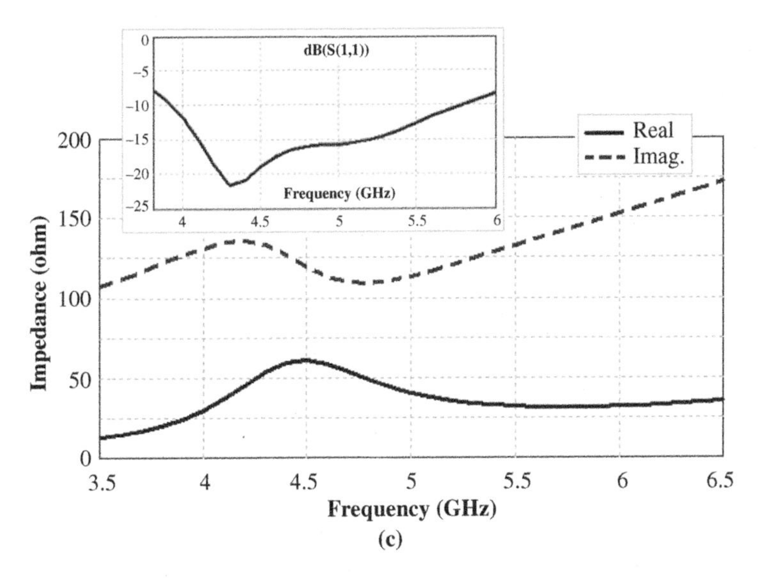


**Figure 11.16** *(Continued)*

**Table 11.1** Desired antenna features in scavenging and transmitting mode

| Specifications | Scavenging antenna | Transmitting antenna |
|---|---|---|
| Operating frequency | 5.8 GHz | Between 3 and 10 GHz |
| Bandwidth (VSWR < 2:1) | >50 MHz (1%) | >500 MHz |
| Pattern | Omni-directional | Omni-directional |
| Input impedance | 2500 Ohm ‖ 170fF, at 5.8 GHz ($10 - j160$ Ohm) | 350 Ohm ‖ 186fF, at 5.8 GHz ($52 - j125$ Ohm) |
| Input port | Balance (differential) | Balance (differential) |
| Dimension | 3 × 1.5 mm | Less than 3 × 1.5 mm |

technology, the biggest available size is around $\lambda/20$ or $ka = 0.204$, where $k$ is the wave number and $a = 1.68$ mm is the radius of the disk surrounding the available area.

Considering the lossy silicon substrate, the final structure of the antenna must meet the minimum required surface current path to have a better efficiency and less loss. In fact, the lossy substrate works as a resistive loading on the main current path and forces the current to rapidly go to zero over a small physical length. The dipole arms are therefore truncated to have a single turn with just four folds in the radiating parts with minimum physical length.

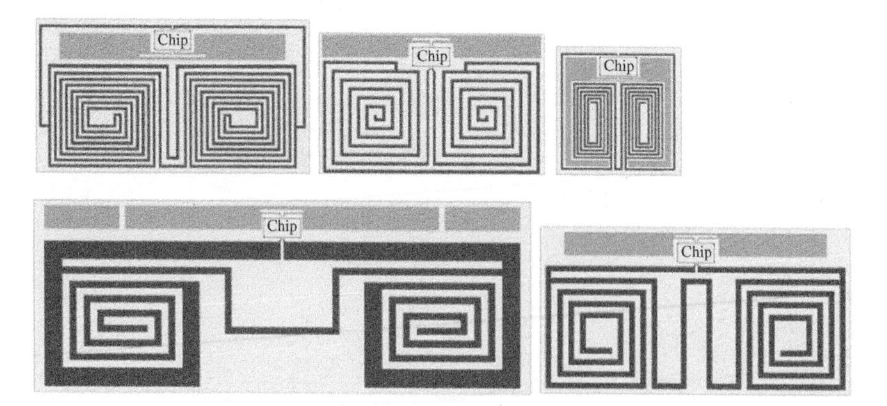


**Figure 11.17** Designed off-chip antennas with scavenging at 915 MHz (from [10], reprinted with permission).

The same structure as a scavenging antenna but mirrored horizontally is used for the UWB antenna operating around 5.5 GHz. It is placed in the empty regions of the scavenging antenna in the same metal layer. At the crossing point, internal vias are utilized to avoid any interception. Note that these structures also can be used as off-chip antennas in order to realize AiP solutions (see Figure 11.17).

## 11.5 AiP Solutions in Low-cost PCB Technology

One of the technologies used to implement AiP solutions is PCB technology. The main benefit of this technology is its lower manufacturing cost when compared to others. However, in terms of packaging capabilities and performance, other technologies, such as low-temperature co-fired ceramic (LTCC) and thin film, are preferable. In this section, PCB technology is used to implement a cost-effective 3D AiP solution for wireless communication systems such as sensor nodes and IoT devices.

### 11.5.1 Introduction to Wireless Sensor Networks and IoT

The recent advances in very large scale integration (VLSI) technology, microelectromechanical systems (MEMS), and wireless communications have driven many researchers towards innovative solutions for wireless sensor networks (WSNs) [53–55]. WSNs are based on the availability of cheap, low power, and miniature embedded processors, radios, sensors, and actuators, often integrated on a single chip. WSNs are composed of different sensor nodes which are connected wirelessly, to feel the physical characteristics of the surrounding environment.

On the other hand, the IoT is a paradigm where billions of devices with embedded intelligence, communication means, and sensing capabilities connect to the internet and share information. In fact, WSNs are the ancestors of the IoT network. The IoT plays an important role in many applications such as smart homes and cities, smart farming, automation, surveillance, medication, etc. [56].

By controlling the flow of information in a smart manner, the concept of ambient intelligence [57, 58] can be achieved in these networks. Ambient intelligence technologies represent a vision of the future where humans will be surrounded by electronic environments, sensitive and responsive to people.

#### 11.5.1.1 Examples of Antennas for IoT Devices

Most antennas used for IoT applications have planar and 2D structures. However, beside these a number of 3D structures have been reported. In [59], a folded miniaturized antenna which operates at 820 MHz was proposed. This antenna benefits from different size reduction techniques such as folding, slots and slits, and inductive loading using vias. Another work presents a transparent double-folded loop antenna that has a low level of fluctuation in realized gain when it is in close proximity to metal [60]. In [61], different fabrication technologies and materials, including paper, textile and 3D printed substrates, are presented to implement substrate-integrated waveguide components and antennas for IoT devices. A 3D miniature antenna working at a ultra-high frequency (UHF) band has been implemented in [62] using liquid metal and additive technologies. In [63], a 3D printed loop antenna was implemented using flexible material (NinjaFlex) for wearable and IoT applications. A multiband microstrip patch antenna was investigated in [64] by introducing metamaterial substrate and loading the ground plane with split-ring resonators (SRRs) that result very good gain and improvement in bandwidth. Another work describes the integration of an electrically small broadside-coupled split-ring resonator (BC-SRR) antenna in a Bluetooth low energy (BLE) module package [65].

### 11.5.2 3D System-in-Package Solutions for Microwave Wireless Devices

From an architectural viewpoint, 3D system-in-package (SiP) technologies for RF and microwave applications can be split into two main categories:

1) The first category makes use of different 2D multilayer technologies such as standard PCB [67], LTCC [68] and thin-film multichip module (MCM) [69, 70]. In order to have SiP integration, different modules are implemented in different technologies and mounted together in a stacked configuration [71, 72].
2) The second category covers technologies that are 3D in nature. Their processing and integration afterwards incorporate some customized non-standard steps, e.g. PCB with flexible materials as a core laminate in its layer buildup [73, 74].

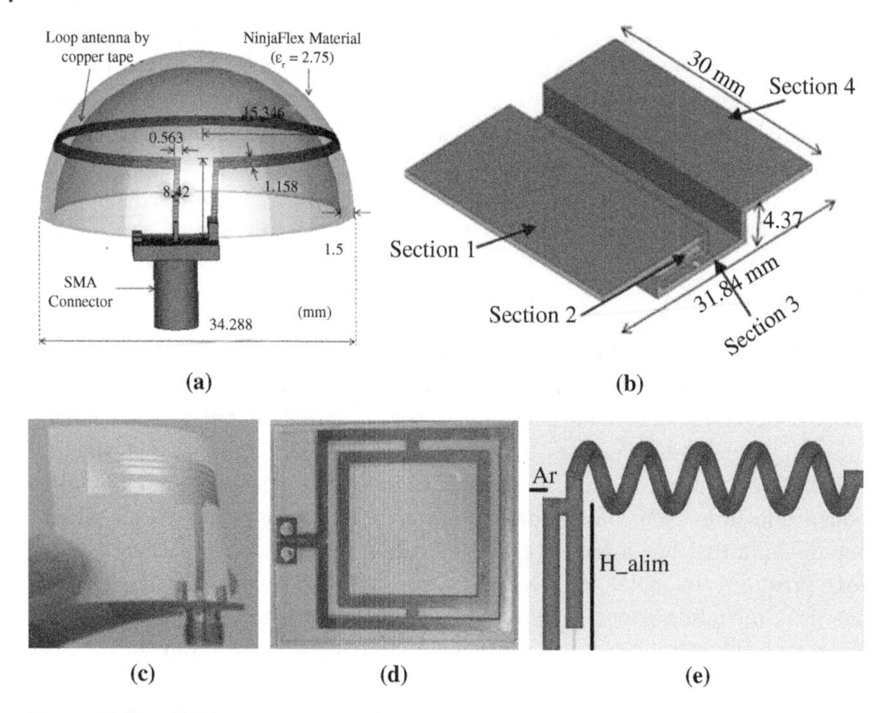


**Figure 11.18** Different antennas which can be used for IoT applications (from [59, 60, 62, 63, 66], © 2018, 2018, 2017, 2015, 2017 IEEE, reprinted with permission).

3D printing techniques enable the integration of an antenna directly onto the package of a wireless device. There are various 3D-printed cubic structures that can be utilized for SiP technology [75]. A 3D tripolar antenna has been developed for operation at 2.4 GHz in [76]. This compact system mitigates multipath and depolarization channel effects and has been specifically designed to integrate with a commercial wireless sensor node. In [77], a 3D-printed balun-fed bowtie antenna working at 4 GHz frequency has been fabricated using acrylonitrile butadiene styrene (ABS) for the substrate and packaging structure.

Three low-cost 3D near-isotropic global system for mobile communications (GSM) antennas, which are good choices for IoT and sensor applications, were implemented in [79–81] using screen printing and 3D printing technology. These antennas have been realized on the packaging of electronics to ensure efficient utilization of available space and lower cost. Also, a fully integrated and 3D packaged wireless sensor for environmental monitoring applications has been reported in [78].

In [82], a fully spherical dipole antenna was fabricated via 3D printing of conductive ink. The volume-filling approach in this work ensures near-optimal bandwidth performance of the small antenna. Another 3D-packaged dipole

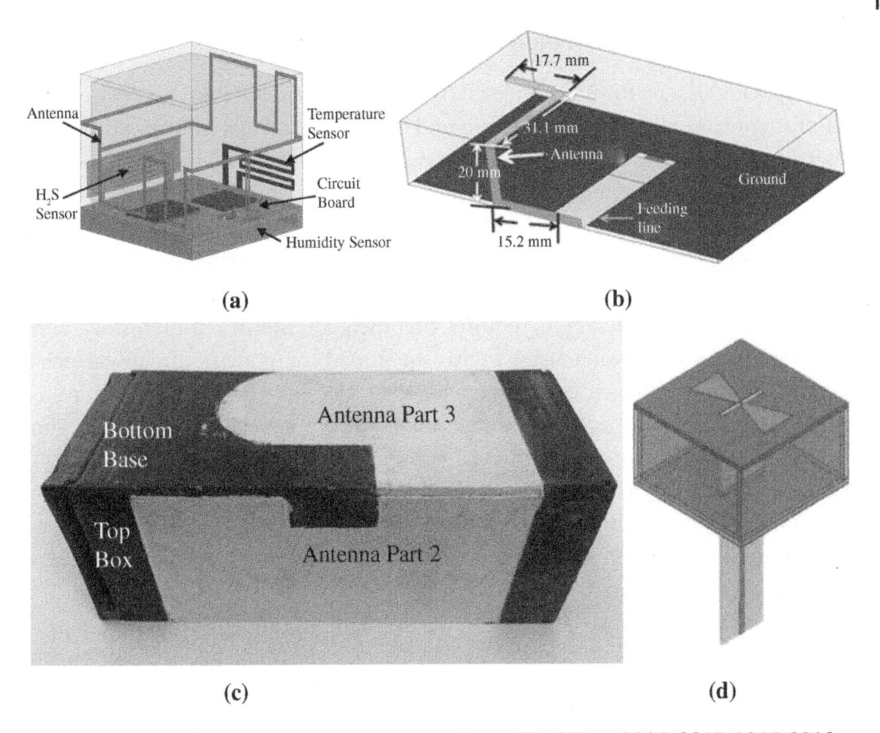


**Figure 11.19** Different 3D printed antennas (from [77–80], © 2016, 2017, 2017, 2018 IEEE, reprinted with permission).

antenna is reported in [83]. An electrically small loop antenna standing on the compact ground is proposed in [84]. The small volume makes this antenna possible for sensor package and IoT applications. In [85], a phased array antenna package is proposed to achieve high-gain beam and broad 3D scanning capability.

Another technique to realize 3D AiP is inkjet printing on flexible materials such as paper, plastics, leather, and even textiles, which can be used as substrates for such antennas [63, 66]. In [86], two high-gain, multidirector Yagi–Uda antennas on a flexible substrate are proposed for use within the 24.5 GHz ISM band.

In [87], a dual circular polarized spherical conformal antenna array was designed based on a 3D printing framework. This array has 23 elements and the capability to scan a beam within the hemispherical coverage. Another work presents a broadband 3D conformal spiral antenna that is based on the standard equiangular spiral geometry [88].

High-impedance surfaces with artificial magnetic conductor (AMC) or frequency selective surfaces (FSS) are becoming increasingly important for the improvement of antenna performance and noise isolation in emerging ultra-miniaturized 3D antenna-integrated packages [89]. In [89] novel miniaturized nanostructured AMCs are demonstrated for 5G and IoT devices.

### 11.5.3 E-CUBE: A 3D SiP Solution

By definition [90], electronic cube (e-CUBE) devices measure parameters in their environment through a sensor function and communicate this information, with rather low data rates, within an *ad hoc* wireless network. The central nodes of the e-CUBEs network connect it to other network services and systems and to users. Envisaged applications are WSNs for health and fitness, distributed intelligence for automotive control, and distributed smart monitoring for aeronautics and space applications.

The nature of the applications requires that the e-CUBEs are miniaturized and cost-effective. This is possible using highly integrated components and novel technologies, which in some cases still have to be developed. Among these technologies, 3D packaging and assembling technology play a central role [91, 92]. The goal is to transfer the SiP approach to RF and microwave applications, which will become a major factor in the next generation of commercialized telecommunication systems. Figure 11.20 shows a schematic and a realized example of the different modules needed in an autonomous sensor node and the integration architecture required to form an e-CUBE.

The systematic goal of e-CUBE design is a 3D topology that has a dipole-like pattern that includes all the electronics inside. This means that each node will have an omni-directional pattern in the azimuth plane, which makes it capable of communicating with the surrounding nodes with low sensitivity to the architecture of the network, i.e. the placement of the wireless nodes.

In the literature, there are three main PCB technologies used to manufacture 3D-like topologies: rigid, flex, and flex-rigid [94–97]. In this section, flex-rigid PCB technology is preferred in order to realize a rigid cube with flat faces. It has been used for different electronic SiP integrations [98–100], antennas for UWB applications [101], and conformal antennas integrated with RF front-ends [102]. The chosen topology introduces a 3D packaging solution which works at the same time as a conformal antenna array. The packaging structure forms an impenetrable cube on the faces of which the antennas of the transceiver system

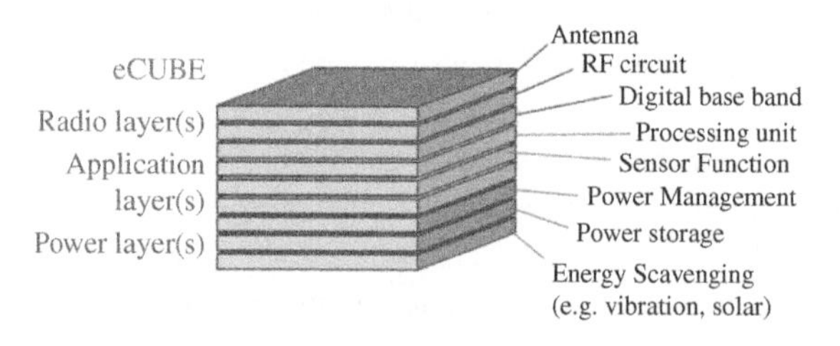

**Figure 11.20** An autonomous bio-sensor as an e-CUBE. (a) Schematic view of a SiP-technology-based e-CUBE. (b) Photograph of a wireless bio-electronic sensor integrated with a 3D SiP approach (from [93], reprinted with permission).

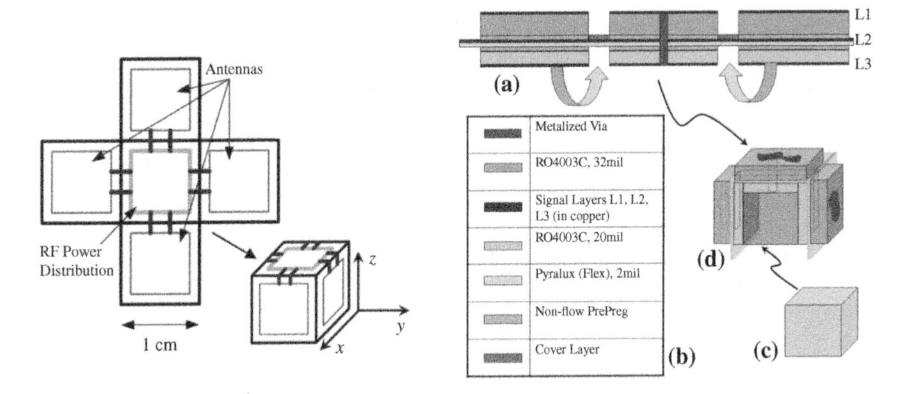


**Figure 11.21** Schematic description of the topology introduced and the layer buildup for the chosen flex-rigid PCB technology. (a) 2D layer buildup. (b) Materials used for different layers and their thicknesses. (c) The metallic cube containing all the electronics needed for the functionality of the e-CUBE. (d) A cross-sectional cut of the finalized e-CUBE (from [103], © 2011 IEEE, reprinted with permission).

are positioned. It is designed in such a way that all the electronics needed for the power generation/distribution and data processing can be placed within the volume of the cube. A schematic overview is given in Figure 11.21.

#### 11.5.3.1 Multilayer Flex-rigid PCB for Antenna Element Design

The layer buildup chosen to implement the 3D antenna array is shown in Figure 11.21. To prevent probable damage at the locations where the PCB is not covered with rigid laminates, i.e. at the corners of the cube, the flexible material is covered with a compatible material. As shown, the folded PCB, which carries the antenna elements on four of its five sides, covers a metallic box. This metallic box acts as an impenetrable electromagnetic shield, which is needed for the electronics in an autonomous wireless node.

As shown in Figure 11.22, the antenna element used on each side of the e-CUBE is a dipole antenna. The main reason for choosing the dipole antenna as the radiating element is its small lateral dimension, i.e. in the $x$ direction. This allows two antenna elements to be put on each surface to yield an array behavior in the $x$–$y$ plane. The RF signal is coupled through a microstrip line to the dipole antenna, which is placed on layer L1. This microstrip line has its signal line on layer L2 and its ground line on layer L3. The fully metal layer L3 also acts as a back plane, reflecting the signal radiated backwards from the dipole towards the front side, and yields a more directive radiation pattern. It also prevents the radiated field from penetrating inside the e-CUBE.

Figure 11.23 shows the simulated return loss and radiation pattern for the optimized dipole antenna.

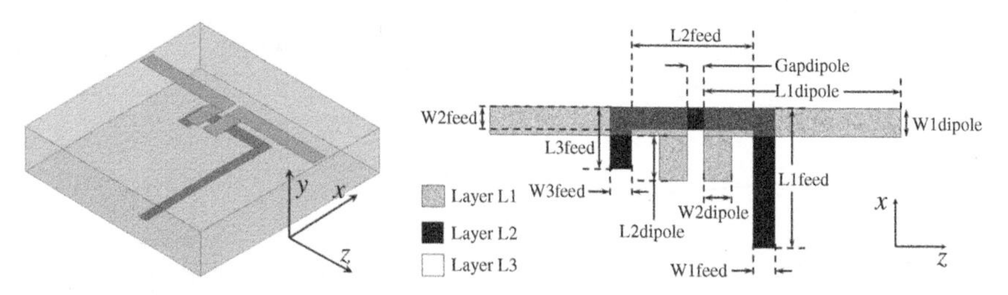


**Figure 11.22** The chosen dipole antenna fed by a via-less balun as the coupling structure: (a) 3D view and (b) top view of the layout (from [103], © 2011 IEEE, reprinted with permission).

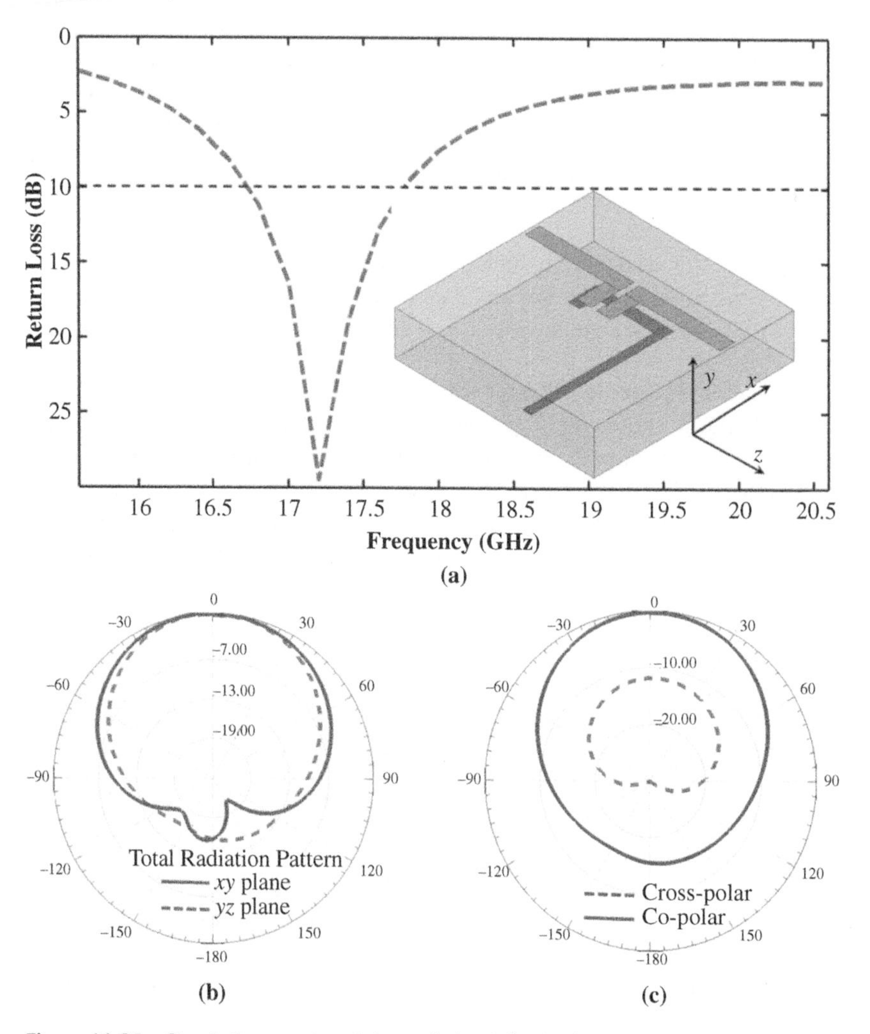


**Figure 11.23** Simulation results of the optimized dipole: (a) return loss, (b) total radiation patterns in the $x$–$y$ and $y$–$z$ planes at 17.2 GHz, and (c) co- and cross-polarization radiation patterns in the $y$–$z$ plane at 17.2 GHz (from [103], © 2011 IEEE, reprinted with permission).

The maximum gain of the dipole is around 4 dBi. To study the polarization purity of the antenna, the cross- and co-polar radiation patterns in the *y*–*z* plane are plotted in Figure 11.23c. As can be deduced, these radiation patterns are not completely symmetrical. This originates from the fact that the geometry of the antenna is not fully symmetrical. Also, the co-polar and cross-polar levels in the broadside direction differ by less than 20 dB. This ratio is improved by the arraying technique.

#### 11.5.3.2 Modular Design of the Antenna Array and Power Distribution Network

To have an omni-directional radiation pattern in the azimuth plane, two antenna elements which minimize the ripple and the grating lobes in the total radiation pattern are required to be placed on each side of the cube, as shown in Figure 11.24a.

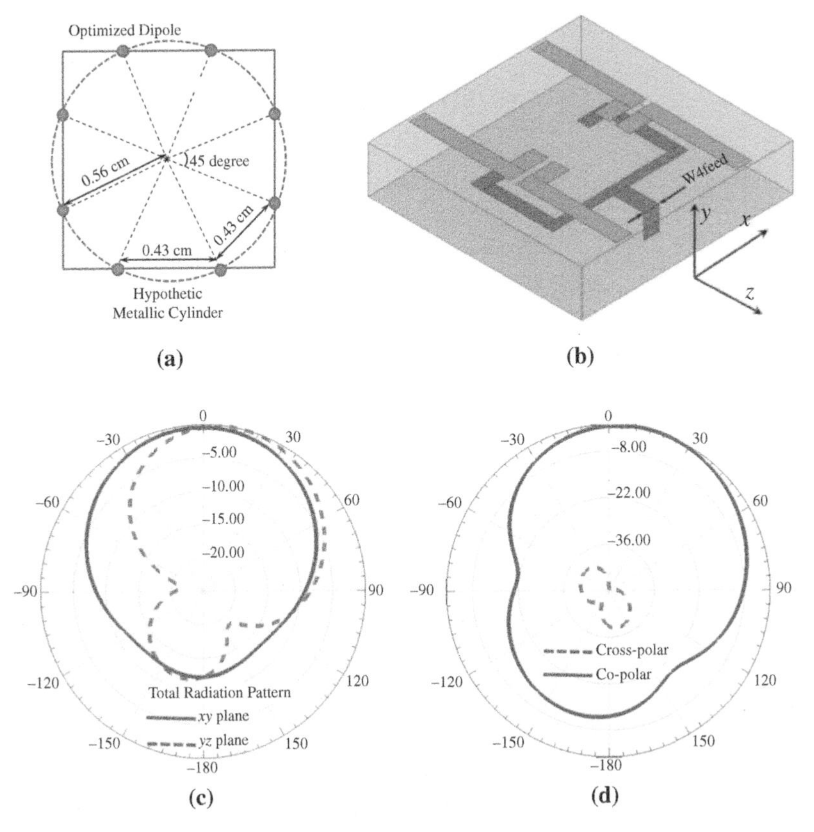


**Figure 11.24** (a) The optimum location for the optimized dipoles to yield minimum ripple and grating lobes in the radiation pattern. (b) 3D view of each face of the e-CUBE. (c) Radiation pattern in the *x*–*y* and *y*–*z* planes. (d) Co- and cross-polarization patterns in the *y*–*z* plane (from [103], © 2011 IEEE, reprinted with permission).

Thanks to the array configuration, the total radiation pattern of the 1 × 2 array has a symmetrical shape in the *x*–*y* plane and the ratio of co-polar to cross-polar patterns in *y*–*z* plane is better than 40 dB in the broadside direction. The pattern is rotated in the *y*–*z* plane due to the radiation from its feeding network. This does not have a destructive effect because it still allows the radiation pattern in the azimuth

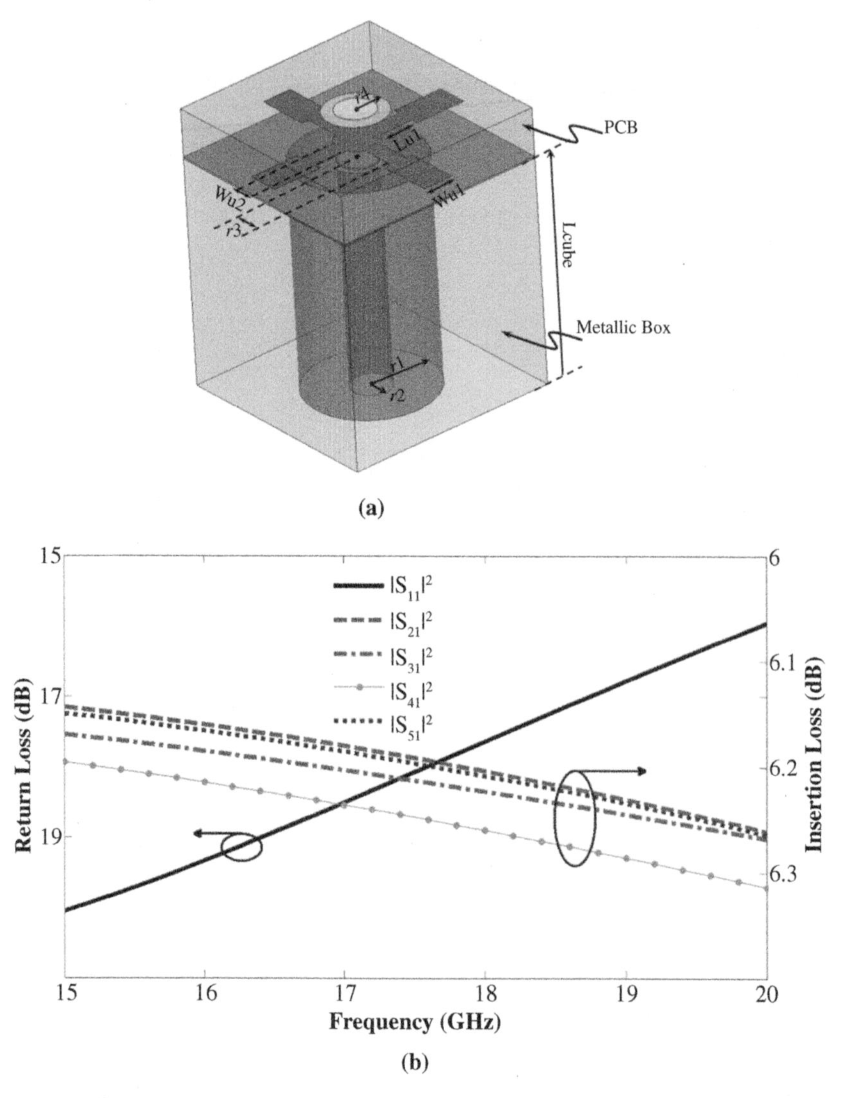


**Figure 11.25** 1 × 4 coaxial-to-microstrip splitter: (a) schematic view and (b) return and insertion losses (from [103], © 2011 IEEE, reprinted with permission).

plane of the final cube ($x$–$y$ plane) to be omni-directional. Moreover, the rotation is in such a direction (+$z$ in Figure 11.24b) that the probable interference between the radiating power from the e-CUBE and its connector or the ground plane on which it will be installed is reduced.

To feed the arrays on the faces of the e-CUBE with a corporate feeding network, a power splitter receiving the power from a coaxial transmission line and distributing it over four different microstrip paths has been designed, as shown in Figure 11.25a. Its outer conductor is fabricated by drilling a cylindrical hole in a metallic cube and the central conductor is the central conductor of a commercial small sub-miniature version A (SSMA) connector. The volume in between the two conductors is filled with Teflon to generate the correct transmission line characteristics. A through-metalized via is needed in the PCB structure for the central conductor of the SSMA connector to go through and then be soldered to the top layer of the PCB.

After being split, the power should feed the 1 × 2 antenna arrays on each of the four surrounding faces. To implement this, a transition has been designed between the microstrip lines on the top face to similar microstrip lines in the same layers on each of the surrounding sides. Hence, as shown in Figure 11.26, a rounded shape in the flex part of the PCB was used to model the finalized structure at the corners. The matching is done using two pieces of microstrip line in series. Note that the signal lines of these microstrip lines are implemented in layer L2 while the ground plane is realized by a face of the metallic box. The dielectric part (RO4003C 20mil) has been removed from underneath the curved part of the PCB. Hence, the dielectric of the curved microstrip line is mainly air. Although the transition has a small curvature radius, the radiation insertion loss is less than 0.2 dB, which is

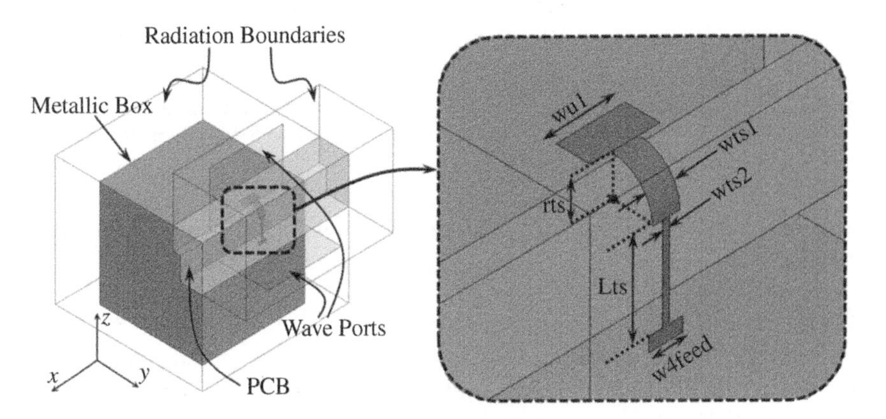


**Figure 11.26** Microstrip-to-microstrip transition from the top face of the cube to one of the surrounding faces (from [103], © 2011 IEEE, reprinted with permission).

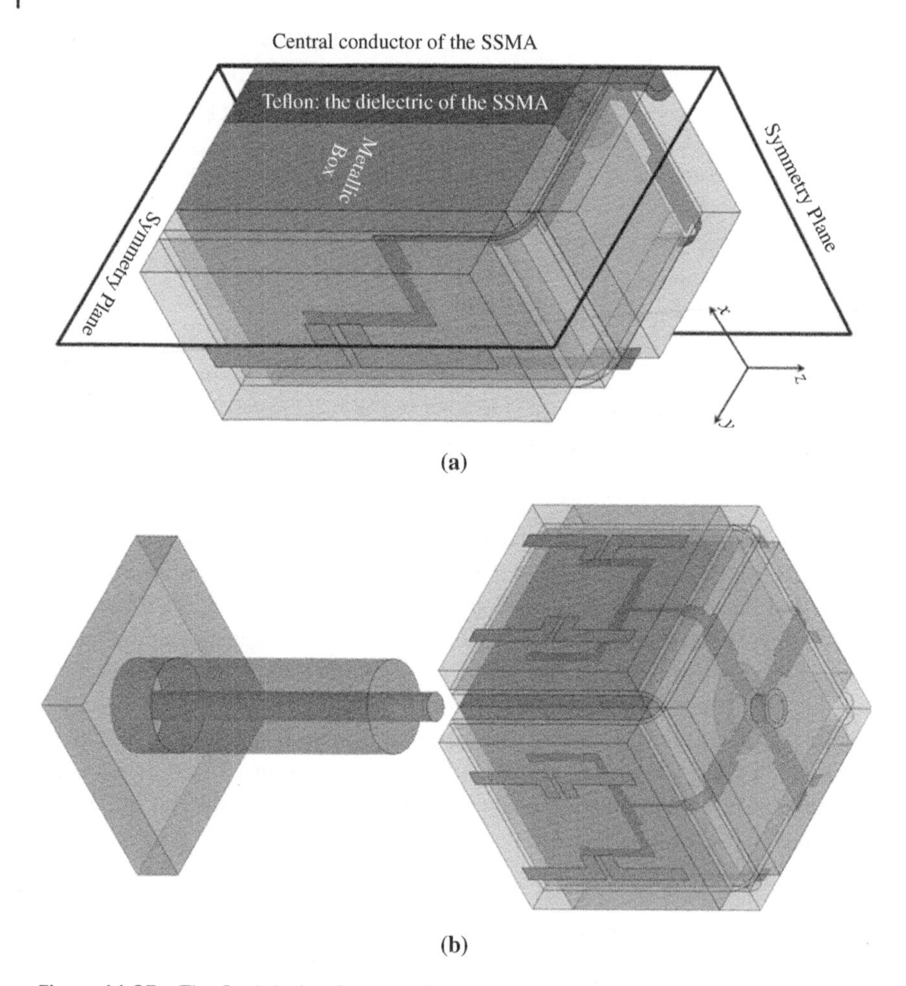


**Figure 11.27** The final design for the e-CUBE antenna: (a) one-quarter of the structure with two symmetry planes and (b) 3D model for the antenna and SSMA connector (from [103], © 2011 IEEE, reprinted with permission).

very promising [103, 104]. A complete 3D model of the e-CUBE conformal antenna array is shown in Figure 11.27.

#### 11.5.3.3 Construction and Measurement Results

Figure 11.28a shows the different parts assembled together to form the final structure: metallic box, SSMA connector, and the flex-rigid PCB. A low-temperature conductive glue has been used to assemble them.

Figure 11.29 shows the simulation versus measurement results for the antenna. The return loss is less than -10 dB at 17.2 GHz, with an acceptable bandwidth.

**Figure 11.28** (a) Manufactured separate parts: SSMA connector, flex-rigid PCB, and metal box. (b) Pictures of the finalized 3D e-CUBE (from [104], © 2011 IEEE, reprinted with permission).

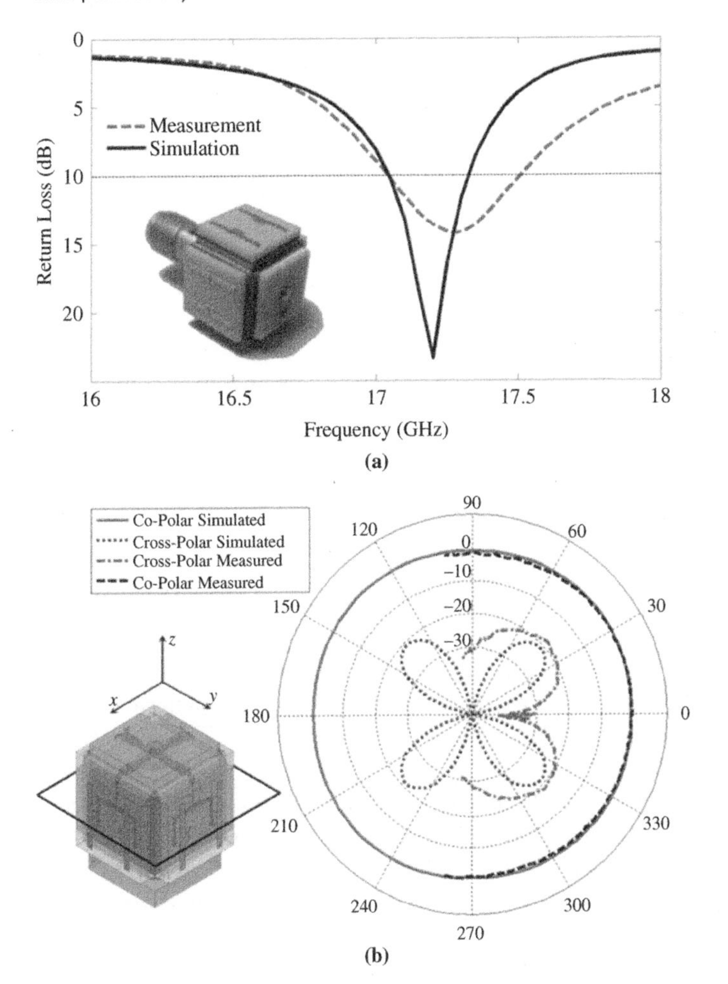


**Figure 11.29** Simulation versus measurement results for the finalized antenna: (a) return loss, (b) radiation patterns in the $x$–$y$ plane, theta = 90, (c) radiation pattern in the $y$–$z$ plane, phi = 90, and (d) peak radiation gain in the $y$–$z$ plane (from [103], © 2011 IEEE, reprinted with permission).

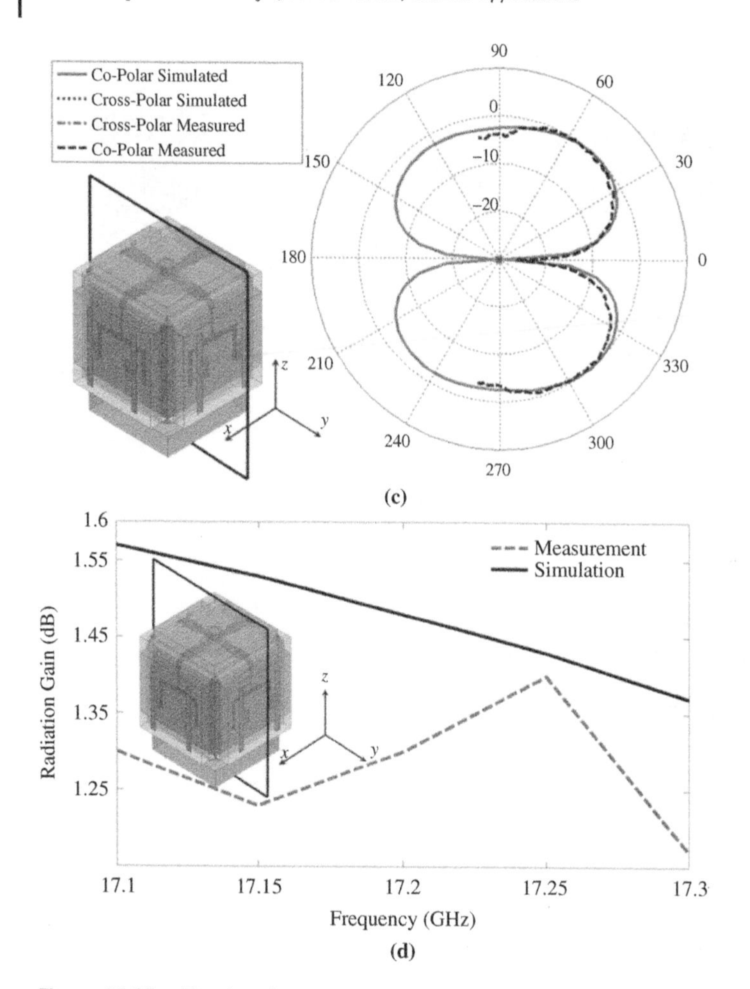


**Figure 11.29** (*Continued*)

The pattern is omni-directional, with a ripple of less than 0.9 dB for the measured pattern in the $x$–$y$ plane. However, the measured cross-polar pattern has an average amplitude higher than the simulated one and degrades at angles around +90° and −90°.

Due to the small dimensions of the antenna under test, the rotating holders commercially available for antenna measurement systems are unsuitable. Hence, the custom measurement system prevented the measurements from being performed in all azimuth angles, which is why the measured patterns are shown from −110° to +110°.

Note that the applicability of the proposed solution is not restricted to microwave frequencies such as 17.2 GHz. The same, or similar, technology can be used to implement similar AiP solutions at millimeter-wave frequencies also. However, by increasing the operation frequency, the sensitivity of the final module characteristics to manufacturing inaccuracies increases. In that case, a sensitivity analysis can be performed in the design stage to foresee the effects of the manufacturing and assembly processes on the behavior of the final solution.

## References


1. L.J. Chu, Physical limitations of omni-directional antennas. *Journal of Applied Physics,* vol. 19, pp. 1163–1175, December 1948.
2. R.E. Collin and S. Rothschild, Evaluation of antenna Q. *IEEE Transactions on Antennas and Propagation,* vol. 12, pp. 23–27, January 1964.
3. R.C. Hansen, Fundamental limitations in antennas. *Proceedings of the IEEE,* vol. 69, pp. 170–182, February 1981.
4. R.F. Harrington, Effect of antenna size on gain, bandwidth and efficiency. *Journal of Research of the National Bureau of Standards,* vol. 64D, no. 1, pp. 1–12, January/February 1960.
5. J.S. McLean, A re-examination of the fundamental limits on the radiation Q of electrically small antennas. *IEEE Transactions on Antennas and Propagation,* vol. 44, pp. 672–676, May 1996.
6. H.A. Wheeler, Fundamental limitations of small antennas. *Proceedings of the IRE,* vol. 35, pp. 1479–1484, December 1947.
7. J. Volakis, C. Chen, and K. Fujimoto, *Small Antennas: Miniaturization Techniques & Applications,* McGraw-Hill, 2010.
8. H.L. Thal, New radiation Q limits for spherical wire antennas. *IEEE Transactions on Antennas and Propagation,* vol. 54, pp. 2757–2763, 2006.
9. R.C. Hansen and R.E. Collin, A new Chu formula for Q. *IEEE Antennas and Propagation Magazine,* vol. 51, no. 5, pp. 38–41, October 2009.
10. K. Mohammadpour-Aghdam, Miniaturized Integrated Antennas for RFID and UWB Applications, PhD Thesis, Leuven, Belgium, 2011.
11. M. Gustafsson, C. Sohl, and G. Kristensson, Illustration on new physical bounds on linearly polarized antennas. *IEEE Transactions on Antennas and Propagation,* vol. 57, no. 5, pp. 1319–1327, May 2009.
12. C. Sohl, M. Gustafsson, and G. Kristensson, Physical limitations on broadband scattering by heterogeneous obstacles. *Journal of Physics A: Mathematical Theory,* vol. 40, pp. 11165–11182, 2007.
13. A.D. Yaghjian and H.R. Stuart, Lower bounds on the Q of electrically small dipole antennas. *IEEE Transactions on Antennas and Propagation,* vol. 58, pp. 3114–3121, October 2010.
14. D.R. Rhodes, On the stored energy of planar apertures. *IEEE Transactions on Antennas and Propagation,* vol. 14, no. 6, pp. 676–683, November 1966.

**15** W. Geyi, P. Jarmuszewski, and Y. Qi, The Foster reactance theorem for antennas and radiation Q. *IEEE Transactions on Antennas and Propagation,* vol. 48, no. 3, pp. 401–408, March 2000.

**16** A.D. Yaghjian and S.R. Best, Impedance, bandwidth, and Q of antennas. *IEEE Transactions on Antennas and Propagation,* vol. 53, pp. 1298–1324, 2005.

**17** S.R. Best, Bandwidth and the lower bound on Q for small wideband antennas. *IEEE Antennas and Propagation Society International Symposium,* Albuquerque, NM, USA, July 2006.

**18** S.R. Best, Low Q electrically small linear and elliptical polarized spherical dipole antennas. *IEEE Transactions on Antennas and Propagation,* vol. 53, no. 3, pp. 1047–1053, March 2005.

**19** H.A. Wheeler, Small antennas. *IEEE Transactions on Antennas and Propagation,* vol. 23, pp. 462–469, July 1975.

**20** A.K. Skrivervik, J.F. Zurcher, O. Staub et al., PCS antenna design: the challenge of miniaturization. *IEEE Antennas and Propagation Magazine,* vol. 43, August 2001.

**21** P. Hallbjorner, Electrically small unbalanced four-arm wire antenna. *IEEE Transactions on Antennas and Propagation,* vol. 52, pp. 1424–1428, June 2003.

**22** E.E. Altshuler, Electrically small self-resonant wire antennas optimized using a genetic algorithm. *IEEE Transactions on Antennas and Propagation,* vol. 50, pp. 297–300, March 2002.

**23** J.P. Gianvittorio and Y. Rahmat-Samii, Fractal antennas: a novel antenna miniaturization technique, and applications. *IEEE Antennas and Propagation Magazine,* vol. 44, pp. 20–36, February 2002.

**24** S.R. Best, On the performance properties of the Koch fractal and other bent wire monopoles. *IEEE Transactions on Antennas and Propagation,* vol. 51, pp. 1292–1300, June 2003.

**25** H. Mosallaei and K. Sarabandi, Magneto-dielectrics in electromagnetics: Concept and applications. *IEEE Transactions on Antennas and Propagation,* vol. 52, no. 6, pp. 1558–1567, June 2006.

**26** A. Buerkle and K. Sarabandi, A wide-band, circularly polarized, magneto-dielectric resonator antenna. *IEEE Transactions on Antennas and Propagation,* vol. 53, pp. 3446–3442, November 2005.

**27** K. Buell, H. Mosallaei, and K. Sarabandi, A wide-band, circularly polarized, magnetodielectric resonator antenna. *IEEE Transactions on Microwave Theory and Techniques,* vol. 54, pp. 135–146, January 2006.

**28** H. Mosallaei and K. Sarabandi, Antenna miniaturization and bandwidth enhancement using a reactive impedance substrate. *IEEE Transactions on Antennas and Propagation,* vol. 52, pp. 2403–2414, September 2004.

**29** A. Erentok, P.L. Luljak, and R.W. Ziolkowski, Characterization of a volumetric metamaterial realization of an articial magnetic conductor for antenna applications. *IEEE Transactions on Antennas and Propagation,* vol. 53, no. 1, pp. 160–172, January 2005.

**30** H.R. Stuart and A. Pidwerbetsky, Electrically small antenna elements using negative permittivity resonators. *IEEE Transactions on Antennas and Propagation,* vol. 54, no. 6, pp. 1644–1653, June 2006.

**31** M. Bahr, A. Boag, E. Michielssen et al., Design of ultra broad-band loaded monopole antennas. *IEEE AP-S International Symposium*, Seattle, WA, June 1994.

**32** A. Boag, A. Boag, E. Michielssen et al., Design of electrically loaded antennas using genetic algorithms. *IEEE Transactions on Antennas and Propagation,* vol. 44, no. 5, p. 687, May 1996.

**33** S.R. Best, A discussion on the properties of electrically small self-resonant wire antennas. *IEEE Antennas and Propagation Magazine,* vol. 46, no. 6, pp. 9–22, December 2004.

**34** S. Seki and H. Hasegawa, Cross-tie slow-wave coplanar waveguide on semi-insulating GaAs substrates. *Electronics Letters,* vol. 17, no. 25, pp. 940–941, December 1981.

**35** K. Mohammadpour-Aghdam, R. Faraji-Dana, and G.A. Vandenbosch, Miniaturized tunable meanderline loaded antenna with Q-factor approaching the lower bound. *XXXth URSI General Assembly and Scientific Symposium*, Istanbul, Turkey, August 2011.

**36** J.H. Apostolos, Meanderline Loaded Antenna. US Patent 5,790,080, August 1998.

**37** J.H. Apostolos, Slow wave meander line having sections of alternating impedance relative to a conductive plate. US Patent 6,313,716 B1, 6 November 2001.

**38** M. Bagheriasl, K. Mohammadpour-Aghdam, and R. Faraji-Dana, Efficient design procedure of dual-mode antennas based on the characteristic modes theory. *IEEE Middle East Conference on Antennas and Propagation*, Beirut, Lebanon, September 2016.

**39** M. Bagheriasl, K. Mohammadpour-Aghdam, and R. Faraji-Dana, Design of a dual-mode meander-line loaded monopole antenna with characteristic mode theory. *11th European Conference on Antennas and Propagation*, Paris, France, March 2017.

**40** F. Molaee-Ghaleh and K. Mohammadpour-Aghdam, Wideband tunable meander-line loaded antenna with dual-mode capability in V/UHF frequencies. *IEEE Asia Pacific Microwave Conference*, Kuala Lampur, Malaysia, November 2017.

**41** H. Huang and T. Li, A spiral electrically small magnetic antenna with high radiation efficiency for wireless power transfer. *IEEE Antennas and Wireless Propagation Letters,* vol. 15, pp. 1495–1498, January 2016.

**42** D.W.K. Ng, T.Q. Duong, C. Zhong et al. (eds), *Wireless Information and Power Transfer*, Wiley & Sons Ltd, 2019.

**43** P. Gadfort and P.D. Franzon, Millimeter-scale true 3-D antenna-in-package structures for near-field power transfer. *IEEE Transactions on Components, Packaging and Manufacturing Technology,* vol. 4, no. 10, pp. 1574–1581, September 2014.

**44** J. Kimionis, M. Isakov, B.S. Koh et al., 3D-printed origami packaging with inkjet-printed antennas for RF harvesting sensors. *IEEE Transactions on Microwave Theory and Techniques,* vol. 63, no. 12, pp. 4521–4532, November 2015.

**45** J. Bito, R. Bahr, J.G. Hester et al., A novel solar and electromagnetic energy harvesting system with a 3-D printed package for energy efficient internet-of-things wireless sensors. *IEEE Transactions on Microwave Theory and Techniques,* vol. 65, no. 5, pp. 1831–1842, May 2017.

**46** C. A. Balanis, *Antenna Theory, Analysis and Design*, 2nd edn, John Wiley & Sons Inc., 1997.

**47** K. Mohammadpour-Aghdam, S. Radiom, R. Faraji-Dana et al., Miniaturized RFID/UWB antenna structure that can be optimized for arbitrary input impedance. *IEEE Antennas and Propagation Magazine,* vol. 54, no. 2, pp. 74–87, July 2012.

**48** J. Rashed and C.-T. Tai, A new class of resonant antennas. *IEEE Transactions on Antennas and Propagation,* vol. 39, no. 9, pp. 1428–1430, September 1991.

**49** T.J. Warnagiris and T.J. Minardo, Performance of a meandered line as electrically small transmitting antenna. *IEEE Transactions on Antennas and Propagation,* vol. 46, pp. 1797–1876, December 1998.

**50** K. Finkenzeller, *RFID Handbook: Fundamentals and Applications in Contactless Smart Cards and Identification*, 2nd edn, Wiley & Sons Ltd, 2003.

**51** A. Babakhani, X. Guan, A. Komijani et al., A 77-GHz phased-array transceiver with on-chip antennas in silicon: Receiver and antennas. *IEEE Journal of Solid-State Circuits,* vol. 41, no. 12, pp. 2795–2806, September 2006.

**52** A. Hajimiri, mm-Wave silicon ICs: Challenges and opportunities. *IEEE Custom Integrated Circuits Conference*, San Jose, CA, September 2007.

**53** F.L. Lewis, Wireless sensor networks. In *Smart Environments: Technologies, Protocols, and Applications*, D. Cook and S. Das (eds), John Wiley, New York, 2004.

**54** Y. Zhang, Y. Gu, V. Vlatkovic et al., Progress of smart sensor and smart sensor networks. *Fifth World Congress on Intelligent Control and Automation,* June 2004.

**55** C. Townsend and S. Arms, Wireless sensor networks: principles and applications. In *Sensor Technology Handbook*, J.S. Wilson (ed.), Elsevier, Oxford, 2005.

**56** S. Cirani, G. Ferrari, M. Picone et al. (eds), *Internet of Things: Architectures,* Protocols and Standards, Wiley & Sons Ltd, 2019.

**57** H. Nakashima, H. Aghajan and J.C. Augusto, *Handbook of Ambient Intelligence and Smart Environments*, Springer, New York, 2010, pp. 3–30.

**58** C. Ramos, J.C. Augusto, J. Carlos et al., Ambient intelligence: the next step for artificial intelligence. *IEEE Intelligent Systems,* vol. 23, pp. 15–18, March 2008.

**59** R.M. Bichara, F.A. Asadallah, J. Costantine et al., A folded miniaturized antenna for IoT devices. *IEEE Conference on Antenna Measurements & Applications*, 2018.

**60** Y. Koga and M. Kai, A transparent double folded loop antenna for IoT applications. *IEEE-APS Topical Conference on Antennas and Propagation in Wireless Communications*, 2018.

**61** M. Bozzi, S. Moscato, L. Silvestri et al., Innovative SIW components on paper, textile, and 3D-printed substrates for the Internet of Things. *Asia-Pacific Microwave Conference*, Nanjing, China, December 2015.

**62** M. Cosker, L. Lizzi, F. Ferrero et al., 3D compact antenna using liquid metal and additive technologies. *European Conference on Antennas and Propagation*, Paris, France, March 2017.

**63** K. Nate and M.M. Tentzeris, A novel 3-D printed loop antenna using flexible NinjaFlex material for wearable and IoT applications. *IEEE 24th Electrical Performance of Electronic Packaging and Systems*, San Jose, CA, USA, October 2015.

**64** I. Aggarwal, M.R. Tripathy, and S. Pandey, Metamaterial inspired multiband slotted antenna for application in IOT band. *Online International Conference on Green Engineering and Technologies*, Coimbatore, India, May 2017.

**65** M. Sano, K. Yamada, and M. Higaki, An electrically small broadside-coupled split-ring resonator antenna integrated in a BLE module package. *IEEE International Workshop on Electromagnetics: Applications and Student Innovation Competition*, 2018.

**66** A. Shamim, 3D inkjet printed flexible and wearable antenna systems. *International Symposium on Antennas and Propagation*, Phuket, Thailand, October 2017.

**67** X. Jiang and H. Shi, Effective die-package-PCB co-design methodology and its deployment in 10 Gbps serial link transceiver FPGA packages. *IEEE MTT-S International Microwave Symposium Digest*, June 2009.

**68** S. Stoukatch, C. Winters, E. Beyne et al., 3D-SIP integration for autonomous sensor nodes. *Electronic Components and Technology Conference*, July 2006.

**69** J.H. Lee, G. Dejean, S. Sarkar et al., Highly integrated millimeter-wave passive components using 3-D LTCC system-on-package (SOP) technology. *IEEE Transactions on Microwave Theory and Techniques,* vol. 53, no. 6, pp. 2220–2229, June 2005.

**70** T. Barbier, F. Mazel, B. Reig et al., A 3D wideband package solution using MCM-D technology for tile TR module. *EGAAS Symposium*, October 2005.

**71** A. Enayati, S. Brebels, W.D. Raedt et al., Vertical vial-less transition in MCM technology for millimeter-wave applications. *Electronic Letters,* vol. 46, no. 4, pp. 287–288, February 2010.

**72** P. Ramm and A. Klumpp, Through-silicon via technologies for extreme miniaturized 3D integrated wireless sensor networks (e-CUBES). *Interconnect Technology Conference*, June 2008.

**73** N. Khan, V.S. Rao, S. Lim et al., Development of 3D silicon module with TSV for system in packaging. *IEEE Transactions on Component Packaging Technology,* vol. 33, no. 1, pp. 3–9, March 2010.

**74** E.C.W.D. Jong, L.A. Ferreira, and P. Bauer, 3D integration with PCB technology. *IEEE Applied Power Electronics Conference and Exposition*, March 2006.

**75** R. Bahr, A. Nauroze, W. Su et al., Self-actuating 3D printed packaging for deployable antennas. *IEEE 67th Electronic Components and Technology Conference*, Orlando, FL, USA, May 2017.

**76** R.A. Ramirez, M. Golmohamadi, J. Frolik et al., 3D printed on-package tripolar antennas for mitigating harsh channel conditions. *IEEE Radio and Wireless Symposium*, Phoenix, AZ, USA, January 2017.

**77** P.B. Nesbitt, H. Tsang, T.P. Ketterl et al., 4 GHz 3D-printed balun-fed bowtie antenna with finite ground plane for gain and impedance matching enhancement. *IEEE 17th Annual Wireless and Microwave Technology Conference,* Clearwater, FL, USA, April 2016.

**78** M.F. Farooqui and A. Shamim, 3D inkjet printed disposable environmental monitoring wireless sensor node. *IEEE MTT-S International Microwave Symposium*, Honolulu, HI, USA, June 2017.

**79** S. Zhen, R.M. Bilal, and A. Shamim, 3D printed system-on-package (SoP) for environmental sensing and localization applications. *International Symposium on Antennas and Propagation*, Phuket, Thailand, 2017.

**80** S. Zhen, K. Klionovski, R.M. Bilal et al., A dual band additively manufactured 3-D antenna on package with near-isotropic radiation pattern. *IEEE Transactions on Antennas and Propagation,* vol. 66, no. 7, pp. 3295–3305, July 2018.

**81** S. Zhen and A. Shamim, A 3D printed dual GSM band near isotropic on-package antenna. *IEEE International Symposium on Antennas and Propagation & USNC/URSI National Radio Science Meeting*, San Diego, CA, USA, July 2017.

**82** J. Adams, S. Slimmer, J. Lewis et al., 3D-printed spherical dipole antenna integrated on small RF node. *Electronics Letters,* vol. 51, no. 9, pp. 661–662, April 2015.

**83** D.F. Hawatmeh, S. LeBlanc, P.I. Deffenbaugh et al., Embedded 6-GHz 3-D printed half-wave dipole antenna. *IEEE Antennas and Wireless Propagation Letters,* vol. 16, pp. 145–148, 2016.

**84** H. Liu, Y. Cheng, and M. Yan, Electrically small loop antenna standing on compact ground in wireless sensor package. *IEEE Antennas and Wireless Propagation Letters,* vol. 15, pp. 76–79, May 2015.

**85** N. Ojaroudiparchin, M. Shen, S. Zhang et al., A switchable 3-D-coverage-phased array antenna package for 5G mobile terminals. *IEEE Antennas and Wireless Propagation Letters,* vol. 15, pp. 1747–1750, February 2015.

**86** B.K. Tehrani, B.S. Cook, and M.M. Tentzeris, Inkjet printing of multilayer millimeter-wave Yagi-Uda antennas on flexible substrates. *IEEE Antennas and Wireless Propagation Letters,* vol. 15, pp. 143–146, May 2015.

**87** Y.F. Wu and Y.J. Cheng, Spherical conformal antenna array based on 3D printing framework. *IEEE MTT-S International Microwave Workshop Series on Advanced Materials and Processes for RF and THz Applications*, Chengdu, China, July 2016.

**88** J.D. Mingo, C. Roncal, and P.L. Carro, 3-D conformal spiral antenna on elliptical cylinder surfaces for automotive applications. *IEEE Antennas and Wireless Propagation Letters,* vol. 11, pp. 148–151, January 2012.

**89** T. Lin, P.M. Raj, A. Watanabe et al., Nanostructured miniaturized artificial magnetic conductors (AMC) for high-performance antennas in 5G, IoT, and smart skin applications. *IEEE 17th International Conference on Nanotechnology*, Pittsburgh, PA, USA, July 2017.

**90** P. Ramm, A. Klumpp, J. Weber et al., The European 3D technology platform (e-CUBES). *Future Fab International,* vol. 34, pp. 103–116, July 2010.

**91** W. Chen, W.R. Bottoms, K. Pressel et al., The next step in assembly and packaging: system level integration in package (SiP). *Hard Copy of the International Technical Roadmap for Semiconductors Report,* 2008.

**92** P. Ramm, A. Klumpp, R. Merkel et al., 3D system integration technologies. *Materials Research Society Symposium*, 2003.

**93** A. Enayati, Antenna-in-package Solutions for Microwave and Millimeter-wave Applications, PhD Thesis, Leuven, Belgium, November 2012.

**94** C. Chiu, J.B. Jang, and R.D. Murch, 24-port and 36-port antenna cubes suitable for MIMO wireless communications. *IEEE Transactions on Antennas and Propagation,* vol. 56, no. 4, pp. 1170–1176, April 2008.

**95** B. Bonnet, P. Monfraix, R. Chiniard et al., 3D packaging technology for integrated antenna front-ends. *European Microwave Technology Conference*, October 2008.

**96** S.Y.Y. Leung, P.K. Tiu, and D.C.C. Lam, Printed polymer-based RFID antenna on curvilinear surfaces. *Electronic Materials. Packaging Confereence*, December 2006.

**97** J. Jung, H. Lee, and Y. Lim, Broadband flexible comb-shape monopole antenna. *IET Microwave Antennas Propagation*, vol. 3, no. 2, pp. 325–332, March 2009.

**98** V. Gjokaj, J. Papapolymerou, J. Albrecht, et al., A novel rigid-flex aperture coupled patch antenna array. *Asia-Pacific Microwave Conference*, Kyoto, Japan, November 2018.

**99** M. Xue, L. Cao, Q. Wang et al., A compact 27 GHz antenna-in-package (AiP) with RF transmitter and passive phased antenna array. *IEEE 68th Electronic Components and Technology Conference*, San Diego, CA, USA, May 2018.

**100** R.N. Das, F.D. Egitto, B. Wilson et al., Development of rigid-flex and multilayer flex for electronic packaging. *Electronic Components Technology Conference*, December 2010.

**101** M. Karlsson and S. Gong, Circular dipole antenna for mode 1 UWB radio with integrated balun utilizing a flex-rigid structure. *IEEE Transactions on Antennas and Propagation*, vol. 57, no. 10, pp. 2967–2971, October 2009.

**102** N. Altunyurt, R. Rieske, M. Swaminathan et al., Conformal antennas on liquid crystalline polymer based rigid-flex substrates integrated with the front-end module. *IEEE Transactions on Advanced Packagaging*, vol. 32, no. 4, pp. 797–808, November 2009.

**103** A. Enayati, S. Brebels, W.D. Raedt et al. G.A.E. Vandenbosch, 3D-antenna-in-package solution for microwave wireless sensor network nodes. *IEEE Transactions on Antennas and Propagation*, vol. 59, no. 10, pp. 3617–3623, October 2011.

**104** A. Enayati, M. Libois, W.D. Raedt et al., Conformal antenna-in-package solution implemented in a 3D flex-rigid multilayer PCB technology. *Asia-Pacific Microwave Conference*, Melbourne, Australia, December 2011.

# Index



*Antenna-in-Package Technology and Applications,* First Edition.
Edited by Duixian Liu and Yueping Zhang.

Printed and bound by CPI Group (UK) Ltd, Croydon, CR0 4YY